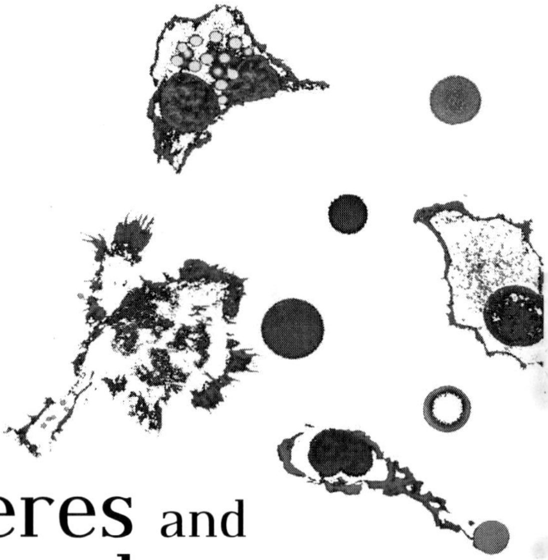

Microspheres and Microcapsules in Biotechnology

GUANGHUI MA
ZHIGUO SU

Design, Preparation, and Applications

Microspheres and Microcapsules in Biotechnology

PAN STANFORD PUBLISHING

Published by

Pan Stanford Publishing Pte. Ltd.
Penthouse Level, Suntec Tower 3
8 Temasek Boulevard
Singapore 038988

Email: editorial@panstanford.com
Web: www.panstanford.com

British Library Cataloguing-in-Publication Data
A catalogue record for this book is available from the British Library.

Microspheres and Microcapsules in Biotechnology: Design, Preparation, and Applications

ISBN 978-981-4316-47-7 (Hardcover)
ISBN 978-981-4364-62-1 (eBook)

Printed in the USA

Contents

Preface

Microspheres and microcapsules are spherical materials having diameters in nano- to micro-size range, with or without active materials encapsulated or on the surface. These spherical materials have very important applications in biotechnology, a rapidly developing field in the 21st century. Various bio-applications require the supply of high-quality microspheres and microcapsules that have controlled sizes, structures, and properties and are made of biocompatible materials.

The available books on polymer microsphere materials usually lack in presenting the role of polymer microsphere materials in bio-applications. On the other hand, numerous biotechnology books have been published that mention the use of microspheres or micro-capsules but lack description on how these materials are made and how to improve their performance.

This situation gave us an idea to prepare a book that can provide the knowledge and information about how to design and prepare microspheres and microcapsules specifically for bio-applications and to build up a bridge between the fields of biotechnology and advanced material technology, so that the scientists and students working in both these fields may be benefitted.

Therefore, this book focuses on design, preparation, and application of new microspheres and new microcapsules in biotechnology, including bio-reaction, bio-separation, bio-formulation, and bio-detection. In bio-reaction, new microspheres or microcapsules are introduced for enzyme immobilization, cell culture, and transplantation. The section on bio-separation discusses chromatographic media for purification of proteins and small bioactive compounds. It also contains information on microspheres for protein refolding and PEG-modified protein separation. As related to both bio-reaction and bio-separation, the solid-phase synthesis of peptides, oligonucleotides, other chemicals, and biomolecules and solid-phase modification of proteins are described. In bio-formulation, new microspheres and microcapsules are designed, synthesized, and used for peptide/protein drug delivery, DNA drug delivery, targeting drug delivery of anti-cancer agents, and brain-targeting drug delivery. The bio-detection section focuses on microspheres for diagnosis and medical imaging.

We have tried to give the knowledge and information about the principles of design, preparation methods, and application results of new microspheres and microcapsules for each bio-application area and also tried to identify problems that need to be studied further.

Contributions to this book have been made by researchers of the State Key Laboratory of Biochemical Engineering, China, and distinguished professors like Jan-Christer Janson of Uppsala University, Sweden; Haruma Kawaguchi *et al.* of Kanagawa University, Japan; Chen Jiang and Xiguo Jiang *et al.* of Fudan University; Xiaojun Ma *et al.* of Dalian Institute of Chemical Physics, Chinese Academy of Sciences; and Chunyin Chen *et al.* of National Center for Nanoscience and Technology, China. We thank all these authors who updated the theory and methodology, discussed deeply, and cooperated to successfully build up a link between the design, preparation, and applications of microspheres and microcapsules.

We hope this book will be helpful to researchers and students who work or will work in the field of material design and preparation and application of microspheres and microcapsules.

Guanghui Ma and Zhiguo Su
Winter 2012

Chapter 1

Microspheres for Enzyme Immobilization

Fei Gao, Yuxia Wang, and Guanghui Ma

State Key Laboratory of Biochemical Engineering, Institute of Process Engineering, Chinese Academy of Sciences, Beijing, P. R. China

ghma@home.ipe.ac.cn

1.1 General Introduction

Modern developments in biotechnology have paved the way for the widespread application of biocatalysis and biotransformation. Nowadays, it is possible to produce a great variety of useful enzymes at a commercially acceptable price. However, industrial application is often hampered by a poor long-term stability and difficulty in recycling, which usually lead to an unacceptable operation cost (Bornscheuer, 2003; Cao, 2005; Sheldon, 2007). These intrinsic drawbacks, in principle, could be often overcome by enzyme immobilization, which always in turn determine the final cost of enzymatic process. Thus, an effective approach and a robust carrier to realize enzyme immobilization become crucial for the acceptability of many enzymatic processes.

In principle, immobilized enzyme, namely the enzyme-carrier complex, consists of two essential components: the noncatalytic structural component, which is designed to facilitate separation and to enhance stability of the catalyst, and the catalytic functional component, which is necessary to convert the substrates to the desired products (Cao *et al.*, 2003; Cao, 2005). The properties of supported enzyme preparations are governed by the properties of both the enzyme and the carrier material (Sheldon, 2007). In

Microspheres and Microcapsules in Biotechnology: Design, Preparation, and Applications
Edited by Guanghui Ma and Zhiguo Su

ISBN 978-981-4316-47-7 (Hardcover), 978-981-4364-62-1 (eBook)
www.panstanford.com

general, coupling enzyme molecules to carriers is a specific kind of modification grafting enzyme with heterogeneous phase. The coupling approach is reflected not only by the irreversibility of the binding or freezing of enzyme conformation, but also by the fact that enzyme performance, such as activity, selectivity, and stability can be improved in comparison with the native enzyme (Cao *et al.*, 2003).

Current methods for enzyme immobilization include adsorption and covalent attachment; structural components for the immobilization include solid beads, porous particles, capsules, membranes, and gels. Unfortunately, none of these methods or carriers could so long solve all the problems and meet all the requirements. The selection criteria for supporting materials, especially for the geometric parameters, are largely depended on the needs of process engineering. Parameters are chosen with the purpose of achieving easy recycling of bioactive catalyst, flexibility of process engineering, and broad applicability to meet various demands (Cao, 2005).

It has been evidenced from time to time that the performance of immobilized enzyme depends not only on the chemical nature of the carriers but also on the geometric properties of the carriers, such as shape, size, porosity, and pore-size distribution (Aparicio *et al.*, 1993; Cao *et al.*, 2003). Accordingly, a great number of synthetic or natural carriers with tailor-made chemical and physical properties were specifically designed for various bio-immobilizations.

Spherical particles, preferably from tens to hundreds of micrometers in size, will be potential supports for industrial bioreactors and the minimal diffusion limitation and will be suitable to quite a lot of industrial applications. In this chapter, we focus on some typical examples of enzyme carriers, for example, porous microspheres, nanoparticles, microcapsules, magnetic beads, and smart gel microspheres.

1.2 Enzyme Supported by Porous Microspheres

A primary requirement for any carrier is the need to have a large surface area, as can be achieved with small particle size or with highly porous materials (Hanefeld *et al.*, 2008). Compared to small carriers, which could lead to arduous separation, porous carriers are more preferred in many areas. In general, it is an accepted fact that high internal specific surface area is necessary to realize high loading capacity. As illustrated in Fig. 1.1, compared to solid beads, porous microspheres usually offer considerable surface area, which promises maximal loading. The immobilization of enzymes on porous materials is very useful in practical applications and for the significantly improved stability of enzyme, especially under relatively severe conditions (Omi *et al.*, 1997).

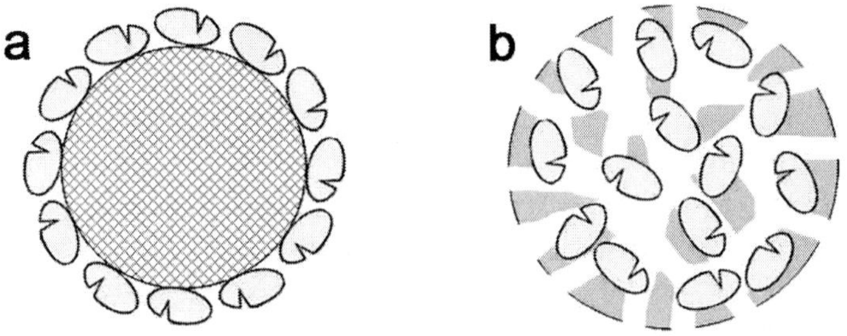

Figure 1.1 (a) Enzyme molecules attached on the surface of a spherical carrier, (b) Enzyme molecules attached on the internal surface of a porous carrier.

Attaching enzyme molecules on the internal surface of porous microspheres, like entrapping enzyme molecules on the inside of porous microspheres, will protect the enzyme molecules from the external environment. Accordingly, the immobilization with porous microsphere could prevent enzyme molecule from external denaturing factors, such as strong sheering within traditional reactor, proteolysis by microbe, and hydrophobic interaction from other supporting materials (Caussette *et al.*, 1998, 1999; Colombié *et al.*, 2001; Bommarius & Karau, 2005). On the other hand, the presence of diffusion constraints with porous carrier often reduces the express of activity (Cao *et al.*, 2003). Pore size plays an important role not only in loading capacity but also in activity retention. Selection of geometric properties for an immobilized enzyme is largely dependent on the peculiarity of certain applications. Loading capacity as well as activity retention must be taken into account when selecting suitable carrier.

Pore-related properties include pore size, pore-size distribution, pore volume, pore structure, and porosity. Pore size plays an important role not only in loading capacity but also in activity retention. Usually, porous microspheres employed as enzyme support have pore-size distribution in a very wide range.

Although inorganic carriers such as silica beads usually feature with mesoporous structure, macroporous structure is the character for many synthetic polymer beads.

In this part, our purpose is not to review all the typical immobilized enzyme systems that are supported by porous microsphere, though we try to give some impressive examples.

1.2.1 Mesoporous Inorganic Microspheres

While nanotechnology-inspired biocatalysis has recently become attractive, materials featured with clearly defined nanoscale structures are already being used as enzyme carriers for decades. A great variety of inorganic particles with mesoporous structures have been widely applied as enzyme carriers, for example, glass beads (Marsh *et al.*, 1973), zeolites (Diaz & Balkus, 1996; Yan *et al.*, 2002), and mesoporous silicas (Springuel-Huet *et al.*, 2001; Wang *et al.*, 2001; Borole *et al.*, 2004; Moelans *et al.*, 2005) such as MCM-41 and SBA-15. Although the most famous mesoporous silicas do not possess spheric morphology, the particle sizes are at micrometer scale, and belong to a kind of broad-scene microsphere.

In 1992, the Mobil Research and Development Corporation (Beck *et al.*, 1992; Kresge *et al.*, 1992) discovered the M41S family of mesoporous silicates (MPS), which possess a narrow pore-size distribution, with pore size in the range of 2–30 nm, and amorphous silica surfaces. Regular mesoporous silicas could be formed by liquid crystal-templating routes, which have been intensively studied for many years. Once the template molecules are removed, the inorganic scaffold possesses pores at a size ranging from 2 to 50 nm (Yiu *et al.*, 2001; Wang *et al.*, 2005). Such pores are well defined, with very narrow pore-size distribution, usually possessing a specific internal area of more the 1000 m^2/g. During the past decade, lots of work has been undertaken in the synthesis of MPS and in the immobilization of enzymes within these supports. These mesoporous structures offer the possibility of adsorbing or entrapping large molecules within their pores (Hudson *et al.*, 2005).

In 1996, Diaz and Balkus first immobilized enzymes with mesoporous MCM-41, which possess pore diameters at 2–8 nm. Since then, research groups around the world have proposed quite a number of methods (Hudson *et al.*, 2005). By adding functional silane precursors, functional groups, for example, amino or carboxylate groups, can be introduced onto the surface to facilitate enzyme immobilization. The feasibility to design the surface groups and to fix the diameter of the pores in the synthesis of mesoporous materials would be surely preferred by enzyme supporting.

However, due to the small pore size, MCM-41 mesoporous materials are restricted to immobilization of enzyme of relatively small sizes (Takahashi *et al.*, 2000; Yiu *et al.*, 2001). In addition, relatively low enzyme loadings and slow loading process were observed, despite these materials possessing specific areas as high as 1000 m^2/g. With the development of carriers, improved enzyme loadings have been reported with SBA-15 materials, which possess a pore size at 15–40 nm (Yiu *et al.*, 2001; Lei *et al.*, 2002; Fan *et al.*, 2003; Kang *et al.*, 2007).

It seems reasonable to believe that confining an enzyme molecule within a space of comparable size, such as nanopores of mesoporous support, could limit the surrounding environment, which jeopardizes the enzyme morphology, thus providing an explanation of enzyme stabilization compared to those involved in macroscopic carriers (Wang *et al.*, 2001; Yan *et al.*, 2006). Bismuto *et al.* (2002) studied the confinement of matrix for immobilized enzyme and revealed that the enhanced stability is mainly due to spatial restriction, which reduces the possibility of deactivation caused by molecule movement. Takahashi *et al.* (2000) also proved that when the average mesopore size matches the molecular diameters of the enzyme, immobilized horseradish peroxidase (HRP) express the highest activity in organic solvent and the best stability.

In porous microspheres, enzyme molecules are attached inside the pores, and a mesoporous structure may be preferable for long-term stability. However, such small pore size will reduce the available surface for an enzyme and bring in severe diffusion constraints for a substrate and a product. Thus, the size of pores is crucial as well as the chemical nature of the carrier (Hanefeld *et al.*, 2008). For these reasons, despite attracting much attention recently, mesoporous materials are still not being used in industrial applications. Recent efforts on mesoporous materials reflect the need to move from the use of model enzyme to more complex and practical enzymes (Hudson *et al.*, 2008). However, it is still a hard job to use mesoporous materials to accommodate the contradictory requirements: relative high loading, considerable stability, and excellent activity retention.

1.2.2 Macroporous Organic Microspheres

While nanometer-sized pores provide a reliable support to enzyme molecules and usually enhanced their stability, mesoporous microspheres suffer from the pore size being of very tiny scale. Many other examples have shown that increase in pore size could not only improve the enzyme loading but also activity retention. Artyomova *et al.* (1986) and Cao *et al.* (2003) proved that the maximum accessible surface for enzyme can be obtained when the pore size for enzyme immobilization should be 5 to 10 times of the size of protein globule.

Compared to inorganic microspheres that usually possess nanometer-sized pores, synthetic organic microspheres usually possess even larger ones (Zhou *et al.*, 2007a,b). In order to reach high enzyme loading and to reduce diffusion constrains, the larger pore size and the more available surface deep inside are much more preferred. As diffusion constraint is pore-size-dependent, use of carrier with large pores should be able to reduce the diffusion limitation, which will certainly translate into high retention of activity (Cao *et al.*, 2003).

A great variety of polymer microspheres have been induced as robust biocatalyst carriers in quite a lot of industrial applications (Bahulekar *et al.*, 1993). Polystyrene (PS) and polymethyl methacrylate (PMMA) resins, which have been widely applied in a variety of industrial processes, are the major species (Omi *et al.*, 1997). Suspension polymerization has been proved as a facile approach to manufacture porous microspheres for enzyme immobilization for a relatively long time (Skovby & Kops, 1990). Phase-separation mechanism, which requires appropriate solvent, is usually employed to produce nanopores in the matrix. Omi *et al.* (1997) employed membrane emulsification technology in porous resin synthesis, manufacturing uniform spherical particles, preferably from tens to hundreds of micrometers in size.

Acrylic resins, such as Eupergit® C, have already been widely used as enzyme carriers (Katchalski-Katzir & Kraemer, 2000). Eupergit® family is a macroporous copolymer of *N,N′*-methylene-bi-methacrylamide, glycidyl methacrylate, allyl glycidyl ether, and methacrylamide, with particle size at 100–250 μm. The manufacturing procedure was based on a novel bead-polymerization method, in which two immiscible organic phases were employed. Because of its structure, Eupergit® is highly hydrophilic and stable over a pH range from 0 to 14 and does not swell or shrink even upon drastic pH changes in this range (Katchalski-Katzir & Kraemer, 2000). For example, Eupergit® C is highly reactor-compatible since almost any common type of reactor, stirred tank, or fixed bed can be used.

Eupergit® C beads bind with proteins via their oxirane groups which react at neutral and alkaline pH with the amino groups of the protein molecules, for example, amino groups of an enzyme to form covalent bonds which are long-term stable within a pH range of 1 to 12. Such carriers can also bind enzyme molecules via their sulfhydril groups or their carboxyl groups in acidic, neutral, and alkaline pH ranges (Turkova *et al.*, 1978). Immobilization with Eupergit® C has been successfully applied to a variety of enzymes for industrial applications. Penicillin amidase on Eupergit® C, for example, maintained 60% of the native activity over more than 800 cycles (Katchalski-Katzir & Kraemer, 2000). Similarly, various porous acrylic resins, such as cross-linked PMMA resins are also used by simple adsorption. For example, the widely used enzyme Candida antarctica lipase B (CALB) is commercially available as Novozyme® 435, which consists of the enzyme adsorbed on a macroporous acrylic resin (Kirk & Christensen *et al.*, 2002; Sheldon, 2007).

Gross's group (Mei *et al.*, 2003; Chen *et al.*, 2007a,b) was the first to employ infrared microspectroscopy to image the distribution of immobilized enzyme molecules in Novozyme® 435. They revealed that although the carrier has a macroporous structure, which is much larger than enzyme globules, the enzyme is located only in the external shell of the beads that have a limited thickness of 80–100 μm. The average pore size of Novozyme® 435 beads is more than 10 times larger than the size of the lipase. They believed that immobilization of lipase molecules on the Lewatit® polymer matrix involves a strong affinity of the enzyme for the matrix or strong protein–protein interactions, which could prevent enzyme to penetrate deep inside the resin beads. Chen *et al.* (2007a,b) investigated and proved that the lipase molecule would distribute evenly in relatively smaller diameter resin beads. However, in cases with larger resin beads with similar pore-size distribution, a nonuniform distribution exists, with most enzyme molecules present in the outer region of resin beads.

Pore size is not only a key factor for loading capacity but is also crucially important for catalysis performances (Frings *et al.*, 1999). For instance, the commercially available carriers Eupergit® C and Eupergit® C 250 possess the same chemical structure but differ mainly in pore size (Katchalski-Katzir & Kraemer, 2000). The pore size of Eupergit® C 250 is approximately 10 times larger than that of Eupergit® C. With enzyme Penicillin G acylase and similar loading, the activity retention of a large-pore carrier is about twice that with small pore size.

Li *et al.* (2010) investigated various polystyrene (PST) microspheres, which are around 50 μm in sizes and much smaller than Novozyme® 435 featuring different pore diameters that were named as giga-, macro-, and mesoporous PST microspheres. The giga- and macroporous PST microspheres were obtained by surfactant-enriched suspension polymerization featuring with reverse-micelle swelling of the oil phase (Zhou *et al.*, 2007b; Li *et al.*, 2010). During their studies, lipase (from *Burkholderia cepacia*) was immobilized by hydrophobic adsorption using similar approach as was used for Novozyme® 435. According to the observations with a laser scanning confocal microscope illustrated in Fig. 1.2, enzyme molecules could easily penetrate to the center of the giga- and macroporous microspheres; however, lipase was adsorbed only in the shell region of the mesoporous microspheres. The thermal stability, storage reliability, and reusability were all proved to be enhanced with an increase in pore size. Similarly, compared to porous microsphere serving as enzyme carrier, large pore size is preferable for final performance and process control.

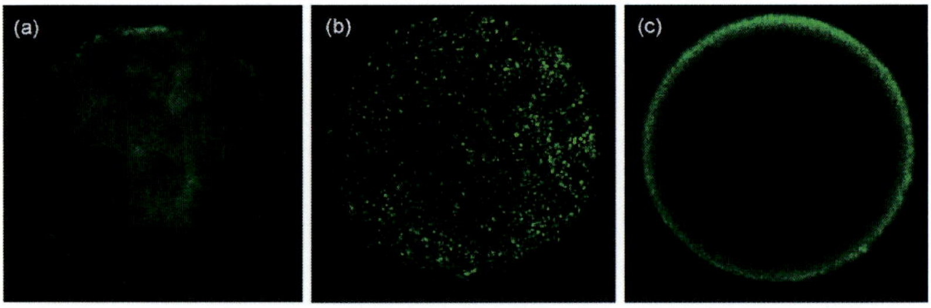

Figure 1.2 Laser confocal images of fluorescamine-lipase immobilized on (a) gigaporous, (b) macroporous, and (c) mesoporous PST microspheres. Reprinted from Li *et al.* (2010), with permission from Elsevier.

For porous carriers with different pore sizes, it is necessary to classify the surface as accessible surface and inaccessible surface. Accessible surface refers to the external surface and the surface where the enzyme molecules can reach, while inaccessible surface refers to the surface where the enzyme molecules cannot reach because of physical restriction (Cao *et al.*, 2003). It is also well accepted that the accessible surface rather than the total surface area is more important to enzyme loading and the catalytic performance. For macroporous carriers, their internal surface is generally 100 times more than the external one; therefore, enzyme molecules that are attached on the external surface can almost be neglected (Kotha *et al.*, 1998). The considerable accessible surface of the organic macroporous carriers, which is their most impressive characteristic, makes them superior to the inorganic mesoporous ones.

However, the preference for large pore size might involve other problems, such as diffusion constraints when the enzyme molecules get entrapped at the center of the microparticles (Ison *et al.*, 1994). The diffusion process of substrate(s) in the porous microspheres seemed to retard the rate of catalysis in large microspheres.

1.3 Spheres at Nanometer Scale

Immobilization of an enzyme on a heterogeneous carrier, which offers a much different microenvironment than from the native state of the enzyme molecule, often leads to loss of over 50% original activity and even more at high enzyme loadings (Cao *et al.*, 2003). Porous materials can afford high enzyme loading but usually suffer from great limitation of diffusion of the substrate. Reducing the size of enzyme-carrier materials is another commonly used strategy to improve loading efficiency of immobilized enzymes.

In the cases of surface attachment, the smaller particles provide the larger surface area for the residence of enzyme molecules, which will lead to the higher enzyme loading (Jia *et al.*, 2003; Vertegel *et al.*, 2004; Kim *et al.*, 2006a; Wang, 2006). There have been extensive studies on the use of micrometer-sized or even larger particles for enzyme immobilization (Omi *et al.*, 1997; Mei *et al.*, 2003; Chen *et al.*, 2007a, 2008). Interest has been growing in the field of employing nanoparticles as carriers for enzyme immobilization ever since nanotechnology emerged (Daubresse *et al.*, 1996; Martins *et al.*, 1996; Caruso & Schüler., 2000; Liao & Chen, 2001; Jia *et al.*, 2003; Kang *et al.*, 2005, 2006).

The considerable specific surface area afforded by nanoparticles is not the only size-dependent feature that is important to the performance of attached enzymes. Vertegel *et al.* (2004) used lysozyme as a model enzyme and demonstrated that the structure and activity of enzyme are strongly dependent upon the size of its carrier, as illustrated in Fig. 1.3. Less perturbation in lysozyme's secondary structure is observed when a protein is adsorbed onto smaller nanoparticles under similar attaching conditions. The structural information strictly correlates with activity recovery for adsorption on the smaller nanoparticles. In short, their pioneering experiment suggested that small nanoparticles, because of their great surface curvature, promote the retention of more native-like protein structure as well as native function compared to their large counterparts.

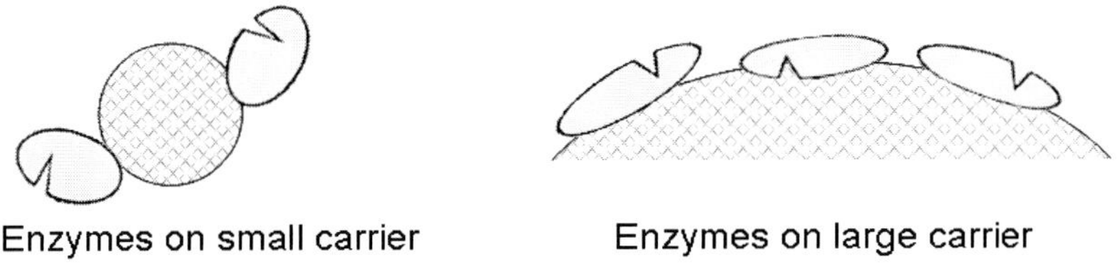

Figure 1.3 Schematic illustration of enzymes adsorption on nanoparticles of different sizes. Reprinted from Gao *et al.* (2009), with permission from John Wiley & Sons.

While explaining the excellent performance brought by nanoparticle support, Wang (2006) concluded that the unique solution behavior of the nanoparticles, that is, the mobility of catalyst in solution, is another crucial factor in determining the activity recovery and explains the higher activities that are usually observed for enzymes attached to nanoparticles than those attached to larger ones. Jia *et al.* (2003) and Wang (2006), with α-chymotrypsin as the model enzyme, demonstrated that the mobile state of the immobilized enzymes is a key factor in their activity recovery. Unlike solid materials of large size, nanoparticles dispersed in a solution are mobile in the form of Brownian motion. In that sense, the enzymes attached to the nanoparticles, as sampled in Fig. 1.4, are not immobilized and are thus different from traditionally immobilized enzymes. Such solution behavior may point to a transitional region between homogeneous catalysis with free native enzymes and a heterogeneous one with immobilized enzymes.

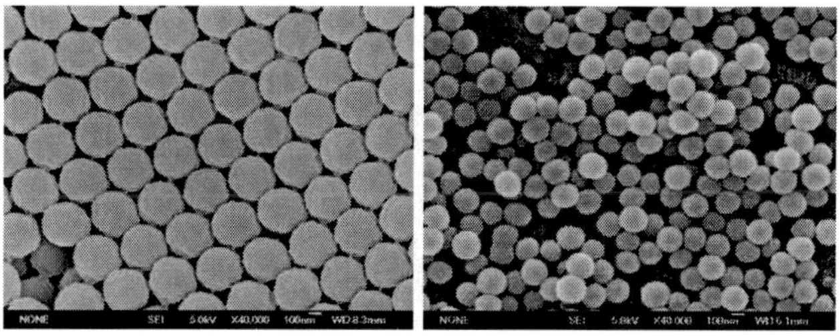

Figure 1.4 Polymer nanoparticles prepared as enzyme carriers. Reprinted from Gao *et al.* (2011), with kind permission from Springer Science+Business Media.

On the other hand, the attachment of enzymes to nanoparticles can also impact the motion of the carriers (Wang, 2006). It was found that the reactions catalyzed by enzymes can also drive the motion of nanoparticles, such as mobility enhancement (Wada *et al.*, 2000; Blum *et al.*, 2001; Lee *et al.*, 2003a). In other words, the enzymes functioned as "nanomotors" on the particles.

The latex particles can be used as carriers for multienzyme system. Zhang *et al.* (2011) explored the potential of simultaneous production of multienzyme system. The production of value-added products from glycerol and xylose was realized by simultaneously recycling and regenerating NAD(H) from two enzymes, glycerol dehydrogenase and xylose reductase, as shown in Fig. 1.5. The shuttling behavior of nanoparticles, featured by the typical Brownian motion, benefits the recycling of cofactor which is necessary for the cofactor-dependant redox enzymatic reactions. Compared to the native enzyme system, the nanoparticle-supported enzymes showed much enhanced stability and reusability.

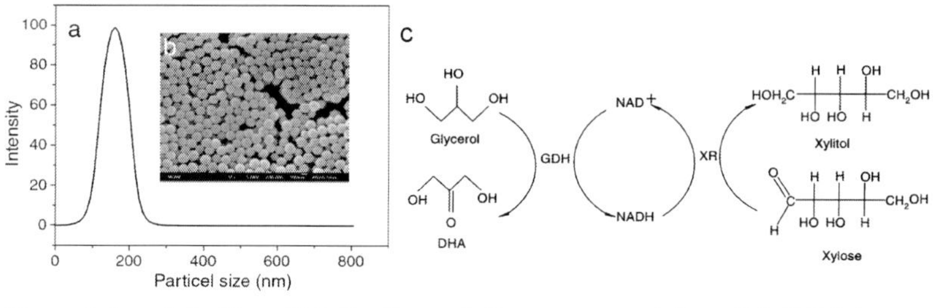

Figure 1.5 (a) Size distribution of the nanoparticles. (b) SEM image of nanoparticles applied for multienzyme supporting. (c) Illustration for enzyme–cofactor–enzyme catalyst. Reprinted from Zhang *et al.* (2011), with permission from Elsevier.

Although such nanoparticle-formed biocatalysts are gifted in action, they induce a lot of inconvenience in application because of their tiny size which always results in difficulty in recycling them, leading to a potential risk to the environment. Commonly applied methods for the separation of nanoparticles, such as ultrafiltration or centrifugation, are time and energy consuming. Also due to their small size, nanoparticles remain suspended in air or water for a long time and can hardly be cleaned up from the environment spontaneously. Although it is still hard to assess the environmental impacts, protective research is necessary to ensure sustainable development of biocatalysts formed from nanoparticles (Colvin, 2003).

1.4 Single-Enzyme Capsule at Nanoscale

In 1997, Wang *et al.* introduced an *in situ* covalent grafting from the surface of enzyme molecules. Based on Wang's approach, Kim and Grate (2003), reported a new form of single-enzyme-centered nanocapsules, which they called single-enzyme nanoparticles (SENs). In SENs, each enzyme molecule is coated by a composite organic/inorganic layer, which is less than a few nanometers in thickness. The preparation of SENs represents a new approach that is different from immobilizing enzymes into mesoporous materials or embedding them in sol-gels, polymers, or bulk composite structures.

Kim *et al.* (2006a) illustrated the construction process of SENs from an enzyme, including enzyme modification and subsequent encapsulation (Fig. 1.6). The construction process of SENs also differs from traditional modification in the preparation of SENs, which begins from the enzyme surface, with covalent reactions to anchor, grow, and cross-link a composite organic/inorganic network around each enzyme molecule. Assisted by a small amount of surfactant, the modified enzymes are first dissolved in hexane phase. This dissolution facilitates the subsequent polymerization in the organic phase. Silane monomers containing both vinyl groups and trimethoxysilyl groups are added to the organic phase. Free radical vinyl polymerization is initiated in this homogeneous solution to produce linear polymers on the enzyme surface. By transferring these polymer-coated enzyme molecules into aqueous phase, the pendant trimethoxysilyl groups are hydrolyzed and condensed with each other, leading to a cross-linked composite network around each enzyme molecule (Kim & Grate, 2003).

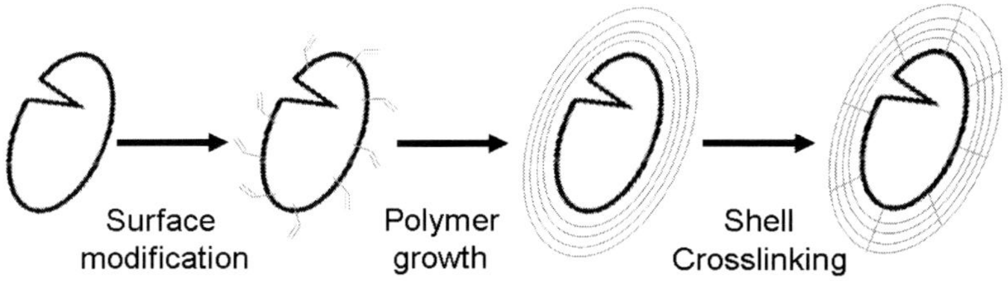

Figure 1.6 Schematic illustration for the synthesis of single-enzyme nanoparticles. Reprinted from Gao *et al.* (2009), with permission from John Wiley & Sons.

Yan *et al.* (2006) proposed a novel two-step *in situ* polymerization procedure that yielded single enzyme nanogels in aqueous solution. In their method, vinyl groups were first generated on the enzyme surface by acryloylation (Yang *et al.*, 1995; Wang *et al.*, 1997), followed by *in situ* polymerization of acrylamides, leading to the formation of enzyme nanogels. The effectiveness of this method has been demonstrated in HRP (Yan *et al.*, 2006), carbonic anhydrase (Yan *et al.*, 2007), and Candida rugosa lipase (CRL) (Ge *et al.*, 2008). In all cases, the enzyme nanogels essentially displayed 80–95% of their original activity. The slight decrease could be mainly attributed to minor mass transfer restriction across the thin, porous, and flexible polymer shell.

The multipoint linkages between the enzyme molecule and hydrophilic polyacrylamide gel shell contribute to the significantly improved stability of enzyme molecules at high temperature and in organic solvents. The activity of the lipase nanogel was almost unchanged for over 500 min while the half-life of the native counterpart was only 30 min at 50°C in aqueous phase (Ge *et al.*, 2008). Moreover, the polyacrylamide shell around the enzyme surface effectively prevents essential water from being replaced by the organic solvent (Klibanov, 2001). All these advantages were sufficient to extend the application of enzymatic catalysis from the aqueous phase to various nonaqueous phases (Ge *et al.*, 2009).

It is believed that enclosing free enzymes in nanocapsules can result in significantly improved stability of the enzymes, and the nanoscale layer of the SEN also does not impose any serious mass-transfer limitation on substrates. However, despite all these advantages, this new technology is still suffering from the drawback of being tiny in scale as was mentioned with nanoparticle carrier in an earlier section of this chapter. Their tiny scale always leads to arduous task in recycling and difficulties in process control. Thus, a more sophisticated approach is required to overcome these drawbacks and to pave the way for industrialization as well.

1.5 Enzyme Supported by Microcapsules

Covalent binding or cross-linking processes often expose enzyme molecules to harsh conditions, which might affect their native structures and activity. On the other hand, immobilization by adsorption could be achieved under mild conditions; however, such immobilization for enzyme is not strong enough to resist changes of microenvironment, such as changes of pH or ionic strength (Cao, 2005). As one of the common choices, encapsulation within microcapsule offers several advantages regarding practical applications, such as mild synthesis condition and relatively higher loading capacity.

Encapsulation of an enzyme is the formation of a membrane-like physical barrier to confine enzyme molecules within a closed compartment (Cao, 2005). What is the difference between entrapment and embedding technology? According to our knowledge, encapsulation will enclose enzyme with a cocoon-like membrane. In this section, we will focus on the enzyme enclosed within capsules that are at micrometer scale. There are several advantages of encapsulation. In such approaches, there is usually no chemical modification involved. Thus deactivation of the native morphology can be effectively avoided during some loading processes. It is also believed that the encapsulation approach provides a universal stage that can integrate more complex functionality with more than one enzyme.

1.5.1 Conventional Microcapsules

Conventional encapsulation is a process by which the enzyme dissolved or dispersed in a solution is further encapsulated in a membrane that forms around the droplets of enzyme solution (Cao, 2005). The encapsulation approach usually employs the process of interfacial solidification in which a membrane or film is formed around the enzyme solution by physical or chemical transformation of the liquid phase around the enzyme solution into the solid phase.

The methods used can be classified into two groups, chemical methods and physical methods. One character of both cases is that the enzyme solution should be immiscible with the phase that is to be solidified. Well-established encapsulation approaches include interfacial polymerization (Chang, 1964), interfacial gelation (Lee *et al.*, 1993), reverse-interfacial gelation (Nigam *et al.*, 1988), interfacial cross-linking (Groboillot *et al.*, 1993), and interfacial deposition (Ozden & Hasirci, 1990). As early as 1960s, Chang (1964) first proposed and developed the "artificial cell" concept, with multienzyme system confined within polymer microcapsules that were fabricated by interfacial polymerization.

Among these methods, interfacial gelation techniques are widely used for a vast variety of enzyme encapsulations as well as whole-cell-associated enzymes (González Siso *et al.*, 1997; Cao, 2005) because such techniques never involve any harsh condition for biomolecules. One characteristic of the usually applied encapsulation approach is that the enzyme molecules encapsulated are free molecules in internal space. It is a hard job to accommodate the usual contradictory requirements: minimum diffusion restriction and maximum loading stability, by the pioneering approach (Kim & Grate, 2003; Cao, 2005). Thus, the difficulties in shell design and permeability control usually restrict its industrial applications. Recently, some more sophisticated techniques for enzyme encapsulation have been developed to overcome the drawback of traditional encapsulation and to meet the needs of different applications (Cao, 2005).

1.5.2 "Ship-in-a-Bottle" Approach

One of the intrinsic drawbacks of the conventional encapsulation is that it is hard to balance loading stability and mass-transfer requirement. Therefore it becomes necessary to control the stability and diffusion resistance of the shell, usually by controlling the pore size and the shell thickness (Cao, 2005). To overcome this dilemma, the post encapsulation and the followed cross-linking approach, which is usually called the "ship-in-a-bottle" approach was proposed and it proved effective (Chang, 1971). The approach has been illustrated in Fig. 1.7, route a.

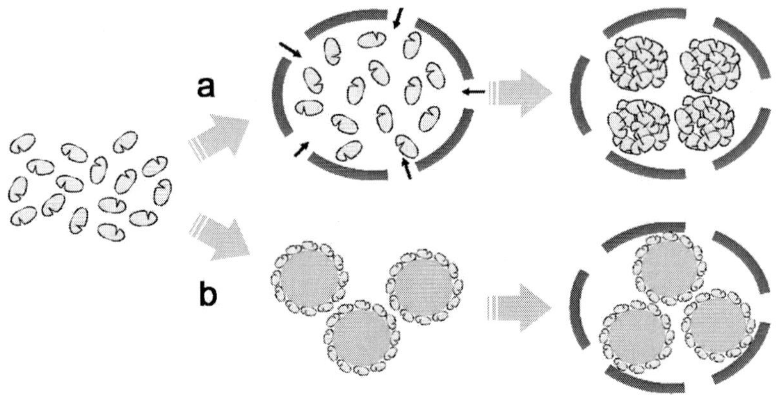

Figure 1.7 Typical multistep encapsulations: (a) Post-loading and cross-linking, (b) Immobilization and then encapsulation.

In the cross-linking approach, the already loaded enzyme molecules aggregate into particles much larger in size. This technique not only loads a large quantity of enzyme but also makes it possible to open large pores in the shell. In the classical technique (Chang, 1964), the semi-permeable membrane allows only small molecules to penetrate and not the cross-linked enzyme aggregates. By using macroporous capsules, this diffusion limitation can easily be overcome (Cao, 2005).

Although the encapsulation cross-linking method has already attracted attention in enzyme immobilization and assembly, it not been industrialized due to the complexity of the fabrication process.

1.5.3 "Fish-in-a-Net" Approach

Another example of an approach in which the enzyme molecules are first supported as nano- or microsized particles and then encapsulated within larger hollow capsules is illustrated in Fig. 1.7, route b. As mentioned before, nanodispersed biocatalysts, such as nanoparticle-formed biocatalysts and single-enzyme capsules, are gifted in actions with enhanced stability and considerable activity recovery, but are dwarfed in practical applications because of their tiny size, which always makes their recycling arduous, resulting in potential risk to the environment. According to Stokes equation, nanodispersed engineered materials tend to remain suspended in aqueous medium or in air for a long time and are therefore not digested by the environment spontaneously. Although the environmental impacts are difficult to assess, protective research is necessary to ensure sustainable development of nanotechnology (Colvin, 2003).

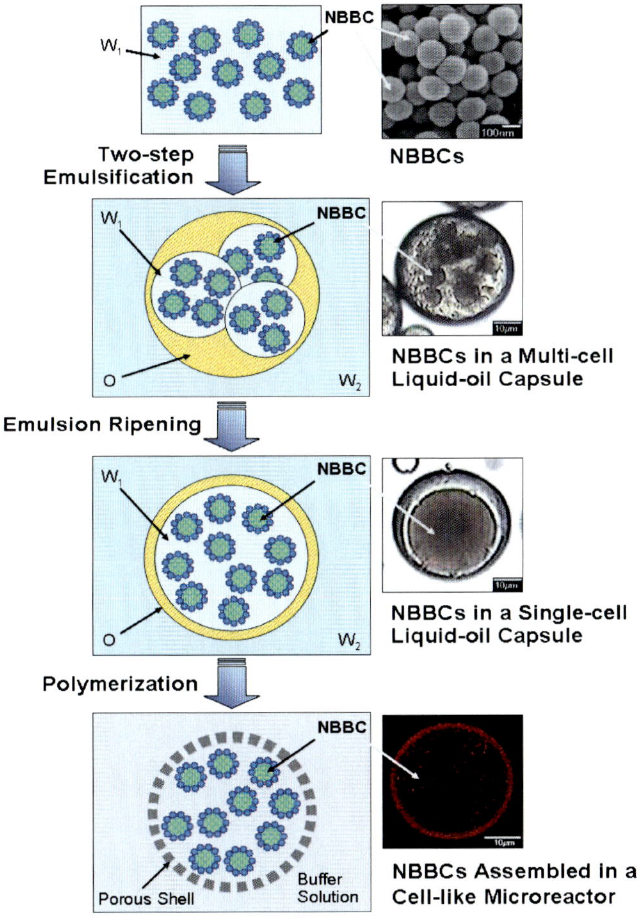

Figure 1.8 Assembly route for cell-like microreactors (CLMRs) with nanoparticles-based biocatalysts (NBBCs) encaged through two-step emulsification, emulsion ripening, and suspension polymerization. Pictures attached aside show the real images under scanning electron microscope, optical microscope, and laser confocal scanning microscope, respectively. Reprinted from Gao *et al.* (2009), with permission from John Wiley & Sons.

Some sophisticated structures, such as porous materials hosting enzyme-carrying nanoparticles (Kim *et al.*, 2006b), cross-linked enzyme aggregates (Lee *et al.*, 2005), and enzyme-attached particles embedded in gel beads (Fan & Lee, 2001), have been developed by applying enzyme modification and fabrication procedures to fulfill practical applications of nanodispersed functional materials.

In our recent study, NBBCs were first constructed and then assembled as CLMRs, using porous microcapsules as containers. The fabrication process is illustrated in Fig. 1.8. The original mobility and activity of nanodispersed biocatalysts were well preserved in the hollow compartments despite the size of the assembled biocatalyst, which is now tens of micrometers and will have little difficulty in recycling. Compared to other hierarchical approaches, for example, hosting enzymatic nanoparticles in mesoporous materials or embedding enzyme-attached particles in gel beads, a remarkable asset of the proposed strategy is that the assembled nanoparticles were still enjoying their original mobility in the form of Brownian motion, which essentially guaranteed native-like performance.

We have developed a novel type of hollow microcapsule to encage pre-constructed nanobiocatalysts, taking into consideration the advantages of the bio-friendly procedure of water-in-oil-in-water (W/O/W) emulsification and the rigid/porous capsule structure obtained from suspension polymerization. We also developed an effective strategy to manipulate the morphology of the W/O/W emulsion globules to prepare single-core water-in-oil (W/O) globules from their multicore precursors (Gao *et al.*, 2009). On the basis of single-core W/O/W emulsion globules, thickness of their oil membranes could be easily controlled, leading to thickness controllable shells of the final capsule products. In addition, phase-separation mechanism is employed to produce nanochannels across the shells. Pores of the microcapsules can be precisely designed, suitable only for rapid exchanging of substrates and products and not allowing the nanoparticle–enzyme conjugates to escape. Such hierarchical scaling-up strategy can well accommodate another pair of contradictories in enzyme support: maximal stability for assembled cores and minimal resistance for molecule diffusion, as is more comprehensive and practical than direct encapsulation.

Figure 1.9 shows the residual activities of α-CT of different forms during recycling. Native enzyme can hardly be reused by simple filtration. In the case with 166 nm NBBCs, biocatalysts were recycled with cellulose acetate membrane (0.1 μm in pore size) and filtered under 0.1 MPa pressure. It costs an hour or more to filter 10 mg NBBCs from 20 mL reaction solution. The residual activity reduced to less than 50% after 20 cycles. The activity loss attributed to denaturation during the separation and redispersion and also attributed to NBBCs being carried away by the membranes. In the case with CLMRs, recycling became more efficient, calling for only a few seconds to separate CLMRs with the same amount of biocatalyst (about 1 g CLMRs) by a membrane filter with a 30 μm pore size. After 100 cycles of utilization, more than 96% activity of the CLMR-assembled NBBCs was still preserved. Such CLMRs brought not only a convenient handling but also an effective protection for delicate enzymes. It is believed that the strategy presented can be universalized to other functional materials in nanoscale, and it is expected to integrate more complex processes in future developments.

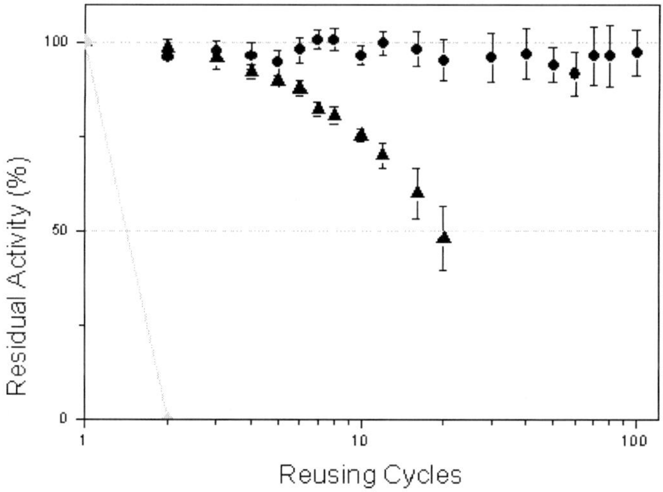

Figure 1.9 Residual activity of α-CT in three different formulations during reusing operation. (◆) Native α-CT, which can hardly be recovered via simple filtration; (Δ) 166 nm NBBCs, which would maintain a stable suspension in liquid; (O) 166 nm NBBCs assembled in 50 μm CLMRs, which would deposit spontaneously after utilization. Reprinted from Gao *et al.* (2009), with permission from John Wiley & Sons.

1.6 Enzyme Supported by Magnetic Carriers

It is time and energy consuming to separate the enzyme-immobilized microspheres from the reaction system. For nanospheres, it is more difficult and almost impossible to separate. Magnetic field–based separations using magnetic particles have attracted people's immense interest in recent two decades (Zeng *et al.*, 2006; Shi *et al.*, 2009). It enables rapid and easy removal of magnetic particle-bound enzymes from complex substrate mixtures. The separation process is easy to realize through manipulation and automation. Fast and cost-effective separation of magnetic carriers from the reaction mixture without filtration or centrifugation makes the magnetic materials useful not only in daily laboratory work but also in production practice.

Magnetic microspheres, compared with other conceivable shapes and large particles, have the advantage of a sufficiently high specific surface area that is available for the immobilization of reactive groups and enzymes. Magnetic microspheres often consist of magnetic cores and a polymer shell. The polymer shell makes it easy to functionalize and immobilize the biomolecules. Magnetic polymer microspheres combine the excellent properties of polymer microspheres (i.e., ease of surface modification and high dispersibility) with the unique magnetic responsibility of magnetic particles. So, magnetic polymer microsphere is a perfect kind of carrier to immobilize enzymes, cells, and other biomolecules.

1.6.1 Design and Preparation of Magnetic Polymer Microspheres

1.6.1.1 Preparation of magnetic polymer microspheres

Good magnetic properties are the essential aspects of the magnetic polymer microspheres. Magnetic properties of the microspheres are greatly influenced by the properties and distribution of magnetic cores inside the polymer microspheres. The magnetic cores should have good response and very low remanence to the applied external magnetic field. Small size and good chemical stability are also the important aspects. Moreover, it should be easy production and have reasonable price (Horak *et al.*, 2007). Superparamagnetic nanoparticles can fulfill the requirements. The magnetic core can be prepared by mechanical (Reshmi *et al.*, 2007), chemical (Bruice & Bruice, 2005), and hydrothermal methods (Carolan *et al.*, 2007).

The chemical method mainly includes coprecipitation with bases and thermolysis of organometallic precursors. The typical process of the coprecipitation method are described as follows (Lei *et al.*, 2009b): $FeCl_3 \cdot 6H_2O$ and $FeCl_2 \cdot 4H_2O$ are first dissolved in deionized water under nitrogen gas with vigorous stirring at high temperature (e.g., 80°C). $NH_3 \cdot H_2O$ and oleic acid are added to the solution. Then the resulting Fe_3O_4 magnetic fluid can be isolated from the solution by a magnet. The resulting Fe_3O_4 is washed several times with deionized water to remove the excess oleic acid. The prepared Fe_3O_4 exhibits zero remanence and coercivity, which proves that the magnetic microspheres have superparamagnetic properties. These properties enable the magnetic microspheres to respond to an applied magnetic field rapidly without any permanent magnetization, which redisperses rapidly when the magnetic field is removed. Shi *et al.* (2010) successfully prepared the Fe_3O_4 microspheres with a solvothermal reaction. The silica-coated magnetic core microspheres showed high saturation magnetization (Fig. 1.10). The resulting magnetic silica microspheres can be quickly

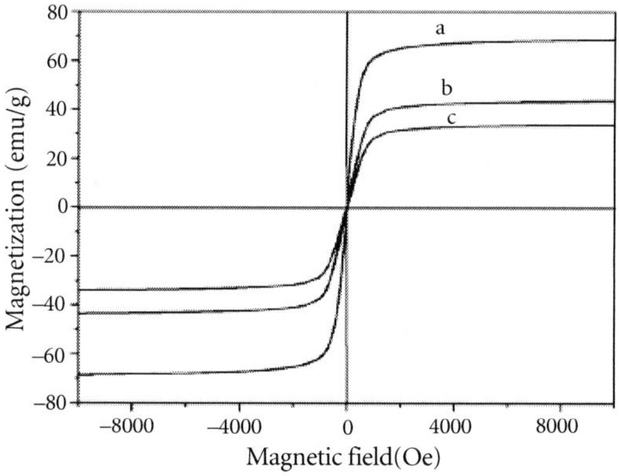

Figure 1.10 The hysteresis loops of (a) Fe_3O_4, (b) $Fe_3O_4/NH_2–SiO_2–10\%$, and (c) $Fe_3O_4/NH_2–SiO_2–10\%$-a at 300 K. *Note*: Fe_3O_4 nanoparticles were about 300 nm in diameter. $NH_2–SiO_2–10\%$ were core-shell microspheres prepared through a solvent thermoreaction. 10% represents the weight percentage of γ-aminopropyltriethoxysilane (APTES) to total silica source. When the total amount of tetraethylorthosilicate (TEOS) and APTES is increased to 0.94 g, the obtained samples are denoted as $Fe_3O_4/NH_2–SiO_2–10\%$-a. Reprinted from Shi *et al.* (2010), with permission from Elsevier.

separated to the wall of the container using a magnet within 1min. The immobilized penicillin G acylase can also be quickly separated and collected

Various methods have been developed to prepare magnetic polymer particles: (i) direct precipitation of iron salt inside pores of the porous polymer microspheres (Ogelstad *et al.*, 1992); (ii) direct polymerization of monomers in the presence of Fe_3O_4 nanoparticles; (iii) post-cross-linking of suspensions of polymer solution droplets with Fe_3O_4 nanoparticle dispersion (Müller-Schulte & Brunner, 1995; Jiang *et al.*, 2005; Liu *et al.*, 2005b); (iv) adsorption by utilizing the opposite charge between polymer particles and magnetic particles. Various preparation methods would provide magnetic polymer microspheres differing in morphology (Fig. 1.11a–d), including magnetic core–polymer shell (Fig. 1.11a), magnetic multicores homogeneously dispersed within the polymer matrix (Fig. 1.11b), magnetic nanoparticles located on the surface of a polymer core ("strawberry" morphology, Fig. 1.11c), "brush" (hair) morphology polymer chains attached to the magnetic core, Fig. 1.11d).

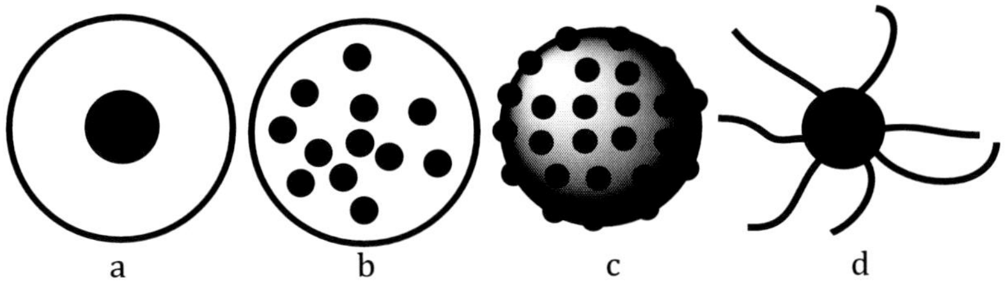

Figure 1.11 Different morphologies of composite magnetic polymer microspheres: (a) single-core, (b) multicore, (c) strawberry, and (d) brush (hair) morphology. Reprinted from Horak *et al.* (2007), with permission from John Wiley & Sons.

1.6.1.2 Introduction of functional groups on magnetic polymer microspheres

As enzyme carriers, the polymer materials of the microspheres should have functional groups to further immobilize enzymes. They are often introduced by copolymerization with functional monomers or modification after polymerization. The functional groups often include amino group, carboxyl group, epoxy group, and aldehyde group, etc.

1.6.1.2.1 *Copolymerization*

The copolymerization methods include suspension polymerization (Liu *et al.*, 2003), emulsion polymerization, (Ding *et al.*, 2001; Liu *et al.*, 2004a), emulsion-templated polymerization, (Liu *et al.*, 2004b), and dispersion polymerization, (Lavayre & Baratil, 1982; Montero *et al.*, 1993). Suspension polymerization, as a simple and easy to scale up method, is widely reported. However, the magnetic polymeric supports made by conventional suspension polymerization were mostly in the size of several ten to hundred micrometers with a very broad size distribution (Cocker *et al.*, 1997; Lee *et al.*, 2003b). The magnetic microspheres were synthesized by the suspension polymerization of GMA, methacrylic acid (MAA), and divinylbenzene (DVB) in the presence of oleic acid-coated Fe_3O_4 nanoparticles. The diameter of the microspheres ranged from 10 μm to 20 μm (Lei *et al.*, 2009a) (Fig. 1.12a). The active epoxy groups were introduced by copolymerizing with GMA. The introduced epoxy groups can covalently immobilize triacylglycerol lipase. The activity yield is up to 63% (±2.3%) and enzyme loading is up to 39(±0.5) mg/g supports. Modified suspension polymerization method was carried out by using special reactor equipped with four vertical stainless steel baffle plates and a four-paddle. In this reaction, the temperature was controlled precisely and large amount of stabilizer was added to prepare the magic microspheres with proper size and distribution. Micron-sized (less than 8 μm in diameter) magnetic poly(methacrylate-divinylbenzene) (PMADVB) spheres were obtained by this modified suspension polymerization (Fig. 1.12b) (Liu *et al.*, 2005a).

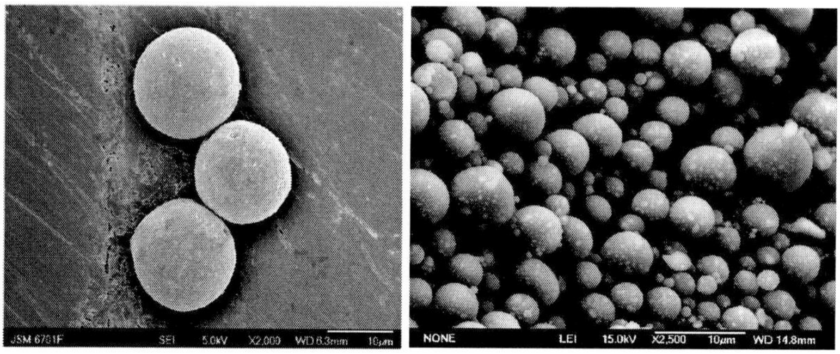

Figure 1.12 SEM images of the prepared microspheres: (a) poly(GMA-DVB-MAA), (b) PMADVB. Reprinted from (a) Lei *et al.* (2009a) and (b) Liu *et al.* (2005a) with permission from Elsevier.

1.6.1.2.2 *Surface modification*

The surface modification methods of the obtained magnetic nanospheres and microspheres often use grafting polymers made by "living" radical polymerization techniques such as nitroxide-mediated polymerization (NMP) (Sonvico *et al.*, 2005), atom transfer radical polymerization (ATRP) (Wan *et al.*, 2005), and reversible addition fragmentation chain transfer (RAFT) (Yuan *et al.*, 2006). They show characteristics of "living" systems, for example, predefined molecular weight, narrow molecular weight distribution, and active polymer end groups. The lifetime of the living radical can be extended up to several hours. So it is possible to control the length of polymers grafted on the nanoparticles. Huang *et al.* (2007) prepared magnetic microspheres with ATRP initiators by coupling the amino group of ethanolamine with the carboxyl group on the precursor particles followed by esterification of the hydroxyl group on the microspheres with 2-bromoisobutyryl bromide (Fig. 1.13).

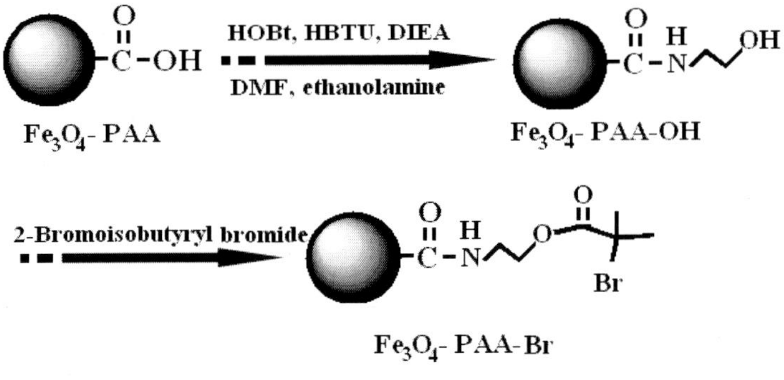

Figure 1.13 Synthetic scheme for forming magnetic microsphere initiators. HOBt, 1-hydroxybenzotriazole hydrate; HBTU, O-Benzotriazole-*N,N,N′,N′*-tetramethyluronium hexafluorophosphate; DIEA, *N,N′*-diisopropyl ethylamine; PAA, poly(acrylic acid).

The carboxylic acid groups of Fe_3O_4–PAA were first coupled with ethanolamine using HBTU as the coupling reagent. The resulting Fe_3O_4–PAA–OH was treated with 2-bromoisobutyryl bromide to form Fe_3O_4–PAA–Br, an ATRP initiator containing magnetic microspheres. Then the grafting polymerization was carried out using Fe_3O_4–PAA–Br microspheres as an initiator and CuBr/1,1,4, 7,7-pentamethyldiethylenetriamine (PMDETA) as the catalyst to graft polymers on the microspheres (Huang *et al.*, 2007). When using polymer mixtures for grafting on Fe_3O_4–PAA–Br, the resulting magnetic microspheres would have multifunctional groups to fulfill multiuse simultaneously. The resulting magnetic microspheres possess poly (glycidyl methacrylate) units, which act as functional groups, and poly (glycerol monomethacrylate) units, which are hydrophilic and make the microspheres dispersible in aqueous solution.

The polyelectrolyte brush poly(sodium 4-styrenesulfonate) (PSStNa) was grafted onto the surface of Fe_3O_4/SiO_2 composite particles by surface-initiated atom transfer radical polymerization (SI-ATRP) using modified magnetic silica as initiator to build more stable assembly. Subsequently, deposition occurs by electrostatic interactions between PSStNa and chitosan layer with opposite charges by introducing a layer-by-layer (LbL) method (Fig. 1.14). The results showed that Fe_3O_4/SiO_2-g-PSStNa nanocomposite microspheres with modified multishells could enhance the stability of both microspheres in solution and the immobilized pectinase. The immobilized enzyme retained >50% of its initial activity over 30 days (Lei *et al.*, 2009b).

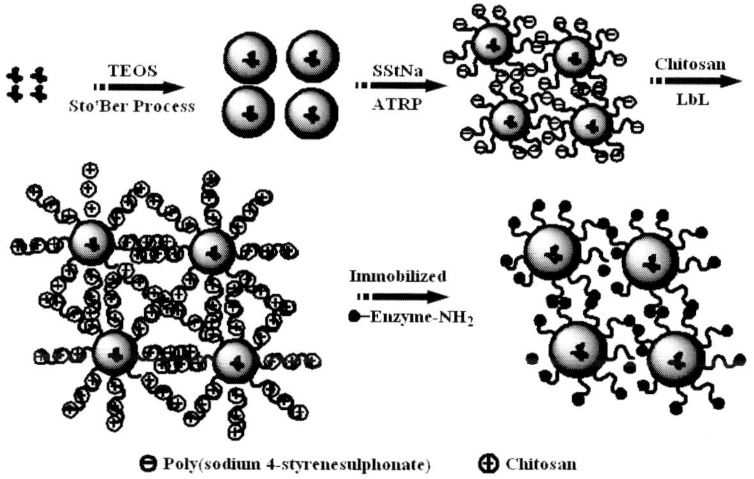

Figure 1.14 Schematic representation of the preparation of Fe_3O_4/SiO_2-g-PSSNa/chitosan particles and pectinase immobilization. Reprinted with permission from Lei *et al.* (2009b). Copyright 2009 American Chemical Society.

Magnetic carbonaceous (MC) microsphere surface was modified by 3-glycidoxypropyltrimethoxysilane (GLYMO) to introduce epoxy group (Fig. 1.15) (Yao *et al.*, 2008) and gave a good immobilization ratio and activity of the enzyme. The amount of trypsin immobilized on the MC microspheres was calculated to be about 37.2 μg/mg, which was higher than the commercial paramagnetic porous glass beads used in Krogh's work (18.1 μg/mg) (Krogh *et al.*, 1999). The enzyme-immobilized magnetic microspheres were successfully applied for fast protein digestion with microwave assistance.

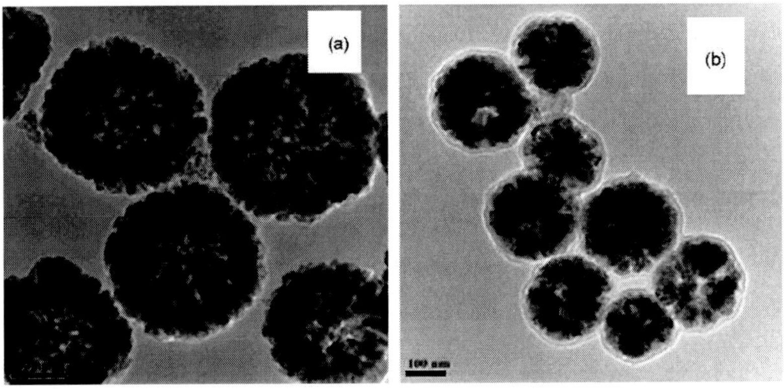

Figure 1.15 The transmission electron microscope (TEM) images of (a) the magnetite microspheres and (b) the magnetic carbonaceous microspheres. Reprinted from Yao *et al.* (2008), with permission from Elsevier.

1.6.2 Property Improvement of the Immobilized Enzymes on Magnetic Polymer Microspheres

Various enzymes including glucoamylase (Bahar & Celebi, 1999), cytochrome C oxidase (Cuyper & Joniau, 1992), β-lactamase (Gao *et al.*, 2003), chymotrypsin (Izmaĭlov *et al.*, 2000, Rebelo *et al.*, 2010), alcohol dehydrogenase (Chen & Liao, 2002; Liao & Chen, 2002), glucose oxidase, galactose oxidase, urease (Varlan & Sansen Ann Van, 1996), neuramidinase (Bilkova *et al.*, 2002), papain (Korecká *et al.*, 2005), Dnase (Rittich *et al.*, 2002), Rnase (Horák *et al.*, 2001), etc. can be immobilized via chemical bonds on coated or embedded magnetic particles. Enzymes immobilized on magnetic supports generally have higher thermal and storage stability facilitating repeated applications but lower activity than free enzymes in solution. The activity of the biomolecule immobilized on these magnetic particles depends on the particular nature of the magnetic microspheres including acidity, surface coverage, and functional group uniformity.

The main disadvantages of those magnetic supports are their large size, low density of surface functional groups, and weak magnetism.

1.6.2.1 Properties of the magnetic polymer microspheres

The morphology, size, polydispersity, and pores of the magnetic polymer microspheres are the important properties when using them as carriers. For magnetic microspheres with too large size, the surface area per mass unit is too small. Furthermore, most functional groups were often buried in the polymer with only a small part localized on the surface of the microspheres during polymerization process. Therefore, it is hardly reported in literature that micron-sized (several microns) magnetic polymeric supports with high density of surface functional groups could be prepared. To overcome this disadvantage, one choice is that highly porous materials with large pore size have to be employed. These materials still have the shortcomings of diffusion restriction. Yong *et al.* (2008) developed a new kind of magnetic porous carrier with active epoxy groups by the copolymerization of glycidyl methacrylate (GMA), ethylene glycol dimethacrylate (EGDMA), and vinyl acetate (VAc) encapsulating oleic acid–coated nanomagnetite (Fe_3O_4) (Fig. 1.16). The average pore diameter and average porosity of the resulting microspheres are 300–500 nm and 40%, respectively. Microspheres with plenty of pores on the surfaces would have greatly increased surface area for the reaction of epoxy groups with the enzyme. This carrier would be suitable for immobilizing enzymes and would also provide a good transmission for substrate and product during the enzymatic reaction. Furthermore, the carrier could form some protections for the immobilized lipase due to the shield functions (such as for

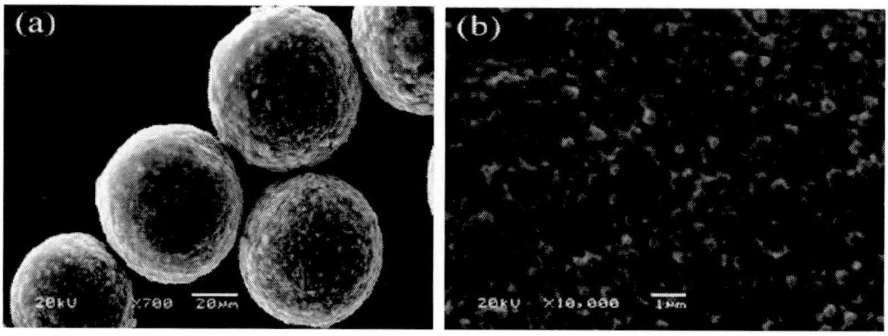

Figure 1.16 SEM micrographs of the magnetic microspheres: (a) Magnification 700×, (b) Magnification 10,000×. Reprinted from Yong *et al.* (2008), with permission from Elsevier.

high temperature, denaturant) of the porous structures. CRL was immobilized by covalent reaction via the amino groups of the lipase and the epoxy groups of the microspheres under mild condition (Fig. 1.17). The CRL activity yield was up to 64.2% and exhibited good thermal stability and reusability.

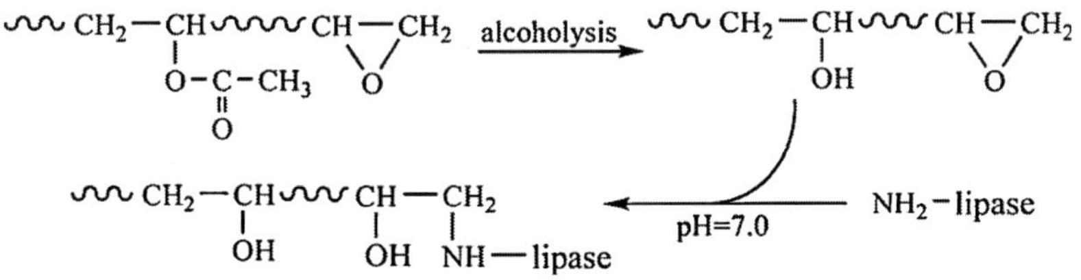

Figure 1.17 Reaction scheme of the magnetic microspheres alcoholysed in NaOH–methanol solution and covalently immobilized lipase on the alcoholysed microspheres. Reprinted from Yong *et al.* (2008), with permission from Elsevier.

The properties of the magnetic microspheres surface have great influence on the enzyme immobilization process. Several magnetic carriers, such as microspheres of various biomaterials encapsulating the magnetic particles and copolymers (Guo *et al.*, 2003) with magnetic particles have been used with good results. However, if the surface polymers are hydrophilic, these microspheres are not able to contact the enzymes, especially those in solution, this leads to reduced enzymes loading amount of the carriers and decreased activity yield. On the other hand, when the enzymes immobilized on hydrophilic materials were employed in organic solution, necessary water could concentrate around enzymes and promote the catalysis reaction greatly (Moreno *et al.*, 1997; Betigeri & Neau., 2002; Szczesna-Antczak *et al.*, 2002).

1.6.2.2 Functional groups on the magnetic polymer microspheres

The amount of functional groups on the magnetic microspheres' surface is another important factor that will greatly affect the loading amount of the enzymes. If it is too low, the loading amount of the enzymes would correspondingly be low. On the contrary, if the amount of the functional groups is too high, the amount of the immobilized enzymes would be high and the activity would also be lowered because of the steric hindrance. Polymer arms grafted on the surface of the magnetic microspheres can improve the loading capacity of the carriers (Huang *et al.*, 2007) and prevent the interaction between the neighboring immobilized enzymes. Arica *et al.* (2000) firstly prepared magnetic PMMA (MPMMA) microspheres by the solvent evaporation method. Then a 6-carbon spacer-arm (i.e., hexamethylene diamine, HMDA) was covalently attached by the reaction of carbonyl groups of PMMA. Glucoamylase was covalently immobilized through the spacer-arm of the MPMMA microspheres by using either carbodiimide (CDI) or cyanogen bromide (CNBr) as a coupling agent (Fig. 1.18). The activity yield

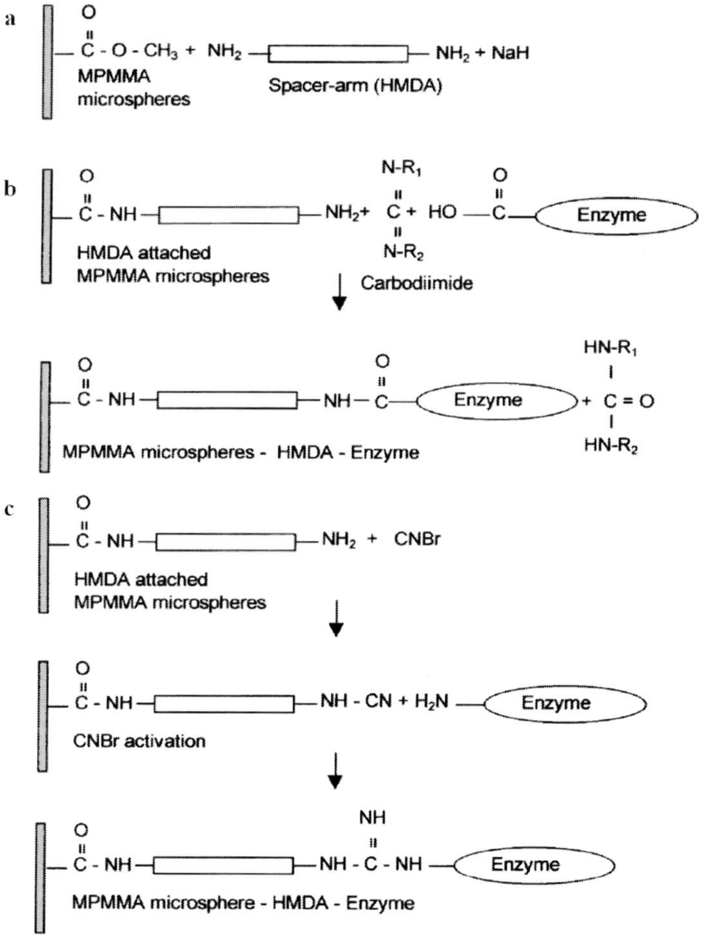

Figure 1.18 Schematic representation of reaction mechanisms: (a) Spacer-arm attachment, (b) immobilization of enzyme via CDI coupling, and (c) immobilization of enzyme via CNBr coupling. Reprinted from Arica *et al.* (2000), with permission from Elsevier.

of the immobilized glucoamylase was greatly improved especially for CNBr coupling (up to 73%). The Km values for immobilized glucoamylases CDI coupling (12.5 g/L dextrin) and CNBr coupling (9.3 g/L dextrin) were higher than that of the free enzyme (2.1 g/L dextrin). Upon immobilization of glucoamylase on the magnetic spacer arm attached microspheres, especially by the CNBr coupling method, yielded a high residual activity and a high operational, thermal, and storage stability.

1.7 Enzyme Supported by Smart Carriers

In recent years, interest has been focused on the design of the smart and intelligent polymeric materials as enzyme carriers to bring some special properties besides traditional protections of immobilized enzymes. Smart materials can undergo rapid and reversible changes in shape, surface characteristics, solubility, formation of an intricate molecular assembly or a sol-to-gel transition, which are triggered by small changes in the environment or small external stimuli, such as temperature, pH, ionic strength, magnetic field, etc. The system is capable of returning to its initial state when the trigger is removed (Galaev & Mattiasson., 1999; Kumar *et al.*, 2007). The advantages of the use of smart polymers for immobilization of enzymes are as follows: (i) the properties can be regulated rapidly just by changing the surroundings including temperature, pH, ionic strength, etc. to fit to special use; (ii) they allow reuse of the biocatalyst after homogeneous catalysis (Roy *et al.*, 2004) and the separation (and reuse) of the immobilized enzymes can be precipitated simply by changing the environment conditions. In our lab, poly(*N*-isopropylacrylamide-*co*-acrylic acid)(P(NIPAM-AA)) microspheres were prepared and used as smart carriers to immobilize trypsin (Fig. 1.19). The hydrophilic/hydrophobic properties of the microspheres can be controlled just by changing the environment temperature. When the digestion reaction is going on, we keep the temperature above the lower critical solution temperature (LCST). The microspheres are hydrophobic, and they can enrich the substrate proteins around the microspheres. This greatly improves the digestion rate. While the digestion reaction finishes, we can change the property of the microspheres to hydrophilic just by lowering the temperature below

LCST. Thus, the digested peptides would break off completely from the microspheres.

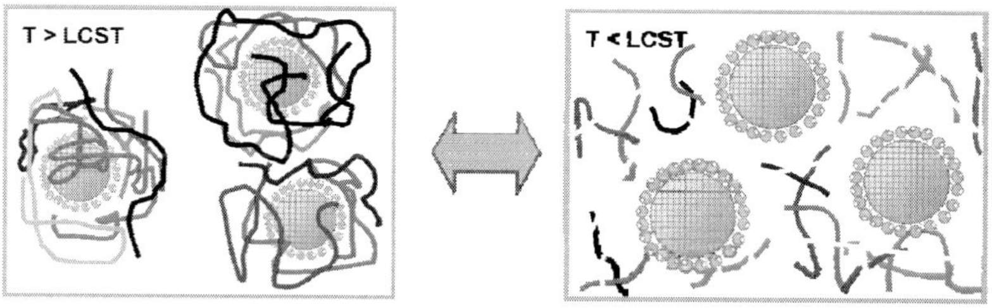

Figure 1.19 Design of thermosensitive P(NIPAM-AA) microspheres to immobilize trypsin used in proteome digestion.

It is a key point that the smart carriers should have rapid and reversible change circle. Microspheres and nanospheres may be one of the most intensively studied subjects because of their small dimension for fast stimuli responses. Particles of polyacrylamide (Pizarro *et al.*, 1997), agarose (Megías *et al.*, 2006), chitosan (Sun *et al.*, 2010; Biró *et al.*, 2008), poly(*N*-isopropylacrylamide) (PNIPAM) (Chen & Hoffman, 1993), and other gel matrix are often used as support materials.

Among these intelligent polymeric materials, *N*-isopropylacrylamide (NIPAM) is the most widely studied thermosensitive polymer. P(NIPAM) has a LCST) of 32°C in water. It becomes hydrophilic below the LCST and hydrophobic above it due to the reversible formation and cleavage of hydrogen bond between the amide groups and the surrounding water molecules (Ding *et al.*, 2000). Comonomers were often used to introduce functional groups which can further react with other molecules to obtain versatile properties and applications.

Various thermoresponsive gels were produced by copolymerization of NIPAM with different acrylate-based comonomers, and some of them have been proposed as support materials for the immobilization of enzymes (Han *et al.*, 1995). Comonomers along with NIPAAm often include acrylic acid, MAA, 2-methyl-2-acrylamidopropane sulfonic acid, trimethyl-acrylamido propyl ammonium 3-methyl-1-vinylimidazolium iodide, sodium acrylate, sodium methacrylate, and 1-(3-sulphopropyl)-2-vinyl-pyridinium betaine (Lee & Shieh, 1999; Opan *et al.*, 2002). Copolymerization of NIPAM with 2-hydroxyethyl methacrylate (HEMA) made suitable materials for immobilization of enzymes because their LCST values were higher than that of pure P(NIPAM) (Bayhan & Tuncel, 1998), which will be consistent with the optimum catalytic reaction temperature of the immobilized enzymes. The presence of hydroxyl groups on the HEMA structure provides functional groups for derivation of the copolymer of P(NIPAM)-HEMA (Arica *et al.*, 1999). The use of MAA, along with NIPAAm, not only changes LCST but also makes the hydrogels responsive to both temperature and pH (Daubresse *et al.*, 1996).

Methods to prepare thermosensitive microspheres include precipitation polymerization emulsion polymerization and suspension polymerization. Huang and coworkers (2004) successfully synthesized P(NIPAM-co-allylamine) and P(NIPAM-co-acrylic acid) nanoparticles by precipitation polymerization method, but the uncontrollable polymerization process resulted in a broad particle-size distribution. Conventional suspension polymerization methods often accompany broad size distribution and are incapable of tuning the particle size during the synthesis. As a result, consistent and repeated environment stimuli responses are hard to achieve, which greatly limits the applications of the particles. Hence, further studies are carried out to prepare monodispersed nanoparticles and microparticles. Monodispersed hollow P(NIPAM) microcapsules were prepared using membrane emulsification technique combined with UV-initiated polymerization by Cheng and coworkers, (2007) but a problem with UV-initiated polymerization is the difficulty of large-scale production because of the limited UV penetrating power. Monodispersed P(NIPAM-co-AAc) hydrogel microsphere were synthesized with membrane emulsification technique by Makino and coworkers (2000), but it is difficult to prepare uniform microspheres with diameters larger than 10 μm when cyclohexane was used as oil phase, resulting from the deformation and break-up of the microspheres during stirring. Therefore, it is valuable to prepare temperature-/pH-responsive monodispersed microsphere with micron size at large scale.

In our lab, monodispersed P(NIPAM-co-AAc) microspheres with tunable pore sizes by copolymerizing with different content of AAc were successfully prepared using combination of SPG membrane emulsification technique and reverse phase suspension polymerization at room temperature with *N,N,N′N′*-tetramethylethylenediamine (TEMED) as an accelerator (Fig. 1.20).

Relationship between the mean pore diameter of P(NIPAM-AA) microspheres and their AA% (w/w) are shown in Fig. 1.21. The nearly linear character will be an important guidance for design of the pore size in the P(NIPAM-AA) microsphere. Thus, we can design and prepare the P(NIPAM-AA) microspheres with needed pore size for different enzyme. Such thermo-/pH-sensitive microspheres with controllable porous structures are expected to meet various requirements in enzyme immobilization.

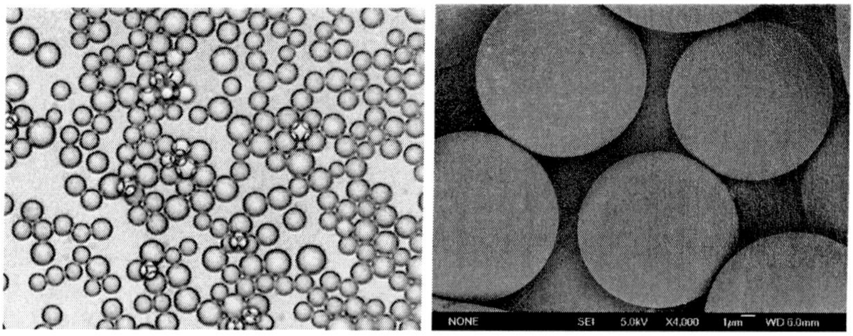

Figure 1.20 (a) Optical micrograph of P(NIPAM-co-AAc) microspheres in water at room temperature; the scale bar is 50 μm. (b) SEM photograph of the dried P(NIPAM-co-AAc) microspheres.

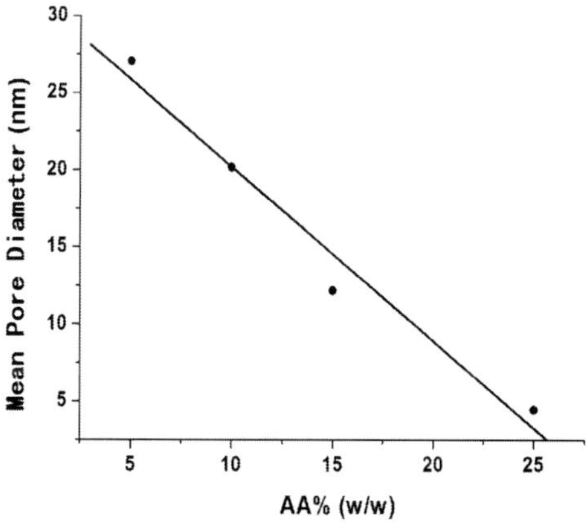

Figure 1.21 Relationship between the mean pore diameter and the AA% (w/w).

Besides thermosensitive microspheres, other stimulate-responsive smart hydrogels were reported in recent years. Smart hydrogels with magnetic character have also been studied.

The mutisensitive hydrogels were prepared by copolymerizing NIPAM, MAA, and *N,N*′-methylene-bis-acryamide (MBA) in the presence of ultraline magnetite particles (average particle diameter of 10–20 nm) prepared by coprecipitation. All the magnetic hydrogel microspheres showed a reversible transition between flocculation and dispersion as a function of temperature (4–40°C) and the thermoflocculated microspheres were separated quickly in the magnetic field. Two different enzymes were immobilized onto these thermosensitive magnetic hydrogel microspheres and retained high activity (Kondo & Fukuda, 1997). For both immobilized trypsin and AGβgal, over 95% of the initial activity was retained after repeated thermocycles (10 times), which shows that the thermosensitive magnetic hydrogel microspheres are novel support materials for enzyme immobilization.

Gai and Wu (2009) prepared a novel reversible pH-triggered release and reloading immobilized enzyme system supported by poly (acrylic acid/*N,N*′-methylene-bisacryl-amide) hydrogel microspheres. The prepared microspheres were regularly spherical with 3.8–6.6 μm diameters and trypsin was immobilized using adsorption method. Loading efficiency of 77.2% was achieved at pH 4.0 in 1 h. Release efficiency of 91.6% was obtained under optimum pH catalysis condition set at 8.0 and trypsin was free in solutions with retention activity of 63.3%, and 51.5% of the released trypsin could be reloaded in 10 min. This study offers a promising alternative for enzyme recovery in biotechnology.

Photo regulation of enzyme activities is of general interest for future bioelectronic devices. "On-off" photo-stimulation of enzyme activities was achieved by covalent attachment of photoisomerizable (photochromic) components to proteins and modification of the enzyme-active site by photoisomerization by the use of photoisomerizable inhibitors. Willner *et al.* (1993) described a novel method by which the activity of α-chymotrypsin can be regulated by its immobilization in a photoisomerizable azobenzene copolymer. The switchable activities of the polymer-immobilized enzyme were attributed to the photostimulated transport of the enzyme substrate across the polymer matrix in its two photostimulated *cis* and *trans* states. This "on" and "off" cycle could be carried out repeatedly in a reversible fashion.

Smart microspheres are increasingly being used to immobilize enzymes. The special response of the smart microspheres to the environment would bring great convenience in the separation of the immobilized enzymes from the reaction system. Therefore, using smart microspheres as enzyme carriers is emerging as a promising field.

1.8 Conclusion and Prospects

In this chapter, we focused on enzyme carriers featured mainly with a spherical structure and at a scale ranging from several nanometers to hundreds of micrometer, including porous microspheres, nanoparticles, microcapsules, and magnetic and smart gel microspheres. Such supporting materials, not only serve as a scaffold on which the protein molecule could be tethered but also provide a unique microenvironment, physically and chemically, which strongly affects the performance of the loaded enzyme, such as the activity, selectivity, and stability.

All immobilization approaches today have to be a compromise between maintaining the native performance of enzyme molecules and achieving feasible control that will meet the requirement of industrial processes. Two apparent trends dominate the development today: first, the use of new reagents and/or carriers, and second, the approaches that take into account the increasing knowledge of enzyme structure and the microenvironment. Thanks to the diversity and broad availability of the carrier, it is always possible to find a robust carrier to meet a unique application.

References

Aparicio, J., & Sinisterra, J. (1993). Influence of the chemical and textural properties of the support in the immobilization of penicillin G-acylase from Kluyvena citrophila on inorganic supports. *Journal of Molecular Catalysis*, **80**(2), 269–276.

Arica, M. Y., Oktem, H. A., *et al.* (1999). Immobilization of catalase in poly (isopropylacrylamide-co-hydroxyethylmethacrylate) thermally reversible hydrogels. *Polymer International*, **48**(9), 879–884.

Arica, M. Y., Yavuz, H., *et al.* (2000). Immobilization of glucoamylase onto spacer-arm attached magnetic poly (methylmethacrylate) microspheres: Characterization and application to a continuous flow reactor. *Journal of Molecular Catalysis B: Enzymatic*, **11**(2–3), 127–138.

Artyomova, A., Voroshilova, O., *et al.* (1986). Macroporous silica in chromatography and immobilization of biopolymers. *Advances in Colloid and Interface Science*, **25**, 235–248.

Bahar, T., & Celebi, S. S. (1999). Immobilization of glucoamylase on magnetic poly (styrene) particles. *Journal of Applied Polymer Science*, **72**(1), 69–73.

Bahulekar, R., Prabhune, A., *et al.* (1993). Immobilization of penicillin G acylase on functionalized macroporous polymer beads. *Polymer*, **34**(1), 163–166.

Bayhan, M., & Tuncel, A. (1998). Uniform poly (isopropylacrylamide) gel beads for immobilization of α-chymotrypsin. *Journal of Applied Polymer Science*, **67**(6), 1127–1139.

Beck, J., Vartuli, J., *et al.* (1992). A new family of mesoporous molecular sieves prepared with liquid crystal templates. *Journal of the American Chemical Society*, **114**(27), 10834–10843.

Betigeri, S. S., & Neau, S. H. (2002). Immobilization of lipase using hydrophilic polymers in the form of hydrogel beads. *Biomaterials*, **23**(17), 3627–3636.

Bilkova, Z., Slováková, M., *et al.* (2002). Oriented immobilization of galactose oxidase to bead and magnetic bead cellulose and poly (HEMA-co-EDMA) and magnetic poly (HEMA-co-EDMA) microspheres. *Journal of Chromatography B: Analytical Technologies in the Biomedical and Life Sciences*, **770**(1–2), 25–34.

Biró, E., Németh, Á. S., *et al.* (2008). Preparation of chitosan particles suitable for enzyme immobilization. *Journal of Biochemical and Biophysical Methods*, **70**(6), 1240–1246.

Bismuto, E., Martelli, P. L., *et al.* (2002). Effect of molecular confinement on internal enzyme dynamics: Frequency domain fluorometry and molecular dynamics simulation studies. *Biopolymers*, **67**(2), 85–95.

Blum, D. J., Ko, Y. H., *et al.* (2001). ATP synthase motor components: Proposal and animation of two dynamic models for stator function. *Biochemical and Biophysical Research Communications*, **287**(4), 801–807.

Bommarius, A. S., & Karau, A. (2005). Deactivation of formate dehydrogenase (FDH) in solution and at gas-liquid interfaces. *Biotechnology Progress*, **21**(6), 1663–1672.

Bornscheuer, U. T. (2003). Immobilizing enzymes: How to create more suitable biocatalysts. *Angewandte Chemie (International Edition)*, **42**(29), 3336–3337.

Borole, A., Dai, S., *et al.* (2004). Performance of chloroperoxidase stabilization in mesoporous sol–gel glass using *in situ* glucose oxidase peroxide generation. *Applied Biochemistry and Biotechnology*, **113**(1), 273–285.

Bruice, T. C., & Bruice, P. Y. (2005). Covalent intermediates and enzyme proficiency. *Journal of the American Chemical Society*, **127**(36), 12478–12479.

Cao, L. (2005). *Carrier-bound Immobilized Enzymes: Principles, Applications and Design*. Wiley-VCH Verlag GmbH: Weinheim, Germany.

Cao, L., Langen, L., *et al.* (2003). Immobilised enzymes: Carrier-bound or carrier-free? *Current Opinion in Biotechnology*, **14**(4), 387–394.

Carolan, N., Forster, R. J., *et al.* (2007). Covalent attachment of ferrocene to soybean peroxidase glycans: Electron transfer mediation to redox enzymes. *Bioconjugate Chemistry*, **18**(2), 524–529.

Caruso, F., & Schüler, C. (2000). Enzyme multilayers on colloid particles: Assembly, stability, and enzymatic activity. *Langmuir*, **16**(24), 9595–9603.

Caussette, M., Gaunand, A., *et al.* (1998). Enzyme inactivation by inert gas bubbling. *Progress in Biotechnology*, **15**, 393–398.

Caussette, M., Gaunand, A., *et al.* (1999). Lysozyme inactivation by inert gas bubbling: Kinetics in a bubble column reactor. *Enzyme and Microbial Technology*, **24**(7), 412–418.

Chang, T. (1964). Semipermeable microcapsules. *Science*, **146**(3643), 524.

Chang, T. (1971). Stabilization of enzyme by microencapsulation with a concentrated protein solution or by crosslinking with glutaraldehyde. *Biochemical and Biophysical Research Communications*, **44**, 1531–1533.

Chen, B., Hu, J., *et al.* (2008). Candida antarctica lipase B chemically immobilized on epoxy-activated micro-and nanobeads: Catalysts for polyester synthesis. *Biomacromolecules*, **9**(2), 463–471.

Chen, B., Miller, E. M., *et al.* (2007a). Effects of macroporous resin size on Candida antarctica lipase B adsorption, fraction of active molecules, and catalytic activity for polyester synthesis. *Langmuir*, **23**(3), 1381–1387.

Chen, B., Miller, M. E., *et al.* (2007b). Effects of porous polystyrene resin parameters on Candida antarctica lipase B adsorption, distribution, and polyester synthesis activity. *Langmuir*, **23**(11), 6467–6474.

Chen, D. H., & Liao, M. H. (2002). Preparation and characterization of YADH-bound magnetic nanoparticles. *Journal of Molecular Catalysis B: Enzymatic*, **16**(5–6), 283–291.

Chen, G., & Hoffman, A. S. (1993). Preparation and properties of thermoreversible, phase-separating enzyme-oligo (*N*-isopropylacrylamide) conjugates. *Bioconjugate Chemistry*, **4**(6), 509–514.

Cheng, C. J., Chu, L. Y., *et al.* (2007). Preparation of monodisperse thermo-sensitive poly (*N*-isopropylacrylamide) hollow microcapsules. *Journal of Colloid and Interface Science*, **313**(2), 383–388.

Cocker, T. M., Fee, C. J., *et al.* (1997). Preparation of magnetically susceptible polyacrylamide/magnetite beads for use in magnetically stabilized fluidized bed chromatography. *Biotechnology and Bioengineering*, **53**(1), 79–87.

Colombié, S., Gaunand, A., *et al.* (2001). Lysozyme inactivation and aggregation in stirred-reactor. *Journal of Molecular Catalysis B: Enzymatic*, **11**(4–6), 559–565.

Colvin, V. L. (2003). The potential environmental impact of engineered nanomaterials. *Nature Biotechnology*, **21**(10), 1166–1170.

Cuyper, M., & Joniau, M. (1992). Binding characteristics and thermal behaviour of cytochrome-C oxidase, inserted into phospholipid-coated, magnetic nanoparticles. *Biotechnology and Applied Biochemistry*, **16**(2), 201–210.

Daubresse, C., Grandfils, C., *et al.* (1996). Enzyme immobilization in reactive nanoparticles produced by inverse microemulsion polymerization. *Colloid & Polymer Science*, **274**(5), 482–489.

Diaz, J. F., & Balkus, K. J. (1996). Enzyme immobilization in MCM-41 molecular sieve. *Journal of Molecular Catalysis B: Enzymatic*, **2**(2–3), 115–126.

Ding, X., Sun, Z., *et al.* (2000). Characterization of Fe_3O_4/poly (styrene-co-*N*-isopropylacrylamide) magnetic particles with temperature sensitivity. *Colloid & Polymer Science*, **278**(5), 459–463.

Ding, X. B., Li, W., *et al.* (2001). Preparation and characterization of magnetic amphiphilic polymer microspheres. *Journal of Applied Polymer Science*, **79**(10), 1847–1851.

Fan, C. H., & Lee, C. K. (2001). Purification of D-hydantoinase from adzuki bean and its immobilization for *N*-carbamoyl–phenylglycine production. *Biochemical Engineering Journal*, **8**(2), 157–164.

Fan, J., Lei, J., *et al.* (2003). Rapid and high-capacity immobilization of enzymes based on mesoporous silicas with controlled morphologies. *Chemical Communications*, (17), 2140–2141.

Frings, K., Koch, M., *et al.* (1999). Kinetic resolution of 1-phenyl ethanol with high enantioselectivity with native and immobilized lipase in organic solvents. *Enzyme and Microbial Technology*, **25**(3–5), 303–309.

Gai, L., & Wu, D. (2009). A novel reversible pH-triggered release immobilized enzyme system. *Applied Biochemistry and Biotechnology*, **158**(3), 747–760.

Galaev, I. Y., & Mattiasson, B. (1999). Smart polymers and what they could do in biotechnology and medicine. *Trends in Biotechnology*, **17**(8), 335–340.

Gao, F., Ma, G. H., *et al.* (2009). Enzyme immobilization, biocatalyst featured with nanoscale structure. In Encyclopedia of Industrial Biotechnology: Bioprocess, Bioseparation, and Cell Technology. John Wiley & Sons, Inc.

Gao, F., Su, Z. G., *et al.* (2009). Double emulsion templated micro-capsules with single hollow cavities and thickness-controllable shells. *Langmuir*, **25**(6), 3832–3838.

Gao, X., Yu, K. M. K., *et al.* (2003). Colloidal stable silica encapsulated nano-magnetic composite as a novel bio-catalyst carrier. *Chemical Communications*, **24**, 2998–2999.

Gao, F., Wang, P., *et al.* (2011). Microencapsulation of bioactive nanoparticles. *Methods in Molecular Biology*, (743), 161–174.

Ge, J., Lu, D., *et al.* (2008). Molecular fundamentals of enzyme nanogels. *The Journal of Physical Chemistry B*, **112**(45), 14319–14324.

Ge, J., Lu, D., *et al.* (2009). Recent advances in nanostructured biocatalysts. *Biochemical Engineering Journal*, **44**(1), 53–59.

González Siso, M., Lang, E., *et al.* (1997). Enzyme encapsulation on chitosan microbeads. *Process Biochemistry*, **32**(3), 211–216.

Groboillot, A., Champagne, C., *et al.* (1993). Membrane formation by interfacial cross-linking of chitosan for microencapsulation of Lactococcus lactis. *Biotechnology and Bioengineering*, **42**(10), 1157–1163.

Guo, Z., Bai, S., *et al.* (2003). Preparation and characterization of immobilized lipase on magnetic hydrophobic microspheres. *Enzyme and Microbial Technology*, **32**(7), 776–782.

Han, J., Park, C. H., *et al.* (1995). Concentrating alkaline serine protease, subtilisin, using a temperature-sensitive hydrogel. *Biotechnology Letters*, **17**(8), 851–852.

Hanefeld, U., Gardossi, L., *et al.* (2008). Understanding enzyme immobilisation. *Chemical Society Reviews*, **38**(2), 453–468.

Horák, D., Rittich, B., *et al.* (2001). Properties of RNase A immobi-lized on magnetic poly(2-hydroxyethyl methacrylate) microspheres. *Biotechnology Progress*, **17**(3), 447–452.

Horak, D., Babi, M., *et al.* (2007). Preparation and properties of magnetic nano- and microsized particles for biological and environmental separations. *Journal of Separation Science*, **30**(11), 1751–1772.

Huang, G., Gao, J., *et al.* (2004). Controlled drug release from hydrogel nanoparticle networks. *Journal of Controlled Release*, **94**(2–3), 303–311.

Huang, J., Han, B., *et al.* (2007). Magnetic polymer microspheres with polymer brushes and the immobilization of protein on the brushes. *Journal of Materials Chemistry*, **17**(36), 3812–3818.

Hudson, S., Cooney, J., *et al.* (2008). Proteins in mesoporous silicates. *Angewandte Chemie International Edition*, **47**(45), 8582–8594.

Hudson, S., Magner, E., *et al.* (2005). Methodology for the immobiliza-tion of enzymes onto mesoporous materials. *The Journal of Physical Chemistry B*, **109**(41), 19496–19506.

Ison, A., Macrae, A., *et al.* (1994). Mass transfer effects in solvent-free fat interesterification reactions: Influences on catalyst design. *Biotechnology and Bioengineering*, **43**(2), 122–130.

Izmaĭlov, A. F, Kiselev, M. V., *et al.* (2000). Alpha-chymotrypsin immobilized on ferromagnetic particles coated with titanium oxide: Production and catalytic properties. *Prikladnaia Biokhimiia I Mikrobiologiia*, **36**(1), 68.

Jia, H., Zhu, G., *et al.* (2003). Catalytic behaviors of enzymes attached to nanoparticles: The effect of particle mobility. *Biotechnology and Bioengineering*, **84**(4), 406–414.

Jiang, D. S., Long, S. Y., *et al.* (2005). Immobilization of Pycnoporus sanguineus laccase on magnetic chitosan microspheres. *Biochemical Engineering Journal*, **25**(1), 15–23.

Kang, K., Kan, C., *et al.* (2005). The properties of covalently immobilized trypsin on soap-free P (MMA-EA-AA) latex particles. *Macromolecular Bioscience*, **5**(4), 344–351.

Kang, K., Kan, C., *et al.* (2006). The immobilization of trypsin on soap-free P (MMA-EA-AA) latex particles. *Materials Science and Engineering: C*, **26**(4), 664–669.

Kang, Y., He, J., *et al.* (2007). Influence of pore diameters on the immobilization of lipase in SBA-15. *Industrial & Engineering Chemistry Research*, **46**(13), 4474–4479.

Katchalski-Katzir, E., & Kraemer, D. M. (2000). Eupergit(R) C, a carrier for immobilization of enzymes of industrial potential. *Journal of Molecular Catalysis B: Enzymatic*, **10**(1–3), 157–176.

Kim, J., & Grate, J. W. (2003). Single-enzyme nanoparticles armored by a nanometer-scale organic/inorganic network. *Nano Letters*, **3**(9), 1219–1222.

Kim, J., Grate, J. W., *et al.* (2006a). Nanostructures for enzyme stabilization. *Chemical Engineering Science*, **61**(3), 1017–1026.

Kim, J., Jia, H., *et al.* (2006b). Single enzyme nanoparticles in nanoporous silica: A hierarchical approach to enzyme stabilization and immobilization. *Enzyme and Microbial Technology*, **39**(3), 474–480.

Kirk, O., & Christensen, M. W. (2002). Lipases from *Candida antarctica*: Unique biocatalysts from a unique origin. *Organic Process Research & Development*, **6**(4), 446–451.

Klibanov, A. M. (2001). Improving enzymes by using them in organic solvents. *Nature*, **409**(6817), 241–246.

Kondo, A., & Fukuda, H. (1997). Preparation of thermo-sensitive magnetic hydrogel microspheres and application to enzyme immobilization. *Journal of Fermentation and Bioengineering*, **84**(4), 337–341.

Korecká, L., Jezová, J., *et al.* (2005). Magnetic enzyme reactors for isolation and study of heterogeneous glycoproteins. *Journal of Magnetism and Magnetic Materials*, **293**(1), 349–357.

Kotha, A., Raman, R. C., *et al.* (1998). Beaded reactive polymers 3 effect of triacrylates as crosslinkers on the physical properties of glycidyl methacrylate copolymers and immobilization of penicillin G acylase. *Applied Biochemistry and Biotechnology*, **74**(3), 191–203.

Kresge, C., Leonowicz, M., *et al.* (1992). Ordered mesoporous molecular sieves synthesized by a liquid-crystal template mechanism. *Nature*, **359**(6397), 710–712.

Krogh, T. N., Berg, T., *et al.* (1999). Protein analysis using enzymes immobilized to paramagnetic beads. *Analytical Biochemistry*, **274**(2), 153–162.

Kumar, A., Srivastava, A., *et al.* (2007). Smart polymers: Physical forms and bioengineering applications. *Progress in Polymer Science*, **32**(10), 1205–1237.

Lavayre, J., & Baratil, J. (1982). Preparation and properties of immobilized lipases. *Biotechnology and Bioengineering*, **24**(4), 1007–1013.

Lee, J., Kim, J., *et al.* (2005). Simple synthesis of hierarchically ordered mesocellular mesoporous silica materials hosting cross-linked enzyme aggregates. *Small*, **1**(7), 744–753.

Lee, K. H., Lee, P. M., *et al.* (1993). Immobilization of aminoacylase by encapsulation in poly-L-lysine-stabilized calcium alginate beads. *Journal of Chemical Technology & Biotechnology*, **57**(1), 27–32.

Lee, B. S., Lee, S. C., *et al.* (2003a). Biochemistry of mechanoenzymes: Biological motors for nanotechnology. *Biomedical Microdevices*, **5**(4), 269–280.

Lee, Y., Rho, J., *et al.* (2003b). Preparation of magnetic ion-exchange resins by the suspension polymerization of styrene with magnetite. *Journal of Applied Polymer Science*, **89**(8), 2058–2067.

Lee, W. F., & Shieh, C. H. (1999). pH-hydrogels. II. Synthesis and swelling behaviors of *N*-isopropylacrylamide-co-acrylic acid-co-sodium acrylate hydrogels. *Journal of Applied Polymer Science*, **73**(10), 1955–1967.

Lei, C., Shin, Y., *et al.* (2002). Entrapping enzyme in a functionalized nanoporous support. *Journal of the American Chemical Society*, **124**(38), 11242–11243.

Lei, L., Bai, Y., *et al.* (2009a). Study on immobilization of lipase onto magnetic microspheres with epoxy groups. *Journal of Magnetism and Magnetic Materials*, **321**(4), 252–258.

Lei, Z., Ren, N., *et al.* (2009b). Fe_3O_4/SiO_2-g-PSStNa polymer nanocomposite microspheres (PNCMs) from a surface-initiated atom transfer radical polymerization (SI-ATRP) approach for pectinase immobilization. *Journal of Agricultural and Food Chemistry*, **57**(4), 1544–1549.

Li, Y., Gao, F., *et al.* (2010). Pore size of macroporous polystyrene microspheres affects lipase immobilization. *Journal of Molecular Catalysis B: Enzymatic*, **66**(1–2), 182–189.

Liao, M. H., & Chen, D. H. (2001). Immobilization of yeast alcohol dehydrogenase on magnetic nanoparticles for improving its stability. *Biotechnology Letters*, **23**(20), 1723–1727.

Liao, M. H., & Chen, D. H. (2002). Characteristics of magnetic nanoparticles-bound YADH in water/AOT/isooctane microemulsions. *Journal of Molecular Catalysis B: Enzymatic*, **18**(1–3), 81–87.

Liu, X. Y., Ding, X. B., *et al.* (2003). Synthesis of novel magnetic polymer microspheres with amphiphilic structure. *Journal of Applied Polymer Science*, **90**(7), 1879–1884.

Liu, X., Guan, Y., *et al.* (2004a). Surface modification and characterization of magnetic polymer nanospheres prepared by miniemulsion polymerization. *Langmuir*, **20**(23), 10278–10282.

Liu, X., Guan, Y., *et al.* (2005a). Immobilization of lipase onto micron-size magnetic beads. *Journal of Chromatography B*, **822**(1–2), 91–97.

Liu, X., Ma, Z., *et al.* (2004b). Preparation and characterization of amino-silane modified superparamagnetic silica nanospheres. *Journal of Magnetism and Magnetic Materials*, **270**(1–2), 1–6.

Liu, G., Yang, H., *et al.* (2005b). Preparation of magnetic microspheres from water-in-oil emulsion stabilized by block copolymer dispersant. *Biomacromolecules*, **6**(3), 1280–1288.

Müller-Schulte, D., & Brunner, H. (1995). Novel magnetic microspheres on the basis of poly (vinyl alcohol) as affinity medium for quantitative detection of glycated haemoglobin. *Journal of Chromatography. A*, **711**(1), 53.

Makino, K., Agata, H., *et al.* (2000). Dependence of temperature-sensitivity of poly(*N*-isopropylacrylamide-co-acrylic acid) hydrogel microspheres upon their sizes. *Journal of Colloid and Interface Science*, **230**(1), 128–134.

Marsh, D. R., Lee, Y. Y., *et al.* (1973). Immobilized glucoamylase on porous glass. *Biotechnology and Bioengineering*, **15**(3), 483–492.

Martins, M., Simoes, S., *et al.* (1996). Development of enzyme-loaded nanoparticles: Effect of pH. *Journal of Materials Science: Materials in Medicine*, **7**(7), 413–414.

Megías, C., Pedroche, J., *et al.* (2006). Immobilization of angiotensin-converting enzyme on glyoxyl-agarose. *Journal of Agricultural and Food Chemistry*, **54**(13), 4641–4645.

Mei, Y., Miller, L., *et al.* (2003). Imaging the distribution and secondary structure of immobilized enzymes using infrared microspectroscopy. *Biomacromolecules*, **4**(1), 70–74.

Moelans, D., Cool, P., *et al.* (2005). Immobilisation behaviour of biomolecules in mesoporous silica materials. *Catalysis Communications*, **6**(9), 591–595.

Montero, S., Blanco, A., *et al.* (1993). Immobilization of Candida rugosa lipase and some properties of the immobilized enzyme. *Enzyme and Microbial Technology*, **15**(3), 239–247.

Moreno, J. M., Arroyo, M., *et al.* (1997). Covalent immobilization of pure isoenzymes from lipase of Candida rugosa. *Enzyme and Microbial Technology*, **21**(8), 552–558.

Nigam, S. C., Tsao, I. F., *et al.* (1988). Techniques for preparing hydrogel membrane capsules. *Biotechnology Techniques*, **2**(4), 271–276.

Ogelstad, J., Berge, A., *et al.* (1992). Preparation and application of new monosized polymer particles. *Progress in Polymer Science (Oxford)*, Pergamon Press Inc, Tarrytown, NY, 1992, **17**(1), 87–161.

Olpan, D., Duran, S., *et al.* (2002). Synthesis and properties of radiation-induced acrylamide–acrylic acid hydrogels. *Journal of Applied Polymer Science*, **86**(14), 3570–3580.

Omi, S., Kaneko, K., *et al.* (1997). Application of porous microspheres prepared by SPG (Shirasu Porous Glass) emulsification as immobilizing carriers of glucoamylase (GluA). *Journal of Applied Polymer Science*, **65**(13), 2655–2664.

Ozden, M. Y., & Hasirci, V. N. (1990). Enzyme immobilization in polymer coated liposomes. *British Polymer Journal*, **23**(3), 229–234.

Pizarro, C., M. Fernández-Torroba, *et al.* (1997). Optimization by experimental design of polyacrylamide gel composition as support for enzyme immobilization by entrapment. *Biotechnology and Bioengineering*, **53**(5), 497–506.

Rebelo, L. P., Netto, C. G. C. M., *et al.* (2010). Enzymatic kinetic resolution of (RS)-1-(Phenyl) ethanols by Burkholderia cepacia lipase immobilized on magnetic nanoparticles. *Journal of the Brazilian Chemical Society*, **21**(8), 1537–1542.

Reshmi, R., Sanjay, G., *et al.* (2007). Immobilization of [alpha]-amylase on zirconia: A heterogeneous biocatalyst for starch hydrolysis. *Catalysis Communications*, **8**(3), 393–399.

Rittich, B., Spanova, A., *et al.* (2002). Characterization of deoxyribonuclease I immobilized on magnetic hydrophilic polymer particles. *Journal of Chromatography B: Analytical Technologies in the Biomedical and Life Sciences*, **774**(1), 25–31.

Roy, I., Sharma, S., *et al.* (2004). Smart biocatalysts: Design and applications. *New Trends and Developments in Biochemical Engineering*, 251–310.

Sheldon, R. A. (2007). Enzyme immobilization: The quest for optimum performance. *Advanced Synthesis & Catalysis*, **349**(8–9), 1289–1307.

Shi, B. F., Wang, Y. S., *et al.* (2009). Aminopropyl-functionalized silicas synthesized by W/O microemulsion for immobilization of penicillin G acylase. *Catalysis Today*, **148**(1–2), 184–188.

Shi, B., Wang, Y., *et al.* (2010). Superparamagnetic aminopropyl-functionalized silica core-shell microspheres as magnetically separable carriers for immobilization of penicillin G acylase. *Journal of Molecular Catalysis B: Enzymatic*, **63**(1–2), 50–56.

Skovby, M. H. B., & Kops, J. (1990). Preparation by suspension polymerization of porous beads for enzyme immobilization. *Journal of Applied Polymer Science*, **39**(1), 169–177.

Sonvico, F., Mornet, S., *et al.* (2005). Folate-conjugated iron oxide nanoparticles for solid tumor targeting as potential specific magnetic hyperthermia mediators: Synthesis, physicochemical characterization, and *in vitro* experiments. *Bioconjugate Chemistry*, **16**(5), 1181–1188.

Springuel-Huet, M. A., Bonardet, J. L., *et al.* (2001). Mechanical properties of mesoporous silicas and alumina-silicas MCM-41 and SBA-15 studied by N_2 adsorption and 129Xe NMR. *Microporous and Mesoporous Materials*, **44**, 775–784.

Sun, S., Zhang, Y., *et al.* (2010) Using of silica particles as porogen for preparation of macroporous chitosan macrospheres suitable for enzyme immobilization. *Kinetics and Catalysis*, **51**(5), 771–775.

Szczesna-Antczak, M., Antczak, T., *et al.* (2002). Catalytic properties of membrane-bound Mucor lipase immobilized in a hydrophilic carrier. *Journal of Molecular Catalysis B: Enzymatic*, **19**, 261–268.

Takahashi, H., Li, B., *et al.* (2000). Catalytic activity in organic solvents and stability of immobilized enzymes depend on the pore size and surface characteristics of mesoporous silica. *Chemistry of Materials*, **12**(11), 3301–3305.

Turkova, J., Blaha, K., *et al.* (1978). Methacrylate gels with epoxide groups as supports for immobilization of enzymes in pH range 3–12. *Biochimica et Biophysica Acta (BBA)-Enzymology*, **524**(1), 162–169.

Varlan, A. R., & Sansen Ann Van, W. (1996). Covalent enzyme immobilization on paramagnetic polyacrolein beads. *Biosensors and Bioelectronics*, **11**(4), 443–448.

Vertegel, A. A., Siegel, R. W., *et al.* (2004). Silica nanoparticle size influences the structure and enzymatic activity of adsorbed lysozyme. *Langmuir*, **20**(16), 6800–6807.

Wada, Y., Sambongi, Y., *et al.* (2000). Biological nano motor, ATP synthase F_0F_1: From catalysis to [gamma][epsilon] c10–12 subunit assembly rotation. *Biochimica et Biophysica Acta (BBA)-Bioenergetics*, **1459** (2–3), 499–505.

Wan, S., Huang, J., *et al.* (2005). Size-controlled preparation of magnetite nanoparticles in the presence of graft copolymers. *Journal of Materials Chemistry*, **16**(3), 298–303.

Wang, P. (2006). Nanoscale biocatalyst systems. *Current Opinion in Biotechnology*, **17**(6), 574–579.

Wang, P., Dai, S., *et al.* (2001). Enzyme stabilization by covalent binding in nanoporous sol–gel glass for nonaqueous biocatalysis. *Biotechnology and Bioengineering*, **74**(3), 249–255.

Wang, P., Sergeeva, M. V., *et al.* (1997). Biocatalytic plastics as active and stable materials for biotransformations. *Nature Biotechnology*, **15**(8), 789–793.

Wang, Y., Yu, A., *et al.* (2005). Nanoporous polyelectrolyte spheres prepared by sequentially coating sacrificial mesoporous silica spheres. *Angewandte Chemie*, **117**(19), 2948–2952.

Willner, I., Rubin, S., *et al.* (1993). Reversible light-stimulated activation and deactivation of alpha-chymotrypsin by its immobilization in photoisomerizable copolymers. *Journal of the American Chemical Society,* **115**(19), 8690–8694.

Yan, A. X., Li, X. W., *et al.* (2002). Recent progress on immobilization of enzymes on molecular sieves for reactions in organic solvents. *Applied Biochemistry and Biotechnology*, **101**(2), 113–129.

Yan, M., Ge, J., *et al.* (2006). Encapsulation of single enzyme in nanogel with enhanced biocatalytic activity and stability. *Journal of the American Chemical Society*, **128**(34), 11008–11009.

Yan, M., Liu, Z., *et al.* (2007). Fabrication of single carbonic anhydrase nanogel against denaturation and aggregation at high temperature. *Biomacromolecules*, **8**(2), 560–565.

Yang, Z., Mesiano, A. J., *et al.* (1995). Activity and stability of enzymes incorporated into acrylic polymers. *Journal of the American Chemical Society*, **117**(17), 4843–4850.

Yao, G., Qi, D., *et al.* (2008). Functionalized magnetic carbonaceous microspheres for trypsin immobilization and the application to fast proteolysis. *Journal of Chromatography A*, **1215**(1–2), 82–91.

Yiu, H. H. P., Wright, P. A., *et al.* (2001). Enzyme immobilisation using SBA-15 mesoporous molecular sieves with functionalised surfaces. *Journal of Molecular Catalysis B: Enzymatic*, **15**(1–3), 81–92.

Yong, Y., Bai, Y. X., *et al.* (2008). Characterization of Candida rugosa lipase immobilized onto magnetic microspheres with hydrophilicity. *Process Biochemistry*, **43**(11), 1179–1185.

Yuan, J. J., Armes, S., *et al.* (2006). Synthesis of biocompatible poly [2-(methacryloyloxy) ethyl phosphorylcholine]-coated magnetite nanoparticles. *Langmuir*, **22**(26), 10989–10993.

Zeng, L., Luo, K. K., *et al.* (2006). Preparation and characterization of dendritic composite magnetic particles as a novel enzyme immobilization carrier. *Journal of Molecular Catalysis B-Enzymatic*, **38**(1), 24–30.

Zhang, Y., Gao, F., *et al.* (2011). Simultaneous production of 1, 3-dihydroxyacetone and xylitol from glycerol and xylose using a nanoparticle-supported multienzyme system with *in situ* cofactor regeneration. *Bioresource Technology*, **102**(2), 1837–1843.

Zhou, W. Q., Gu, T. Y., *et al.* (2007a). Synthesis of macroporous poly (glycidyl methacrylate) microspheres by surfactant reverse micelles swelling method. *European Polymer Journal*, **43**(10), 4493–4502.

Zhou, W. Q., Gu, T. Y., *et al.* (2007b). Synthesis of macroporous poly (styrene-divinyl benzene) microspheres by surfactant reverse micelles swelling method. *Polymer*, **48**(7), 1981–1988.

Chapter 2

Microspheres for Cell Culture

Weiqing Zhou, Guanghui Ma, and Zhiguo Su

State Key Laboratory of Biochemical Engineering, Institute of Process Engineering, Chinese Academy of Sciences, Beijing, P. R. China

ghma@home.ipe.ac.cn

2.1 Introduction

Microspheres with particle sizes of 60–300 μm have been successfully used as carriers of adherent cell culture. Cell culture system based on microspheres is called "microcarrier culture" due to the extremely important role of microspheres. Since Van Wezel first introduced diethyl amino ethyl (DEAE) dextran microspheres in the culture of adherent animal cells in 1967, microspheres have achieved many important applications in cell culture. The microcarrier technology has developed a number of biological products with vital practical and commercial value in the duration of forty years. These products include vaccines, enzymes, hormones, antibodies, interferons, and nucleic acids. Compared to traditional roller bottle cell culture, microcarrier culture combines the excellences of both the monolayer cell culture and suspension culture and provides many significant advantages (Handbooks from GE Healthcare, 2005). Some of these advantages are as follows:

(i) The small spherical particles provide large surface-to-volume ratio, which enhances production. For a given quantity of cells or products, microcarrier cultures require much less space than other types of monolayer cultures. Large numbers of cells

Microspheres and Microcapsules in Biotechnology: Design, Preparation, and Applications
Edited by Guanghui Ma and Zhiguo Su

ISBN 978-981-4316-47-7 (Hardcover), 978-981-4364-62-1 (eBook)
www.panstanford.com

can be cultured in small volumes (more than 10^{11} cells/L) and fewer culture vessels are required. This also greatly reduces the requirement of labor.

(ii) Microcarrier culture require less culture media than other monolayer or suspension cultures. Superior yields as high as 100 times have been achieved in various cell systems including chicken fibroblasts (Giard *et al.*, 1977; Mered *et al.*, 1980), pig kidney cells (Meignier, 1979), fish cells (Nicholson, 1980), Chinese hamster ovary cells (Crespi & Thilly, 1981), human fibroblasts (Griffiths & Thornton, 1980), primary monkey kidney cells (van Wezel, 1972), and transformed mouse fibroblasts (Pharmacia Fine Chemicals, 1979). This reduced requirement for media significantly reduces the cost of production which is particularly high in the use of serum additives.

(iii) The culture environment is homogeneous and has all the advantages of suspension culture. It is favorable for enlarging production and controlling culture conditions such as temperature, pH, CO_2, and PO_2. Monitoring and sampling microcarrier systems are simpler than any other existing technique for producing large numbers of adherent cells. Excellent control offers more convenience for process design and optimization.

(iv) Microcarrier culture provides simplified, systematic, and automated procedures. Separation of cells from the culture medium is simple. As compared to culture in several hundred roller bottles, the risk of contamination is greatly reduced when producing cells in a single microcarrier culture bioreactor (Crespi & Thilly, 1981).

In the microcarrier technique of cultivation, cells attach and spread on solid microspheres suspended in growth medium and gradually grow and propagate on the surface of microspheres or in the pores of macroporous structures. A large number of microcarriers are commercially available and their properties are optimized. Many matrix materials have been used in preparation of microcarriers, including dextran, collagen, gelatin, glass, polystyrene, polyacrylamide, and cellulose. The surface properties such as charge group capacity, coating materials, and particle size have been optimized to enhance cell growth. In recent years, the application of microcarriers in tissue engineering has drawn many attentions although it is still far from clinical application. In this chapter, the requirement and design of microspheres in cell culture will be reviewed. The latest development of microcarriers in three-dimensional (3-D) cell culture, tissue engineering, and regenerative medicine will also be introduced.

2.2 Characteristics of Microcarriers for Cell Culture

Microcarriers were first applied to culture-adherent cells for large-scale production by van Wezel in 1967. Since then, microcarrier technique has received wide attentions. Many types of microcarriers are commercially available and have been successfully used for producing biological products (Tree *et al.*, 2001). Due to the importance to cell culture, biological products and tissue engineering researches on microcarriers are still hot. The main existing commercially available microcarriers are listed in Table 2.1. According to their properties, they can be classified as ionic and nonionic, smooth and macroporous, etc.

During several decades of development, researchers have summarized characteristics of a good microcarrier (van Groot, 1995): (i) nontoxic to cells; (ii) good adhesion properties or good cell entrapment in microcarrier; (iii) slightly higher density than culture media (1.03–1.05 g/mL); (iv) uniform particle size distribution; (v) clear and transparent optical properties, easy to observe cell growth; (vi) enough nutrient supplication at center of porous matrix; (vii) autoclavable, batch-to-batch consistency, and mechanical stability. Surface properties are most important requirements of excellent microcarriers as they determine cell attachment and growth. This section will discuss effects of surface properties, such as charge density, chemical nature, roughness, wettability, and rigidity, on cell culture in detail.

Table 2.1 Main commercially available microcarriers and their physico-chemical parameters

Name and material	Manufacture	Size (μm)	Surface area (cm^2/g)	Beads (/g)	Density (g/mL)	Porous	Material
Dextran							
Cytodex 1	GE HC, USA	147–248	4400	4.3×10^6	1.03	–	Dextran matrix with substituted *N,N*-dimethylaminoethyl groups
Cytodex 2	GE HC, USA	135–200	3300	4.1×10^6	1.04	–	Dextran matrix with a surface layer of *N,N,N*-trimethyl-2-hydroxyaminopropyl groups
Hillex	SoloHill, USA	150–210	350	5.0×10^5	1.10	–	Dextran matrix with treated surface
Plastic							
Plastic coated	SoloHill, USA	150–210	380	4.5×10^5	1.02–1.04	–	Plastic coated with denatured collagen
Plastic plus coated	SoloHill, USA	150–210	380	4.5×10^5	1.02–1.04	–	Plastic coated with denatured collagen
Biosilon	Nunc, Denmark	160–300	255	5.1×10^4	1.05	–	Polystyrene
Cytoline 2	GE HC, USA	400–2500	>1000	–	1.03	+	Polyethylene and silica
Glass							
Glass coated	SoloHill, USA	150–210	380	4.5×10^5	1.02–1.04	–	Plastic with glass coating
Cellulose							
Cytopore 1	GE HC, USA	200–280	~1200	–	1.03	+	Cellulose
Gelatin							
Cultispher G	Percell Biolitica, Sweden	130–380	n/a	1.0×10^6	1.04	+	Cross-linked porcine gelatin
Cultispher S	Percell Biolitica, Sweden	130–380	n/a	8.0×10^5	1.04	+	Cross-linked porcine gelatin
Collagen							
Cytodex 3	GE HC, USA	141–211	2700	3.0×10^6	1.04	–	Dextran matrix with thin layer of denatured pig skin-derived collagen
Cellagen beads	MP Biometicals, USA	100–400	–	–	–	+	Highly cross-linked bovine collagen type 1

Source: Reprinted from Malda *et al.* (2006), with permission from Elsevier.

2.2.1 Effect of Surface Charge and Charge Density of Microcarriers

Most animal cells have a negative surface charge, and they can easily attach on positive surface by electrostatic interaction. Some microcarriers, especially conventional plastic or glass particles, have a negative surface. In this case, divalent cations need to be added in culture system, which act as a "bridge" (Maroudas, 1975). Cells attach on microcarriers by "negative charge–bridge–negative charge" interaction. With the development of microcarriers, researchers are more willing to choose positive microcarriers. It should be noted that compared to the polarity, the charge density is a more critical factor for cell attachment and spreading. Many studies have indicated that a optimal charge density is necessary for cell culture (Hakoda & Shiragami, 2000; Ishida *et al.*, 1985). If the charge density is too low, cells can adhere to the surface of microcarriers and begin to grow, but the cells soon fall off from the microcarriers. If the electrostatic charge is too great, cell will rapidly attach to the surface but will not spread and grow. Furthermore, high electrostatic interactions prevent release of viable cells from the surface while production of some types of vaccines requires that cells be released from the substrate in a living state. Levine *et al.* (1978) reported that DEAE-Sephadex particles with a charge density of 2.0 mmol/g had no toxic effects and provided a higher cell concentration, whereas with a DEAE-substitution of 3.5 or 0.9 mmol/g, poor results were obtained. This was the preliminary work for the development of "Cytodex" microcarriers. The concentration of microcarriers in cell culture system is another crucial factor. The use of microcarrier concentration over 1 g/L was found to result in loss of inoculum, poor growth, and cell detachment (van Wezel, 1977). Kiremitci and Piskin (1990) also reported that the microcarriers significantly inhibited the cell culture when used in large amounts.

The optimal charge density varies according to different matrix materials and charge groups. The nature of positively charged groups determines the optimal charge density (Reuveny *et al.*, 1982). For baby hamster kidney (BHK) cells on carriers with a primary amine, the optimal charge was 0.56 mmol/g, while for carriers with a tertiary amine, the optimal charge was found to be 1.80 mmol/g. Other interesting results were that one kind of charged group was not suitable for culturing all kind of cells (Reuveny, 1983; Reuveny *et al.*, 1983, 1985). Thus, different cells attached to substrates by different ligands or adherence forces. These findings showed that the aim to develop a "general microcarrier, optimal for all purposes," was not realistic.

Besides the charge density and the nature of positively charged groups, Masashi Kunitake presented another view that pKa of microcarriers had a great impact on cell growth (Sakata *et al.*, 2000). They cultivated mouse L-929 cells on animated poly(γ-methyl-L-glutamate) (PG) and cross-linked poly(ε-lysine) (PL) particles containing variable amounts of primary amino groups. The cell growth was strongly related with the apparent pKa of the particles but not on their charge group content. The microspheres with the apparent pKa of 7.4–7.6 were most suitable for cell culture. However, most researchers optimize the properties of microspheres based on the charge density and the nature of positively charged groups instead of the apparent pKa of microcarriers.

2.2.2 Effect of Hydrophobicity (Wettability) of Microcarriers

Cell growth requires hydrophilic environment, while researchers have found that appropriate increase of surface hydrophobicity of microspheres improves cell attachment and growth (Horbett *et al.*, 1985; van Wachem *et al.*, 1987; Ikada, 1994; Tamada & Ikada, 1994). Reuveny *et al.* (1983) investigated the effects of four alkyldiamines with different alkyl chain length [NH_2–$(CH_2)_n$–NH_2; n = 2, 4, 6, and 8] on cell growth of BHK cells. Longer the alkyl chain, stronger the hydrophobicity. The experimental results showed that butyl- and hexyldiamine modified particles provided the most optimal environment.

Kunitake modified PG-based [poly(γ-methyl L-glutamate)] particles with three *n*-alkyl terminal groups and compared cell growth on the alkylated PG particles with unmodified PG and hydroxyl terminated PG particles (Sakata *et al.*, 2000). They found that cells grew better on the particles modified with longer alkyl chains. PG microcarrier (PG-C12) with the longest alkyl chain provided highest cell growth. However, the most hydrophilic hydroxyl-terminated PG (PG-OH) particles revealed the lowest cell growth compared with other PG particles. The reverse effect of hydrophilic hydroxyl-terminated particles for cell attachment was also confirmed by Horbett and Schway (1988) and Massia and Hubbell (1991). They found that the cells' attachment reduced dramatically on the hydrophilic hydroxyethylmethacrylate and ethylmetha-crylate (HEMA-EMA) film and the polyethylene terephthalate (PET) film. The main reason was that the hydrophilic surface prevented the adsorption of proteins which could promote cell adhesion, such as fibronectin, vitronectin, collagen, and laminin. These results suggested that a moderately hydrophobic surface could also be used for cell culture compared with charged surfaces (van der Valk, 1993).

2.2.3 Effect of Particle Size on Cell Attachment and Growth

The diameter of commercially available microspheres mainly range from 10 μm to 5 mm. We can choose microcarriers with appropriate particle size according to the type of bioreactor. For example, smaller microcarriers are suitable for stirred tanks, whereas the larger ones fit the applications in fluidized and packed beds due to their higher sedimentation rates. The conventional size of microcarriers is 60–300 μm. But some researchers revealed that when the surface area of the microcarriers was normalized, they performed equally well even when they were of different sizes (small: average diameter 38–75 μm; intermediate sized: 75–150 μm; large: 150–300 μm) (Hu & Wang, 1987). Besides particle size, a very narrow size distribution is more important for good mixing in the reactor and an equal sedimentation of the beads which provides a similar environment for cell growth (Handbooks from GE Healthcare, 2005).

2.3 Surface Modification of Microcarrier

The ideal material for cell culture should (i) support rapid cell attachment and spreading; (ii) support high-density cell growth; (iii) not interfere with the secretion of metabolic products; and (iv) allow cells to be easily detached (Kato *et al.*, 2003; Varani, 1985). However, a cell cannot attach and grow on conventional matrix material such as dextran or polystyrene unless after appropriate surface modification. There are two basic strategies: (i) covalently linking a strongly charged surface moiety such as DEAE to a neutral substrate (van Wezel, 1967); (ii) coating the surface with molecules promoting cell adhesion such as collagen, fibronectin, and synthetic peptides designed from the attachment supporting domains of extracellular matrix molecules (Ruoslahti & Pierschbacher, 1986; Varani *et al.*, 1989; Varani *et al.*, 1993). After appropriate surface modification, many matrix materials have been successfully applied in cell culture (Varani, 1985; Nilsson, 1989).

2.3.1 Surface Modification with Charged Groups

A variety of microcarriers have been prepared by 2-diethylaminoethylchloride hydrochloride (DEAE) amination modified method. In addition, some other methods have been developed to make surface charged for different matrix materials by using various modification molecules.

Trimethylamine was used to modify the copolymer of styrene and divinylbenzene (CP-TMA) (Varani *et al.*, 1998). The CP-TMA microcarriers were unique in that they combined the advantages of both charged surface and protein coated microcarriers. Charged surfaces allow cells to attach to the surface rapidly, but cell propagation is slower. The protein coated surfaces provide good conditions for cell spread, while the process of cell attachment is rather slower than charged surface. Both the rates of cell attachment and propagation were high on the CP-TMA microcarriers. At the end of the culture, cells could be easily released from the microcarriers by exposure to trypsin/ethylene diamine tetraacetic acid (EDTA), which was another significant advantage.

A large organic polymer molecule, polyethyleneimine (PEI) is used as a modification reagent for cell culture since it has a high density of amino groups. PEI is well known as a DNA transfection reagent. Its role of an efficient and convenient attachment factor is much less recognized (Baek *et al.*, 2011). PEI is more suitable for adherence of weakly anchoring cells (Vancha *et al.*, 2004). Surface modification with PEI incorporated hydrophobic groups was also confirmed to be highly effective in the attachment of several types of cells (Bledi *et al.*, 2000).

Copolymerization with ionic monomer is another way for preparation of charged polymeric microcarriers. Polyethylene glycol (PEG) has excellent biocompatibility and is not toxic for cells. A PEG-based macromonomer, polyethylene glycol methacrylate

(PEGMA), was selected as a matrix for mouse fibroblast cells. A cationic molecule, *N*-[3-(dimethylamino)propyl]methacrylamide (DMAPM), was copolymerized with PEGMA because cells could not attach to PEGMA particles (Esra *et al.*, 2007). DMAPM provided higher cationic charge for poly(PEGMA-DMAPM) hydrogel beads, and an effective cell attachment and growth up to 3.5×10^6 cells/mL were obtained on these copolymer microcarriers.

Low-temperature plasma surface modification can effectively improve the surface compatibility, wettability, and biological properties of materials. After appropriate modification, the chemical structure, surface energy, and surface charge can be optimized to increase cell attachment and growth. Many polymeric materials such as polystyrene and polypropylene have been modified to be good charged microcarriers by this method (Vladkova, 2010).

Besides the positive charged surface modification method, treatment with sulfuric acid can improve the number of negative charged groups of the surface. Poly(2-hydroxyethyl methacrylatey) (PHEMA) modified by sulfuric acid was used to culture endothelial cells although the original material had a hydrophilic hydroxyl-terminated surface which was nonadhesive to cells (Hannan & McAuslan, 1987). However, the attachment strength was not so strong for the negative charged surface therefore its applications are much less than positive charged methods.

2.3.2 Surface Modification with Proteins

In general, charged-surface microcarriers allow rapid attachment through electrostatic interactions (Ginsburg, 1987; Swartz, 1993; Plantefaber & Hynes, 1989). However once cells attached, they usually spread and proliferate more slowly (Varani *et al.*, 1989; Varani *et al.*, 1995). Furthermore, it is often difficult to harvest living cells from charged surfaces. To overcome these shortcomings, scientists proposed to coat the surface by proteins from extracellular matrix such as collagens, proteoglycans, fibronectin, laminin, elastin, and chondronectin. It was found cells grew more rapidly than those attached on charged microcarriers, and readily detached (and, therefore, retained high viability) from protein-coated materials.

Whole proteins such as collagen can be immobilized to the surface, providing the cell with a substrate that more closely resembles the living conditions in tissues (Tamada & Ikada, 1994).

The effects of protein attachment factors on cell attachment and growth have been well investigated (Kleinmann *et al.*, 1981). The collagen-based microcarrier systems have been widely applied in biotechnology. The selection criteria for protein, matrix, and immobilization methods have been well established.

Different proteins including recombinant collagen, gelatin, synthetic collagen-derived peptides, and native collagen were compared as microcarrier surface coating molecules (Dame & Varani, 2008). It was found that the ability of recombinant collagen and native collagen was similar for fibroblast attachment and spreading, which was more rapid than that on gelatin. Gelatin is the most common surface-coating material. Native gelatin was in turn better than the recombinant gelatin fragment. Fibroblast could not attach to the surface coated by the synthetic peptide with valine or lysine residues. The concentration of collagen was also a crucial factor. Higher concentration of collagen was more favorable for initial attachment and overall growth.

Bueno *et al.* (2007) combined a cationic moiety of PEI on recombinant collagen–coated dextran microspheres in order to improve the attachment rate. The content of PEI was high enough to boost initial attachment but not too high to be a growth inhibitor. They found that a combination of protein and charged surface modification efficiently promoted initial cell attachment and proliferation without compromising cell yields.

Polysaccharide particles, such as dextran and alginate, are commonly used matrix. The properties of matrix keep developing. Gröhn *et al.* (1997) reported collagen-coated Ba^{2+}-alginate as a matrix for microcarriers. The preparation procedure of alginate microspheres was easy. The particle size and pore size were conveniently controlled by varying alginate concentration, molecular weight, and the concentration of the divalent cations (Yamagiwa *et al.*, 1995). Further important advantages are the biocompatibility of purified alginates and their common use as encapsulation and implantation materials (Klöck *et al.*, 1994; Zimmermann *et al.*, 1992). Alginates are conventionally cross-linked by Ca^{2+} ions. Ba^{2+} ions provide a more chemically stable matrix. Furthermore, cells with good activity could be harvested without using lytic enzymes to detach cells from collagen-coated Ba^{2+}-alginate microcarriers. This was important for sensitive cells and for *in vitro* tissue reconstruction.

In addition to the surface-coating method, collagen and other extracellular matrix (ECM) molecules can also been incorporated into microspheres by copolymerization method. The proteins could be added in a reaction mixture containing monomers and be polymerized in skeleton (Civerchia-Perez *et al.*, 1980; Carbonetto *et al.*, 1982; Woerly *et al.*, 1993), or be mixed with polymerized polymer in appropriate solvents and attached on surface after removing the solvents (Stol *et al.*, 1985).

2.3.3 Modification with Smaller Biologically Active Functional Groups

With deepening understanding of cell-surface interaction, smaller biologically active functional groups have been used to modify surfaces in addition to whole proteins (Massia & Hubbell, 1989; Schnaar *et al.*, 1978; Blackburn & Schnaar, 1983). These small active groups include oligopeptides, saccharides, or glycolipids. Among these groups, the oligopeptides have been widely investigated. These short amino acid sequences appear to bind to receptors on cell surfaces and promote cell attachment. For example, the cell-binding domain of fibronectin contains the tripeptide, RGD (Arg-Gly-Asp). The matrix coating with oligopeptides containing RGD sequence promoted the cell-binding activity of fibronectin, which revealed the importance of RGD sequence in the adhesion of cells (Pierschbacher & Ruoslahti, 1984). A large number of ECM proteins (fibronectin, collagen, vitronectin, thrombospondin, tenascin, laminin, and entactin) contain the RGD sequence. The sequences YIGSR (Tyr-Ile-Gly-Ser-Arg) and IKVAV (Ile-Lys-Val-Ala-retinol dehydrogenase) in laminin also have cell-binding activity and mediate adhesion in certain cells.

Due to the important role of RGD in cell adhesion, synthetic RGD-containing peptides have been immobilized to various matrix including polytetrafluoroethylene (PTFE), PET, polyacrylamide, polyurethane (PEU), polycarbonate urethane, PEG, polyvinyl alcohol (PVA), polylactide (PLA), and poly(*N*-isopropylacrylamide-co-*N*-*n*-butylacrylamide) substrates (Saltzman, 2000). The addition of RGD dramatically promotes cell adhesion and spreading. However, the surface modified with RGD are only favorable for cell containing the adhesion receptors which recognize RGD sequence, so an appropriate peptide should be selected according to the nature of cells.

In addition to synthetic peptides, other biologically active molecules can also be recognized by cell receptors. It was found that the homopolymers of amino acid and covalently bound amine groups enhanced cell attachment for certain cells (Kikuchi *et al.*, 1992; Massia & Hubbell, 1992). Some saccharides like lactose and *N*-acetylglucosamine were observed to promote the adherence and proliferation of cells (Saltzman, 2000).

2.4 Macroporous Microcarriers for Cell Culture

Solid microcarriers have been widely used in cell culture; however, the surface area provided by solid microcarrier is small and the final cell density is limited. Cells attached on surface are still subject to the impact of shear force. Therefore, researchers propose using macroporous microspheres for cell culture (Nilsson, 1986). Their average pore size is between 30 μm and 400 μm. As the mean diameter of single cells in suspension is about 10 μm, the large pores allow easy access to cells to the interior of macroporous microcarriers. The commercially available macroporous microcarriers are listed in Table 2.1 (Gotoh *et al.*, 1993). Due to larger growth space and surface area, macroporous microspheres provide higher cell densities which offer many other advantages for cell culture compared to smooth microcarriers (Looby & Griffiths, 1990). One main advantage is that the microcarrier concentration can be reduced evidently. Only one-fifth amount of macroporous microcarriers is needed for certain cell density compared to charged dextran microcarriers, which is consistent with the ratio of their surface area (Reuveny, 1985). The second advantage of macroporous microcarriers is that they protect cells against the shearing damage since most cells grow inside the pores. Thus, a higher stirring speed can be used, that is, 80–100 rpm, which is beneficial for transfer of O_2 and nutrients (Xiao *et al.*, 1999; Vournakis & Runstadler, 1989). Another advantage is that macroporous microcarriers allow cells to grow in three dimensions at high densities which decrease the need for external growth factors. This makes it easier to use low serum, serum-free, and even protein-free media, which is very important for downstream processing, and of course, cutting costs. The high cell density also promotes the stability and increases the lifetime of cells. Macroporous microcarriers are not only suitable to culture anchorage-dependent cells but also anchorage-independent cells, while solid microcarriers are only suitable to culture anchorage-dependent cells.

However, macroporous microcarriers also have some shortcomings. Due to their higher cell density and porous structure, cell count and harvesting are more difficult. In addition, the high density makes it harder to effectively infect all cells simultaneously, especially using non-lytic viruses. The pore size should be large enough to reduce the diffusion restrict of nutrients for cells in the interior of pores. The main advantages and disadvantages of macroporous microcarriers are listed in Table 2. 2.

Table 2.2 Advantages and disadvantages of macroporous microcarriers

Advantages
High surface area/volume ratio
High cell density
Homogeneous system
Scale-up potential
Anchorage-dependent as well as suspension cells
Easy separation media/carriers
Different bioreactor systems
Long-term cultures
Disadvantages
Biomass estimation difficult
Harvesting incomplete
Diffusion problems
Production of lytic viruses difficult
Complicated washing procedure

Source: Reprinted from van Groot (1995), with permission from Springer-Verlag Dordrecht.

2.5 Three-Dimensional Cell Culture Using Microcarrier Technology

Microcarriers provide 3-D environment by two methods as shown in Fig. 2.1: (i) cells grow in interstices between particles; (ii) cells grow in macropores of porous particles. Three-dimensional culture systems provide an environment more similar to that *in vivo* and allow cells with good viability, phenotype, and functions (Hoffman, 1993). Therefore, many researchers have focused on applications of 3-D cell culture in tissue engineering and regenerative medicine.

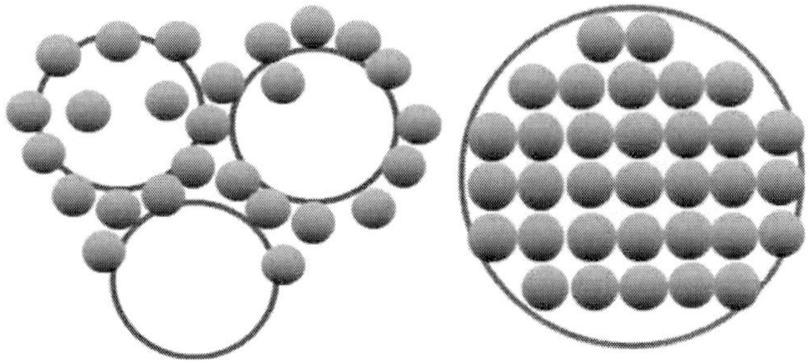

Figure 2.1 Schematic drawing of 3-D cell culture.

2.5.1 Three-Dimensional Microcarrier Culture in Tissue Engineering

Three-dimensional microcarrier culture provides sufficient numbers of living cells with desired phenotype for analysis purposes and cell transplantation, which is important for tissue engineering (Tebb *et al.*, 2006). Some microcarriers appropriate for 3-D culture are also used in cell therapy method, that is, after a period of culture time, the cell-seeded microcarriers are injected directly into the body. Cells cultured on carriers exchange nutrients and metabolic waste by diffusion before the formation of new blood vessels. Cells existing at the internal part of conventional 3-D scaffolds may be dead due to the diffusion limit. The diffusion limit decreases dramatically in microcarrier culture especially macroporous microcarriers, which makes it more favorable for tissue engineering. Compared to transplant operation, this injection method of cell-seeded microcarriers is more convenient, reduces suffering, and saves costs for patients. Microcarriers for tissue engineering must have good biocompatibility, must be completely biodegradable, and must have nontoxic degradation products on cells. Materials that have already been used in clinical applications can meet these requirements.

(i) Cell-seeded microcarriers as *in vitro* models

Microcarrier culture systems have already been used to study the response of cells to the external environment, including biological factors, chemical reagents, and mechanical and physical effects. These investigations of biology of tissue cells in an *in vitro* model are well controllable and have obtained important information including effects of oxygen tension, growth factors, trace elements, microgravity, etc. Some results showed that cells cultured on microcarriers could resemble many features of the original tissue, which was a dramatic advantage for applications in tissue engineering.

The culture of chondrocytes on microcarriers has been extensively studied in tissue engineering (Freed *et al.*, 1993). It was observed that viable chondrocyte cells were maintained on collagen-coated dextran microcarriers for more than four months in 3-D culture while it was significantly shorter for monolayer cultures (Malda *et al.*, 2003a,b). In addition, the 3-D microcarrier system maintained the cell phenotype similar to the original tissue. However, chondrocytes cultured in monolayer changed its phenotype, which was not suitable for further applications in tissue engineering (Frondoza *et al.*, 1996). Other microcarriers like dextran and cross-linked collagen were also confirmed to be good for obtaining viable bone-forming osteoblasts with right phenotype (Howard *et al.*, 1983; Shima *et al.*, 1988; Sautier *et al.*, 1992). However, the reason for such difference between 3-D microcarrier culture and monolayer culture was still unclear. It might relate to the 3-D system provided an environment more similar to their natural growing conditions (Bouchet *et al.*, 2000; Shikani *et al.*, 2004).

(ii) Microcarriers as cell delivery systems

Compared to the conventional transplant method, the injectable cell therapy is much friendlier for patients. Microcarriers are useful delivery vehicles for tissue engineering and have indicated their unique advantages for repair of bone and cartilage defects (Malda & Frondoza, 2006). Various microcarriers such as collagen-coated dextran microspheres, poly(L-lactide) (PLLA) microspheres, and gelatin microspheres have achieved good therapeutic effects for treatment of degenerated or wounded tissue (Demetriou *et al.*, 1988), restoration of tissue function (Moscioni *et al.*, 1989), promotion of wound healing (Voigt *et al.*, 1999; Bayram *et al.*, 2005; LaFrance & Armstrong, 1999), and therapy for Parkinson's disease (Liu *et al.*, 2004).

The therapeutic cells were seeded in microcarriers. After a period of culture, the cellular aggregates formed and further grew to injectable aggregates. The cell-seeded aggregates were directly injected into the defective site (Fig. 2.2). The size of aggregates could be controlled by regulating different culture conditions, including mixing intensity, oxygen tension, duration of culture, and other factors. Smaller aggregates containing several microcarriers that covered viable cells provided structure suitable for injection. If the culture time was long enough, large aggregates would develop into a tissue-like material which was appropriate for repairing bigger defects (Malda & Frondoza, 2006). Cellular aggregates cultured on microcarriers were a principle method to improve cell density and metabolic activity. The morphology of cell and interaction between cells in aggregates was similar to those *in vivo*. This kind of 3-D structure helped them to maintain their functions and bioactivities.

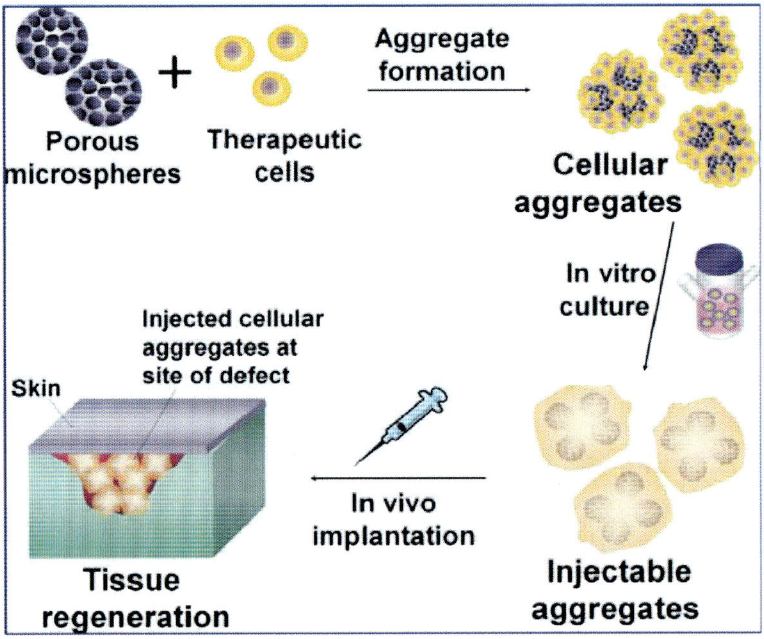

Figure 2.2 Schematic illustration for the formation and culture of cellular aggregates for injectable therapy. Reprinted from Chung *et al.* (2009), with permission from Mary Ann Liebert.

Microspheres not only act as cell-seeded carrier but also as a delivery system for bioactive factors. The bioactive factors embedded in microspheres will influence cell differentiation and improve therapeutic effects (Newman & McBurney, 2004; McBurney & Rogers, 1982). Compared to soluble bioactive factors, microencapsulation provides several advantages. It has a prolonged effect, which avoids repetition of administration of drugs and also helps to deliver the bioactive factors to the target location. This helps in cell differentiation and in selecting appropriate phenotype (Eldridge *et al.*, 1991; Miller *et al.*, 1977).

2.5.2 Preparation and Surface Modification of Microcarriers for Tissue Engineering

Various microcarrier substrates have been used including modified dextran, alginate, plastic, and synthetic polymers like modified poly(lactide-co-glycolide) (PLGA) (Tebb *et al.*, 2006; Overstreet *et al.*, 2003). The surface of these matrixes is coated with collagens or gelatin to enhance cell attachment and growth. Some microspheres prepared by degradation materials like gelatin and collagen are even more appropriate for tissue engineering (Kwon & Peng, 2002; Gröhn *et al.*, 1997). Listed here are some main matrix materials and their modification methods.

(i) Synthetic polyesters

Microcarriers prepared by synthetic polyesters including PLA, polyglycolide (PGA), PLGA, and PLLA have good mechanical properties, processibility, biocompatibility, and biodegradability. Surface modification of the polyester microspheres is still required to improve their adherence for cells. There are many surface modification methods. One effective strategy is to immobilize bioactive molecules like fibronectin (Fn) and RGD peptide sequence (Suh *et al.*, 2000) to form a biomimetic interface. Another method is covalent immobilization of an amine-terminated dendrimer. Thissen *et al.* (2006) obtained the maximum proliferation rate of sheep articular cartilage chondrocytes on the PLGA particles modified by the second method compared to the carriers modified by other methods. Another advantage of the modified PLGA particles was that chondrocytes cultured on these microcarriers maintained their activity as *in vivo*, which demonstrated that the modified PLGA particles could be used as cell carriers for tissue engineering.

(ii) Gellan gels

Gellan gum is an edible microbial gum and is an FDA-approved natural polysaccharide. Cell culture microcarriers are prepared with this novel, fully transparent hydrogel. Same to other polysaccharide materials, gellan gels have no affinity to cells due to their extremely hydrophilic nature, which renders specific functionalities to the cells (Tan *et al.*, 2005). Many ways have been developed to improve the cell adhesion on hydrogel surfaces including enzymatic catalysis, end functionalization, or ultraviolet (UV) activation (Hern & Hubbell, 1998; Chen *et al.*, 2003). Surface coating is also a widely used method. The gellan particles were covalently coated with gelatin, and good culture results were obtained for fibroblasts and osteoblasts which were appropriate for cell delivery (Wang *et al.*, 2008). The detailed coating method is shown in Fig. 2.3. The final gelatin-grafted-gellan microspherical cell microcarriers were called "TriG" microcarriers. The TriG microcarriers were confirmed to be favorable for cell adhesion and proliferation, especially for human dermal fibroblasts (hDFs) and human fetal osteoblasts (hFOBs). Ideal cell viability and great potentials for tissue engineering were indicated by various characterization methods such as optical microscopy, field emission scanning electron microscopy, immunofluorescent staining, and specific histobiochemical indications.

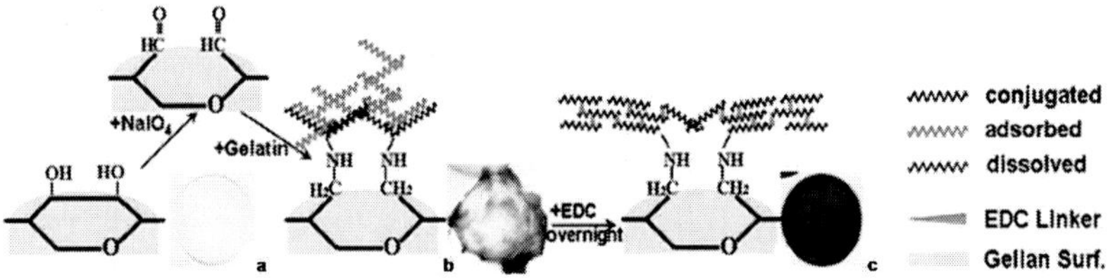

Figure 2.3 Schematic diagram of coating the gellan microspheres by gelatin: (a) gellan microspheres containing numbers of hydroxyl groups; (b) gelatin conjugated on the surface of gellan particles; (c) adsorbed gelatin molecules cross-linked with EDC to improve the stability of the coating. Reprinted from Wang *et al.*, (2008), with permission from Elsevier.

(iii) Chitosan

Chitosan is a widely used biomaterial, and it is a natural material for cell culture. It is nontoxic, nonirritant, and has good biocompatibility and high positive charge density. It is worth noting that chitosan has a similar chemical composition like glycosaminoglycan (GAG), which is an important component of ECM. Macroporous chitosan microspheres with positive charge provide a 3-D environment for cell growth and help in maintaining the cell functions. Chitosan microspheres can be applied in cell therapy especially for liver cells, chondrocytes, and endothelial cells, which are respectively suitable for repair of liver, cartilage, and skin.

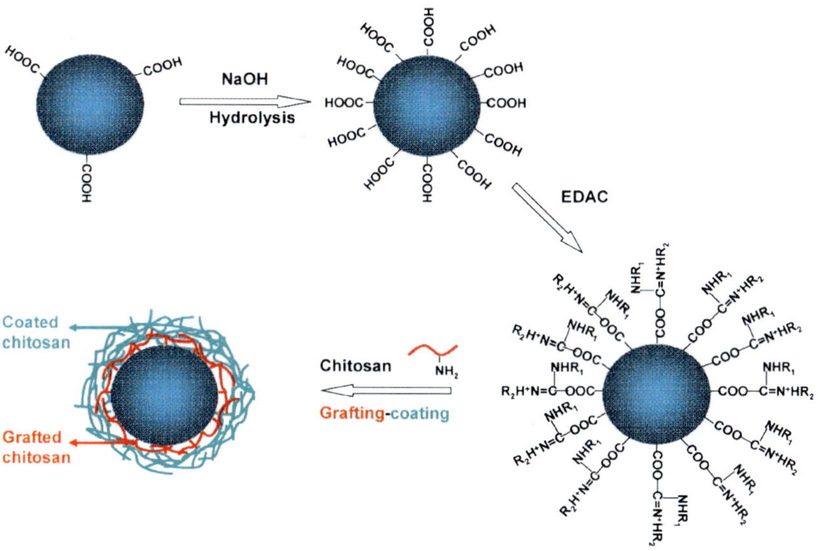

Figure 2.4 The fabrication of chitosan-coated PLLA particles. Reprinted from Lao *et al.* (2008), with permission from Elsevier.

Chitosan is a good coating material because it has a lot of amino groups which can easily interact with the matrix. Many methods can be used in chitosan coating such as covalent bonding (Zhu *et al.*, 2004), layer-by-layer assembly (Liu *et al.*, 2005), plasma treatment, and grafting–coating (Ding *et al.*, 2004). These chitosan-coated microspheres have acted as cell delivery carriers and injectable scaffolds. Here, an example of chitosan-coated PLLA microspheres is given. The PLLA surface was hydrolyzed to produce abundant carboxyl groups, and then chitosan molecules were covalently grafted on the microspheres through carboxyl–

amino interaction as shown in Fig. 2.4 (Lao *et al.*, 2008). Unbonded chitosan remained to form a coating. The coated particles were applied to culture rabbit auricular chondrocytes, which showed good promotion of cell attachment and proliferation. The chondrocytes also well maintained their secretion function especially on the particles with higher chitosan amount.

(iv) Bioceramics

According to the action mechanism of the materials with tissues, bioceramics can be divided into active bioceramics and inactive bioceramics. Inactive bioceramics mainly refer to ceramic materials with chemical stability and good biocompatibility. They have high mechanical strength and are wear resistance. These materials include alumina, single-crystal ceramics, zirconia ceramics, and glass ceramics. Active bioceramics are also known as biodegradable ceramics which usually contain hydroxyl groups and are appropriate for tissue engineering. Bioactive glass ceramics (calcium phosphate systems), hydroxyapatite ceramics, and tricalcium phosphate ceramics are the main active bioceramics. Microspheres composed of active bioceramics have been applied in cell delivery especially for bone repair. Calcium titanium phosphate and hydroxyapatite microcarriers were used in growth of osteoblast bone-forming cells and bone-marrow stromal cells. Good propagation and proliferation were observed. These microcarriers also showed promotion of bone-cell function and bone bonding. However, the density of ceramic materials is usually higher, and the collisions of particles and particles with vessel walls may make cells lose their activity. In order to solve these problems, hollow bioceramic particles were prepared (Doctor *et al.*, 2002). The density of hollow bioceramic particles was slightly higher than or equal to the density of the liquid medium, which could efficiently prevent strong collisions.

2.5.3 Highly Open Porous Microcarriers for Three-Dimensional Cell Culture

Large particles with a highly porous structure have been found to be appropriate for 3-D cell culture. The pore size of highly porous microspheres is about tens of microns, usually larger than twenty microns. The highly porous structure facilitates nutrients and oxygen in and out of the particles and allows high cell density (Fig. 2.5). Various methods have been used for fabricating this structure. Among them, the porogen-leaching method was most extensively used. Porogens were mixed with a polymer/solvent mixture and they were leached out with solvent for formation of pores. Various porogens, such as salts (Mikos *et al.*, 1993), carbohydrates (McGlohorn *et al.*, 2003), and hydrocarbon waxes (Ma *et al.*, 2003), could be chosen according to the requirements of different polymer systems. Other methods including emulsion/freeze drying (Whang *et al.*, 1995), expansion in supercritical fluid (Butler *et al.*, 2001), 3-D-guided ink-jet printing (Park *et al.*, 1998), gas-forming method (Kim *et al.*, 2006), and high interphase emulsion (Li & Benson, 1996; Li *et al.*, 2000) were also reported. Most of the methods have applied highly open porous microspheres in preparation of bulk polymers. Those cited here are methods used in synthesis of microspheres.

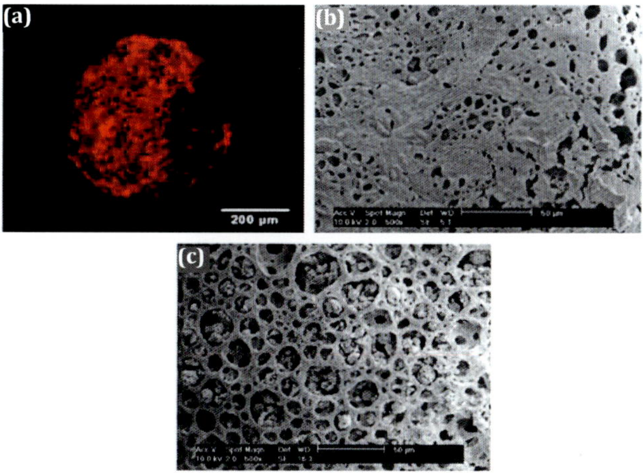

Figure 2.5 Cells cultured on highly porous microspheres: (a) confocal image of cells cultured after 1 day, (b) SEM image of cells cultured after 1 day, and (c) SEM image of cell proliferated within the pores after 7 days. Reprinted from Kim *et al.* (2006), with permission from Elsevier.

(i) Gas foaming method

Polymer particles such as PLGA and PLA particles with highly open porous structures were fabricated by gas foaming method. This preparation strategy combined a double emulsion method and a solvent-evaporation process as shown in Fig. 2.6 (Kim *et al.*, 2006). The primary W/O emulsion solution was re-emulsified into an aqueous solution of PVA for further formation of a $W_1/O/W_2$ double emulsion and solvent evaporation. Carbon dioxide and ammonia gas were gradually generated by the solvent-removing method, and open porous particles with highly interconnected pores were prepared. The size and pore diameter of the particles could be controlled by varying the preparation conditions. For example, when increasing the PLGA concentration, the size of particles increased and the pore diameter decreased (Fig. 2.7), which indicated that the viscosity of organic phase had an important influence on the porous structure. Among these particles, those with a particle size of approximately 175 μm and an average pore diameter of approximately 29 μm were confirmed to be appropriate for cell culture and injectable delivery. Bovine articular chondrocytes were seeded in the particles and the microcarrier–cell aggregate had good cell proliferation and expression of cartilage-specific phenotype.

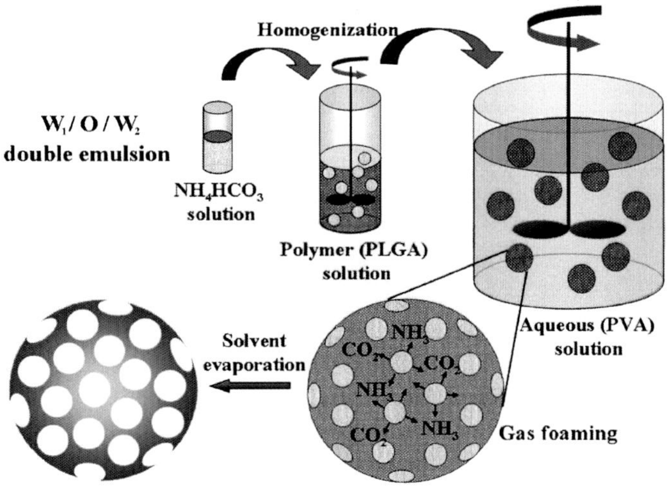

Figure 2.6 Schematic illustration of gas forming method to produce highly porous particles. Reprinted from Chung *et al.* (2008), with permission from Mary Ann Liebert.

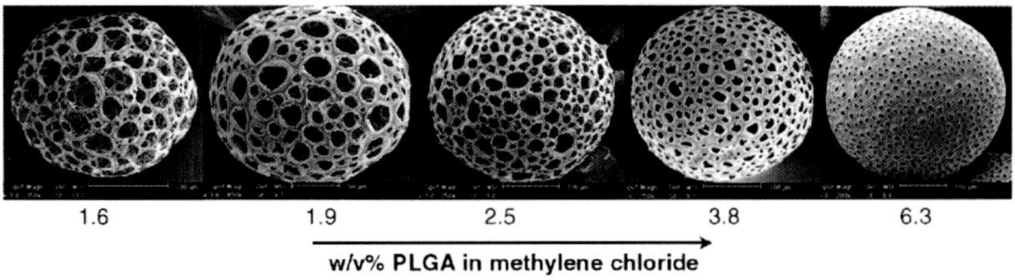

Figure 2.7 Porous PLGA particles prepared by various process conditions. Reprinted from Chung *et al.* (2008), with permission from Mary Ann Liebert.

(ii) High internal phase emulsion method

Bulk polymer with highly open porous structure was fabricated by using a high-internal phase emulsion (HIPE) method in 1985 (Barby & Haq, 1985). In 1996, Li and Benson developed this method and successfully produced HIPE particles (Li & Benson, 1996; Li *et al.*, 2000). The particles prepared by HIPE method usually had a high porosity of 70% to 90%, which showed a highly permeable structure with a low density.

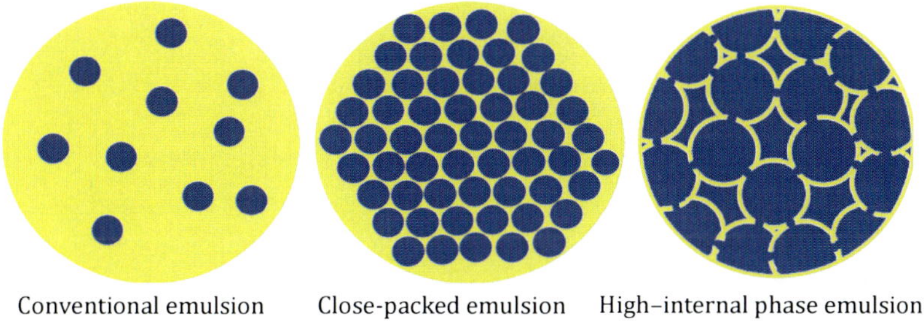

Figure 2.8 Changing from conventional emulsion to high-internal phase emulsion. Reprinted from Benson (2003), with permission from International Scientific Communications, Inc.

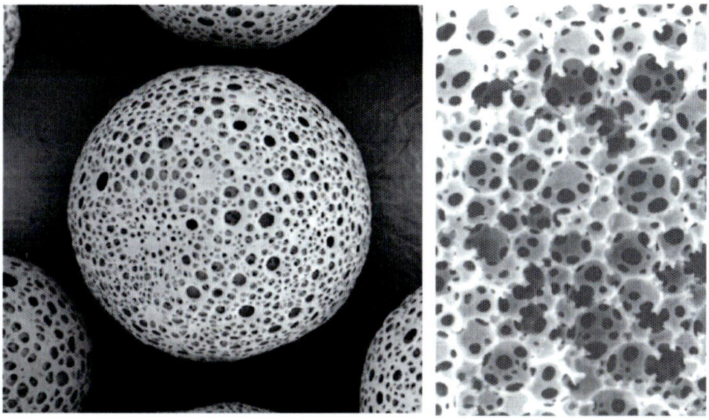

Figure 2.9 SEM images of HIPE particles and its internal structure. Reprinted from Benson (2003), with permission from International Scientific Communications, Inc.

HIPE is actually an emulsion with a large internal-phase volume. As shown in Fig. 2.8, the water phase is dispersed in the oil phase in the initial stage. As its concentration increases, the consistency of the mixture changes to a thick liquid. When the content of water becomes 74%, the aqueous droplets get packed closely in oil. As the water content further increases above 74%, the high-internal phase emulsion, also called concentrated emulsion, comes into being. As for the high volume fraction of internal aqueous phase, the water droplets are separated by the oil film. The high-internal phase emulsion is then suspended in a large amount of water for stability, and microspheres of HIPE are formed after polymerization. The particles prepared by this method have interconnected cavities which can provide a 3-D environment for cell culture. Hydrophobic and hydrophilic microspheres with highly porous structure can been prepared by the HIPE method (Fig. 2.9).

A large number of surfactants have to be used in preparing double emulsion, especially HIPE. To avoid this problem, Song *et al.* (2007) used an amphiphilic copolymer, monomethoxy-poly(ethylene glycol)-b-poly-DL-lactide (PELA), to prepare highly porous microspheres by double-emulsion method. They found that there was no need to add an emulsifier in the system because the amphiphilic property of PELA made the oil–water interface stable. PELA was dissolved in ethyl acetate (EA). Deionized water (inner water phase, W_1) was mixed with the oil phase to form a pre-emulsion. The pre-emulsion was dispersed in outer water phase (W_2) to form $W_1/O/W_2$ double emulsion. The emulsion structure was solidified after removing EA by redispersing the double emulsion in a large amount of water. Microspheres with large pores formed when the volume ratio of inner water phase to oil phase and the solidification rate were suitable (Fig. 2.10).

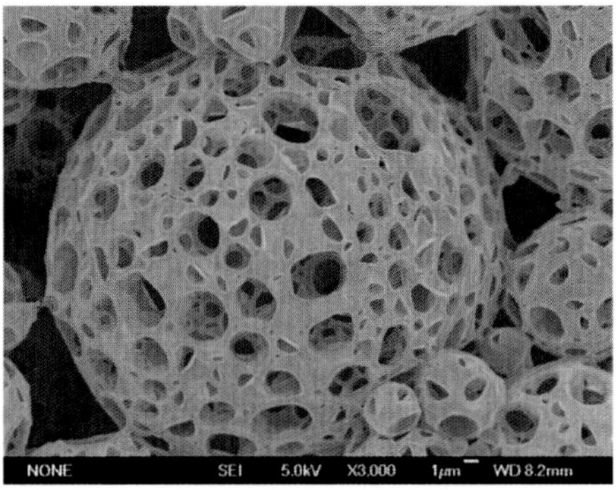

Figure 2.10 Highly porous PELA microspheres prepared by double emulsion method. Reprinted from Song *et al.* (2007), with permission from China Science Publishing & Media Ltd.

(iii) Phase-removing method

Another widely used method to produce highly porous microspheres is phase-removing method. For example, porous collagen beads were prepared by combining an alginate phase with collagen beads and then removing the alginate phase (Tebb *et al.*, 2006). The porosity of the resulting particles is large enough to allow cell growth. In the preparation process, first, the collagen solution was mixed with alginate (Suh *et al.*, 2000). The mixture was then added dropwise into $CaCl_2$ solution to form spherical gel particles. The collagen–alginate particles were then stabilized either by immersing in a Na-CHES (2-(*N*-cyclohexylamino) ethane-sulfuric acid) buffer containing L-lysine or in Na-CHES buffer containing glutaraldehyde for 16 hours. The alginate within the gel particles was liquefied with sodium citrate solution. The resulting particle sizes were readily controlled from 100 μm to 1500 μm. The pore size was adjusted by the concentration of alginate. Further cross-linking with glutaraldehyde (GA) improved the mechanical strength and durability of collagen particles, which made them more appropriate for cell therapy and tissue engineering.

2.6 Summary and Prospects

Microcarrier system will keep developing because of its importance in cell culture. The design of its structure and properties has been widely investigated. Various commercial microcarriers, including smooth, macroporous, positively charged, and protein-modified, have been applied in different cell cultures to produce valuable biological technology products. Related bioreactors, monitoring equipment, and culture media have been developed. Furthermore, microcarrier technology has been demonstrated to be a powerful research tool in the field of tissue engineering and regeneration medicine. Microcarrier cultures can provide sufficient cell numbers with appropriate phenotype to assist the repair or regeneration of damaged or degenerated tissue, which is a significant bottleneck in tissue engineering. Moreover, highly open macroporous microcarriers have been successfully used in 3-D cell culture to support cell proliferation and differentiation, which closely resembles that in *in vivo* environment and can be applied in cell therapy. In future, larger-scale production, higher cell density, and higher yield of target products having bioactivity resembling *in vivo* environment will be the areas of development of microcarrier technique.

References

Baek, N. S., Lee, J. H., Kim, Y. H., *et al.* (2011). Photopatterning of cell-adhesive-modified poly(ethyleneimine) for guided neuronal growth. *Langmuir*, **27**, 2717–2722.

Barby, D., & Haq, Z. (1985). U.S. Pat. 4522953.

Bayram, Y., *et al.* (2005). The cell-based dressing with living allogenic keratinocytes in the treatment of foot ulcers: A case study. *British Journal of Plastic Surgery*, **58**, 988–996.

Benson, J. R. (2003). Highly porous polymers. *American Laboratory*, **35**, 44–52.

Blackburn, C. C., & Schnaar, R. L. (1983). Carbohydrate-specific cell adhesion is mediated by immobilized glycolipids. *The Journal of Biological Chemistry*, **258**, 1180–1188.

Bledi, Y., Domb, A. J., & Linial, M. (2000). Culturing neuronal cells on surfaces coated by a novel polyethyleneimine-based polymer. *Brain Research Protocols*, **5**, 282–289.

Bouchet, B., *et al.* (2000). Beta-1 integrin expression by human nasal chondrocytes in microcarrier spinner culture. *Journal of Biomedical Materials Research*, **52**, 716–724.

Bueno, E. M., Laevsky, G., & Barabino, G. A. (2007). Enhancing cell seeding of scaffolds in tissue engineering through manipulation of hydrodynamic parameters. *Journal of Biotechnology*, **1293**, 516–531.

Butler, R., Davies, C. M., & Cooper, A. I. (2001). Emulsion templating using high internal phase supercritical fluid emulsions. *Advanced Materials*, **13**, 1459–1463.

Carbonetto, S. T., Gruver, M. M., *et al.* (1982). Nerve fiber growth on defined hydrogel substrates. *Science*, **216**, 897–899.

Chen, T., Embree, H. D., Brown, E. M., Taylor, M. M., & Payne, G. F. (2003). Enzymecatalyzed gel formation of gelatin and chitosan: Potential for *in situ* applications. *Biomaterials*, **24**, 2831–2841.

Chung, H. J., Kim, I. K., Kim, T. G., & Park, T. G. (2008). Highly open porous biodegradable microcarriers: *In vitro* cultivation of chondrocytes for injectable delivery. *Tissue Engineering: Part A*, **14**, 607–615.

Chung, H. J., & Park, T. G. (2009). Injectable cellular aggregates prepared from biodegradable porous microspheres for adipose tissue engineering. *Tissue Engineering: Part A*, **15**, 1391–1400.

Civerchia-Perez, L., Faris, B., *et al.* (1980). Use of collagen-hydroxyethyhethacrylate hydrogels for cell growth. *Proceedings of the National Academy of Sciences of the United States of America*, **77**, 2064–2068.

Crespi, C. L., & Thilly, W. G. (1981). Continuous cell propagation using low-charge microcarriers. *Biotechnology and Bioengineering*, **23**, 983–993.

Dame, M. K., & Varani, J. (2008). Recombinant collagen for animal product-free dextran microcarriers. *In Vitro Cellular & Developmental Biology — Animal*, **44**, 407–414.

Demetriou, A., *et al.* (1988). Transplantation of microcarrier-attached hepatocytes into 90% partially hepatectomized rats. *Hepatology*, **8**, 1006–1009.

Ding, Z., Chen, J. N., Gao, S. Y., Chang, J. B., Zhang, J. F., & Kang, E. T. (2004). Immobilization of chitosan onto poly-L-lactic acid film surface by plasma graft polymerization to control the morphology of fibroblast and liver cells. *Biomaterials*, **25**, 1059–1067.

Doctor, J., *et al.* (2002). Evaluating microcarriers for delivering human adult mesenchymal stem cells in bone tissue engineering. *Developmental Biology*, **247**, 505–514.

Eldridge, J. H., Staas, J. K., Meulbroek, J. A., McGhee, J. R., Rice, T. R., & Gilley, R. M. (1991). Biodegradable microspheres as a vaccine delivery system. *Molecular Immunology*, **28**, 287–294.

Esra, C., Gürpinar, A., Onur, M. A., & Tuncel, A. (2007). Polyethylene glycol-based cationically charged hydrogel beads as a new microcarrier for cell culture. *Journal of Biomedical Materials Research Part B: Applied Biomaterials*, **80B**, 406–414.

Freed, L., *et al.* (1993). Cultivation of cell–polymer cartilage implants in bioreactors. *Journal of Cellular Biochemistry*, **51**, 257–264.

Frondoza, C., *et al.* (1996). Human chondrocytes proliferate and produce matrix components in microcarrier suspension culture. *Biomaterials*, **17**, 879–888.

Giard, D. J., Thilly, W. G., Wang, D. I., & Levine, D. W. (1977). Virus production with a newly developed microcarrier system. *Applied and Environmental Microbiology*, **34**, 668–672.

Ginsburg, I. (1987). Cationic polyelectrolytes: A new look at their possible role as opsinins, as stimulators of the respiratory burst in leukocytes, in bacteriolysis and as modulators of immune complex disease. *Inflammation*, **11**, 489–495.

Gotoh, T., Honda, H., Shiragami, N., & Unno, H. (1993). A new type porous carrier and its application to culture of suspension cells. *Cytotechnology*, **11**, 35–40.

Griffiths, B., Thornton, B., & McEntee, I. (1980). Production of Herpes viruses in microcarrier cultures of human diploid and primary chick fibroblast cells. *European Journal of Cell Biology*, **22**, 606.

Gröhn, P., Klöck, G., & Zimmermann, U. (1997). Collagen-coated Ba^{2+}-alginate microcarriers for the culture of anchorage-dependent. *Mammalian Cells*, **22**, 970–975.

Hakoda, M., & Shiragami, N. (2000). Effects of ion exchange capacities on attachment and growth of anchorage-dependent HeLa cell. *Bioprocess Engineering*, **23**, 523–527.

Handbooks from GE Healthcare (2005). *Microcarrier Cell Culture: Principles and Methods*. (pp. 12–14), Sweden: Amersham Biosciences.

Hannan, G., & McAuslan, B. (1987). Immobilized serotonin: A novel substrate for cell culture. *Experimental Cell Research*, **171**, 153–163.

Hern, D. L., & Hubbell, J. A. (1998). Incorporation of adhesion peptides into nonadhesive hydrogels useful for tissue resurfacing. *Journal of Biomedical Materials Research*, **39**, 266–276.

Hoffman, R. M. (1993). To do tissue culture in two or three dimensions? That is the question. *Stem Cells*, **11**, 105–111.

Horbett, T. A., & Schway, M. B. (1988). Correlations between mouse 3T3 cell spreading and serum fibronectin adsorption on glass and hydroxyethylmethacrylate-ethylmethacrylate copolymers. *Journal of Biomedical Materials Research*, **22**, 763–793.

Horbett, T. A., Schway, M. B., & Ratner, B. (1985). Hydrophilic-hydrophobic copolymers as cell substrates: Effect on 3T3 cell growth rates. *Journal of Colloid and Interface Science*, **104**, 28–39.

Howard, G. A., *et al.* (1983). Bone cells on microcarrier spheres. *JAMA*, **249**, 258–259.

Hu, W. S., & Wang, D. I. (1987). Selection of microcarrier diameter for the cultivation of mammalian cells on microcarriers. *Biotechnology and Bioengineering*, **30**, 548–557.

Ikada, Y. (1994). Surface modification of polymers for medical applications. *Biomaterials*, **15**, 725–736.

Ishida, M., Akaike, T., Kuroyanagi, Y., & Seno, M. (1985). *Japanese Journal of Polymer Science and Technology*, **42**, 725–730.

Kato, D., Takeuchi, M., Sakurai, T., *et al.* (2003). The design of polymer microcarrier surfaces for enhanced cell growth. *Biomaterials*, **24**, 4253–4264.

Kikuchi, A., Kataoka, K., *et al.* (1992). Adhesion and proliferation of bovine aortic endothelial cells on monoamine- and diamine-containing polystyrene derivatives. *Journal of Biomaterials Science, Polymer Edition*, **3**, 253–260.

Kim, T. K., Yoon, J. J., Lee, D. S., & Park, T. G. (2006). Gas foamed open porous biodegradable polymeric microspheres. *Biomaterials*, **27**, 152–159.

Kiremitci, M., & Piskin, E. (1990). Cell adhesion to the surfaces of polymeric beads. *Biomaterials: Artificial Cells and Artificial Organs*, **18**, 599–603.

Kleinmann, H., Klebe, R., & Martin, G. R. (1981). Role of collagenous matrices in the adhesion and growth of cells. *The Journal of Cell Biology*, **88**, 473–485.

Klöck, G., Frank, H., Houben, R., Zekorn, T., Horcher, A., Siebers, U., *et al.* (1994). Production of purified alginates suitable for use in immunoisolated transplantation. *Applied Microbiology and Biotechnology*, **40**, 638–643.

Kwon, Y. J., & Peng, C. A. (2002). Calcium-alginate gel bead cross-linked with gelatin as microcarrier for anchorage-dependent cell culture. *BioTechniques*, **33**, 212–214.

LaFrance, M. L., & Armstrong, D. W. (1999). Novel living skin replacement biotherapy approach for wounded skin tissues. *Tissue Engineering*, **5**, 153–170.

Lao, L. H., Tan, H. P., Wang, Y. J., & Gao, C. Y. (2008). Chitosan modified poly(L-lactide) microspheres as cell microcarriers for cartilage tissue engineering. *Colloids and Surfaces B: Biointerfaces*, **66**, 218–225.

Levine, D. W., Thilly, W. G., & Wang, D. I. C. (1978). New microcarriers for the large scale production of anchorage-dependent mammalian cells. *Advances in Experimental Medicine and Biology*, **100**, 15–23.

Li, N. H., & Benson, J. R. (1996). Polymeric microbeads and method of preparation. U.S. Pat. 5583162.

Li, N. H., Benson, J. R., & Kitagawa, N. (2000). Polymeric microbeads and method of preparation. U.S. Pat. 6100306.

Liu, J. Y., Hafner, J., *et al.* (2004). Autologous cultured keratinocytes on porcine gelatin microbeads effectively heal chronic venous leg ulcers. *Wound Repair and Regeneration*, **12**, 148–156.

Liu, Y. X., He, T., & Gao, C. Y. (2005). Surface modification of poly(ethylene terephthalate) via hydrolysis and layer-by-layer assembly of chitosan and chondroitin sulfate to construct cytocompatible layer for human endothelial cells. *Colloids and Surfaces B: Biointerfaces*, **46**, 117–126.

Looby, D., & Griffiths, B. (1990). Immobilization of animal cells in porous carrier culture. *TIBTECH*, **8**, 204–209.

Ma, Z., Gao, C., Gong, Y., & Shen, J. (2003). Paraffin spheres as porogen to fabricate poly(L-lactic acid) scaffolds with improved cytocompatibility for cartilage tissue engineering. *Journal of Biomedical Materials Research Part A*, **67B**, 610–317.

Malda, J., & Frondoza, C. G. (2006). Microcarriers in the engineering of cartilage and bone. *Trends in Biotechnology*, **24**, 299–304

Malda, J., *et al.* (2003a). Expansion of bovine chondrocytes on microcarriers enhances redifferentiation. *Tissue Engineering*, **9**, 939–948.

Malda, J., *et al.* (2003b). Expansion of human nasal chondrocytes on macroporous microcarriers enhances redifferentiation. *Biomaterials*, **24**, 5153–5163.

Maroudas, N. G. (1975). Adhesion and spreading of cells on charged surfaces. *Journal of Theoretical Biology*, **49**, 417–427.

Massia, S. P., & Hubbell, J. A. (1989). Covalent surface immobition of Arg-Gly-Asp- and Tyr-Ile-Gly-Ser-Arg-containing peptides to obtain well-defined cell-adhesive substrates. *Analytical Biochemistry*, **187**, 292–301.

Massia, S. P., & Hubbell, J. A. (1991). Human endothelial cell interactions with surface-coupled adhesion peptides on a nonadhesive glass substrate and two polymeric biomaterials. *Journal of Biomedical Materials Research Part A*, **25**, 223–242.

Massia, S. P., & Hubbell, J. A. (1992). Immobilized amines and basic amino acids as mimetic heparin-binding domains for cell surface proteoglycan-mediated adhesion. *The Journal of Biological Chemistry*, **267**, 10133–10141.

McBurney, M. W., & Rogers, K. A. (1982). Isolation of male embryonal carcinoma cells and their chromosome replication patterns. *Developmental Biology*, **89**, 503–508.

McGlohorn, J. B., Grimes, L. W., Webster, S. S., Burg, K. J. L. (2003). Characterization of cellular carriers for use in injectable tissue-engineering composites. *Journal of Biomedical Materials Research*, **66A**, 441–449.

Meignier, B. (1979). Cell culture on beads used for the industrial production of foot-and-mouth disease virus. *Developments in Biological Standardization*, **42**, 141–145.

Mered, B., Albrecht, P., & Hopps, H. E. (1980). Cell growth optimization in microcarrier culture. *In Vitro*, **16**, 859–865.

Mikos, A. G., Sarakinos, G., Leite, S. M., Vacanti, J. P., & Langer, R. (1993). Laminated three-dimensional biodegradable foams for use in tissue engineering. *Biomaterials*, **14**, 323–230.

Miller, R. A., Brandy, J. M., Cutwright, D. E. (1977). Degradation rates of resorbable implants (polylactates and polyglycolates): Rate modification with changes in PLA/PGA copolymer ratios. *Journal of Biomedical Materials Research*, **11**, 711–719.

Moscioni, A., *et al.* (1989). Human liver cell transplantation. Prolonged function in athymic-Gunn and athymic-analbuminemic hybrid rats. *Gastroenterology*, **96**, 1546–1551.

Newman, K. D., & McBurney, M. W. (2004). Poly(DL lactic-co-glycolic acid) microspheres as biodegradable microcarriers for pluripotent stem cells. *Biomaterials*, **25**, 5763–5771.

Nicholson, B. L. (1980). Growth of fish cell lines on microcarriers. *Applied and Environmental Microbiology*, **39**, 394–397.

Nilsson, K. (1989). Microcarrier culture. *Biotechnology & Genetic Engineering Reviews*, **6**, 403–439.

Nilsson, K., Buzsaky, F., & Masboch, K. (1986). Growth of anchorage-dependent cells on macroporous microcarriers. *Nature Biotechnology*, **4**, 989–990.

Overstreet, M., Sohrabi, A., Polotsky, A., Hungerford, D., & Frondoza, C. G. (2003). Collagen microcarrier spinner culture promotes osteoblast proliferation and synthesis of matrix proteins. *In Vitro Cellular and Developmental Biology*, **39**, 228–234.

Park, A., & Wu, B., Griffith, L. G. (1998). Integration of surface modification and 3D fabrication techniques to prepare patterned poly(Llactide) substrates allowing regionally selective cell adhesion. *Journal of Biomaterials Science, Polymer Edition*, **9**, 89–110.

Pharmacia Fine Chemicals, Uppsala, Sweden. (1979). Save with Cytodex TM 1. *Separation News*, **4**, 5.

Pierschbacher, M. D., & Ruoslahti, E. (1984). Cell attachment activity of fibronectin can be duplicated by small synthetic fragments of the molecule. *Nature*, **309**, 30–33.

Plantefaber, L. C., & Hynes, R. O. (1989). Changes in integrin receptors on oncogenically transformed cells. *Cell*, **56**, 281–290.

Reuveny, S. (1983). Microcarriers for culturing mammalian cells and their applications. *Advances in Biotechnological Processes*, **1**, 1–26.

Reuveny, S., Corett, R., Freeman, A., Kotler, M., & Mizrahi, A. (1985). Newly developed microcarrier culturing systems — An overview. *Developments in Biological Standardization*, **60**, 243–253.

Reuveny, S., Mizrahi, A., Kotler, M., Freeman, A. (1983). Factors effecting cell attachment, spreading, and growth on derivatized microcarriers:

II introduction of hydrophobicelements. *Biotechnology and Bioengineering*, **25**, 2969–2980.

Reuveny, S., Silberstein, L., Shahar, A., Freeman, A., & Mizrahi, A. (1982). DE-52 DE-53 cellulose microcarriers. I. Growth of primary and established anchorage-dependent cells. *In Vitro*, **18**, 92–98.

Ruoslahti, E., & Pierschbacher, M. D. (1986). Arg-Gly-Asp: A versatile cell recognition signal. *Cell*, **44(4)**, 517–518.

Sakata, M., Kato, D., Uchida, M., Todokoro, M., Mizokami, H., Furukawa, S., Kunitake, M., & Hirayama, C. (2000). Effect of pKa of polymer microcarriers on growth of mouse L cell. *Chemistry Letters*, 1056–1057.

Saltzman, M. W. (2000). Cell interactions with polymers. *Principles of Tissue Engineering*. Second Edition (pp. 221–235), Academic Press.

Sautier, J., *et al.* (1992). Mineralization and bone formation on microcarrier beads with isolated rat calvaria cell population. *Calcified Tissue International*, **50**, 527–532.

Schnaar, R. L., Weigel, P. H., *et al.* (1978). Adhesion of hepatocytes to polyacrylamide gels derivitized with *N*-acecylglucosamine. *The Journal of Biological Chemistry*, **253**, 7940–7951.

Shikani, A. H., *et al.* (2004). Propagation of human nasal chondrocytes in microcarrier spinner culture. *American Journal of Rhinology*, **18**, 105–112.

Shima, M., *et al.* (1988). Microcarriers facilitate mineralization in MC3T3-E1 cells. *Calcified Tissue International*, **43**, 19–25.

Song, W., Ma, G. H., & Su, Z. G. (2007). Preparation of macroporous polymer microspheres by double-emulsion method. *Chinese Journal of Process Engineering*, **7**, 1029–1034.

Stol, M., Tolar, M., *et al.* (1985). Poly(2-hydroxyethyl methacrylatea)-collagen composites which promote muscle cell differentiation *in vitro*. *Biomaterials*, **6**, 193–197.

Suh, J. K. F., & Matthew, H. W. T. (2000). Application of chitosan-based polysaccharide biomaterials in cartilage tissue engineering: A review. *Biomaterials*, **21**, 2589–2598.

Swartz, M. A. (1993). Signaling by integrins; implications for tumorigenesis. *Cancer Research*, **53**, 1503–1506.

Tamada, Y., & Ikada, Y. (1994). Fibroblast growth on polymer surfaces and biosynthesis of collagen. *Journal of Biomedical Materials Research*, **28**, 783–789.

Tan, J., Gemeinhart, R. A., Ma, M., & Saltzman, W. M. (2005). Improved cell adhesion and proliferation on synthetic phosphonic acid-containing hydrogels. *Biomaterials*, **26**, 3663–3671.

Tebb, T. A., Tsai, S. W., Glattauer, V., White, J. F., Ramshaw, J. A. M., & Werkmeister, J. A. (2006). Development of porous collagen beads for chondrocyte culture. *Cytotechnology*, **52**, 99–106.

Thissen, H., Chang, K. Y., Tebb, T. A., Tsai, W. B., Glattauer, V., Ramshaw, J. A. M., & Werkmeister, J. A. (2006). Synthetic biodegradable microparticles for articular cartilage tissue engineering. *Journal of Biomedical Materials Research A.*, **77A**, 590–598.

Tree, J. A., Richardson, C., & Fooks, A. R. (2001). Comparison of large-scale mammalian cell culture systems with egg culture for the production of influenza virus A vaccine strains. *Vaccine*, **19**, 3444–3450

van der Valk, P., van Pelt, A., *et al.* (1983). Interaction of fibroblasts and polymer surfaces: Relationship between surface free energy and fibroblast spreading. *Journal of Biomedical Materials Research*, **17**, 807–817.

van Groot, C. A. M., (1995). Microcarrier technology, present status and perspective, *Cytotechnology*, **18**, 51–56.

van Wachem, P. B., Hogt, A. H., *et al.* (1987). Adhesion of cultured human endothelial cells onto methacrylate polymers with varying surface wettability and charge. *Biomaterials*, **8**, 323–328.

van Wezel, A. L. (1967). Growth of cell strains and primary cells on microcarriers in homogeneous culture. *Nature*, **216**, 64–65.

van Wezel, A. L. (1972). New trends in the preparation of cell substrates for the production of virus vaccines. *Progress in Immunobiological Standardization.* **5**, 187–192.

van Wezel, A. L. (1977). The large scale cultivation of diploid cell strains in microcarrier culture. Improvement of microcarriers. *Developments in Biological Standardization*, **37**, 143–147.

Vancha, A. R., Govindaraju, S., Parsa, K. V. L., Jasti, M., González-García, M., & Ballestero, R. P. (2004). Use of polyethyleneimine polymer in cell culture as attachment factor and lipofection enhancer. *BMC Biotechnology*, **4**, 23–34.

Varani, J. (1985). Substrate-dependent differences in growth and biological properties of fibroblasts and epithelial cells grown in microcarrier culture. *Journal of Biological Standardization*, **13**, 67–76.

Varani, J., Fligiel, S. E. G., Inman, D. R., Beals, T. F., & Hillegas, W. J. (1995). Modulation of adhesive properties of DEAE dextran with laminin. *Journal of Biomaterials Research*, **23**, 993–997.

Varani, J., Fligiel, S. E. G., Inman, D. R., Helmreich, D. L, Bendelow, M. J., & Hillegas, W. J. (1989). Substrate-dependent differences in production of extracellular matrix molecules by squamous carcinoma cells and diploid fibroblasts. *Biotechnology and Bioengineering*, **33**, 1235–1241.

Varani, J., Inman, D. R., Fligiel, E. G. S., & Hillegas, W. J. (1993). Use of recombinant and synthetic peptides as attachment factors for cells on microcarriers. *Cytotechnology*, **13**, 89–98.

Varani, J., Piel, F., Josephs, S., Beals, T. F., & Hillegas, W. J. (1998). Attachment and growth of anchorage-dependent cells on a novel, charged-surface microcarrier under serum-free conditions. *Cytotechnology*, **28**, 101–109.

Vladkova, T. G. (2010). Surface engineered polymeric biomaterials with improved biocontact properties. *International Journal of Polymer Science*, **2010**, 1–22.

Voigt, M., *et al.* (1999). Cultured epidermal keratinocytes on a microspherical transport system are feasible to reconstitute the epidermis in full-thickness wounds. *Tissue Engineering*, **5**, 563–572.

Vournakis, J. N., & Runstadler P. W. (1989). Microenvironment: The key to improved cell culture products. *Biotechnology*, **7**, 143–145.

Wang, C. M., Gong, Y. H., Lin, Y. M., Shen, J. B., & Wang, D. A. (2008). A novel gellan gel-based microcarrier for anchorage-dependent cell delivery. *Acta Biomaterialia*, **4**, 1226–1234.

Whang, K., Thomas, C. H., Healy, K. E., & Nuber, G. A. (1995). A novel method to fabricate bioabsorbable scaffolds. *Polymer*, **36**, 837–842.

Woerly, S., Maghami, G., *et al.* (1993). Synthetic polymer derivatives as substrata for neuronal cell adhesion and growth. *Brain Research Bulletin*, **30**, 423–432.

Xiao, C. Z, Huang, Z. C., Li, W. Q., Hu, X. W., Qu, W. L., Gao, L. H., & Liu, G. Y. (1999). High density and scale-up cultivation of recombinant CHO cell line and hybridomas with porous microcarrier Cytopore. *Cytotechnology*, **30**, 143–147.

Yamagiwa, K., Kozawa, A. T., & Ohkawa, A. (1995). Effects of alginate composition and gelling conditions on diffusional and mechanical properties of calcium-alginate gel beads. *Journal of Chemical Engineering of Japan*, **28**, 462–467.

Zhu, Y. B., Gao, C. Y., & He, T. (2004). Endothelium regeneration on luminal surface of polyurethane vascular scaffold modified with diamine and covalently grafted with gelatin. *Biomaterials*, **25**, 423–430.

Zimmermann, U., Klöck, G., Federlin, K., Hannig, K., Kowalski, M., Bretzel, R. G., Horcher, A., Entenmann, H., Siebers, U., & Zekorn, T. (1992). Production of mitogen contamination free alginates with variable ratios of mannuronic to guluronic acid by free flow electrophoreses. *Electrophoresis*, **13**, 269–274.

Chapter 3

Microcapsules for Cell Transplantation: Design, Preparation, and Application

Guojun Lv, Ying Zhang, Mingqian Tan, Hongguo Xie, and Xiaojun Ma

Dalian Institute of Chemical Physics, Chinese Academy of Sciences, Dalian, P. R. China

maxj@dicp.ac.cn

3.1 Introduction

Microcapsule for *in situ* cell transplantation involves the envelopment of tissues or cells in semipermeable membranes (Fig. 3.1). Microcapsule has been shown to be efficacious in mimicking the cell's natural microenvironment and thereby improving the efficiency of the production of different metabolites and therapeutic agents. During the past few decades, many procedures to fabricate microcapsules have been described. Most of these procedures form a documentation of the characterization of the microcapsules. Unfortunately, many procedures show an extreme lab-to-lab variation and many results cannot be adequately reproduced. Many aspects, such as biomaterials, methods for preparation, design, and characterization of microcapsules, can no longer be neglected, especially since new clinical trials with microencapsulated therapeutic cells have been initiated and the industrial application of microcapsules is growing (Fig. 3.2a,b).

In this chapter, first, we have discussed how to design, prepare, and characterize microcapsules for *in situ* cell transplantation and have introduced novel approaches to produce and characterize microcapsules in clinical application. Second, we have discussed about the adequacy of using the versatility of cell encapsulation

Microspheres and Microcapsules in Biotechnology: Design, Preparation, and Applications
Edited by Guanghui Ma and Zhiguo Su

ISBN 978-981-4316-47-7 (Hardcover), 978-981-4364-62-1 (eBook)
www.panstanford.com

technology for clinical application. In addition, we have also presented the pitfalls of the currently applied methods and have provided recommendations for standardization to avoid lab-to-lab variations.

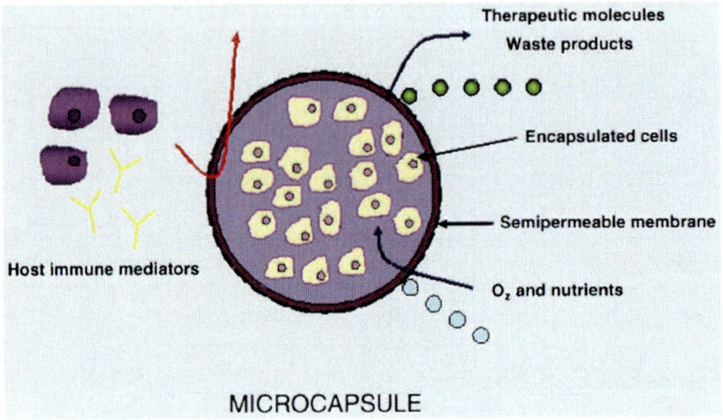

Figure 3.1 Diagram of microcapsules for cell transplantation (1).

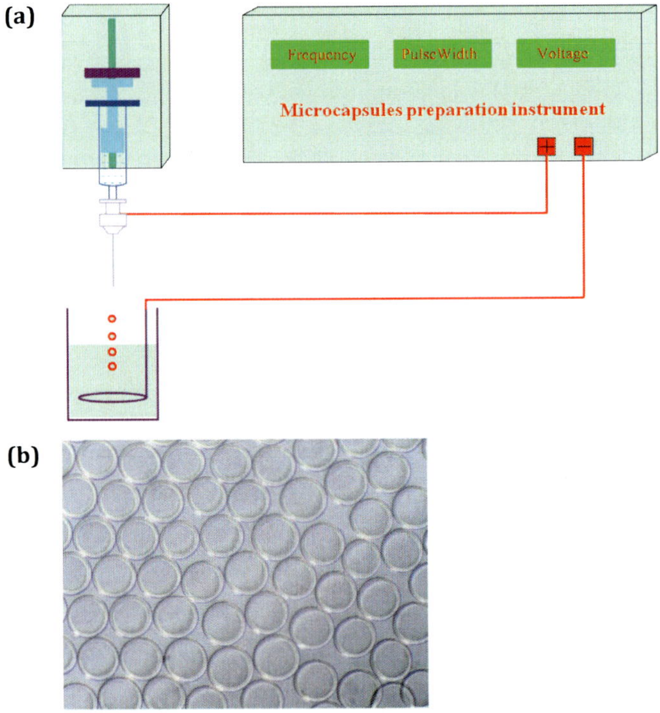

Figure 3.2 (a) The basic equipment of electrostatic droplet. (b) Picture of alginate poly-L-lysine alginate (APA) microcapsules.

3.2 Cell Encapsulation Technology

A very high level of *in vivo* biocompatibility is essential, assuming that the final aim of the encapsulation device is to protect the enclosed cellular tissue from the host's immune response. It is necessary to improve our knowledge about the biomaterial and device properties and to optimize and characterize each of the steps related to cell encapsulation technology, from alginate extraction and purification to cell selection, elaboration, large-scale production, characterization, storage, and administration of the microcapsules.

Biomaterials in Cell Microencapsulation

The most often described microencapsulation system is based on an alginate core surrounded by a polycation layer which, in turn, is covered by an outer alginate membrane. The polycation membrane forms a semipermeable membrane, which improves the stability and biocompatibility of the microcapsule. Many different materials are employed to encapsulate cells. Among them, alginates are the most studied and characterized for cell encapsulation technology. Alginates create three-dimensional (3-D) structures when they react with multivalent ions. Divalent cations, such as calcium, barium, and strontium, cooperatively bind between G-blocks of adjacent alginate chains, creating interchain bridges which cause gelling of the aqueous alginate solutions. Different polycations, such as poly-L-lysine (PLL) (2), poly-L-ornithine (PLO) (3), chitosan (4), lactose-modified chitosan (5, 6), and photopolymerized biomaterials (7) have been employed to cover the alginate matrix.

Alginates are certainly the most frequently employed biomaterials for cell immobilization due to their abundant source, easy gelling properties, and apparent biocompatibility. Although the suitability of other natural and synthetic polymers is under investigation (8, 9), none has reached the same level of performance as alginates. Among problems and details about alginates in cell microencapsulation, more detailed and in-depth knowledge will lead to the production of "transplantation-grade alginates".

As natural polysaccharides, alginates exist in brown seaweeds and bacterium (10), and their compositions vary depending upon the source from which they are isolated (11). Alginates are a family of unbranched binary copolymers of $1 \rightarrow 4$ linked β-D-mannuronic acid (M) and α-L-guluronic acid (G) of widely varying compositions and sequential structures. Standardizing these differences is of paramount importance since they have a significant impact on some of the alginate gel properties including biocompatibility, stability, mechanical resistance, permeability, biodegradability, and swelling behavior.

A key element in the validation of alginate for implantation purposes is the efficient purification process to monitor and remove all its contaminants, which include endotoxins, certain proteins, and polyphenols. In addition, the industrial processes used for extracting alginates from seaweeds could introduce further contaminants into the raw alginates. Some of these impurities can be dangerous for humans as reported by the World Health Organization (12) and can possibly accumulate in the body (13). According to FDA requirements for device implantations, the content of endotoxins must be below 350 EU per patient. As the chemical properties of endotoxins are very similar to alginates, their removal has been a challenging task. For Good Medical Practice (GMP), alginates must be characterized by validated methods and every product batch must be characterized and documented with its laboratory certificate. Characterization parameters for alginates to be used in biomedical and tissue engineered medical products are now thoroughly described in the American Society for Testing and Materials (ASTM) guide F 2064 of the ASTM Book of Standards. Other polymers, in addition to sodium alginate, have been successfully applied in microencapsulation research.

Alginate purification: In the last few years, several research groups have developed their own in-house protocols for alginate purification (14–19). The first published method described by De Vos P *et al.* used a free-flow electrophoresis technique (20), but since it was difficult and expensive, it was abandoned in favor of chemical extraction procedures. Even, the first comparative evaluation of some of these in-house alginate purification protocols was published (21). Results from this study showed that in general all of the studied purification methods reduced the amounts of endotoxins and polyphenols but were less effective in eliminating proteins. Overall, the results of this study reflected that currently employed methods to purify alginates may not be efficient enough to completely remove contaminating and potentially immunogenic species. Recently, Professor Ma's group used a set of alginate purification protocols which can effectively eliminate proteins. (www.lbme.dicp.ac.cn). Moreover, it has been demonstrated that purifying the alginate induces a number of changes in the polymer's characteristics. Alginate hydrophilicity was shown to increase by 10% to 40% following purification by different methods in correlation with a decrease in protein and polyphenol content. This increased hydrophilicity correlated with lower immunogenicity of the alginate gel. In this study, reducing the contamination level of the alginate also correlated with an increased solution viscosity, a property that influences the morphology of the final microcapsule.

The composition of the alginate is another critical issue to be considered. In fact, alginate composition regulates some main properties of the alginate gels, including stability, biocompatibility, and permeability. Since the variability in the chemical character of monomer units is virtually infinite, application of well-characterized polymers in terms of chemical composition is critical in order to understand and control the microcapsule properties.

In addition to the chemical composition, the molecular weight characteristics should be part of the conventional documentation. This includes identification of weight number (M_w) and average molecular weight (M_n) and polydispersity (M_w/M_n). The molecular weight averages and the molecular weight distribution can be measured by various techniques. Conventionally, static light-scattering, viscometry, and size-exclusion chromatography (SEC) are used to determine the molecular weight averages. The molecular weight distribution is most typically determined by SEC, although the mass spectrometry techniques have been advancing to assess the molecular weight distribution of synthetic (22) and natural (23) polymers. Molecular weight characteristics are linked to the viscosity and other rheological properties of the polymer solution, which are important for the process of microcapsule formation. The rheological properties are affected by temperature and concentration of the polymer and by ionic strength in case of polyelectrolytes, which should all be specified for the materials applied in encapsulation. There are some additional items that should be documented in specific applications, that is, for medical application, it is mandatory to have information on the degree of purity of the polymer.

Chitosan is a cationic linear polysaccharide composed essentially of $\beta(1 \rightarrow 4)$ linked glucosamine units together with some proportion of *N*-acetylglucosamine units. Chitosan occurs rarely in nature and is generally obtained by extensive deacetylation of chitin, a homopolymer of β(1→4) linked *N*-acetyl-D-glucosamine present in the shells of crustaceans, mollusks, the cell walls of fungi, and the cuticle of insects. The structure of chitosan depends on whether the deacetylation reaction was carried out under homogeneous or heterogeneous conditions. This structural difference is very important in determining many properties of chitosan, such as solubility. Chitosan obtained by a heterogeneous procedure is not soluble in water, whereas water-soluble chitosan can be prepared by homogeneous deacetylation of chitin.

Chitosan purification: Chitosan is a chelating polymer that is very effective for the binding of transition and post-transition metal ions due to the primary amino groups distributed along its chain. Therefore, the industrial processes used for extracting chitosan could further introduce toxic metals, such as mercury and other contaminants into the raw chitosan. Removal of such content has been a challenging task. For GMP, chitosan must be characterized by validated methods and every product batch must be characterized and documented with its laboratory certificate as alginate.

Solubility of chitosan: The properties of chitosan depend not only on its average degree of acetylation (DA) but also on the distribution of the acetyl groups along the main chain in addition to the molecular weight (24–26). The deacetylation, usually done in the solid state, gives an irregular structure due the semicrystalline character of the initial polymer. The role of the microstructure of the polymer is clearly shown when a fully deacetylated chitin is reacetylated in solution; the critical value of chitosan DA to achieve insolubility in acidic media is then greater than 60%. In addition, solubility at neutral pH has also been claimed for chitosan with DA around 50% (24). Recently, a water-soluble form of chitosan at neutral pH was obtained in the presence of glycerol 2-phosphate (27–30). Stable solutions were obtained at pH 7–7.1 and at room temperature.

Characterization DA of chitosan: The characterization of a chitosan sample requires the determination of its average DA. Various techniques, in addition to potentiometric titration (31), have been proposed, such as IR (32–34), elemental analysis, an enzymatic reaction (35), UV-Vis (36), 1H liquid-state NMR (37), and solid-state 13C NMR (38–40). The fraction of $-NH_2$ in the polymer (which determines the DA) can be obtained by dissolution of neutral chitosan in the presence of a small excess of HCl on the basis of stoichiometry followed by neutralization of the protonated $-NH_2$ groups by NaOH using pH or conductivity measurements. These techniques and the analysis of the data obtained have been described earlier (31). Presently, we consider that 1H NMR is the most convenient technique for measuring the acetyl content of soluble samples.

Characterization molecular weight of chitosan: Another important characteristic to consider for these polymers is the molecular weight and its distribution. To choose a solvent for chitosan characterization, various systems have been proposed, including an acid at a given concentration for protonation together with a salt to screen the electrostatic interaction. The solvent is also important when molecular weight has to be calculated from intrinsic viscosity using the Mark–Houwink relation. Absolute M values were obtained from SEC, with on-line viscometer and light-scattering detectors to allow determination of the Mark–Houwink parameters and also the relation between the molecular radius of gyration (Rg) and molecular weight. We proposed average values for the Mark–Houwink parameters within portions of the total range of DA covered.

Cells Selection

For cell-contained microcapsules, selection of the cells depends upon the intended application, such as the secretion of bioactive substance like neurotransmitter, cytokine, chemokine, growth factor, growth-factor inhibitor, angiogenic factor, or the metabolism of a toxic agent, or the release of an immunizing agent, Epo or glucose, and insulin.

Cell encapsulation technology has in part failed to reach clinical approval so far mainly due to the high immunogenicity of the encapsulated cells, which eventually evoke an inflammatory reaction in the microenvironment surrounding the micro-devices that lead to suffocation and death of the encapsulated cells (41–44). The key issue to overcome this problem could be to use cells that can downregulate or reduce this immune response (43). One promising solution to reduce host immune reaction is by administering anti-inflammatory drugs along with the therapeutic system (45,46). Another approach under study to reduce host immune reaction is to replace the cell lines commonly used for cell encapsulation with naive cells, such as stem cells. Human mesenchymal stem cells (hMSCs) show promising properties as the cells of choice for cell microencapsulation and cell-based therapy. MSCs improve the biocompatibility of the microcapsules *in vivo* and can serve as a platform for continuous long-term delivery of therapeutic factors, including potent cancer therapies.

Method of Preparing Microcapsules Containing Live Cells

There are various methods available for preparing artificial cells containing live cells for therapy. For preparation of the classic alginate–polylysine–alginate (APA) membrane, the live cells are suspended in a matrix of natural polymer alginate (1.5%). The viscous polymer-cell suspension is passed through a needle using a syringe pump. Sterile compressed air, passed through a coaxial needle, is then used to shear the droplets coming out of the tip of the needle. The droplets are allowed to gel for 15 minutes in a gently stirred ice-cold solution of solidifying chemicals, such as $CaCl_2$ (1.4%). After gelation in $CaCl_2$, the beads are then washed with 4-(2-hydroxyethyl)-1-piperazineethanesulfonic acid (HEPES) (0.05% in HEPES, pH 7.20), coated with polylysine (0.1% for 10 minutes), and washed again in HEPES (0.05% in HEPES, pH 7.20). The resultant microcapsules are then coated by reaction with alginate (0.1% for 10 minutes) and washed with appropriate chemicals to dissolve their inner core content. For this step, a 3.00% citrate bath (3.00% in 1:1 HEPES-buffer saline, pH 7.20) is often used. The microcapsules formed can then be stored at 4°C in minimal solution (10% cell nutrient to 90% water).

For modulating the properties of microcapsules, a balance has to be maintained among the physical properties of microcapsule membranes so as to support the entrapped cells' survival. The mass transport properties of a membrane are critical since the influx rate of molecules, essential for cell survival, and the outflow rate of metabolic waste ultimately determine the viability of the entrapped cells. Ordinarily, the desired microcapsule permeability is determined by the molecular weight cutoff (MWCO) and is application-dependent. The MWCO is the maximum molecular weight of a molecule that is allowed passage through the pores of the microcapsule membrane. For transplantation, the MWCO must be high enough to allow passage of nutrients but low enough to reject antibodies and other immune molecules. We will discuss and outline how to modulate MWCO in Section 3.3 (Design and Elaboration of Microcapsules for Cell Transplantation).

Large-scale production of artificial cell microcapsules: If large-scale preparation of sterile microcapsules containing cells is required, automated encapsulators such as the Inotech Encapsulator produced by Biosystems International Inc (Rockville, Md, USA) and DICP Large-Scale Encapsulator (Fig. 3.3) produced by Professor Ma's group can be used (www.lbme.dicp.ac.cn). The DICP Large-Scale Encapsulator is based on the principle that a laminar liquid jet is broken into equal-sized droplets by an imposed vibration. Furthermore, an electrostatic charge may be applied to each microcapsule so that a spray of droplets is produced and a more homogeneous set of microcapsules result. This type of microencapsulator can manufacture large numbers of superior quality microcapsules and can facilitate the procedures described in the methods outlined above.

Figure 3.3 DICP Large-Scale Encapsulator.

Ideally, the preparation of microcapsules would be scaled up to 10 L to allow single-batch production of enough cell-contained microcapsules to treat 100 patients. However, for commonly used methods and large-scale generators mentioned above, the highest extrusion throughput allowing adequate droplet generation varies between 10 mL/h and 360 mL/h. In addition, the damage imparted by energy dissipation at the nozzle tip for high-viscosity alginate solutions at these flow rates has not been investigated.

Recently, a scaleable and low-shear alternative for generating alginate beads by emulsion and internal gelation was set up by Poncelet (47). This immobilization processes is modified by Corinne (48). Their emulsion process was adapted from the Poncelet method (47) by changing the buffer system, the processing times, and the bead recovery method (Fig. 3.4). The "HEPES emulsion" refers to the initial process using 10 mM HEPES process buffer with 12 min emulsification and 8 min acidification, while "MOPS emulsion" refers to the optimized process using 60 mM 3-(*N*-morpholino) propanesulfonic acid (MOPS) process buffer, 3 min emulsification, and 1 min acidification. Cells from trypsinized cultures were washed and resuspended in medium as a concentrated cell stock (10.5 fold). One volume of this stock was added to 9 volumes of alginate and 0.5 volumes of a 500 mM $CaCO_3$ suspension to obtain 1.5–2% final alginate concentration and $1–2.5 \times 10^6$ cells/mL alginate, depending on the experiment. First, 20 mL of light mineral oil was agitated at 500 rpm for 2 min in a 100 mL microcarrier spinner flask (Step 1). A 10.5 mL volume of the alginate, cells, and $CaCO_3$ mixture was then added and emulsified for 3 or 12 min (Step 2). Internal Ca^{2+} release and bead gelling was triggered by adding 40 mL glacial acetic acid dissolved in 10 mL mineral oil (Step 3). After 1 or 8 min acidification, 40 mL HEPES buffered saline solution mixed with 10% medium was added to neutralize the pH (Step 4). The agitation was ceased 1 min later and the mixturet 20 mL medium was centrifuged for 3 min at 630 g (Step 5). The oil and solution above the beads were removed by aspiration, followed by two washes with medium, centrifugation, and aspiration. The beads were filtered (Step 6) on a 40 mm nylon cell strainer and transferred into the desired solution with a spatula.

After gelation, the beads were then washed with HEPES (0.05% in HEPES, pH 7.20), coated with polylysine (0.1% for 10 min), and washed again in HEPES (0.05% in HEPES, pH 7.20). The resultant microcapsules were then coated by reaction with alginate (0.1% for 10 min) and washed with appropriate chemicals to dissolve their inner core content. For this step a 3.00% citrate bath (3.00% in 1:1 HEPES-buffer saline, pH 7.20) is often used. The microcapsules formed can then be stored at 4°C in minimal solution (10% cell nutrient to 90% water).

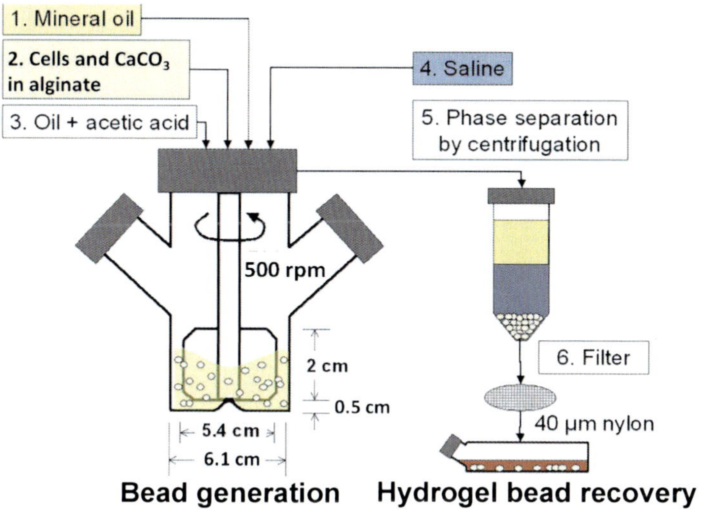

Figure 3.4 Schematic illustration of the emulsion process adapted to mammalian cell immobilization (48).

Storage Conditions and Methods for Microcapsules

Storage of encapsulated cells for transport or in the time period between manufacture and application is mandatory for almost all fields of encapsulation. Determination of suitable conditions for storage of microcapsules plays an important role in microencapsulation research since it is broadly accepted that microcapsule characteristics and functionalities are often very sensitive to environmental parameters, such as temperature, humidity, osmotic pressure, storage solution, or solvent (49–52).

It is mandatory to choose those storage conditions that ensure that the optimal performance is maintained. The adequacy of a storage condition depends on the field of application in which the microcapsule characteristic should be maintained.

Here is an example for microcapsules cold storage and rewarming. Cell-contained microcapsules were rinsed twice and resuspended in freshly prepared cold (0°C) and modified University of Wisconsin (UW) solution. The composition of the modified UW solution has been previously described (53,54). Microcapsules (106 cells in 40 mL UW solution) were allowed to settle to the bottom of 50 mL screw cup polycarbonate tubes and left undisturbed at 0°C up to 120 h. After that, the suspensions were washed twice with a rinse solution and sedimented in warm Krebs–Henseleit resuspension (KHR) media. The microcapsules were subsequently incubated (120 min, 37°C, 2–3 × 10^6 cells/mL) in KHR media under carbogen atmosphere in a Dubnoff metabolic shaker.

Cryopreservation of microcapsules: Briefly, freezing was carried out in L15 medium supplemented with 2 mmol/L glutamine, 4 μg/mL bovine insulin, 50 IU/mL penicillin, 50 μg/mL streptomycin, 0.1% bovine serum albumin and containing 10% FCS and 16% dimethyl sulfoxide (DMSO). Aliquots of microcapsules were distributed in 1.8 mL freezing vials and transferred at −20°C for 12 min and then at −80°C for 2 h before being plunged and stored in liquid nitrogen. In some experiments, the caspase inhibitor benzyloxycarbonyl-Val-Ala-DL-Asp-fluoromethylketone (ZVAD-fmk) was added to the freezing medium (50 μmol/L). For thawing, vials were placed in a water bath at 37°C until ice disappeared; microcapsules were gently pipetted, and aliquots were transferred into L15 medium containing 0.8 mol/L glucose. After a 2 min incubation period at room temperature, microcapsules were suspended in the culture medium, plated, and maintained at 37°C for 24 h as described above.

Subzero nonfreezing preservation of microcapsules: Cryopreservation has been suggested as the best technique for storage of cell-contained microcapsules; however, cell recovery and viability post-preservation is insufficient. Recently, a subzero nonfreezing storage protocol that uses UW solution and 8% (W/V) 1,4-butanediol (BDL) as cryoprotectant agent has been applied. With this method, we could maintain viable and functional microencapsulated cells at −4°C for up to 120 h, without damage due to ice crystal formation. This much time is sufficient for transportation of microcapsules.

Briefly, microcapsules were subzero nonfreezing preserved up to 120 h in modified UW solution with 8% 1,4-butanediol (1,4-BDL) (in 40 mL UW solution + 8% 1,4-BDL). After that microcapsules were warmed to 0°C and were then washed once with the rinse solution containing 4% of 1, 4-BDL and later, twice with rinse solution with no additives. Then, the microcapsules were incubated (120 min) in KHR media under carbogen atmosphere in a Dubnoff metabolic shaker.

3.3 Design and Elaboration of Microcapsules for Cell Transplantation

Although important advances have already been made in cell encapsulation technology, in the coming years the obstacles encountered during the optimization process of the technology should be overcome to come closer to its clinical application. There are some critical aspects that should be carefully taken into consideration for the technology. A compilation of important microcapsule properties is provided in recent reviews (55,56).

3.3.1 Biocompatibility

Biocompatibility is defined as the ability of a biomaterial to perform with an appropriate host response in a "specific application." Biocompatibility of microcapsules and their biomaterials' components is a critical issue if the long-term efficacy of this technology is aimed. Usually, a fully biocompatible system is considered to be a system made of membranes which elicit no or very minimal foreign body reaction. The host response is potentially serious and can cause deleterious effects to the clinical implementation of the technology.

As previously mentioned, a key element in the validation of alginate for implantation purposes is the efficient purification process to monitor and remove all its contaminants which include endotoxins, certain proteins, and polyphenols. Several experiments have demonstrated that implantation surgery can activate a nonspecific response against implanted cells. Moreover, although it has been described as a transient response, it is difficult to avoid as it cannot be solved by chemical modification of the microcapsule. In order to overcome this obstacle, the use of transient immunosuppressive protocols have been recently proposed (57).

In addition, transplantation of encapsulated cells is associated with the release of bioactive proteins from the host such as fibrinogen, thrombin, histamine, and fibronectin. These factors induce an influx of granulocytes, basophils, mast cells, and macrophages during the first few days after implantation. Mast cells and macrophages produce bioactive factors such as IL-1β, TNF-α, TGF-β, and histamine, which stimulate cells in the capsules. Within two weeks, basophils and granulocytes gradually disappear from the graft site while macrophages and some fibroblasts remain attached to an average of 2–10% of the microcapsules. These attached macrophages remain activated and contribute to the deleterious circle of activation.

Another important issue to take into consideration is the diameter of the microcapsule as it could influence the immune response against microcapsules. An interesting approach using agarose microcapsules of different diameters developed by Sakai *et al.* revealed that cellular reaction to the smaller capsules was much lower than that to the larger capsules (58).

Finally, an important issue to take into consideration is the surface properties of the microcapsule as they could influence the immune response against microcapsules. An interesting advance to predict biocompatibility was recently reported by de Vos and Professor Ma's group where the measurement of the electrical charge of the surface by means of zeta potential was found to predict the interfacial reactions between the biomaterial and the surrounding tissue (17,59,60).

Thus, most approaches developed to minimize the immune response to hydrogels are focused on preventing protein adsorption and cellular adhesion to microcapsules through the (i) encapsulation of cells into biologically inert hydrogels, (ii) modification of the microcapsules' surface with biologically inert materials, (iii) anchorage of biologically inert molecules to the cell membrane.

Biocompatible materials are often used to modify the surface properties of certain cell-encapsulating hydrogels and reduce the immune response. Certain hydrogels, which facilitate the microcapsules forming process and provide the desirable physical properties, may induce severe immune responses. Alginate molecules are often used to coat microcapsules containing polycations, such as poly-L-lysine, chitosan, and poly(ornithine). Poly-L-lysine, chitosan, and poly(ornithine) allow for efficient control of pore size and mechanical properties but induce mild to severe immune responses (61). Overlaying alginate molecules strongly associate with polycations to form polyelectrolyte complexes and decrease the immune response, as depicted in Fig. 3.5. The polycationic surfaces are also modified by chemically coupling PEG molecules to poly-L-lysine or chitosan. Like alginate molecules,

PEG molecules sterically inhibited the protein adsorption and cell adhesion onto poly-L-lysine/or presenting hydrogels *in vivo* (62, 63). These surface treatments improved the therapeutic efficacy of islet cells by reducing fibrosis around the microcapsules. The presence of the PEG layers on the microcapsules membrane did not affect the cells' ability to secrete therapy molecules while providing a thin biochemical and biophysical barrier to the immune system for almost 3 months (64–67). Individually coated islets with thin PEG layers have shown excellent preclinical results, and the technology is currently in phase I/II clinical trials (www.clinicaltrials.gov). Combining this cell-membrane coating technique with cell encapsulation techniques using these biocompatible polymers may further minimize the immune response and extend the life of transplanted cells.

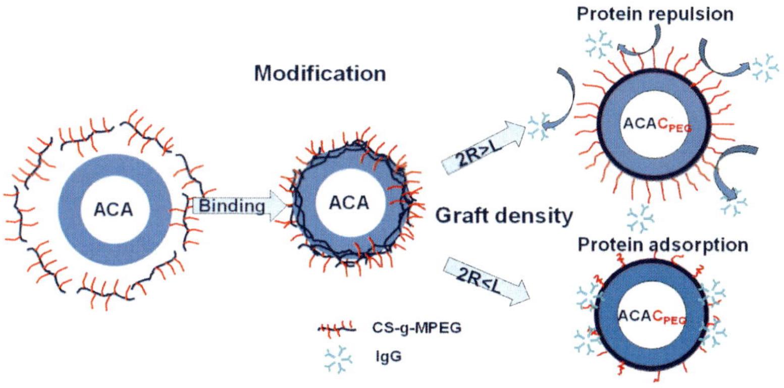

Figure 3.5 Design chitosan-g-MPEG-modified alginate/chitosan hydrogel microcapsules to resist protein adsorption. R: Radius of gyration (Rg) of PEG chain. L: Distance of two neighbored PEG chain.

3.3.2 Searching for the Optimal Transplantation Site

Determining the optimal site for systemic drug delivery is a matter of intense research. Figure 3.6 represents the main implantation sites employed in cell encapsulation technology.

Several factors have to be taken into consideration such as biocompatibility, oxygen transfer limitations, and mechanical resistance. In general, intraperitoneal implantations result in poorer functionality of the devices due to an increased inflammatory response that takes place in the peritoneum. Microcapsules implanted intraperitoneally showed decreased viability and function (such as insulin secretion rates) than microcapsules implanted subcutaneously or under the kidney capsule. Furthermore, the immune response against microcapsules implanted intraperitoneally was more severe compared to those implanted subcutaneously or under the kidney capsule. Microcapsules that were implanted into the striatum and the subcutaneous space were found to be more stable (6 months) in comparison to the peritoneal cavity (2 months).

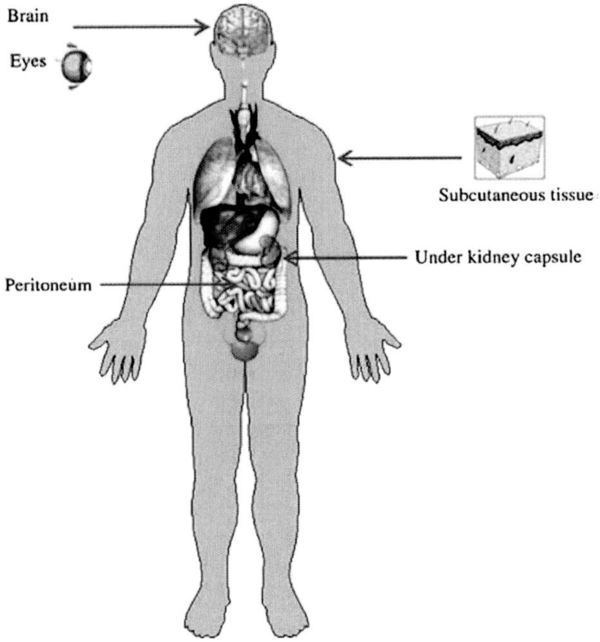

Figure 3.6 Main implantation sites employed in cell encapsulation technology (68).

3.3.3 Internal Oxygen Mass Transfer Limitations

Aggregating into multicellular spheroids within alginate-poly-L-lysine-alginate (APA) microcapsules is important in maintaining the cellular viability and specific functions. However, in the absence of a vascular network, cells in the core of large-sized spheroids are gradually necrotic because of oxygen transfer limitations. Different solutions have been proposed to overcome this obstacle. On one hand, Sakai *et al.* developed small-size microcapsules, less than 100 μm in diameter, to improve oxygen transfer into the microcapsules where cell viability was observed not to be affected by the small size of the capsules (69). On the other hand, Khattak *et al.* included synthetic oxygen carriers (perfluorocarbons) in alginate gels to improve oxygen supply and transport, and found an improvement in cell viability and metabolic activity due to a reduction in anaerobic glycolysis which resulted in an increase in glucose consumption/lactate production efficiency (70).

Presently, an effective method to increase oxygen mass transfer in the microcapsules is to prepare microcapsules embedded with 3-D fibrous scaffolds for cell culture (71). In this study, a novel APA microcapsule embedded with 3-D fibrous scaffolds (called APA-FS) was proposed to eliminate cellular necrosis by regulating cells to form multi-small spheroids. Comparing with the conventional APA microcapsules, the cells within APA-FS organized into multi-small spheroids. The size of these spheroids depended on the concentration of fibrous scaffolds embedded within the microcapsules. In the APA-FS embedded with 5% (v/v) fibrous scaffolds, the average size of cellular spheroids was controlled below 100 μm and the cellular viability was increased by 50% than the control. The results showed that the improved cellular viability might be attributed to the increased oxygen transfer in the core of these spheroids.

3.3.4 Microcapsule Permeability and MWCO

The mass transport properties of an encapsulation membrane are critical since the influx rate of substance essential for cell survival and the efflux rate of therapeutic products will ultimately determine the extent of encapsulated cell viability. Moreover, membrane pore size must be carefully designed and controlled to avoid the undesired entrance of immune system components from the host. The metabolic requirements of different cell types are diverse and, hence, in principle optimal membrane permeability depends on the choice of cells (72).

For assessing and expressing the diffusion and permeability of microcapsules, two factors are very important. The first factor that is relevant for expressing the diffusion and permeability of microcapsules is the rate of solute diffusion, which is reflected in the mass transfer, permeability, and diffusion coefficients. The second one is represented by the membrane exclusion properties considering minimum size of a solute completely excluded by the capsule membrane. This is usually referred to as the exclusion limit or MWCO.

Selection of solute for quantifying permeability is the first step. Many experimental techniques to assess the permeability properties of microcapsules have been described during the past two decades. These techniques are comprehensively compiled in a number of review articles, for example, by Schuldt and Hunkeler (73) and Uludag *et al.* (72). Which technique is most suitable for a specific application depends on the type of solute that is applied for measuring permeability. The selection of the technique starts with choosing the solute type, which typically involves proteins, dextrans, and pullulans. In the medical field, the permeability of microcapsules to IgG is considered as the most important criterion since it is assumed to predict the immunoprotective properties of the microcapsules after implantation in humans.

Extensive efforts are being made to control the size of the membrane surface pores to allow the transport of oxygen, nutrients, therapeutic drug molecules, and cellular wastes, while preventing the transport of immunogenic molecules and cells. Specifically, the diameter of surface pores for calcium cross-linked alginate microcapsules commonly varies from 5 nm to 200 nm (74). This pore size prevents the diffusion of large molecules, like fibrinogen (M_W of 3.41×10^5 g/mol), but allows the diffusion of small molecules, like albumin (M_W of 6.9×10^4 g/mol) (75). This pore size is further controlled by forming polyelectrolyte complex layers on the gel surface. Exposing alginate to polycations such as poly-L-lysine reduces the diffusivity of small molecules, like albumin or hemoglobin (M_W of 6.8×10^4 g/mol) up to 50%. The ability of poly-L-lysine to reduce the effective surface pore diameter is dependent on the molecular weight of poly-L-lysine and the exposure time. Decreasing the molecular weight of poly-L-lysine molecules and extending exposure time leads to a significant decrease in the average diameter of pores on the gel surfaces (76). Coating alginate gels with poly-L-ornithine (PLO) further reduced the permeability of a 75 kDa protein 10 times over poly-L-lysine (77). These approaches may be broadly applicable to cell encapsulation with other anionic polyelectrolyte biomaterials.

3.3.5 Mechanical Integrity and Stability

The mechanical properties of cell-contained microcapsules are highly important to ensure the persistent therapeutic efficacy of transplanted cells. The assessment of microcapsule mechanical properties is important, not only to determine the durability of microcapsules during production and transportation, but also as an indication of the microcapsule membrane integrity during transplantation procedure when long-term clinical studies have been carried out. Microcapsules may also lose their mechanical stiffness and structural integrity over time because of the several intrinsic and extrinsic factors exposing the encapsulated cells to the hostile immune system.

Various techniques are available to improve the mechanical properties and stabilities of microcapsules. The mechanical properties of the gel are commonly controlled with the polymer concentration and molar ratio between polymers and cross-linking molecules. Increasing the polymer concentration and shortening the distance between the cross-links leads to an increase of the mechanical stiffness (78). However, covalently cross-linked polymers become more brittle and susceptible to failure under mechanical loading exerted by neighboring tissues, as polymers are made increasingly stiff through increasing the cross-linking densities (79). In contrast, calcium cross-linked alginate microcapsules allow increases in both the stiffness and the toughness with increased cross-link density, unlike other polymer formed from covalent cross-linking.

The long-term mechanical stability of hydrogels is enhanced by modifying gel surfaces with polyelectrolyte complexes and cross-linkable molecules. These complexes formed with various polycations including poly-L-lysine, poly(ethyleneimine), and poly-L-ornithine (PLO) significantly increase the gel stiffness and toughness (80–82).

One of the most commonly applied assays to quantify mechanical resistance is the osmotic pressure test in which microcapsules are exposed to various deleterious solutions with the aim to quantify the swelling of microcapsules. The advantage of this technique is that it is nonlaborious and readily available in all laboratories.

Another assay that is nowadays more commonly applied is evaluation of the physical integrity of the microcapsules by using a surface texture analyzer (83,84). With this technique a specific force is placed on the microcapsules. The quantity of deformation or rupture of microcapsules is applied as a measure for the mechanical stability of the microcapsules.

3.3.6 Surface Properties of Microcapsules

The surface properties of microcapsules determine their functional performance. It is the site that is responsible for the biocompatibility and it also determines the diffusion properties.

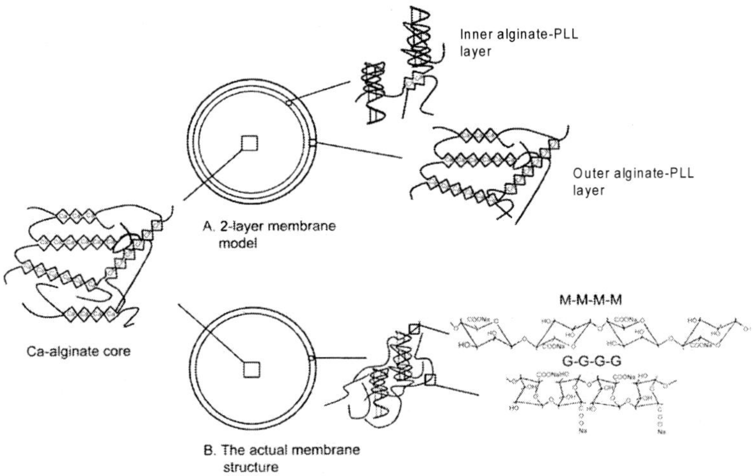

Figure 3.7 The considered and the actual structure of alginate-PLL capsules Reprinted from Ref. 55, with permission from Elsevier.

In order to provide a deep insight into the structure of alginate-PLL microcapsules a physico-chemical analysis of the microcapsules has been performed by applying X-ray photoelectron spectroscopy (20). This technique allows for identification of the chemical groups on the surface of the microcapsule on an atomic level. Up to now the microcapsule was assumed to be composed of a core of calcium-alginate which is enveloped by a membrane composed of two layers, that is, an inner layer of alginate-PLL and an outer layer of calcium-alginate. In subsequent studies, Fourier transform infrared spectroscopy, X-ray photoelectron spectroscopy (XPS), and confocal microscopy were applied to study the structure of the alginate-PLL capsule membrane

(13,15–18,20). From confocal images and from electron microscopy pictures it was visualized that the PLL penetrates the alginate core, forming an alginate-PLL complex of about 30 mm, depending on the exposure time to PLL. It was found that the microcapsules were not composed of a generally considered three-layer system of alginate-polycation, and an outer alginate layer but only of an alginate-core surrounded by an alginate-polycation shell (18). Figure 3.7 shows the actual structure of alginate-PLL capsules (19).

3.4 Application of Cell Encapsulation Technology

Many diseases are closely tied with deficient or subnormal metabolic functions. Hypoparathyroidism, hemophilia, Parkinson's disease, diabetes mellitus, and hepatic failure belong to this kind of degenerative and disabling disorders. It is impossible even to mimic the precise regulation and the complex roles of the hormone, factor, or enzyme that is not produced by the body. Many studies contributed by highly experienced research groups are now shedding light on the main challenges of cell encapsulation technology. One requirement that is of paramount importance to implement this technology for the treatment of chronic diseases is the long-term production of therapeutic products from the encapsulated cells.

3.4.1 Diabetes

Diabetes is a global disease with an incidence rate of about 3–5% of the population. According to the update statistics of the WHO, 150 million people around the world suffer from diabetes mellitus at present. The figure will be duplicated by 2025. Until recently, transplantation of encapsulated pancreatic islets for the treatment of diabetes has been the most common application of cell encapsulation technology.

In 1980, Lim and Sun implanted microencapsulated xenograft islet cells into rats and the microencapsulated islets corrected the diabetic state for several weeks (85). Since then, there has been considerable progress toward understanding the biological and technological requirements for successful transplantation of encapsulated cells in experimental animal models, including rodents and nonhuman primates. Bio-artificial pancreatic constructs based on islet microencapsulation could eliminate or reduce the need for immunosuppressive drugs and offer a possible solution to the shortage of donors, as it may allow for the use of animal islets or insulin-producing cells engineered from stem cells (86–88) (Fig. 3.8).

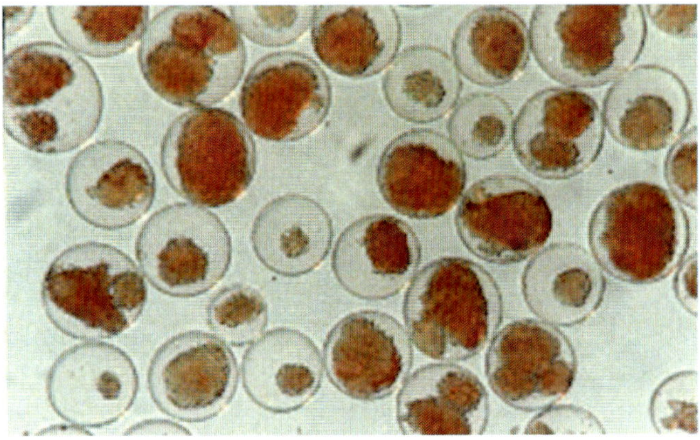

Figure 3.8 Islets encapsulated in alginate microcapsules.

Based on the promising results obtained in allo- and xeno-transplantation approaches (89–92), a pilot clinical trial has recently been initiated by Calafiore *et al.* (93,94). Microencapsulated human islets were implanted into 10 non-immunosuppressed patients with type 1 diabetes. Data of two patients have been published although treatment with exogenous insulin was not totally suspended. The measurement of several parameters indicated that pancreatic islets remained metabolically active. In addition, as presented at the XV International Workshop on Bioencapsulation and COST 865 meeting in Vienna, the clinical study is ongoing and improvements are being observed due to an important advance in achieving a cell source of higher quality. It could be concluded that the correct isolation and treatment of cells shows to be of outstanding importance if a successful clinical outcome is desired.

Diabetes is a metabolic disorder characterized by hyper-glycemia resulting from defects in insulin secretion, insulin action, or both. Current research efforts toward therapy of type 1 diabetes are aimed at developing approaches for restoration of regulated insulin supply. Nowadays, researchers assessed the safety and clinical activity of alginate-encapsulated porcine islets in a nonhuman primate model of streptozotocin-induced diabetes. They noted worsening of the disease in control animals. Six out of eight control monkeys required increased doses of daily insulin. In contrast, six of the eight islet-transplanted monkeys had reduced insulin requirements. After islet transplantation, individual blood glucose values varied and one monkey was weaned off insulin for 36 weeks (93).

In the past few years, the renewed interest in porcine islet xenotransplantation has generated some controversy about the human clinical trials carried out. The study by Living Cell Technologies Ltd with the Diabecell® device provided evidence of improvement in glycemic control individuals and showed no evidence of porcine viral retroviral infection. Moreover, they reported evidence of residual, viable, encapsulated porcine islets being retrieved from a patient 9.5 years after transplantation (95). However, this approach has been criticized by the International Xenotransplantation Association as being premature and potentially risky (96). Recent progress in the use of closed, porcine endogenous retroviruses free (PERV-free) herds or advances in immunoisolation may help to improve the formulations. In fact, a new open-label investigation about the safety and effectiveness of Diabecell® in patients with diabetes type 1 is currently recruiting patients (NCT00940173).

Much work is clearly needed before microencapsulated cell therapy for diabetes can be advanced to the clinic. The challenges center on generation of an abundant source of regulated insulin-producing cells and some aspects of the cell-based encapsulation methods that should be improved in order to increase the transplant longevity and functional performance of the capsules *in vivo*.

3.4.2 Stem Cell Technology

An increasing interest in which microencapsulation of stem cells could be a promising therapeutic strategy is tissue replacement therapy. Some *in vivo* studies have been developed such as the implantation of encapsulated bone marrow mesenchymal stem cells to improve the formation of the osseous and cartilaginous architecture (97,98) and the transplantation of encapsulated embryonic stem cells (ESC) in order to overcome the rejection that may take place when transplanting ESCs into recipients with different major histocompatibility antigens (99).

Last but not least, *in vivo* approaches employing genetically modified stem cells to improve their characteristics include the works in the field of bone formation. On the one hand, marrow-derived mesenchymal stem cell encapsulation transfected with the BMP-2 gene was found to induce bone formation, and on the other hand, encapsulated human mesenchymal progenitor stem cells transfected with the Sox-9 gene revealed chondrogenic differentiation (100,101). Although stem cell encapsulation technology presents an interesting alternative to the current EC therapies, several issues need to be addressed by stem cell encapsulation technology to become a realistic proposal for clinical application. Most of the *in vivo* studies have been developed using immunodeficient animals. Therefore, *in vivo* assays are required to be carried out in immunocompetent recipients to study the effects derived from the activation of the host's immune system after implantation.

3.4.3 Bone and Cartilage Defect

Bone defects resulting from trauma and tumor resection are common clinical problems. Bone tissue usually has the ability to regenerate. However, the repair attempt fails in most cases when a defect of critical size needs to be bridged. The standard tissue used currently is autologous tissue, which is usually harvested from the iliac crest of the patient. Although autografting has been a major treatment, it has several limitations including patient pain, cost, and limited supply. As an alternative, allografting has been studied due to its abundant source. However, its drawbacks, including the uncertainty of biocompatibility and disease transmission, have limited its use (102).

To overcome these drawbacks, researchers are considering alternative therapies involving the use of mesenchymal stem cells (MSCs). MSCs are multipotent progenitor cells that can be isolated from bone marrow, muscle tissue, umbilical cord blood, peripheral blood, and other tissues (103–105). They have the capability to differentiate into multiple tissue-forming cell lines, such as osteoblasts and chondrocytes, which contribute to the regeneration of bone and cartilage. Recently, some works have showed that microcapsules could create a 3D microenvironment that would provide a niche for stem cell growth and differentiation. In this respect, Endres *et al.* have confirmed that these hMSC were able to differentiate along the chondrogenic lineage *in vitro* when microencapsulated in Ca-alginate microcapsules and stimulated with TGF-β3 (106). The size of these microcapsules is in the injectable range (mean diameter of 500–700 μm), making this administration easier.

3.4.4 Neurological Diseases

Human neurological disorders such as Parkinson's disease (PD), Huntington's disease (HD), amyotrophic lateral sclerosis (ALS), stroke, and spinal cord injury (SCI) are caused by a loss of neurons and glial cells in the brain or spinal cord. Recently, the transplantation of cells to the brain and/or spinal cord has been pursued as a potential curative treatment for a broad spectrum of human neurological diseases (107,108). Promising results have been achieved in clinical trials, but there is much work to be done before cell-based therapies can be practiced extensively.

Recently an interesting therapeutic application employing genetically engineered cells, including the immobilization of growth factor–producing fibroblast for delivery of therapeutic products in spinal cord injuries (SCIs). This pathology results in the disruption of ascending and descending axons that produce a devastating loss of motor and sensory function. Several strategies have been used to provide trophic and antiapoptotic molecules that can change the environment of the injured central nervous system. In this sense, various studies (109,110) have pointed out that primary fibroblasts that were genetically modified to produce brain-derived neurotrophic factor (BDNF) survived in the injured spinal cord of Sprague-Dawley rats, rescuing axotomized neurons, promoting regeneration, and contributing to the recovery of locomotor function.

In a word, these results suggest that improvements in cell microencapsulation systems that release higher concentrations of neurotrophin are required to stimulate regeneration of axotomized brainstem neurons. In future efforts, there is a need to systematically approach each potential clinical indication with emphasis on optimizing the cell sources, determining which cells are most beneficial, and identifying the optimal post-injury timing, transplant location, and dosage of cells.

3.4.5 Cancer

Microencapsulation of recombinant cells is also a novel alternative approach to tumor gene therapy. Tumor growth is dependent on the formation of blood vessels to supply nutrients and eliminate metabolic wastes. Inhibiting neovascularization may retard tumor growth and lead to tumor dormancy. Endostatin, a 20 kDa C-terminal globular domain of collagen XVIII, is one of the potent angiogenesis inhibitors. It strongly inhibits angiogenesis by inducing endothelial cell apoptosis. A number of studies have demonstrated that endostatin can suppress the growth of primary and metastatic lesions in multiple murine tumor models. Combination therapy of androgen-independent prostate cancer, using a prostate-restricted replicative adenovirus together with antiangiogenic therapy did not induce acquired drug resistance. At present, the primary approach in endostatin treatment is administration of recombinant protein. Because of the expense and inconvenience of repeated administration of endostatin, it is necessary to develop alternative therapeutic strategies to bring antiangiogenic therapy into the clinic. Microencapsulation of recombinant cells represents a novel alternative therapeutic approach in which endostatin could be sustained and delivered long term by recombinant cells entrapped into microcapsules. It has been proved that microencapsulation could favor the long-term secretion of endostatin from genetically engineered cells and allow the treatment of malignant tumors (111,112).

Two independent groups treated glioma models of cancer with encapsulated xenogenically derived cell lines, genetically modified to secrete endostatin (113). As shown in Figs. 3.9 and 3.10, one of the most potent antiangiogenic drugs can directly induce apoptosis in tumor cells. Both groups reported that local delivery of endostatin significantly inhibited tumor growth. Nevertheless, local delivery of drugs might not be feasible in the treatment of many tumors, especially for metastasis. In this respect, Chinese hamster ovary (CHO) cells that were engineered to secrete human endostatin encapsulated in APA microcapsules were transplanted into peritoneal cavities of mice bearing subcutaneous B16 melanoma (114). The results proved that the intraperitoneally implanted microencapsulated cells could significantly inhibit the subcutaneous growth of melanoma in mice. Using a similar approach, Cirone *et al.* (113)

tested angiostatin to treat a murine model of melanoma/breast cancer. When angiostatin was delivered systemically by implanting microencapsulated cells that were genetically modified to express angiostatin, tumor growth in the recipient animals was almost suppressed by >90% within 3 weeks of post-tumor induction and survival was 100% compared to 100% mortality in the untreated or mock-treated controls.

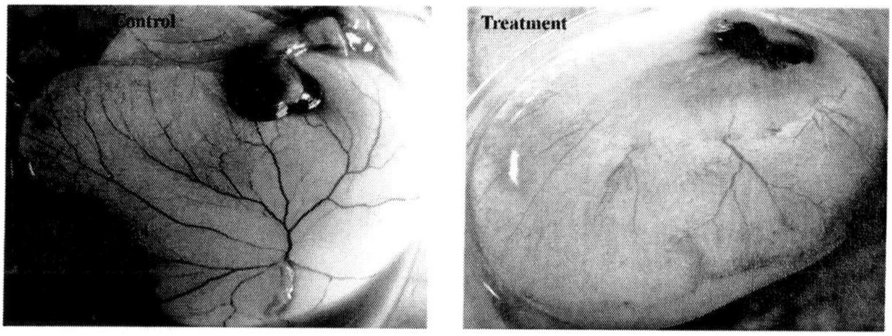

Figure 3.9 Encapsulated CHO-endo cells–inhibited angiogenesis in chicken chorioallantoic membrane.

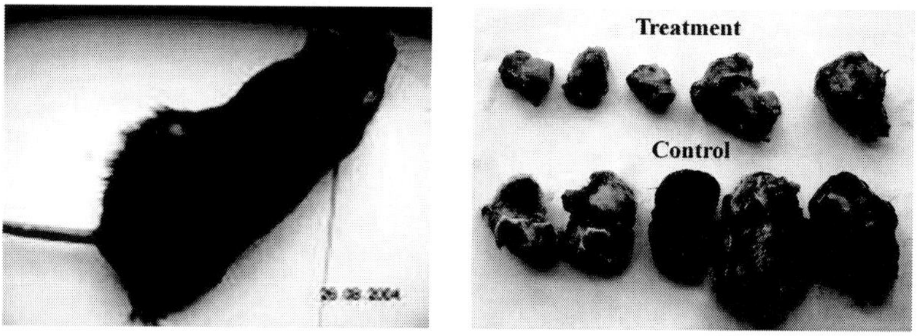

Figure 3.10 Implantation of encapsulated CHO-endo cells in the peritoneal cavity inhibiting tumor growth.

Currently, simultaneous delivery of cytokines and antiangiogenic drugs is also being deeply explored for cancer therapy. The best results were obtained with IL-2-secreting cells and angiostatin-secreting cells microencapsulated in separate microcapsules and implanted at different times post-tumor induction. Thus, the use of a combined strategy with microcapsules containing cells engineered to release different molecules and with anti-tumor properties targeting multiple pathways opens up some new possibilities in the treatment of cancer.

3.5 Future Directions

The primary purpose of this chapter is to provide an overview of current applications of cell microencapsulation. With advances in genetics, biomaterials science, pharmaceutical technology, biology, and chemical engineering, this technology can be largely improved in the next few decades. Future research will focus on the development of a technically advanced microcapsule technology to satisfy the demands of the GMP guide lines for large-scale transplantation and transportation. This ambitious goal requires integrated effort of scientists with different areas of expertise.

In spite of the tremendous growth of the industrial and clinical application of microencapsulation in the past decade, it is still difficult to define the requirement of microcapsules for long-term functionality of the enveloped cells or bioactive components. A standard for the microencapsulation technology is needed to measure specific characteristics. Fortunately, a COST 865 action has been elected to define standardized protocols for characterization and standardization of microcapsule properties for a given application. In a word, it can be expected that the importance of cell microencapsulation technology will continue to increase in the future due to its benefits in offering a living-cell delivery system.

References

1. Rosa, Ma H., Gorka, O., Ainhoa, M., & Jos, L. P. (2010). Microcapsules and microcarriers for *in situ* cell delivery. *Advanced Drug Delivery Reviews*, **62**, 711–730.
2. Wang, X. L., Wang, W., Ma, J., Guo, X., Yu, X. J., & Ma, X. J. (2006). Proliferation and differentiation of mouse embryonic stem cells in APA microcapsule: A model for studying the interaction between stem cells and their niche. *Biotechnology Progress*, **22**, 791–800.
3. Thanos, C. G., Calafiore, R., Basta, G., Bintz, B. E., Bell, W. J., Hudak, J., *et al.* (2007). Formulating the alginate-polyornithine biocapsule for prolonged stability: Evaluation of composition and manufacturing technique. *Journal of Biomedical Materials Research, Part A*, **83**, 216–224.
4. Baruch, L., & Machluf, M. (2006). Alginate-chitosan complex coacervation for cell encapsulation: Effect on mechanical properties and on long-term viability. *Biopolymers*, **82**, 570–579.

5. Marsich, E., Borgogna, M., Donati, L., Mozetic, P., Strand, B. L., Salvador, S. G., *et al.* (2008). Alginate/lactose-modified chitosan hydrogels: A bioactive biomaterial for chondrocyte encapsulation. *Journal of Biomedical Materials Research, Part A*, **84**, 364–376.

6. Donati, I., Haug, J., Scarpa, T., Borgagna, M., Draget, K. I. B., & Paoletti, S. (2007). Synergistic effects in semidilute mixed solutions of alginate and lactose-modified chitosan (chitlac). *Biomacromolecules*, **8**, 957–962.

7. Baroli, B. (2006). Photopolymerization of biomaterials: Issues and potentialities in drug delivery, tissue engineering, and cell encapsulation applications. *Journal of Chemical Technology and Biotechnology*, **81**, 491–499.

8. Sakai, S., Kawabata, K., Ono, T., Ijima, H., & Kawakami K. (2005). Development of mammalian cell-enclosing subsieve-size agarose capsules (b100 μm) for cell therapy. *Biomaterials*, **26**, 4786–4792.

9. Cellesi, F., Weber, W., Fussenegger, M., Hubbell, J. A., & Tirelli, N. (2004). Towards a fully synthetic substitute of alginate: Optimization of a thermal gelation/chemical cross-linking scheme ("tandem" gelation) for the production of beads and liquid-core capsules. *Biotechnology and Bioengineering*, **6**, 740–749.

10. Govan, J. R. W., Fyfe, J. A. M., & Jarman, T. R. (1981). Isolation of alginate-producing mutants of Pseudomonas fluorescens, Pseudomonas putida and Pseudomonas mendocina. *Journal of General Microbiology*, **125**, 217–220.

11. Smidsrød, O., & Skjåk-Bræk, G. (1990). Alginate as immobilization matrix for cells. *Trends in Biotechnology*, **8**, 71–78.

12. Kim, K., Liu, X. Y., Zhang, Y., Cheng, J., Wu, X. Y., & Sun, Y. (2008). Elastic and viscoelastic characterization of microcapsules for drug delivery using a force-feedback MEMS microgripper. *Biomedical Microdevices*, **11**, 421–427.

13. Bunger, C. M., Gerlach, C., Freier, T., Schmitz, K. P., Pilz, M., Werner, C., *et al.* (2003). Biocompatibility and surface structure of chemically modified immunoisolating alginate-PLL capsules. *Journal of Biomedical Materials Research*, **67**, 1219–1227.

14. Lacík, I. (2006). Polymer chemistry in diabetes treatment by encapsulated islets of Langerhans: Review to 2006. *Australian Journal of Chemistry*, **59**, 508–524.

15. De Vos, P., Van Hoogmoed, C. G., Van Zanten, J., Netter, S., Strubbe, G. H., & Busscher, H. J. (2003). Long-term biocompatibility, chemistry,

and function of microencapsulated pancreatic islets. *Biomaterials*, **24**, 305–312.

16. De Vos, P., Andersson, A., Tam, S. K., Faas, M. M., & Halle, J. P. (2006). Advances and barriers in mammalian cell encapsulation for treatment of Diabetes. *Immunology, Endocrine & Metabolic Agents in Medicinal Chemistry*, **6**(2), 139–153.
17. De Vos, P., De Haan, B. J., Kamps, J. A. A. M., Faas, M. M., & Kitano, T. (2007). Zeta-potentials of alginate-PLL capsules: A predictive measure for biocompatibility? *Journal of Biomedical Materials Research*, **80**, 813–819.
18. Van Hoogmoed, C. G., Busscher, H. J., & De Vos, H. J. (2003). Fourier transform infrared spectroscopy studies of alginate-PLL capsules with varying compositions. *Journal of Biomedical Materials Research*, **67**, 172–178.
19. Tam, S. K., Dusseault, J., Polizu, S., Ménard, M., Hallé, J. P., & Yahia, L. H. (2005). Physicochemical model of alginate-poly-L-lysine microcapsules defined at the micrometric/nanometric scale using ATR-FTIR, XPS, and ToF-SIMS. *Biomaterials*, **26**, 6950–6961.
20. De Vos, P., Hoogmoed, C. G., & Busscher, H. J. (2002). Chemistry and biocompatibility of alginate-PLL capsules for immunoprotection of mammalian cells. *Journal of Biomedical Materials Research*, **60**, 252–259.
21. Williams, D. F. (1987). Summary and definitions. In *Progress in biomedical engineering: Definition in biomaterials* (Vol. 4, pp. 66–71). Amsterdam: Elsevier Science Publisher BV.
22. Montaudo, G., Samperi, F., & Montaudo, M. S. (2006). Characterization of synthetic polymers by MALDI-MS. *Progress in Polymer Science*, **31**, 277–357.
23. Schnoll-Bitai, I., Ullmer, R., Hrebicek, T., Rizzi, A., & Lacik, I. (2008). Characterization of the molecular mass distribution of pullulans by matrix-assisted laser desorption/ionization time-of-flight mass spectrometry using 2,5-dihydroxybenzoic acid butylamine (DHBB) as liquid matrix. *Rapid Communications in Mass Spectrometry*, **22**, 2961–2970.
24. Kubota, N., & Egnchi, Y. (1997). Facile preparation of water-soluble *N*-acetylated chitosan and molecular weight dependence of its water-solubility. *Polymer Journal*, **29**, 123–127.
25. Aiba, S. (1991). Studies on chitosan: Evidence for the presence of random and block copolymer structures in partially *N*-acetylated chitosans. *International Journal of Biological Macromolecules*, **13**, 40–44.

26. Rinaudo, M. (1989). Solution properties of chitosan. *Chitin and Chitosan. Sources, Chemistry, Biochemistry, Physical Properties and Applications* (pp. 71–86). London and New York: Elsevier.

27. Chenite, A., Chaput, C., Wang, D., Combes, C., Buschmann, M. D., Hoemann, C. D., *et al.* (2000). Novel injectable neutral solutions of chitosan form biodegradable gels *in situ*. *Biomaterials*, **21**, 2155–2161.

28. Chenite, A., Buschmann, M., Wang, D., Chaput, C., & Kandani, N. (2001). Rheological characterization of thermogelling chitosan/glycerol-phosphate solutions. *Carbohydrate Polymers*, **46**, 39–47.

29. Molinaro, G., Leroux, G. C., Damas, J., & Adam, A. (2002). Biocompatibility of thermosensitive chitosan-based hydrogels: An *in vivo* experimental approach to injectable biomaterials. *Biomaterials*, **23**, 2717–2722.

30. Cho, J., Heuzey, M. C., Bégin, A., & Carreau, P. J. (2005). Physical gelation of chitosan in the presence of b-glycerophosphate: The effect of temperature. *Biomacromolecules*, **6**, 3267–3275.

31. Rusu-Balaita, L., Desbrières, J., & Rinaudo, M. (2003). Formation of a biocompatible polyelectrolyte complex: Chitosan-hyaluronan complex stability. *Polymer Bulletin*, **50**, 91–98.

32. Miya, M., Iwamoto, R., & Yoshikawa S. (1980). IR spectroscopic determination of CONH content in highly deacylated chitosan. *International Journal of Biological Macromolecules*, **2**, 323–324.

33. Baxter, A., Dillon, M., & Anthony Taylor, K. D. (1992). Improved method for I.R. determination of the degree of *N*-acetylation of chitosan. *International Journal of Biological Macromolecules*, **14**, 166–169.

34. Doisy, J. G., & Roberts, G. A. F. (1985). Evaluation of infrared spectroscopic techniques for analyzing chitosan. *Macromolecular Chemise*, **186**, 1671–1677.

35. Pelletier, A., Limier, I., Sygusch, J., Chornet, E., & Overend, R. P. (1990). Chitin/chitosan transformation by thermo-mechano-chemical treatment including characterization by enzymic depolymerization. *Biotechnology and Bioengineering*, **36**, 310–315.

36. Muzzarelli, R. A. A. (1985). Determination of the degree of acetylation of chitosans by first derivative ultraviolet spectrophotometry. *Carbohydrate Polymers*, **5**, 461–472.

37. Rinaudo, M., Gey, C., & Milas, M. (1992). Substituent distribution on O, *N*-carboxymethylchitosans by 1H and 13C NMR. *International Journal of Biological Macromolecules*, **14**, 122–128.

38. Saito, H., Tabeta, R., & Ogawa, K. (1987). High-resolution solid-state 13C NMR study of chitosan and its salts with acids: Conformational

characterization of polymorphs and helical structures as viewed from the conformation-dependent 13C chemical shifts. *Macromolecules*, **20**, 2424–2430.

39. Raymond, L., Morin, F. G., & Marchessault, R. H. (1993). Degree of deacetylation of chitosan using conductometric titration and solid-state NMR. *Carbohydrate Research*, **246**, 331–336.
40. Heux, L., Brugnerotto, J., Desbrières, J., Versali, M. F., & Rinaudo, M. (2000). Solid state NMR for determination of degree of acetylation of chitin and chitosan. *Biomacromolecules*, **1**, 746–751.
41. Orive, G., Hernández, R. M., Gascón, A. R., Calafiore, R., Chang, T. M. S., De Vos, P., *et al.* (2004). History, challenges and perspectives of cell microencapsulation. *Trends in Biotechnology*, **22**, 87–92.
42. Orive, G., Tam, S. K., Pedraz, J. L., & Hallé, J. P. (2006). Biocompatibility of alginate–poly-L-lysine microcapsules for cell therapy. *Biomaterials*, **20**, 3691–3700.
43. Goren, A., Goren, E., Baruch, L., & Machluf, M. (2010). Encapsulated human mesenchymal stem cells: A unique hypoimmunogenic platform for long-term cellular therapy. *FASEB Journal*, **24**, 22–31.
44. Groot, M. de., Schuur, T. A., & Van Schilfgaarde, R. (2004). Causes of limited survival of microencapsulated pancreatic islet grafts. *Journal of Surgical Research*, **121**, 141–150.
45. Bunger, C. M., Tiefenbach, B., Jahnke, A., Gerlach, C., Freier, T., Schmitz, K. P., *et al.* (2005). Deletion of the tissue response against alginate-PLL capsules by temporary release of co-encapsulated steroids. *Biomaterials*, **26**, 2353–2360.
46. Figliuzzi, M., Plati, T., Cornolti, R., Adobati, F., Fagiani, A., Rossi, L., *et al.* (2006). Biocompatibility and function of microencapsulated pancreatic islets. *Acta Biomaterialia*, **2**, 221–227.
47. Poncelet, D., Lencki, R., Beaulieu, C., Halle, J. P., Neufeld, R. J., & Fournier, A. (1992). Production of alginate beads by emulsification/internal gelation. I. Methodology. *Applied Microbiology and Biotechnology*, **38**, 39–45.
48. Corinne, A. H., Raghuram, K., Kiang, R. L. J., Mocinecová, D., Hu, X. K., Johnson, J. D., *et al.* (2011). Pancreatic cell immobilization in alginate beads produced by emulsion and internal gelation. *Biotechnology and Bioengineering*, **108**, 424–434.
49. Gibbs, B. F., Kermasha, S., Alli, I., Muligan, C., & Anon, A. (2002). Constant air quality humidity and temperature control during soft capsule manufacturing. *Chemical Plants Processing*, **1**, 22–23.

50. Vinogradova, O. I. (2004). Mechanical properties of polyelectrolyte multilayer microcapsules. *Journal of Physiology*, **16**, 32–36.

51. Gupta, V., Verma, S., Nanda, A., & Nanda S. (2005). Osmotically controlled drug delivery. *Drug Delivery Technology*, **5**, 68–76.

52. Pirvu, C. (2005). Chemical characterization and applications of microcapsules. *Farmacia*, **53**, 61–68.

53. Rodríguez, J. V., Mamprin, M. E., Mediavilla, M. G., & Guibert, E. E. (1998). Glutathione movements during cold preservation of rat hepatocytes. *Cryobiology*, **36**, 236–244.

54. Guibert, E. E., Almada, L. L., Mamprin, M. E., Bellarosa, C., Pizarro, M. D., Tiribelli, C., *et al.* (2009). Subzero nonfreezing storage of rat hepatocytes using UW solution and 1,4-butanediol. II- functional testing on rewarming and gene expression of urea cycle enzymes. *Annals of Hepatology*, **8**, 129–133.

55. De Vos, P., Bučko, M., Gemeiner, P., Navrátil, M., Švitel, J., Faas, M., *et al.* (2009). Multiscale requirements for bioencapsulation in medicine and biotechnology. *Biomaterials*, **30**, 2559–2570.

56. Wilson, J. T., & Chaikof, E. L. (2008). Challenges and emerging technologies in the immunoisolation of cells and tissues, *Advanced Drug Delivery Reviews*, **60**, 124–145.

57. Blas, P., Giovagnoli, S., Schoubben, A., Ricci, M., Rossi, C., Luca, G., *et al.* (2006). Preparation and *in vitro* and *in vivo* characterization of composite microcapsules for cell encapsulation. *International Journal of Pharmaceutics*, **324**, 27–36.

58. Sakai, S., Mu, C. J., Kawabata, K., Hashimoto, I., & Kawakami, K. (2006). Biocompatibility of subsieve-size capsules versus conventional-size microcapsules. *Journal of Biomedical Materials Research, Part A*, **78**, 394–398.

59. Xie, H. G., Zheng, J. N., Li, X. X., Liu, X. D., Zhu, J., Wang, F., *et al.* (2010). Effect of surface morphology and charge on the amount and conformation of fibrinogen adsorbed onto alginate/chitosan microcapsules. *Langmuir*, **26**, 5587–5594.

60. Xie, H. G., Li, X. X., Lv, G. J., Zhu, J., Xie, W. Y., Luxbacher, T., *et al.* (2010). Effect of surface wettability and charge on protein adsorption onto implantable alginate-chitosan-alginate microcapsule surfaces. *Journal of Biomedical Materials Research Part A*, **92**, 1357–1365.

61. Ponce, S., Orive, G., Hernández, R., Gascón, A. R., Pedraz, J. L., De Haan, B. J., *et al.* (2006). Chemistry and the biological response against immunoisolating alginate-polycation capsules of different composition. *Biomaterials*, **27**, 4831–4839.

62. Sawhney, A. S. (1992). Poly(ethylene oxide)-graft-poly(L-lysine) copolymers to enhance the biocompatibility of poly(L-lysine)-alginate microcapsule membranes. *Biomaterials*, **13**, 863–870.

63. Sawhney, A. S., & Hubbell, J. A. (1993). Interfacial photopolymerization of poly(ethylene glycol)-based hydrogels upon alginate poly(L-lysine) microcapsules for enhanced biocompatibility. *Biomaterials*, **14**, 1008–1016.

64. Cruis, G. M., Hegre, O. D., Lamberti, F. V., Hegre, S. R., Scharp, D. S., & Hubbell, J. A. (1999). *In vitro* and *in vivo* performance of porcine islets encapsulated in interfacially photopolymerized poly(ethylene glycol) diacrylate membranes. *Cell Transplant*, **8**, 293–306.

65. Sawhney, A. S., Pathak, C. P., & Hubbell, J. A. (1994). Modification of Islet of Langerhans surfaces with immunoprotective poly(ethylene glycol) coatings via interfacial photopolymerization. *Biotechnology and Bioengineering*, **44**, 383–386.

66. Cruise, G. M., Hegre, O. D., Scharp, D. S., & Hubbell, J. A. (1998). A sensitivity study of the key parameters in the interfacial photopolymerization of poly(ethylene glycol) diacrylate upon porcine islets. *Biotechnology and Bioengineering*, **57**, 655–665.

67. Zheng, J. N., Xie, H. G., Yu, W. T., Liu, X. D., Xie, W. Y., Zhu, J., *et al.* (2010). Chitosan-g-MPEG-Modified alginate/chitosan hydrogel micro-capsules: A quantitative study of the effect of polymer architecture on the resistance to protein adsorption. *Langmuir*, **26**, 17156–17164.

68. Murua, A., Portero, A., Orive, G., Hernández, R. M., De Castro, M., & Luis Pedraz, J. (2008). Cell microencapsulation technology: Towards clinical application. *Journal of Controlled Release*, **132**, 76–83.

69. Sakai, S., Hashimoto, I., & Kawakami, K. (2006). Development of alginate-agarose subsieve-size capsules for subsequent modification with a polyelectrolyte complex membrane. *Biochemical Engineering Journal*, **30**, 76–81.

70. Khattak, S. F., Chin, K. S., Bhatia, S. R., & Roberts, S. C. (2007). Enhancing oxygen tension and cellular function in alginate cell encapsulation devices through the use of perfluorocarbons. *Biotechnology and Bioengineering*, **96**, 156–166.

71. Huang, X. B., Wang, J. Z., Xie, H. G., Zhang, Y., Wang, W., Yu, W. T., *et al.* (2010). Microcapsules embedded with three-dimensional fibrous scaffolds for cell culture and tissue engineering. *Tissue Engineering: Part C*, **16**, 1023–1032.

72. Uludag, H., De Vos, P., & Tresco, P. A. (2000). Technology of mammalian cell encapsulation. *Advanced Drug Delivery Reviews*, **42**, 29–64.

73. Schuldt, U., & Hunkeler, D. (2000). Characterization methods for microcapsules. *Minerva Biotechnologica*, **12**, 249–264.
74. Gombotz, W. R., & Wee, S. F. (1998). Protein release from alginate matrices. *Advanced Drug Delivery Reviews*, **31**, 267–285.
75. Tanaka, H., Matsumura, M., & Veliky, I. A. (1984). Diffusion characteristics of substrates in Ca-alginate gel beads. *Biotechnology and Bioengineering*, **26**, 53–58.
76. Goosen, M. F. A., O'Shea, G. M., Gharapetian, H. M., Chou, S., & Sun, A., M. (1985). Optimization of microencapsulation parameters — Semipermeable microcapsules as a bioartificial pancreas. *Biotechnology and Bioengineering*, **27**, 146–150.
77. Darrabie, M. D., William F., Kendall, J., & Emmanuel, C. (2005). Characteristics of poly-L-ornithine-coated alginate microcapsules. *Biomaterials*, **26**, 6846–6852.
78. Weeks, T. S., Adolf, D., & Mccoy, J. D. (1999). Cohesive failure in partially cured epoxies. *Macromolecules*, **32**, 1918–1922.
79. Kong, H. J., Wong, E., & Mooney, D. J. (2003). Independent control of rigidity and toughness of polymeric hydrogels. *Macromolecules*, **36**, 4582–4588.
80. Raamsdonk, V., Cornelius, R. M., Brash, J. L., & Chang, P. L. (2002). Deterioration of polyamino acid-coated alginate microcapsules *in vivo*. *Journal of Biomaterials Science, Polymer Edition*, **13**, 863–884.
81. Raamsdonk, V., & Chang, P. L. (2001). Osmotic pressure test: A simple, quantitative method to assess the mechanical stability of alginate microcapsules. *Journal of Biomedical Materials Research*, **54**, 264–271.
82. Chang, S. J., Lee, C. H., Hsu, C. Y., & Wang, Y. J. (2002). Biocompatible microcapsules with enhanced mechanical strength. *Journal of Biomedical Materials Research*, **59**, 118–126.
83. Thano, C. G., Bintz, B. E., & Emerich, D. F. (2007). Stability of alginate–polyornithine microcapsules is profoundly dependent on the site of transplantation. *Journal of Biomedical Materials Research A*, **81**, 1–11.
84. Rosinski, S., Grigorescu, G., Lewińska, D., Ritzén, L. G., Viernstein, H., Teunou, E., *et al.* (2002). Characterization of microcapsules: Recommended methods based on round-robin testing. *Journal of Microencapsulation*, **19**, 641–659.
85. Lim, F., & Sun, A. M. (1980). Microencapsulated islets as bioartificial endocrine pancreas. *Science*, **210**, 908–910.
86. Efrat, S. (2008). Beta-cell replacement for insulin-dependent diabetes mellitus. *Advanced Drug Delivery Reviews*, **60**, 114–123.

87. Calafiore, R., & Basta, G. (2007). Artificial pancreas to treat type 1 diabetes mellitus. *Methods in Molecular Medicine*, **140**, 197–236.

88. Jones, P. M., Courtney, M. L., Burns, C. J., & Persaud, S. J. (2008). Cell-based treatments for diabetes. *Drug Discovery Today*, **13**, 888–893.

89. Black, S. P., Constantinidis, I., Cui, H., Tucker-Burden, C., Tucker-Burden, C. J., & Safley, S. A. (2006). Safley, Immune responses to an encapsulated allogeneic islet β-cell line in diabetic NOD mice. *Biochemical and Biophysical Research Communications*, **340**, 236–243.

90. Schneider, S., Feilen, P. J., Brunnenmeier, F., Minnemann, T., Zimmermann, H., Zimmermann, U., *et al.* (2005). Long-term graft function of adult rat and human islets encapsulated in novel alginate-based microcapsules after transplantation in immunocompetent diabetic mice. *Diabetes*, **54**, 687–693.

91. Elliott, R. B., Escobara, L., Tan, P. L. J., Garkavenko, O., Calafiore, R., Basta, P., *et al.* (2005). Intraperitoneal alginate-encapsulated neonatal porcine islets in a placebo-controlled study with 16 diabetic cynomolgus primates. *Transplantation Proceedings*, **37**, 3505–3508.

92. Dufrane, D., Rose-Marie, G., Alain, S., Yves, G., & Pierre, G. (2006). Six-Month Survival of microencapsulated pig islets and alginate biocompatibility in primates: Proof of concept. *Transplantation*, **81**, 1345–1353.

93. Calafiore, R., Basta, G., Luca, G., Lemmi, A., Racanicchi, L., Mancuso, F., *et al.* (2006). Standard technical procedures for microencapsulation of human islets for graft into nonimmunosuppressed patients with Type 1 diabetes mellitus. *Transplantation Proceedings*, **38**, 1156–1157.

94. Calafiore, R., Bast, G., Luca, G., Lemmi, A., Montanucci, M. P., Calabrese, G., *et al.* (2006). Microencapsulated pancreatic isl*et al*lografts into nonimmunosuppressed patients with type 1 diabetes: First two cases. *Diabetes Care*, **29**, 137–138.

95. Elliott, R. B., Escobar, L., Tan, P. L. J., Muzina, M., Zwain, S., & Buchanan, C. (2007). Live encapsulated porcine islets from a type 1 diabetic patient 9.5 yr after xenotransplantation. *Xenotransplantation*, **14**, 157–161.

96. Grose, S. (2007). Critics slam Russian trial to test pig pancreas for diabetics. *Nature Medicine*, **13**, 390–391.

97. Cai, X. X., Lin, Y. F., Ou, G. M., Luo, E., Man, Y., Yuan, Q., *et al.* (2007). Ectopic osteogenesis and chondrogenesis of bone marrow stromal stem cells in alginate system. *Cell Biology International*, **31**, 776–783.

98. Kaigler, D., Krebsbach, P. H., Wang, Z., West, E. R., Horger, K., & Mooney, D. J. (2006). Transplanted endothelial cells enhance orthotopic bone regeneration. *Journal of Dental Research*, **85**, 633–637.

99. Dean, S. K., Yulyana, Y., Williams, G., Sidhu, K., & Tuch, B. E. (2006). Differentiation of encapsulated embryonic stem cells after transplantation. *Transplantation*, **82**, 1175–1184.

100. Ding, H. F., Liu, R., Li, B. G., Lou, J. R., Dai, K. R., & Tang, T. T. (2007). Biologic effect and immunoisolating behavior of BMP-2 gene-transfected bone marrow-derived mesenchymal stem cells in APA microcapsules. *Biochemical and Biophysical Research Communications*, **362**, 923–927.

101. Babister, J. C., Tare, R. S., Green, D. W., Inglis, S., Mann, S., & Oreffo, R. O. C. (2008). Genetic manipulation of human mesenchymal progenitors to promote chondrogenesis using "bead-in-bead" polysaccharide capsules. *Biomaterials*, **29**, 58–65.

102. Stevens, B., Yang, Y. Z., Mohandas, A., Stucker, B., & Nguyen, K. T. (2008). A review of materials, fabrication methods, and strategies used to enhance bone regeneration in engineered bone tissues. *Journal of Biomedical Materials Research Part B: Applied Biomaterials*, **85**, 573–582.

103. Bieback, K., Kern, S., Klüter, H., & Eichler, H. (2004). Critical parameters for the isolation of mesenchymal stem cells from umbilical cord blood. *Stem Cells*, **22**, 625–634.

104. Zuk, P. A., Zhu, M., Mizuno, H., Huang, J., Fetrell, W., Katz, A. J., *et al.* (2001). Multilineage cells from human adipose tissue: Implications for cell based therapies. *Tissue Engineering*, **7**, 211–228.

105. Kimelman, N., Pelled, G., Helm, G. A., Huard, J., Schwarz, E. M., & Gazit, D. (2007). Review: Gene and stem cell-based therapeutics for bone regeneration and repair. *Tissue Engineering*, **13**, 1135–1150.

106. Endres, M., Wenda, N., Woehlecke, H., Neumann, K., Ringe, J., Erggelet, C., *et al.* (2010). Microencapsulation and chondrogenic differentiation of human mesenchymal progenitor cells from subchondral bone marrow in Ca-alginate for cell injection. *Acta Biomaterialia*, **6**, 436–444.

107. Laguna, G. R., Tyers, P., & Barker, R. A. (2008). The search for a curative cell therapy in Parkinson's disease. *Journal Neurological Sciences*, **265**, 32–42.

108. Winkler, C., Kirik, D., & Björklund, A. (2005). Cell transplantation in Parkinson's disease: How can we make it work? *Trends in Neuroscience*, **28**, 86–92.

109. Kim, D., Schallert, T., Liu, Y., Browarak T., Nayeri, N., Tessler, A., *et al.* (2001). Transplantation of genetically modified fibroblasts expressing BDNF in adult rats with subtotal hemisection improves specific motor

and sensory functions. *Neurorehabilitation and Neural Repair*, **15**, 141–150.

110. Liu, Y. D., Kim, D., Himes, B. T., Chow, S. Y., Schallert, T., Murray, M., *et al.* (1999). Transplants of fibroblasts genetically modified to Express BDNF promote regeneration of adult rat rubrospinal axons and recovery of forelimb function. *Journal of Neuroscience*, **19**, 4370–4387.

111. Read, T. A., Sorensen, D. R., Mahesparan, R., Enger, P., Timpl, R., Olsen, B. R., *et al.* (2001). Local endostatin treatment of gliomas by microencapsulated producer cells. *Nature Biotechnology*, **19**, 29–34.

112. Joki, T., Machluf, M., Atala, A., Zhu, J. H., Seyfried, N. T., Dunn, I. F., *et al.* (2001). Continuous release of endostatin from microencapsulated engineered cells for tumor therapy. *Nature Biotechnology*, **19**, 35–39.

113. Cirone, P., Bourgeois, J. M., & Chang, P. L. (2003). Antiangiogenic cancer therapy with microencapsulated cells. *Human Gene Therapy*, **14**, 1065–1077.

114. Teng, H., Zhang, Y., Wang, W., Ma, X. J., & Fei, J. (2007). Inhibition of tumor growth in mice by endostatin derived from abdominal transplanted encapsulated cells. *Acta Biochimica et Biophysica Sinica*, **39**, 278–284.

Chapter 4

Microspheres for the Separation and Refolding of Proteins with an Emphasis on Particles Made of Agarose

Jan-Christer Janson

Department of Physical and Analytical Chemistry, Uppsala Biomedical Center, Uppsala University, Uppsala, Sweden

Jan-Christer.Janson@ytbioteknik.uu.se

4.1 Introduction

The first attempts to use microspheres, in a wider sense of the term, for protein separation were based on batch adsorption to a variety of inorganic materials. The most frequently used were alumina hydrates ($Al_2O_3 \cdot nH_2O$), such as Bauxite, Al-silicates (bentonites), Mg-silicates (soapstone etc.), silicas ($SiO_2 \cdot nH_2O$), Ca- and Zn-carbonates, Ca-phosphate, iron oxides (hematite, Fe_2O_3), and Al,Mg-silicate (Florisil). Sometimes, these adsorbents were mixed with filter aids and packed in columns and were used in chromatography. An early review of these techniques was published by Turba in 1954 (1).

The first applications of organic polymer microspheres for protein separation were based on the weak cation exchanger Amberlite™ IRC-50, a co-polymer of methacrylic acid and divinyl benzene (2–4). However, its hydrophobic character was an obstacle for use with high–molecular weight proteins. This problem was not solved until about 30 years later when products based on hydrophilized polystyrene divinyl benzene copolymers were introduced (5).

The first useful chromatographic media that were used for high–molecular weight proteins were not spherical and, therefore, cannot be formally called microspheres. Thus cellulose ion

Microspheres and Microcapsules in Biotechnology: Design, Preparation, and Applications
Edited by Guanghui Ma and Zhiguo Su

ISBN 978-981-4316-47-7 (Hardcover), 978-981-4364-62-1 (eBook)
www.panstanford.com

exchangers first introduced in 1954 were microgranular (fibrous) (6,7), and hydroxyapatite, introduced by Tiselius *et al.* in 1956, was microcrystalline (8). Later, spherical cellulose media were introduced by Determann *et al.* in 1969 (9) and put to the market in 1977 under the trade name Sephacel by Pharmacia Fine Chemicals AB (now GE Healthcare Bio-Sciences AB), Uppsala, Sweden. Sometime later, Motozato *et al.* (10) developed a similar product and marketed it in 1984 under the trade name Cellulofine by Chisso Corporation, Kumamoto, Japan.

In 1959, Porath and Flodin (11) introduced cross-linked dextran for gel filtration (size-exclusion chromatography, SEC) of water-soluble substances. The product Sephadex® was launched in the market the same year by Pharmacia Fine Chemicals AB, Uppsala, Sweden. Review articles covering the history of the development of Sephadex are available in (12,13).

In 1961, Polson (14) reported on the use of columns packed with granulated agar gels (5–15%) for the separation of a variety of high–molecular weight proteins by molecular sieving (an early name for SEC). In 1962, Hjertén and Mosbach (15) introduced microspheres of cross-linked polyacrylamide for protein separation by molecular size, which were launched a few years later in the market as Bio-Gel P by Bio-Rad Laboratories, Richmond, CA, USA.

In 1964, Hjertén (16) pioneered in reporting the preparation and use of spherical particles made of agarose for molecular sieve chromatography of proteins and particles such as viruses. Soon after, several companies introduced agarose-based microspheres for chromatography in the market (Mann Research Inc., New York, USA, Seravac Ltd., Maidenhead, England, and Pharmacia Fine Chemicals AB, Uppsala, Sweden).

Microspheres made of porous glass were introduced by Haller in 1965 (17), and later microspheres for protein separation made of porous silica were introduced by Unger *et al.* (18), Regnier and Noel (19), and Hashimoto *et al.* (20).

Microspheres based on cross-linked dextran have got many favorable properties that make them suitable as size-exclusion media and as matrices and ligand carriers for protein separation, for example, for the manufacturing of ion exchangers (21). However, at low dextran concentrations [4–6% (w/w)], which are required to produce microspheres possessing pore sizes large enough to allow the separation of high–molecular weight proteins, the rigidity of the microspheres is inadequate, restricting the use of particles smaller than approximately 100 μm.

This drawback has to a large extent been compensated for by the design of composite gels formed by copolymerization of allyl dextran with *N,N*-methylenebisacrylamide, such as in the Sephacryl™ series of media (22), or by grafting dextran chains to microspheres made of agarose gel polymer networks, such as in the Superdex™ series of media (23).

4.2 Design and Preparation of Agarose Microspheres for Protein Chromatography

4.2.1 Agarose Raw Material

The starting material for the preparation of agarose microspheres originates from the marine polysaccharide agar (also named agar–agar). Agar is extracted from red sea weeds (Rhodophycae), primarily belonging to the genera Gelidium and Gracilaria harvested along temperate seashores all around the globe.

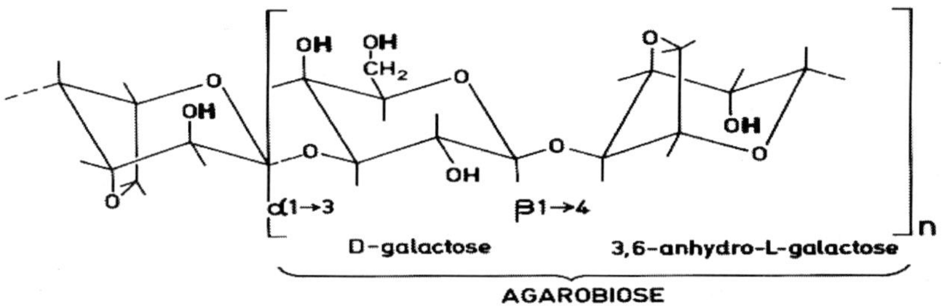

Figure 4.1 Molecular structure of the agarose disaccharide unit agarobiose. The alternative disaccharide 3,6-anhydro-L-galactose-1,3-D-galactose is called neo-agarobiose. The symbol outside the bracket, *n*, is approximately 350 (the average molecular weight of agarose is around 120,000).

Studies performed by Araki (24) concluded that agarose is the ideal, noncharged component of agar built up from alternating 1,3-linked-D-galactose and 1,4-linked 3,6-anhydro-L-galactose (Fig. 4.1). According to Duckworth and Yaphe (25), agar is a mixture of related polysaccharides, substituted to various degrees with sulfate esters, pyruvic acid, and methyl groups. The original conception that agar is composed of a neutral component, agarose, and a charged component called agaropectin has now been abandoned. The relationship between the agar concentration and the corresponding gel pore size was investigated by Ackers and Steere in 1962 (Fig. 4.2) (26).

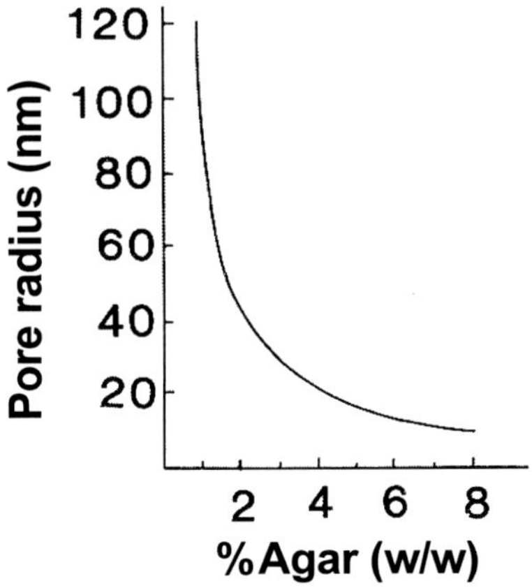

Figure 4.2 Effective pore radii of agar gels as a function of gel concentration. Adapted from Ref. 26.

Today, agarose is defined as the low-charge fraction of agar obtained either by precipitating its high-charge fraction, using positively charged reactants such as cetylpyridinium chloride (27), or by precipitating its low-charge fraction, using polyethylene glycol (28). Agarose with extremely low concentrations of sulfate groups have been prepared by Hjertén (29), who used adsorption of the more charged polysaccharide fraction to an anion exchanger, and by Låås (30), who used reduction with lithium aluminium hydride. The latter process does not only remove the sulfate groups but also the carboxylate groups producing a virtually charge-free agarose material.

A typical industrial scale production process for agarose can be summarized as follows:

1. Extraction of agar from red sea weeds in a stirred 3000 L vessel at 95°C
2. Precipitation of charged polysaccharides (formerly called agaropectin)

3. Precipitate removal by centrifugation in a disc separator at 75°C
4. Gel setting by cooling (~2% agarose)
5. Gel fragmentation, often by freezing/thawing, and washing in a basket centrifuge
6. Final washing on a sprayed continuous conveyor belt filter
7. Hydraulic pressing out of water to 6% agarose gel concentration
8. Drying in air at maximum 80°C to a moisture level <7%
9. Milling to a powder in a hammer mill
10. Analysis
11. Several 20 kg batches are mixed to 150 kg lots in a double-cone mixer
12. Final analysis and packaging

A typical raw material specification for agarose intended for the manufacturing of microspheres for chromatography media is summarized in Table 4.1.

Table 4.1 Typical raw material specification for agarose intended for the production of chromatographic media

Property	Specification
Appearance	White to faintly yellow powder
Gel strength of 1.5% solution	Min. 1700 g/cm^2
Viscosity of 6% solution, 80°C	600–1000 cps
Gelling temperature of a 1.5% solution	40–43°C
A_{500} of a 1.5% solution and gel	60°C, max. 0.015, 10°C, max. 0.300
Electroendosmosis (-mr)	0.08 ± 0.02
Sulfate	Max. 0.30%
Pyruvic acid	Max. 0.05%
Residue on ignition	Max. 0.5%
Loss on drying	Max. 7%
Microbial contamination	Max. 10^3/g
pH in the agarose solution	5.5–8.0

4.2.2 Methods for Preparing Agarose Microspheres

4.2.2.1 Granulation

The first agar and agarose gel particles used for chromatography were suspensions prepared by granulation of gels set in bulk (14,27). Also some early commercially available agarose gel preparations were of granular (irregular) type such as those at the time produced by Mann Research Inc., New York, USA.

4.2.2.2 Pearl condensation

This technique is based on spraying hot agar solution under pressure on ice-cold ether layered on top of ice-cold water. The agar solution is adjusted to a temperature of 65°C and poured into a vessel equipped with a tube made of glass with an orifice of approximately 0.5 mm. The agar solution was forced to pass through the orifice at a pressure of 2–3 kg/cm^2 obtained from a nitrogen gas tank. Agar drops condensed immediately on contact with the ice-cold ether layered on top of ice-cold water and could be collected after sedimentation (31).

4.2.2.3 Suspension gelation (water-in-oil emulsification)

The first pearl-condensed agarose microspheres intended for chromatography were prepared by Hjertén (16) by emulsification of a water solution of agarose in an organic liquid phase containing dissolved suitable stabilizer. Because of the fact that the emulsification principle is the same as for all subsequent published methods, it is relevant to add a somewhat more detailed description of the procedure used by Hjertén. Thus, the agarose was dissolved in distilled water to the desired concentration by boiling in a round-bottom flask equipped with a reflux condenser. For agarose concentrations higher than 3%, a pressure boiler was used. The dissolved agarose was allowed to cool to approximately 50°C and added to another round-bottom flask, equipped with a suitable stirrer, containing an organic liquid phase (toluene and carbon tetrachloride) and the dissolved stabilizer, and adjusted to the same temperature. The stirring was started immediately. After 1 minute of stirring, the suspension of agarose droplets was cooled using tap water by pouring it into the glass beaker in which the reactor flask was kept. After transferring the agarose microspheres to a Büchner funnel, they were washed by ethyl ether to free them of excess reactants and the sedimented suspension was finally washed with distilled water.

Agarose microspheres for chromatography have been produced using water/oil emulsification by several commercial manufacturers ever since the middle of the 1960s, but no details of the manufacturing processes have been published. However, one may be assured that ethyl ether is not used as a washing solvent on an industrial scale. A comparatively recent and detailed study of the effect of various processing conditions for the preparation of porous agarose beads by water/oil emulsification was published in 2005 by Mu *et al.* (32).

4.2.2.4 Membrane emulsification

In a recent report, a totally different approach for the production of agarose gel microspheres was introduced by the group headed by Professor Guanghui Ma at the Institute of Process Engineering, Chinese Academy of Sciences, Beijing, China (33). In this process, a pre-emulsified water-in-oil suspension of agarose solution droplets was forced, by using excessive pressure, to pass through a membrane with suitable pore-size distribution into a cooling vessel. The microspheres so produced have got a much more favorable homogeneous particle-size distribution than what is possible to achieve by normal-stirred batch-type water-in-oil emulsification.

4.2.3 Cross-Linking Methods

Agarose owes its gel-forming ability to the existence of hydrogen bond–forming groups on the polysaccharide backbone. The most important being the intramolecular ether bond between carbons 3 and 6 in the 3,6-anhydro-L-galactose residue (Fig. 4.1), which contributes to the formation of the agarose chain double helix and is a prerequisite for spontaneous gel setting (Fig. 4.3). It has been shown that the double helix is held together by hydrogen bonds via water molecules located inside the helix structure, utilizing one ether bond and one alcohol hydroxyl in one disaccharide residue in each of the two helix structures (34). During the gel formation process, the agarose double helices aggregate through another set of hydrogen bonds into a three-dimensional network of fibers surrounding large voids of water forming the interconnected pores through which the proteins diffuse without much resistance (Fig. 4.4).

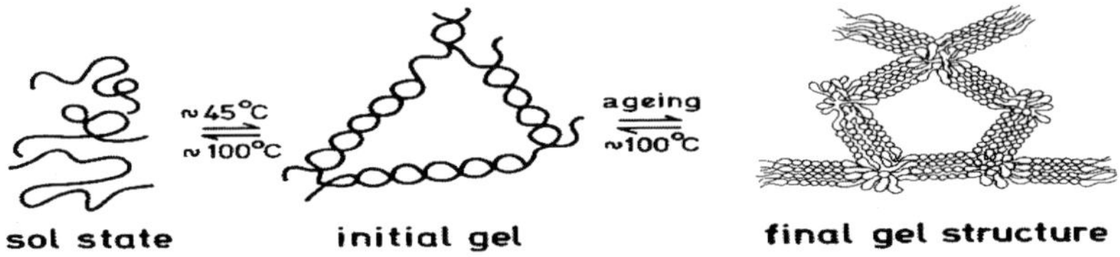

Figure 4.3 Schematic representation of the gelling process of agarose.

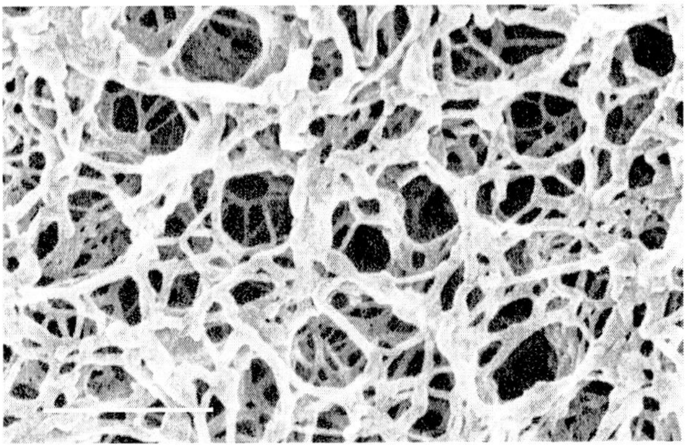

Figure 4.4 Scanning electron micrograph of a 4% agarose gel. The white bar represents 500 nm. Photo adapted with permission from Anders Medin, Uppsala Biomedical Center, Uppsala, Sweden, 1989.

4.2.3.1 Cross-linking with dibromopropanol and epichlorohydrin

This inherent and spontaneous physical cross-linking of the galactan polymers of agarose is very sensitive to environmental factors such as high temperature and the presence of structure-breaking reagents. Examples of these are high concentrations of urea, guanidine hydrochloride, and certain chaotropic ions, such as isothiocyanate, perchlorate, and iodide, all leading to gel dissolution. Also, substitution with charged groups results in the formation of soluble reaction products. This means that it is impossible to make ion exchangers based on neat agarose gels. This is why Porath and coworkers in 1971 introduced covalent cross-links in agarose microspheres in order to stabilize the already existing physically cross-linked structures (35,36). These cross-links are confined to the conservation of the helix structures inside an individual fiber and do not involve inter-fiber reactions. Thus, the cross-linking will not affect the original agarose gel pore dimensions. Dibromopropanol and bisepoxides were used as cross-linking agents. Under alkaline conditions that are required during the cross-linking reaction, dibromopropanol gives rise to epibromohydrin, which reacts with alcohol hydroxyls in the agarose polymer backbone. The work by Porath *et al.* was utilized for the introduction of the Sepharose™ CL series of products by Pharmacia Fine Chemicals AB, Uppsala, Sweden, in 1975.

4.2.3.2 Cross-linking with divinyl sulphone

In a subsequent report by Porath and coworkers, it was shown that there exists an optimal length of the cross-linking reagent in order to achieve maximum agarose gel rigidity. Thus, cross-linking of a 2% agarose gel with divinyl sulphone gave rise to a 20-fold increase of the flow rate through a chromatographic column of a given bed volume. The increase in rigidity was dependent on the amount of cross-linker used, and an optimization of the reaction in this respect was required. The cross-linking with divinyl sulphone did not affect the pore-size distribution of the gel, indicating that the cross-linking took place inside the helix aggregates and not between them. Using divinyl sulphone as cross-linker, it is possible to make and use agarose microspheres for chromatography at a concentration as low as 0.5%. Such gels are extremely porous and allow penetration of tobacco mosaic virus, which is approximately 300 nm long. Particles as large as 70S ribosomes eluted at the total liquid volume of the column (37).

4.2.3.3 Cross-linking in the presence of salt

Studies on the structure of agarose gels, published by Laurent in 1967, showed that the agarose gel fiber itself contains as much as 50% water (38). Data showing around 65% intra-fiber-bound water have later been reported by Axelsson (39). This fact triggered the idea of performing the cross-linking in the presence of high concentrations of polar organic solvents (40) or in the presence of a cosmotropic salt such as sodium sulfate. The latter discovery finally led to the development and market introduction of the Sepharose™ Fast Flow series of chromatography products in 1985. A variation of this cross-linking method is described in a United States Patent Application Publication from 2008 (41). The salt binds water much more strongly than the agarose fiber, which forces its polymer helix aggregates to contract and its individual polysaccharide polymers to become more accessible to the cross-linking reagents. Also, the high salt suppresses the electrostatic repelling of even the slightest negatively charged polymers in the gel fibers. The result is a more efficient locking of the helices in the fiber, leading to a more rigid gel product.

4.2.3.4 Cross-linking using a combination of reagents

Another procedure aimed at obtaining a more rigid agarose gel product is based on the consecutive use of two complementary bi- or polyfunctional cross-linking reagents, one possessing a short carbon chain between the reacting groups (e.g., epoxy) and another possessing a long carbon chain (42,43). Using this concept, the Superose™ series of small-diameter agarose microspheres was developed and launched in the market by Pharmacia Fine Chemicals AB, Uppsala, Sweden, in 1982.

4.2.3.5 Cross-linking of pre-activated agarose polymers

In a more recent development, the unorthodox approach was taken to modify the agarose polymers while still in solution before gel setting, using bifunctional cross-linkers containing one active and one inactive moiety. After the gel particles (microspheres) were formed by emulsifying the solution of the modified agarose polymers and after gel setting, the inactive reactant moiety was activated and allowed to cross-link the agarose fiber polymer chains. In this way, enhanced rigidity could be achieved (44). The high-flow agarose base matrix so obtained is used for the production of the Capto™ and Capto™ ImpRes series of agarose gel media manufactured by GE Healthcare Bio-Sciences AB, Uppsala, Sweden.

4.2.4 Superporous Agarose Gel Media

In late 1980s, Professor Klaus Mosbach at the Department of Applied Biochemistry, University of Lund, Sweden, and his pupils were interested in the development of gelatin particles that were suitable for the growth of anchorage-dependent animal cells (45). These particles were prepared by a double water-in-oil emulsification procedure and had large cavities that increased the surface area for cell growth. Under the electron microscope, large interconnecting cavities could be observed in these particles. During a discussion with Mosbach and his pupils in 1987, the author of this chapter put forward the idea that such particles could be produced using agarose instead of gelatin but for another purpose, namely the design of a new type of media for protein chromatography. For such a media, one can anticipate that a part of the chromatographic flow would pass through the interior of each particle, which would significantly improve the chromatographic efficiency. In the publication that followed, the above mentioned discussion is referred to on page 232 of reference (46). Superporous agarose is also described in the corresponding patent (47). A few years later Afeyan *et al.* (48) introduced the technique of perfusion chromatography using PS-DVB media based on the same concept.

4.2.5 Advantages of Agarose as a Matrix for Protein Chromatography

Adsorption chromatography is based on the solutes being reversibly attached to a surface. The larger the surface area, the higher is the binding capacity of the adsorbent. The higher the surface-to-volume ratio, the higher is the capacity per unit volume. The highest surface-to-volume ratios are obtained in gels, and among the hydrophilic gels, agarose is considered to be closest to an ideal matrix for protein adsorbents. The main advantage of an agarose gel is its hydrophilicity, which is due to the presence of hydroxyl groups, and its macroporosity, which is based on a three-dimensional network of fibers resulting in a highly accessible interconnected system of pores (Fig. 4.4). Schematically, such a system can be simplified as consisting of a three-dimensional array of randomly oriented infinitely long straight rods as shown in Fig. 4.5. Using suitable cross-linking agents, the gel rigidity can be enhanced to a practical level for preparative chromatography but cannot match totally rigid gels such as those based on silica or organic polymers, for example, polystyrene/divinyl benzene (PS/DVB) (49).

Figure 4.5 Schematic visualization of an agarose gel simplified as consisting of a three-dimensional array of randomly oriented infinitely long straight rods.

In Fig. 4.6, a schematic representation of the difference in macrostructure of gels based on silica or PS/DVB and agarose are shown. The most striking difference is the relative pore volume: silica and PS/DVB contain 45–55% (v/v) solids whereas the agarose gel fiber volume accounts for only 5–10% (v/v) solids. As a consequence, the total available specific surface area for protein binding differs as well. Thus, porous silica with nominal pore diameters of 30 nm and 50 nm have got specific surface areas of 150 m^2/mL and 83 m^2/mL, respectively (50) while a 4% (w/w)

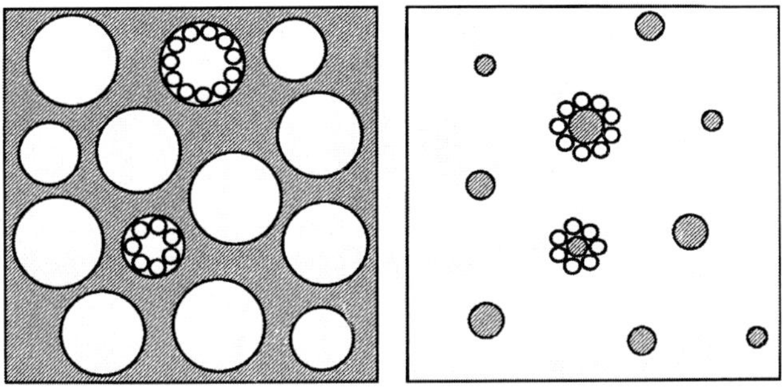

Figure 4.6 Schematic representation of the difference in macrostructure of gels based on silica or PS/DVB (left) and agarose (right) at similar apparent pore size. The most striking difference is the relative pore volume where silica and PS/DVB contain 45–55% (v/v) solids whereas the agarose gel fiber volume accounts for only 5–10% (v/v). The small circles represent protein molecules adsorbed to the solid phase (hatched area).

agarose gel network has got a total surface area of approximately 28 m^2/mL packed chromatographic bed according to the calculations shown in the following example:

Prerequisites

The total gel volume in a packed chromatographic bed of agarose microspheres is independent of particle size at constant void volume.

As a simplification, the agarose microsphere is considered to be composed of a random homogeneous network of straight gel-fiber cylinders with a radius of 0.0048 μm as illustrated in Fig. 4.5.

The total agarose gel fiber volume (in μm^3) can be obtained from the approximation that the fiber network consists of one fiber cylinder with a radius of 0.0048 μm.

The fiber concentration in a 4% (w/w) agarose gel is 9.5% (v/v) due to an appreciable amount of intra-fiber-bound water (65%) (39).

Example

For a 1 mL packed bed of agarose microspheres with 30% interstitial (void) volume,

Gel volume = $7 \cdot 10^{11}$ μm^3

For a 1 mL packed bed of 4% (w/w) agarose microspheres,

Total fiber volume = $0.095 \cdot 7 \cdot 10^{11}$

= $6.65 \cdot 10^{10}$ μm^3

Thus, for a 1 mL packed bed of 4% (w/w) agarose microspheres,

Total fiber surface area = $6.65 \cdot 10^{10} \cdot 2/0.0048$

= $27.7 \cdot 10^{12}$ μm^2

≈ 28 m^2

Not only the specific surface area but also the relative pore volume and the gel pore-size distribution are important chromatographic parameters to consider (21), particularly in SEC as this will affect the shape of the selectivity curve as well as the relative protein diffusion coefficients in adsorption chromatography. One way to estimate the apparent relative pore volume and the pore-size distribution of chromatographic gel media is by correlating the elution behavior of well-defined fractions of a neutral and an inert linear homopolymer, such as dextran. The resulting calculated distribution coefficients, K_d and K_{av}, are then plotted against the hydrodynamic (viscosity) radius calculated for each dextran fraction using the equation $R_\eta = 0.0271\ M_p^{0.498}$ as suggested by Hagel (51). In Fig. 4.7, the apparent pore-size distributions of a variety of common chromatographic media for proteins obtained in this way are shown.

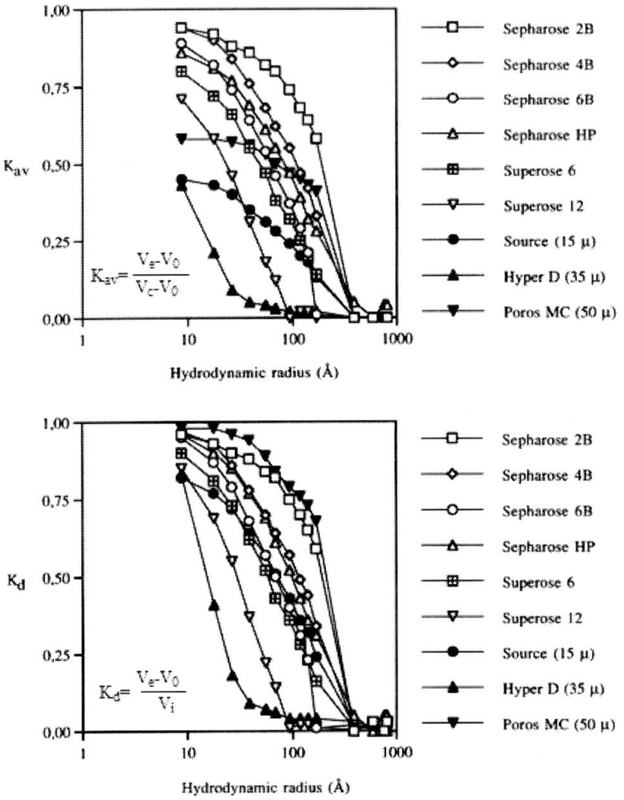

Figure 4.7 Distribution coefficients K_{av} and K_d plotted against the hydrodynamic (viscosity) radius of several dextran fractions for a variety of gel media frequently used in protein chromatography. V_e = elution volume, V_0 = void volume, V_c = geometrical bed volume, and V_i = inner liquid volume in the gel phase. Data on the dextran fractions used can be found in (51).

The apparent relative pore volume and pore-size distribution curves for the agarose gel media (Sepharose 2B, Sepharose 4B, Sepharose 6B, Sepharose HP, Superose 6, and Superose 12) are on top in comparison with the polymer-based macroreticular media (Source, Hyper D and Poros MC). The reason for this becomes clear when we study the schematic representation of macro gel structures shown in Fig. 4.6. It is striking that the 2% agarose gel Sepharose 2B has got a pore-size distribution curve similar to that of Poros MC, a perfusion chromatography gel material. The data might be interpreted such that it should be possible to get perfusion effects also in this agarose gel if the rigidity of the gel fiber network was high enough (52).

4.2.6 Recent Developments of Agarose Microspheres for Protein Separation

Figure 4.8 shows the timeline for the market introduction of some selected agarose gel media intended for protein separation.

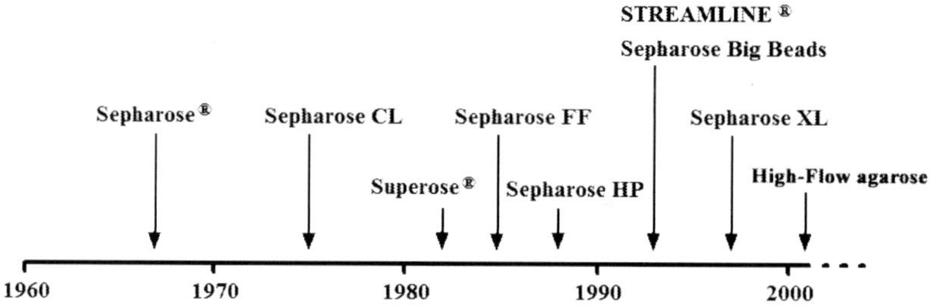

Figure 4.8 Timeline for the market introduction of some well-known agarose gel products.

These media have all primarily been used as base matrices for the synthesis of ion exchangers, hydrophobic interaction media, and affinity chromatography media. Thus, in 1968, Cuatrecasas *et al.*, developed affinity chromatography based on the use of Sepharose 4B for the coupling of affinity ligands (53), using the CNBr-activation technique introduced by Axén *et al.* (54). As a consequence, in 1971, Pharmacia Fine Chemicals AB, Uppsala, Sweden, introduced CNBr-activated Sepharose, which became the economically most important product of the company a few years later. Another example of an early breakthrough in protein separation was the introduction of Sepharose Fast Flow in 1984, which for the first time allowed the application of ion-exchange chromatography at industrial scale in which a 200 L column of DEAE Sepharose Fast Flow was used as the first step to purify the recombinant human growth hormone by KabiVitrum AB (now Pfizer AB), Stockholm, Sweden.

Recent developments in agarose microspheres for protein separation are Capto™ Q and Capto S (90 μm average particle diameter), Capto MMC and Capto adhere (75 μm), and Capto Q ImpRes and Capto SP ImpRes (36–44 μm) chromatography media (GE Healthcare Bio-Sciences AB, Uppsala, Sweden). These are based on the "high-flow" type of agarose mentioned above. The high rigidity of these microspheres enables the production of particles of much smaller diameters, which results in enhanced column efficiency. In order to increase the functional surface area, Capto Q and Capto S ion-exchange media are equipped with dextran surface extenders grafted to the mantle of the cross-linked agarose gel fibers. Another important development is the introduction of multimodal protein adsorbents combining several complementary binding mechanisms such as ion–ion attraction, hydrogen bond formation, and hydrophobic interaction (55–57).

Besides having a potential for higher binding selectivity, multimodal ligands offer binding at higher ionic strengths than conventional monomodal ion exchangers. This is of particular importance in capture situations in which large volumes of cell culture supernatants obtained after animal cell culture or fermentation of the methylotropic yeast *Pichia pastoris* have to be handled. An example of this is the process of binding of a supernatant containing recombinant human serum albumin to a conventional cation exchanger such as CM Sepharose Fast Flow. The process normally requires a three-fold dilution of the supernatant, but when the multimodal ion exchanger Capto MMC is used for the same purpose, no dilution is required (58). In Fig. 4.9, the molecular structure of Capto MMC ligand is shown.

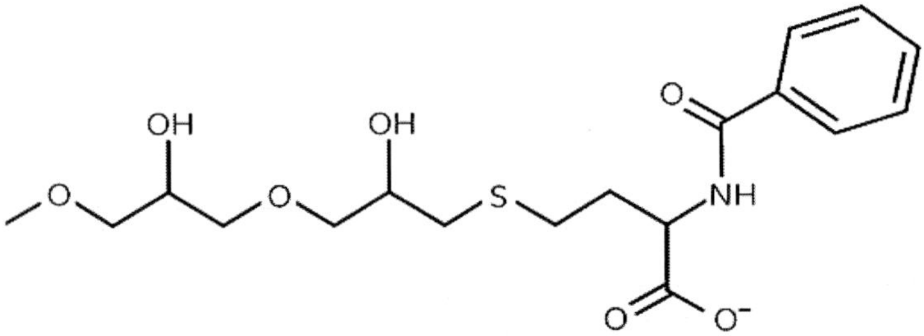

Figure 4.9 Molecular structure of the multimodal weak ion-exchange ligand of Capto™ MMC. In addition to ion–ion interactions with the negatively charged carboxylate group, hydrophobic interactions might occur with the phenyl group and hydrogen-bond interactions with the carbonyl (hydrogen-bond acceptor) and the amido hydrogen (hydrogen-bond donor).

Another example is the use of the strong anion exchanger Capto adhere. Capto adhere possesses multimodal functionality, giving a different selectivity compared to traditional anion exchangers. It is designed for the purification of monoclonal antibodies, particularly in combination with the Protein A–based high-flow agarose affinity medium MabSelect SuRe. When used in flow-through mode, Capto adhere contributes to the removal of leached Protein A, aggregates, host-cell proteins, nucleic acids, and viruses by passing the antibodies directly through the column and adsorbing the contaminants (59). In Fig. 4.10, the molecular structure of the Capto adhere ligand is shown.

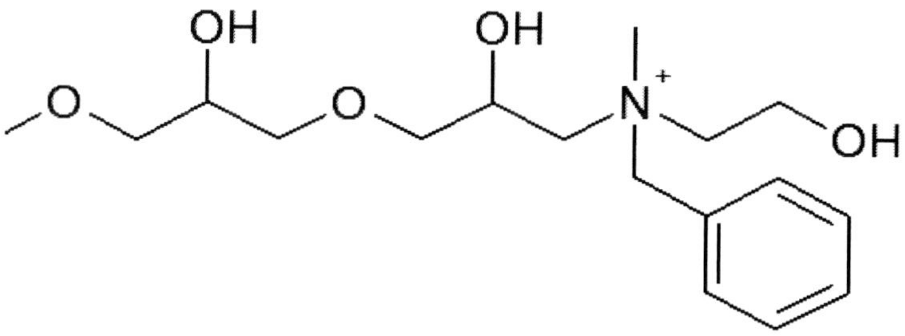

Figure 4.10 Molecular structure of the multimodal strong anion exchange ligand of Capto™ adhere: *N*-benzyl-*N*-methylethanolamine. In addition to the ion–ion interaction with the tertiary amino group, there is also a possibility to utilize hydrophobic interaction with the benzyl group.

4.3 The Use of Microspheres in Protein Refolding

Overexpression of recombinant proteins in bacterial systems, observed particularly when using *Escherichia coli* (*E. coli*) as host, often yields high concentrations of product in the form of insoluble aggregates, called inclusion bodies, composed of misfolded protein. Many methods have been developed to solubilize and refold recombinant proteins into their native structures. However, it is not possible to predetermine the best method to refold a given protein. However, this has to be optimized experimentally by a trial-and-error procedure and by using some general guidelines from the literature.

Before refolding is possible, the inclusion bodies are dissolved in denaturing solutions, such as 8 M urea or 6 M guanidine hydrochloride. The principle approach then is to remove the denaturants under circumstances that facilitate transformation to the native structure using an optimized refolding buffer individually designed for every protein. The critical problems are misfolding and aggregation, respectively. How to cope with these limitations in detail is out of the scope of this chapter and is discussed in specialized texts (60–64).

The use of microspheres in the refolding process is based on two different alternative approaches. SEC media are used only as a physical tool to transfer the denatured protein from the denaturant into the refolding buffer under controlled conditions and can be compared to an ultrafiltration cell used in diafiltration/dilution mode. Adsorption chromatography media, such as ion exchangers, hydrophobic interaction media and affinity media, on the other hand, are primarily used as a means to prevent aggregation allowing higher protein concentrations during the refolding process, but also in order to impose a control of the rate of transfer to the refolding buffer.

4.3.1 Refolding Using Size-Exclusion Microspheres

Protein refolding by SEC can be performed according to two different principles. In one of these, the denatured protein, dissolved in denaturant-containing buffer, is applied to a desalting type of SEC column, such as Sephadex G-25, which has been pre-equilibrated with a descending concentration gradient of the denaturant. When the sample solution containing the denatured protein migrates down the column, the polypeptide is confined to the column-void compartment whereas the denaturant can occupy the whole liquid volume of the column. The consequence is that the protein, during its migration down the column, will be surrounded by a continuously decreasing denaturant concentration. When leaving the column, the protein is expected to be entirely in the final refolding buffer and to have attained its native structure. By regulating the flow rate through the column, the rate of refolding can be controlled (65). The procedure is illustrated in Fig. 4.11.

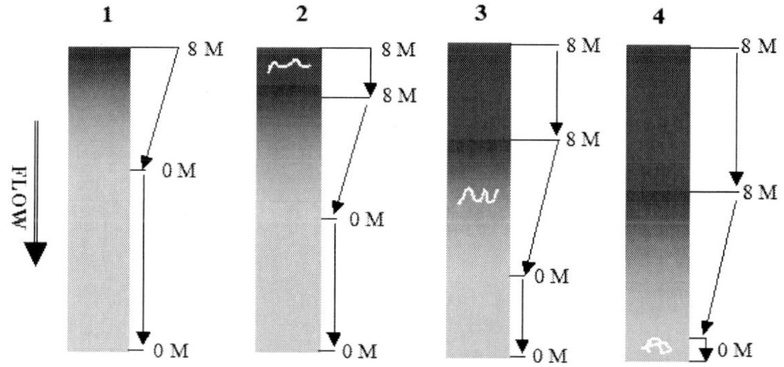

Figure 4.11 Flow scheme of gradient SEC for refolding. (1) The column is equilibrated in refolding buffer followed by a descending urea gradient. (2) Loading of the sample dissolved in 8 M urea. (3) During sample elution, the urea gradient moves downward as well. However, the protein moves faster than the gradient and enters areas with gradually lowering denaturant concentration. (4) At the end of the column, the protein has passed through the whole urea gradient and leaves the column refolded. The key design parameters for this refolding procedure are the shape of the denaturant concentration gradient and the flow rate, both affecting the kinetics of the refolding process (65).

An alternative way of protein refolding using SEC is by applying a gel media that is porous enough to allow the correctly folded protein to enter the internal network of the column bed particles. In doing so, the nonfolded or aggregated versions of the protein, by occupying a larger molecular volume, will move ahead of the target protein, causing its purification to occur in parallel to the refolding process. As an early example of this approach, Werner *et al.* (66) used a Superdex 75 HR 10/30 column equilibrated in a refolding buffer having the following composition: 20 mM HEPES, pH 6.8, containing 150 mM NaCl, 3.3 mM Na_2EDTA, and 0.1% Tween 20. The target proteins (1–10 mg) were dissolved in 1–2 mL of 50 mM Tris–HCl, pH 8.5, containing 50 mM dithiothreitol (DTT), 200–500 mM NaCl, and 6–8 M Gu–HCl and were directly loaded to the column and eluted, resulting in around 70% active protein. A similar approach was used by Batas and Chaudhuri (67) in which they applied a Sephacryl S-100 column loaded with up to 80 mg/mL protein. Average protein recoveries of around 60%, with specific activities in the range 80–100% were obtained for the proteins tested.

4.3.2 Solid-Phase (On-Column) Refolding

The most serious problem to solve during protein refolding is spontaneous aggregation caused by too high protein concentration. Traditionally, this problem has been partially compensated for by extensive dilution. However, in recent years solid-phase (on-column) protein refolding has become more and more popular. Adsorption of the denatured protein molecules onto a solid support, which separates them from each other during refolding, is an efficient means to avoid aggregation. In principle, all kinds of protein adsorbents can be used that allow binding in the presence of strong denaturants, such as 8 M urea or 6–8 M guanidine hydrochloride. Adsorption to ion exchangers is restricted to the use of the noncharged urea while hydrophobic interactions and biospecific affinity interactions can be applied in both denaturants. For a recent comprehensive review of the subject, the reader is referred to the text by Su *et al.* (64).

4.3.2.1 Refolding using binding to ion-exchange media

Successful attempts have been made to refold proteins after adsorption to ion-exchange media in the presence of 8 M urea. The principle is shown in Fig. 4.12. After adsorption of the denatured protein, the denaturant concentration in the mobile phase is gradually allowed to decrease and the protein starts to fold. When the protein is totally surrounded by the refolding buffer and is expected to have attained its correct, native conformation, it is eluted from the column by increasing the ionic strength of the mobile phase, either stepwise or gradually. A certain degree of purification can then be achieved as well.

Refolding by gradual decrease of the denaturant concentration can also be combined, in a dual process, with gradient elution of the refolded protein from the ion exchanger, either by an increase in ionic strength of the mobile phase or by gradual change of pH. The latter principle also facilitates the formation of correct disulfide bridges. In Fig. 4.13, a schematic representation of a dual-gradient ion-exchange chromatography refolding procedure for lysozyme with a descending urea-concentration gradient combined with an ascending pH gradient as suggested by Li *et al.* (68) is shown. In Fig. 4.14, the corresponding chromatogram obtained by applying this dual gradient principle for lysozyme is shown.

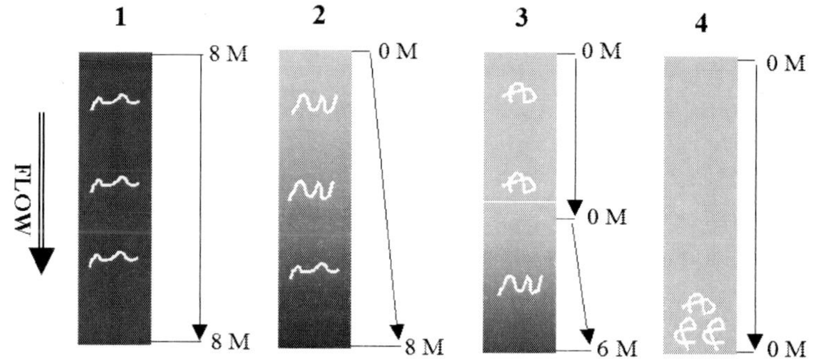

Figure 4.12 Schematics of a refolding process based on binding to an ion exchanger. (1) The denatured protein is adsorbed to the ion exchanger pre-equilibrated with denaturation buffer containing 8 M urea. (2) The protein refolding is induced by a decrease in urea concentration. (3) The still adsorbed protein continues to refold as the gradient of urea concentration decreases as the refolding buffer moves down the column. (4) The refolded proteins are eluted from the column as a consequence of the addition of a high ionic strength buffer (68).

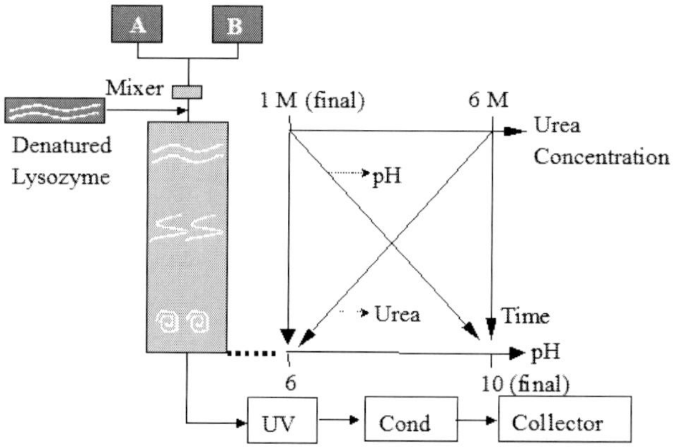

Figure 4.13 Schematic representation of a dual-gradient ion-exchange chromatography refolding procedure for lysozyme with a descending urea concentration gradient combined with an ascending pH gradient. The protein is adsorbed to the ion exchanger in the presence of buffer at pH 6 containing 6 M urea. Buffer at pH 10 containing 1 M urea is gradually introduced into the column inducing protein refolding and desorption from the ion exchanger simultaneously. Adapted from Ref. 68.

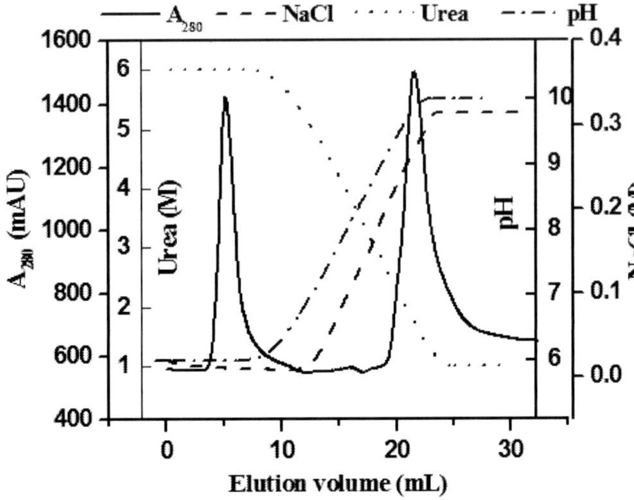

Figure 4.14 Example of a dual gradient ion-exchange refolding procedure. Lysozyme dissolved in denaturing buffer at pH 6.2 containing 6 M urea is adsorbed to a HiTrap™ SP Sepharose™ HP column equilibrated in the same buffer. Refolding buffer at pH 10 containing 1 M urea is allowed to gradually replace the denaturing buffer. Refolding of the adsorbed protein occurs as a consequence of the decrease in urea concentration and an increase in pH. At some point in the gradient, the protein is desorbed from the ion exchanger as a consequence of the increase in salt concentration. The left peak in the chromatogram represents the absorbance of DTT, and the right peak represents the absorbance of the refolded lysozyme. Adapted from Ref. 68.

4.3.2.2 Refolding using binding to hydrophobic media

When using hydrophobic interaction media for solid-phase refolding, there is always a risk that the binding of the denatured protein will occupy residues that are necessary for a correct polypeptide folding. Nevertheless, there are a few successful examples worth mentioning. Thus, hydrophobic interaction binding was used to refold recombinant human interferon in one step. The refolding yield was twice as high as that obtained using dilution or dialysis (69). Also reversed-phase high-performance liquid chromatography has been used as a refolding tool. Thus, recombinant human interleukin-2, expressed as inclusion bodies in *E. coli*, resulted in increases in recovery and specific activity of 9 and 14 fold (70).

Binding by hydrophobic interaction in the presence of glycerol has been used to refold lysozyme at high protein concentration (71). A very mild hydrophobic ligand of polyethylene glycol immobilized onto agarose microspheres was used successfully for refolding of the recombinant elongation factor G from *Staphylococcus aureus* (72).

4.3.2.3 Refolding using binding to affinity media

An alternative solid-phase (on-column) refolding technique is applicable when the proteins are expressed as histidine tail fusions (His-tagged). The technique is based on the fact that the histidine tail–fused protein binds to the chelated copper, nickel, or zinc ion in the presence of high concentrations of the denaturants urea and guanidine hydrochloride. In the first published application, *N*-terminally His-tagged prion proteins were immobilized on nickel-NTA agarose microspheres (73). Following refolding and oxidation, the proteins were depleted of their histidine tail using thrombin without cleaving the prion protein polypeptide chain.

In another example Rogl *et al.* (74) refolded and purified His-tagged recombinant chloroplast membrane proteins and solubilized it using 8 M urea on a Ni^{2+}-saturated chelating Sepharose Fast Flow column. After washing for several volumes with refolding buffer containing 10 mM imidazole in order to displace loosely bound protein, the membrane proteins were finally eluted in one step using 0.3 M imidazole in the same buffer. Noncharged detergent Triton X-100 (0.1%) was added to the elution buffer in order to keep the membrane proteins in solution.

4.3.2.4 Refolding using binding to a combination of media

Solid-phase (on-column) refolding can also be performed using a combination of different adsorptive interactions. One such example is refolding of recombinant nonstructural protein 3 (NS3) a potential target protein for hepatitis C virus therapy (74). Because it has a histidine tag, partially purification and refolding were initially performed after adsorption to a Ni^{2+}-saturated Chelating Sepharose Fast Flow column in the presence of 8 M urea. This step separated the denatured NS3 from other impurities present in the inclusion body. It also refolded the protein to its intermediate state. A further refolding and purification were followed either after adsorption to DEAE Sepharose Fast Flow in the presence of 8 M urea or by SEC

(Superdex 200 HR 10/30 or Superdex 75 HR 10/30 prepacked columns). Comparing the results from the second on-column refolding step demonstrated that SEC was the best choice. Thus, the latter combination gave a total activity recovery of around as much as 90% (75).

4.4 Conclusion and Prospects

Microspheres have had a profound impact on the development of biotechnology because of their importance for the manufacturing of chromatographic separation media for both low- and high-molecular weight compounds. For the development of chromatographic media for proteins, microspheres based on polysaccharides, such as cellulose, dextran, and agarose have been particularly important. Because of their inherent hydrophilicity, combined with neutral inertness and ease of derivatization, microspheres based on rigid, covalently cross-linked agarose have been the dominating preparative tool for both laboratory and industrial scale protein separation and purification during the past 25 years. Recent developments in new techniques for emulsification and cross-linking of agarose microspheres will most probably allow them to maintain this position during a foreseeable future.

References

1. Turba, F. (1954). *Chromatographische Metoden in der Protein-Chemie* (pp. 1–358). Berlin, Göttingen, Heidelberg: Springer-Verlag.
2. Paleus, S., & Neilands, J. B. (1950). Preparation of cytochrome C with the aid of ion exchange resin. *Acta Chemica Scandinavica*, **4**, 1024–1030.
3. Hirs, C. H. W., Stein, W. H., & Moore, S. (1951). Chromatography of proteins. Ribonuclease. *Journal of the American Chemical Society*, **73**, 1893.
4. Boardman, N. K., & Partridge, S. M. (1955). Separation of neutral proteins on ion exchange resins. *Biochemical Journal*, **59**, 543–552.
5. Ugelstad, J., Söderberg, L., Berge, A., & Bergström, J. (1983). Monodisperse particles — a step forward for chromatography. *Nature*, **303**, 95–96.

6. Sober, H. A., & Peterson, E. A. (1954). Chromatography of proteins on cellulose ion exchangers. *Journal of the American Chemical Society*, **76**, 1711–1712.
7. Peterson, E. A., & Sober, H. A. (1956). Chromatography of proteins. I. Cellulose ion-exchange adsorbents. *Journal of the American Chemical Society*, **78**, 751–755.
8. Tiselius, A., Hjertén, S., & Levin, Ö. (1956). Protein chromatography on calcium phosphate columns. *Archives of Biochemistry and Biophysics*, **65**, 132–155.
9. Determann, H., Meyer, N., & Wieland, T. (1969). Ion exchanger from pearl-shaped cellulose gel. *Nature*, **223**, 499–500.
10. Motozato, Y., & Hirayama, C. (1984). Preparation and properties of cellulose spherical particles and their ion exchangers. *Journal of Chromatography*, **298**, 499–507.
11. Porath, J., & Flodin, P. (1959). Gel filtration, a method for desalting and group separation. *Nature*, **183**, 1657–1659.
12. Janson, J. C. (1987). On the history of the development of Sephadex®. *Chromatographia*, **23**, 361–369.
13. Flodin, P. (1998). The Sephadex® story. *Polymer Engineering and Science*, **38**, 1220–1228.
14. Polson, A. (1961). Fractionation of protein mixtures on columns of granulated agar. *Biochimica et Biophysica Acta*, **50**, 565–567.
15. Hjertén, S., & Mosbach, R. (1962). Molecular-sieve chromatography of proteins on columns of cross-linked polyacrylamide. *Analytical Biochemistry*, **3**, 109–118.
16. Hjertén, S. (1964). The preparation of agarose spheres for chromatography of molecules and particles. *Biochimica et Biophysica Acta*, **79**, 393–398.
17. Haller, W. (1965). Chromatography on glass of controlled pore size. *Nature*, **206**, 693–696.
18. Unger, K. K., Kern, R., Minou, M. C., & Krebs, K. F. (1974). High performance gel permeation chromatography with a new type of silica packing material. *Journal of Chromatography*, **99**, 435–443.
19. Regnier, F. E., & Noel, R. J. (1976). Glycerolpropylsilane bonded phases in the steric exclusion chromatography of biological macromolecules. *Journal of Chromatography Science*, **14**, 316–320.
20. Hashimoto, T., Sasaki, M., Aiura, M., & Kato, Y. (1978). High-speed aqueous gel-permeation chromatography. *Journal of Polymer Science, Part A: Polymer Physics*, **16**, 1789–1801.

21. Flodin, P., Johansson, O., & Carlsson, S. (1966). U.S. Patent No. 3,275,576 and U.S. Patent No., 3,277,025. Washington, DC: Patent and Trademark Office.

22. Hagel, L., Lundström, H., Andersson, T., & Lindblom, H. (1989). Properties, in theory and practice, of novel gel filtration media for standard liquid chromatography. *Journal of Chromatography*, **476**, 329–344.

23. Kågedal, L., Engström, B., Ellegren, H., Lieber, A. K., Lundström, H., Sköld, A., & Schenning, M. (1991). Chemical, physical and chromatographic properties of Superdex 75 prep grade and Superdex 200 prep grade gelfiltration media. *Journal of Chromatography*, **537**, 17–32.

24. Araki, C. (1956). Structure of the agarose constituent of agar–agar. *Bulletin of the Chemical Society of Japan*, **29**, 543–544.

25. Duckworth, M., & Yaphe, W. (1971). The structure of agar. Part I. Fractionation of a complex mixture of polysaccharides. *Carbohydrate Research*, **16**, 189–197.

26. Ackers, G. K., & Steere, R. L. (1962). Restricted diffusion of macromolecules through agar-gel membranes. *Biochimica et Biophysica Acta*, **59**, 137–149.

27. Hjertén, S. (1962). Chromatographic separation according to size of macromolecules and cell particles on columns of agarose suspensions. *Archives of Biochemistry and Biophysics*, **99**, 466–475.

28. Russel, B., Mead, T. H., & Polson, A. (1964). A method for preparing agarose. *Biochimica et Biophysica Acta*, **86**, 169–174.

29. Hjertén, S. (1971). Some new methods for the preparation of agarose. *Journal of Chromatography*, **61**, 73–80.

30. Låås, T. (1972). Charge-free agar. *Journal of Chromatography*, **66**, 347–355.

31. Bengtsson, S., & Philipson, L. (1964). Chromatography of animal viruses on pearl-condensed agar. *Biochimica et Biophysica Acta*, **79**, 399–406.

32. Mu, Y., Lyddiatt, A., & Pacek, A. W. (2005). Manufacture by water/oil emulsification of porous agarose beads: Effect of processing conditions on mean particle size, size distribution and mechanical properties. *Chemical Engineering and Processing*, **44**, 1157–1166.

33. Zhou, Q. Z., Wang, L. Y., Ma, G. H., & Su, Z. G. (2007). Preparation of uniform-sized agarose beads by microporous membrane emulsification technique. *Journal of Colloid and Interface Science*, **311**, 118–127.

34. Arnott, S., Fulmer, A., Scott, W. E., Dea, I. C. M., Moorhouse, R., & Rees, D. A. (1974). The agarose double helix and its function in agarose gel structure. *Journal of Molecular Biology*, **90**, 269–284.

35. Porath, J., Janson, J. C., & Låås, T. (1971). Agar derivatives for chromatography, electrophoresis and gel bound enzymes. I. Desulphated and reduced cross-linked agar and agarose in spherical bead form. *Journal of Chromatography*, **60**, 167–177.

36. Porath, J., & Janson, J. C. (1976). Stabilized agar product and method for its stabilization, U.S. Patent No. 3,959,251. Washington, DC: Patent and Trademark Office.

37. Porath, J., Låås, T., & Janson, J. C. (1975). Rigid agarose gels cross-linked with divinyl sulphone (DVS). *Journal of Chromatography*, **103**, 49–62.

38. Laurent, T. C. (1967). Determination of the structure of agarose gels by gel chromatography. *Biochimica et Biophysica Acta*, **136**, 199–206.

39. Axelsson, I. (1978). Characterization of proteins and other macromolecules by agarose gel chromatography. *Journal of Chromatography*, **152**, 21–32.

40. Janson, J. C. (1969). Increased gel rigidity. Unpublished observation.

41. Berg, H., Brännholm, M., Buckley, D., Busson, P., Hagvall, A., Holmgren, E., Ihre, H., Larsson, A., & Lindström, D. (2008). Method of preparing a separation matrix, U.S. Patent Application Publication No. US 2008/0154031 Al. Washington, DC: Patent and Trademark Office.

42. Andersson, T., Carlsson, M., Hagel, L., Pernemalm, P.-Å., & Janson, J. C. (1985). Agarose-based media for high-resolution gel filtration of biopolymers. *Journal of Chromatography*, **326**, 33–44.

43. Pernemalm, P. Å., Carlsson, M., & Lindgren, G. (1987). Polysaccharide cross-linked separation material and its preparation, U.S. Patent No. 4,665,164. Washington, DC: Patent and Trademark Office.

44. Berg, H. (2003). Process for the production of a porous cross-linked polysaccharide gel and its use as a gel filtration media and in chromatography. U.S. Patent No. 6,602,990 Bl. Washington, DC: Patent and Trademark Office.

45. Nilsson, K., Buzsaky, F., & Mosbach, K. (1986). Growth of anchorage-dependent cells on macroporous microcarriers. *Nature Biotechnology*, **4**, 989–990.

46. Gustavsson, P. E., & Larsson, P.-O. (1996). Superporous agarose, a new material for chromatography. *Journal of Chromatography A*, **734**, 231–240.

47. Larsson, P. O. (1998). Superporous polysaccharide gels, U.S. Patent No. 5,723,601. Washington, DC: Patent and Trademark Office.
48. Afeyan, N. B., Gordon, N. F., Mazsaroff, I., Varady, L., Fulton, S. P., Yang, Y. B., & Regnier, F. E. (1990). Flow-through particles for the high-performance liquid chromatographic separation of biomolecules: Perfusion chromatography. *Journal of Chromatography*, **519**, 1–29.
49. Unger, K. K., Anspach, B., & Giesche, H. (1984). Optimum support properties for protein separations by high-performance size exclusion chromatography. *Journal of Pharmaceutical and Biomedical Analysis*, **2**, 139–151.
50. Hagel, L., Östberg, M., & Andersson, T. (1996). Apparent pore size distributions of chromatography media. *Journal of Chromatography A*, **743**, 33–42.
51. Hagel, L. (1988). Pore size distributions. In P. L. Dubin (Ed.), *Aqueous size exclusion chromatography* (pp. 119–155). Amsterdam: Elsevier.
52. Janson, J. C., & Jönsson, J. Å. (2011). Introduction to chromatography. In J. C. Janson (Ed.), *Protein purification: principles, high resolution methods and applications* (pp. 25–50). New York: Wiley.
53. Cuatrecasas, P., Wilchek, M., & Anfinsen, C. B. (1968). Selective enzyme purification by affinity chromatography. *Proceedings of the National Academy of Sciences of the United States of America*, **61**, 636–643.
54. Axén, R., Porath, J., & Ernback, S. (1967). Chemical coupling of peptides and proteins to polysaccharides by means of cyanogen halides. *Nature*, **214**, 1302–1303.
55. Johansson, B. L., Belew, M., Eriksson, S., Glad, G., Lind, O., Maloisel, J. L., & Norrman, N. (2003). Preparation and characterization of prototypes for multi-modal separation media aimed at capturing negatively charged biomolecules at high salt conditions. *Journal of Chromatography A*, **1016**, 21–33.
56. Johansson, B. L., Belew, M., Eriksson, S., Glad, G., Lind, O., Maloisel, J. L., & Norrman, N. (2003). Preparation and characterization of prototypes for multi-modal separation media aimed at capturing positively charged biomolecules at high salt conditions. *Journal of Chromatography A*, **1016**, 35–49.
57. Carredano, E., Johansson, B. L., & Maloisel, J.-L. (2010). Multi-modal ion exchange chromatography resins, U.S. Patent Application Publication, No. US2010/0009867 A1. Washington, DC: Patent and Trademark Office.

58. Belew, M., Li, M. Y., Zhang, W., & Caldwell, K. (2008). Purification of recombinant human serum albumin (rHSA) produced by genetically modified Pichia pastoris. *Separation Science and Technology*, **43**, 3134–3153.

59. Eriksson, K., Ljunglöf, A., Rodrigo, G., & Brekkan, E. (2009). MAb contaminant removal with a multimodal anion exchanger, A platform step to follow Protein A. *BioProcess International*, **7**, 52–56.

60. Clark, E. D. (2001). Protein refolding for industrial processes. *Current Opinion in Biotechnol*, **12**, 202–207.

61. Li, M., Su, Z. G., & Janson, J. C. (2004). *In vitro* protein refolding by chromatographic procedures. *Protein Expression and Purification*, **33**, 1–10.

62. Chow, M. K. M., Abdullah A., Kate, A. A., Fulton, F., Fernando, T., Batty, L. C., ... Buckle, A. M. (2006). The REFOLD database: A tool for the optimization of protein expression and refolding. *Nucleic Acids Research*, **34** (Database issue), D207–D212.

63. Jungbauer, A., & Kaar, W. (2007). Current status of technical protein refolding. *Journal of Biotechnology*, **128**, 587–596.

64. Su, Z. G., Lu, D., & Liu, Z. (2011). Refolding of inclusion body proteins from *E. coli*. In J. C. Janson (Ed.), *Protein Purification: Principles, High Resolution Methods and Applications* (pp. 319–338). New York: Wiley.

65. Gu, Z., Su, Z. G., & Janson, J. C. (2001). Urea gradient size-exclusion chromatography enhanced the yield of lysozyme refolding. *Journal of Chromatography A*, **918**, 311–318.

66. Werner, M. H., Clore, G. M., Gronenborn, A. M., Kondoh, A., & Fisher, R. J. (1994). Refolding proteins by gel filtration chromatography. *FEBS Letters*, **345**, 125–130.

67. Batas, B., & Chaudhuri, J. B. (1996). Protein refolding at high concentration using size-exclusion chromatography. *Biotechnology and Bioengineering*, **50**, 16–23.

68. Li, M., Zhang, G. F., & Su, Z. G. (2002). Dual gradient ion-exchange chromatography improved refolding yield of lysozyme. *Journal of Chromatography A*, **959**, 113–120.

69. Guo, L. (2001). Simultaneous purification and renaturation of recombinant human interferon alpha expressed by *E. coli* by highperformance hydrophobic interaction chromatography. *Chinese Journal of Chromatography*, **19**, 301–303.

70. Ling, M., Xu, X., Shi, F., Zhu, Y., & Long, N. (1997). Refolding of recombinant human interleukin-2 by reverse phase high performance liquid chromatography. *Chinese Journal of Biotechnology*, **13**, 180–183.

71. Wang, F., Liu, Y., Ma, G. H., & Su, Z. G. (2009). Glycerol-assisted hydrophobic interaction chromatography improving refolding of recombinant human granulocyte colony-stimulating factor. *Applied Biochemistry and Biotechnology*, **159**, 634–641.

72. Li, J. J., Venkataramana, M., Wang, A. Q., Sanyal, S., Janson, J. C., & Su, Z. G. (2005). A mild hydrophobic interaction chromatography involving polyethylene glycol immobilized to agarose media refolding recombinant Staphylococcus aureus elongation factor G. *Protein Expression and Purification*, **40**, 327–335.

73. Zahn, R., von Schroetter, C., & Wüthrich, K. (1997). Human prion proteins expressed in Escherichia coli and purified by high-affinity column refolding. *FEBS Letters*, **417**, 400–404.

74. Rogl, H., Kosemund, K., Kühlbrandt, W., & Collinson, I. (1998). Refolding of Escherichia coli produced membrane protein inclusion bodies immobilised by nickel chelating chromatography. *FEBS Letters*, **32**, 21–26.

75. Li, M., Poliakov, A., Danielson, U. H., Su, Z. G., & Janson, J. C. (2003). Refolding of a recombinant full-length non-structural (NS3) protein from hepatitis C virus by chromatographic procedures. *Biotechnology Letters*, **25**, 1729–1734.

Chapter 5

Microspheres for Separation of Bioactive Small Molecules

Fangling Gong, Guifeng Zhang, and Zhiguo Su

State Key Laboratory of Biochemical Engineering, Institute of Process Engineering, Chinese Academy of Sciences, Beijing, P. R. China

zgsu@home.ipe.ac.cn

5.1 Introduction

Liquid chromatography (LC) is one of the most effective separation techniques to separate a mixture of compounds and has been widely used to identify, quantify or purify the individual components of the mixture for analytical or preparative purpose. During chromatographic separation, analytes are separated based on the difference in their interactions with the stationary phase, the ratio/composition of the solvent used, and the flow rate of the mobile phase. Differential interaction of analytes with mobile and stationary phase results in the different retention behaviors for various compounds and leads to their selectivity (Table 5.1). Thus, the stationary phase plays an important role in achieving highly efficient separation. Both amorphous particles and microspheres can be used as stationary phase. Microspheres are the most widely used substances because of their many distinct advantages. According to the interaction between an analyte and the stationary phase, LC can be categorized into different types as listed in Table 5.1. Proper choice of LC and an eluting solvent are crucial for the successful separation of a target compound from a complex mixture (1).

Microspheres and Microcapsules in Biotechnology: Design, Preparation, and Applications
Edited by Guanghui Ma and Zhiguo Su

ISBN 978-981-4316-47-7 (Hardcover), 978-981-4364-62-1 (eBook)
www.panstanford.com

Table 5.1 Liquid chromatographic method and applications

Method	Mechanism	Application
Reversed phase (RPC)	Polarity	++++
Normal phase (NP)	Polarity	+
Hydrophobic interaction (HIC)	Dispersive interactions Salt gradients	+
Hydrophilic interaction (HILIC)	Hydrophilicity	+++
Size exclusion (SEC)	Molecular size (hydrodynamic diameter)	+++
Ion exchange (IEC)	Electrostatic interaction	++++
Affinity (AC)	Selective non-covalent binding	++
Metal chelate affinity (MCAC)	Specific binding with metal ion	+

Note: +, the range of application areas.

Bioactive small molecules (BSM) are generally referred as low-molecular weight compounds with bioactivities and may range from simple compounds to small peptides, including, but not limited to, metabolites, peptides, nucleotides and their derivatives, bioactive natural products, and their analogues (2). BSM have been widely used in many fields, which has resulted in their increased demand in recent years (3). In most cases, BSM from different resources exist in complex mixtures of different compounds and have to be separated from a very large number of contaminants. High purity is of great importance for BSM to be used for injection, chemical reagents, especially for investigation on various aspects of BSM interactions to probe specific biochemical processes. Effective separation of BSM is an essential process for characterization and analysis, product development, and drug discovery, etc. However, many BSM can now be easily purified as they generally show multiple forms, diverse effects, and complicated interaction between them (4).

Generally, column chromatography is the commonly used method for purifying BSM from various resources and can also be used for target compound identification and quantitative analysis. However, many BSM cannot be easily and fully separated using the common chromatographic column because the target molecules and their derivatives or analogues show high similarity in physicochemical properties, such as hydrophilicity/hydrophobicity, polarity, molecular weight, stability, and solubility. This presents a new challenge to a chemical or biochemical engineer for developing more efficient chromatographic media or separation techniques. In recent years, many novel microspheres have been designed and used as stationary phases that give improved resolution, simplicity, speed, ease of scale-up, and possibility of continuous operation. Some "smart" microspheres with multi-functional properties have been also designed. In this chapter, recent advances in design and application of novel microspheres for chromatographic separation of BSM, especially peptides and natural products, is briefly reviewed.

5.2 Chromatographic Stationary Phase

5.2.1 The Requirements of Microspheres to Be Used as Stationary Phases

The stationary phase packed in the column plays a key role for chromatographic separation and through which the mobile phase flows, distributing the analyte between the mobile phase and the stationary phase. According to the chemical composition, the stationary phases can generally be grouped into inorganic types and families of cross-linked organic polymers. Most classical stationary phases are inorganic types including porous glass and porous oxides, such as silica, alumna, and magnesia, and they may possess a crystalline and amorphous bulk structure (5). Organic types include some porous or nonporous polymer microspheres, polymer-coated microspheres, polymer-grafted microspheres, and others. Theoretically, microspheres to be used as stationary phase should posses the following characteristics:

(i) Mechanical strength: Column packing process is normally carried out at high pressure, increasing from 20 to 200 bar according to the column design and dimensions. The mechanical strength should keep the microsphere's sphericity and should not produce crashed debris during the packing process and the chromatographic separation.

(ii) Chemical stability: The microsphere to be used as stationary phase should not only remain chemically stable and insoluble under alkaline or acidic environments and in organic solvents, but also remain chemically inert to avoid any reactions with the analytes or affect the characteristics, such as molecular structure, chirality and bioactivity, of the analytes in any way.

(iii) Thermal stability: Polymer-based microspheres should exhibit a high thermal stability so that they can be used at elevated temperatures without swelling and shrinkage.

(iv) Biological stability: The microsphere should be resistant to micro-organisms or enzymes and should remain dynamically stable for samples such as cell lysates, tissue extracts, or body fluids.

(v) Controllable pore size and its distribution: The controllable pore sizes and their distribution in porous microspheres is the unique feature, which makes this stationary phase different from inorganic stationary phase. Pore size and its distribution is crucial for chromatographic separation, especially for gel filtration chromatography, which is based on the difference between molecular sizes.

(vi) Selective adsorption: For adsorption chromatography, the microsphere should show high target compound adsorption and low or no impurity adsorption. An ideal microsphere should show different adsorption properties for different compounds in a multi-component mixture with extreme complexity.

(vii) Column capacity: Generally, column capacity is related to the density of adsorbed layer or quantity of adsorbed compound per gram of microsphere, which can be increased by changing the specific surface area or porosity, the pore size, and the distribution of the pores on the microspheres (6).

In fact, it is difficult to obtain microspheres possessing all properties mentioned above. For example, the main limitation of inorganic microsphere-based stationary phases is their low chemical stability. The main limitations of polymer-based microsphere stationary phases are their low mechanical strength and swelling properties. In recent years, great efforts have been made to improve the properties of microsphere to be used as stationary phases.

5.2.2 Inorganic Stationary Phases

In liquid chromatography, silica-based stationary phases are the most popular inorganic stationary phases because of their favorable physical characteristics, such as high mechanical strength, appropriate surface area, and narrow mesopore size distributions (7). Multi-component mixtures with extreme complexity often require rapid separation techniques with increased resolution for preparative or analytical purpose. To meet the ever increasing demand for high efficiency and rapid chromatographic separations, novel micro- and nanoseparation strategies, two- or more-dimensional separation, optimization in column packing, development of columns packed with sub-2 μm particle sizes, and variation in temperature have been carried out and the resolution has been improved to some extent. However, most of these improvements are only limited to the resolution and require extensive time for analysis, ultra-high pressure, or sophisticated instrumentation (8). Moreover, most of these trials are carried out using classical stationary phases, since silica accounts for about 90% of the chromatographic supports, especially in high-performance liquid chromatography (HPLC). Although there are various columns available in the market, many stationary phases are nominally identical if they belong to the same chemical classification, such as hydrocarbon-bonded silicas. The differences in packing material are limited to the range of ligand type, end-capping, type of silica, residual silanols, density of bonded groups, pore size, and so on (9). Silica gel has several limitations including low thermal and pH stability and peak tailing, especially for the separation of basic compounds, as listed in Table 5.2. For acidic compounds, the separation can also be affected by the residual silanol due to electrostatic exclusion phenomena and hydrogen bond, since the coverage of silanol groups in chemically bonded phases is less than 60% due to steric hindrance and other effects. The restricted pH range (2–8) is also a limiting factor for separating basic or acidic compounds (10).

Table 5.2 Comparison of chemical and thermal stability of substrates and reversed-phase stationary phases

Phase	Chemical stability		Thermal stability
	Low pH	Intermediate and high pH	
Silica substrate	+	–	++
RP covalently bonded phases	+++	+++	++
Polymer-coated	+	+	+
Zirconized or titanium-grafted and polymer-coated	++	++	++
Polymer	++++	++++	+++
RP modified phases	++++	++++	+++
Alumina substrate	+++	+++	?
RP covalently bonded phases	+	+	?
Polymer-coated	?	?	?
Zirconia substrate	++++	++++	++++
RP covalently bonded phases	+	+	+
Polymer-coated	++++	++++	++++

Note: +, the range of chemical or thermal stability of microspheres used as stationary phase.

Source: Reprinted from Ref. 11, with permission from Elsevier BV.

5.2.3 Polymer-Based Stationary Phases

To overcome the drawbacks of silica-based bonded LC columns for separating acidic and basic compounds, the polymer-based microspheres, which include polymer-coated microspheres and synthetic polymer microspheres, can be used as a substitute for the packing material (stationary phase). Polymer-coated microspheres usually consist of an inorganic substrate on which a polymer layer is deposited, and they provide a large variety of chromatographic properties combined with the increased chemical stability. The reduction of unwanted residual substrate groups, such as silanols, may improve the pH stability and reproducibility, especially for separating basic compounds. Polymers that are used for coating inorganic microspheres include polymers with hydrocarbon backbone, polyethers, polysaccharides, polyamines, polynucleotides, polyamides, polypeptides, polysiloxanes, and so on. The other polymer-based stationary phases are the synthetic polymer microspheres, including highly cross-linked polymer or copolymer microspheres. The microspheres are composed of a network of much smaller particles, providing a polymer phase with a continuous pore structure and resulting in higher specific surface area and increased mechanical strength. The pore size and pore morphology of microsphere can be controlled through particle preparation technology. Polymers that have been used for microsphere synthesis include poly(styrene-divinylbenzene) (PS-DVB), polyvinylalcohol (PVA), polyvinylesters (PVE), poly(alkyl methacrylate-glycerol dimethacrylate), poly(ethylene-*co*-acrylic acid) (PEA), etc.

Compared with crystalline and amorphous inorganic particles, polymer-based microsphere and polymer-coated microspheres show many advantages including high specific surface area, controllable pore size and pore morphology, chemical stability, and high column capacity. Many polymer-based and polymer-coated microspheres have been successfully applied in purification of BSM, such as peptides, nucleotides, organic acids, isoflavones, and synthetic compounds (11). A brief comparison of different microspheres is listed in Table 5.2.

5.3 Application of Microspheres in Separation of BSM

Separation of BSM has become increasingly important in recent years due to the rapidly growing importance of BSM and their wide applications. The applications of peptides as therapeutic reagents, diagnostic markers, drug transporters, diagnostic reagents, bioactive research compounds, structural probes, modifier, and analytical standards are closely related to their high purity, which can be achieved by high performance separation. Selective separation of low-abundance peptide is of great importance for diagnosis when the peptide is used as a marker in a complex mixture. Effective separations of the target peptides are constantly demanded since the synthesis provides not only the target peptide but also impurities (12, 13). In proteomics, the analysis of a proteome of an organism, tissue, or a single cell is often associated with peptide separation, which has been a challenge to researchers for many years. In many cases, the physico-chemical similarity of peptides (charge, isoelectric point, hydrophobicity, size) makes it difficult to separate them fully. To increase the resolution, column capacity, and column stability, many novel microspheres have been designed as stationary phase and some have been successfully applied for large-scale separation. Bioactive natural products and/or their analogues are another kind of BSM. With the development of Traditional Chinese Medicine, the effective components of the bioactive natural products and their analogues have been extensively used for curing diseases, such as heart disease, cerebrovascular disease, and others. Purification of these effective components is also an important key technology.

Polymeric microspheres have been used for the separation of bioactive natural products in recent years. But these original materials are semi-rigid porous particles and hence have limited mechanical stability (14). This necessitates their use in a large particle size format under low-pressure and low-flow conditions. Their large particle size and inherent low performance make them unsuitable for high pressure liquid chromatography to achieve highly purified products. To overcome this limitation, some polymer-coated microspheres based on silica have been developed and are being used for high-performance separations of bioactive natural products. However, such silica-based polymer-coated microspheres have limited chemical stability which can reduce the lifetime of the column packing and the pH range at which the purification can be performed. To address these issues, cross-linked polymer or copolymer microspheres have been developed for analytical separations. Much work has been carried in the area of polymer chemistry with a view to design polymeric media (stationary phase), which can be used for HPLC.

5.3.1 Hydrophobic Microspheres Used for BSM Separation

Among the cross-linked copolymer microspheres, PS-DVB microspheres have some excellent properties, such as high mechanical strength, chemical stability, and hydrophobic property, and have become one of the most popular polymer-based particles for reversed-phase HPLC separation of biomolecules, which has gained acceptance because of the use of the PS-DVB microspheres in the analysis of bioactive natural products. The performance of the PS-DVB microspheres used as stationary phase depends on their size, distribution, pore size, and specific surface area. A number of commercial large-scale macroporous PS-DVB products for process chromatography have been introduced, but they have been restricted to large inefficient particles with a broad particle-size distribution. However, they have demonstrated the suitability of this type of PS-DVB microspheres for preparative and process scale purification of synthetic peptides and bioactive natural products. Thus there is a need for a range of highly efficient PS-DVB microspheres, in which the particle size, the pore size, and the distribution of the particles, and the specific surface area of the particles have been optimized for the purification of BSM.

Porous uniform PS-DVB microspheres of several tens micrometer, which is a size between the sizes of particles used in HPLC and in low performance liquid chromatography (LPLC), can be used as reversed-phase chromatographic separation media. The mentioned separation system has similar separation efficiency to HPLC and lower work pressure than HPLC, which enable us to obtain the pure product more easily. Therefore, it is highly expected to prepare porous uniform PS-DVB microspheres with several tens micrometer as chromatographic separation media.

Ma *et al.* have prepared uniform PST, PS-DVB, and PDVB microspheres that can be used as LC media to obtain high separation efficiency (15). They used a specific technique, membrane-emulsification technique, to obtain uniform monomer droplets, which were polymerized further to form uniform microspheres. The effect of the preparation condition on the stability of the droplets/microspheres has been studied. It is necessary to select the appropriate emulsifier or/and stabilizer during the solidification of the polymer particles. In their study, sodium dodecyl sulfate (SDS) and poly(vinyl alcohol) (PVA) were used as emulsifiers or/and stabilizers and a series of uniform-sized porous PS-DVB microspheres with large specific surface area were prepared. One of prepared uniform PS-DVB microspheres with microporous structure is shown in Fig. 5.1.

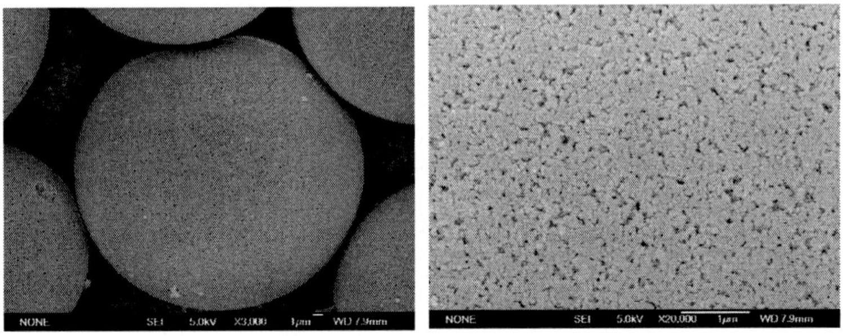

Figure 5.1 Uniform PS-DVB microspheres with microporous structure.

Sun *et al.* (16) used the synthesized uniform microspheres as a hydrophobic packing media to separate natural bioactive product Icariin. During the experiment, a comparison was done of the column chromatographic performance of the PS-DVB microspheres and a commercial silica-based C18. Figure 5.2 showed the influence of the mobile phase composition on the separation and recovery of Icariin. For the C18 column, best result was obtained when 50% methanol was used for washing and 70% methanol was used for eluting. Under this condition, the purity of Icariin was 63% with a recovery of 50%. While for the PS-DVB column, the best mobile phase composition was 55% methanol washing and 60% methanol eluting, and the obtained purity of Icariin was 92% with a recovery of 99.9%. The results demonstrated that PS-DVB microsphere media was superior to the C18 media for obtaining high product purity and recovery. Figure 5.3 shows the profiles of chromatographic separation of Icariin by the PS-DVB and C18 columns under their optimized mobile phase conditions. The peak marked 1 indicates the Icariin fraction in both figures. Results showed that the PS-DVB column produced a better separation in a shorter time and under lower operating pressure than that for the C18 column. Sun *et al.* (17) also used the uniform PS-DVB microspheres as a RPC packing media to separate natural bioactive product Paclitaxel. Under the optimized operation conditions, Paclitaxel was purified simply and efficiently. The purity obtained ranged from 0.46% to more than 98% with a total recovery of more than 71%. The results further confirmed the feasibility and advantage of these uniform porous PS-DVB microspheres as chromatographic separation media. Qin *et al.* (18) studied the production of PS-DVB microspheres on a large scale. Some important influential factors related to the size and morphology of the microspheres, such as the shear force, pipeline linear velocity, and pressure of membrane emulsification, were investigated. Using the optimized technique, PS-DVB microspheres with the coefficient of variation (CV) of diameters being close to 10% were successfully prepared. The size of the polymeric microspheres was controlled between 10 μm and 100 μm and the specific surface area was tunable between 100 m^2/g and 500 m^2/g. Figure 5.4 shows three kinds of prepared PS-DVB microspheres with different particle sizes.

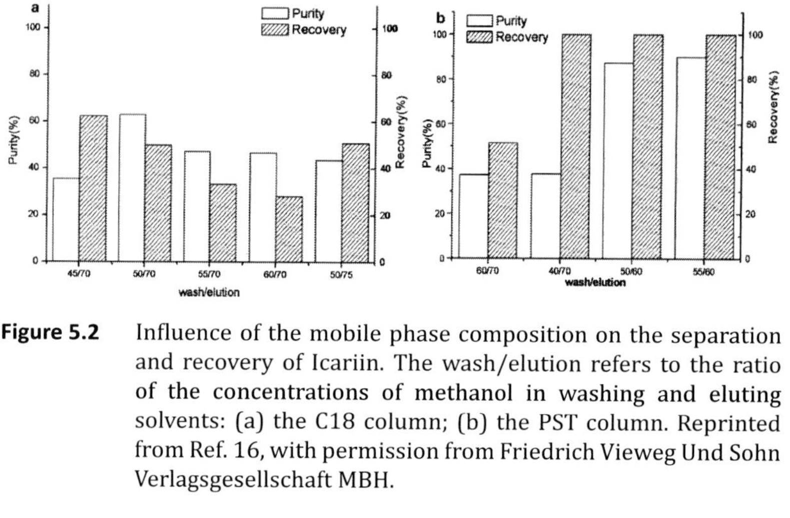

Figure 5.2 Influence of the mobile phase composition on the separation and recovery of Icariin. The wash/elution refers to the ratio of the concentrations of methanol in washing and eluting solvents: (a) the C18 column; (b) the PST column. Reprinted from Ref. 16, with permission from Friedrich Vieweg Und Sohn Verlagsgesellschaft MBH.

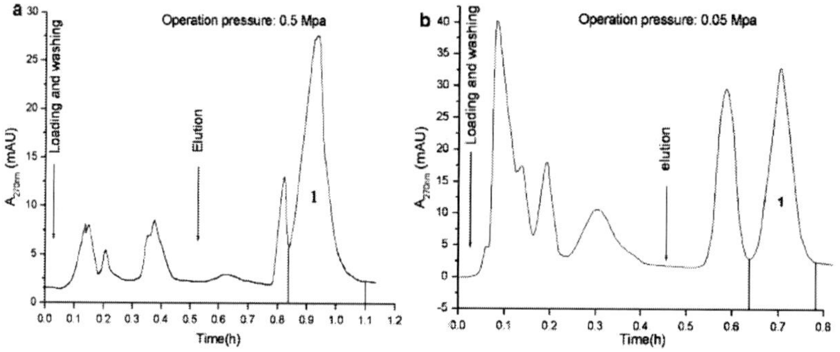

Figure 5.3 Profile of chromatographic purification of Icariin from crude extract by (a) the C18 column and (b) the PST column under their optimized mobile phase conditions. The eluent was detected at 270 nm. Peak 1 represented Icariin. Reprinted from Ref. 16, with permission from Friedrich Vieweg Und Sohn Verlagsgesellschaft MBH.

Figure 5.4 As-prepared PS-DVB microspheres with different particle sizes.

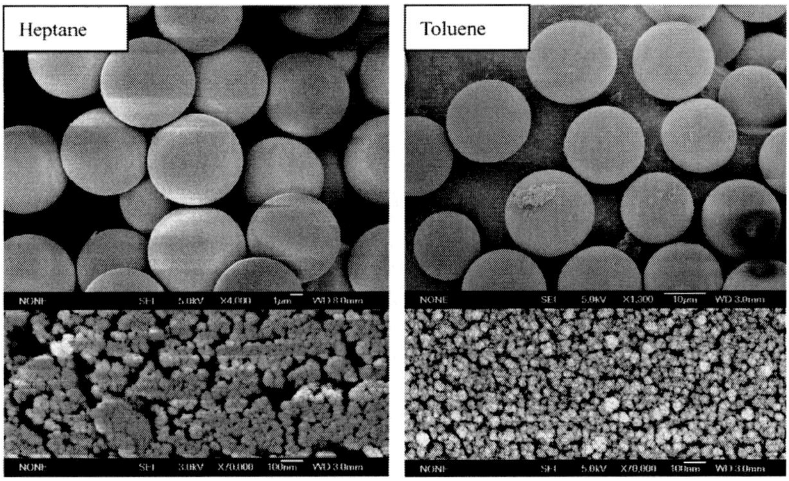

Figure 5.5 As-prepared PDVB microspheres with different specific surface areas. Reprinted from Ref. 19, with permission from Academic Press.

To be used as high efficiency polymeric media in separation chromatography, the pore size and the specific surface area of microspheres are two most important performance parameters. Hao *et al.* (19) studied the porogen effects in synthesis of uniform PDVB microspheres with high surface areas. A new method for preparing uniform PDVB microspheres was developed by combining membrane emulsification, suspension polymerization, and post-cross-linking techniques. They found that the physico-chemical properties of porogens had a great influence on the size distribution and porous features of microspheres. The low aqueous solubility of porogen facilitated preparation of uniform emulsions and microspheres, and high aqueous solubility led to polydispersed emulsions and low microsphere yields. Such aqueous solubility effects can be compensated by adding a low-molecular weight polystyrene (LPST) as co-stabilizer in porogen, thus improving the uniformity of microspheres. Moreover, different affinities of porogens for copolymers demonstrated various contributions to specific surface areas of microspheres in suspension polymerization, especially post-cross-linking. Solvating porogen requires a much higher addition than non-solvating porogen to obtain equal specific surface areas in polymerization, but it also has more potential to enhance the specific surface area in post-cross-linking. Two kinds of uniform microspheres were obtained with high specific surface areas, up to 706.6 m^2/g by heptane and 937.5 m^2/g by toluene, as shown in Fig. 5.5.

Hao *et al.* (20) also studied the relationship between heterogeneous structures and phase-separation behaviors in synthesis of uniform PDVB microspheres by *in situ* tracking photo-transmittance of polymerization system. It is shown that one of the decisive factors is the porogen type. Non-solvating porogen induced fast decline of transmittance and early phase separation, which often formed macropores or hollow structure, while solvating porogen led to much slower declines of transmittance and later-phase separation, which usually formed micropores structure. Besides, the gelling point of the polymerization system during the phase-separation process was found crucial to influence movements of phases in polymerization and the final heterogeneous structures of the microspheres. The later the gelling point occurred, the more heterogeneous structures, such as hollow or microvoid structure, were formed, while an early gelling point facilitated fabrication of small pores or non-porous structure. For example (as shown in Fig. 5.6), for non-solvating porogen as heptane, liquid paraffin, and hexadecane, as the phase separation between the porogen and the growing polymer accelerated, both pore size and nuclei on the microsphere's skeleton enlarged and macropores, multivoids, and hollow structures formed. While for solvating porogen as toluene, under a lower phase-separation rate, the formed pores and nuclei were much smaller than that of non-solvating porogen. Summing-up the main points of this study, the pore size and specific surface areas of microspheres can be controlled through the synthesizing process of polymeric microspheres.

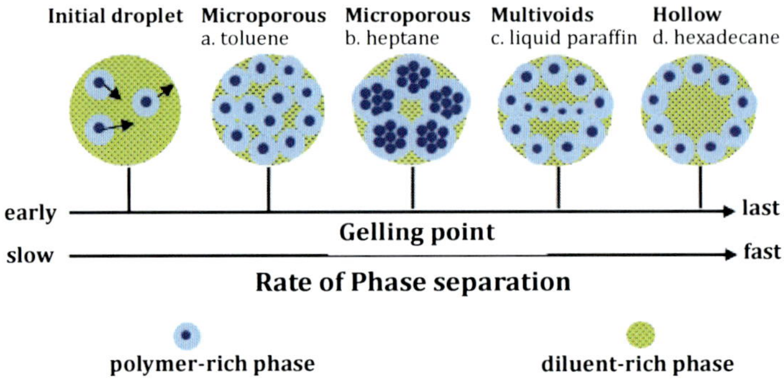

Figure 5.6 Schematic diagram of the formation of heterogeneous structure in phase separation. Reprinted from Ref. 20, with permission from Elsevier.

Zhang *et al.* (21) used the uniform PDVB microspheres (PDVB-SPG) prepared by SPG membrane emulsification as the separation media in reversed-phase liquid chromatography to purify polydatin, resveratrol, and anthraglycoside B from polygonum cuspidatum Sieb. et Zucc. The results were compared with those of a commercial silica-based C18 media. The uniform porous PDVB microspheres columns demonstrated better separation compared with the C18 column. Figure 5.7 showed the profile of HPLC chromatograms of polydatin, resveratrol, and anthraglycoside purified by PDVB-SPG column. After the chromatographic process and further crystallization, polydatin, resveratrol, and anthraglycoside B were purified from 7.5%, 1.3%, and 2.4% to 98.8%, 98.2%, and 98.6%, respectively, with total recoveries of 81.3%, 77.4%, and 80.2%. Lloyd *et al.* (22) prepared macroporous PST-DVB microspheres and used them for purification of synthetic peptides and recombinant proteins. Results demonstrated that the PST-DVB particles were mechanically stable and hence were able to operate at the required high linear velocities. Lloyd *et al.* found that a 10 nm pore size was suitable for synthetic peptides and a 30 nm pore size for recombinant proteins and the narrow particle-size distribution enabled maximum resolution. Wang *et al.* (23) prepared uniform macroporous PDVB microspheres with a mean pore size of 14 nm and a specific surface area of 520 m^2/g. As-prepared PDVB microspheres were packed into semi-preparative columns (300 × 10 mm I.D.) and were tested by a mixture of peptides on HPLC. As shown in Fig. 5.8, four standard peptides, oxytocin, angiotensin II, angiotensin I, and insulin, were completely separated by PDVB column, especially angiotensin II and I as there was only slight difference in the structure of the two amino-acid residues. The resolution of the purification system was higher than 1.5, and the theoretical plate number reached 21,000/m. The result further confirmed that the macroporous microspheres would prove to be a promising reversed-phase stationary phase for the preparation and process purification of peptides.

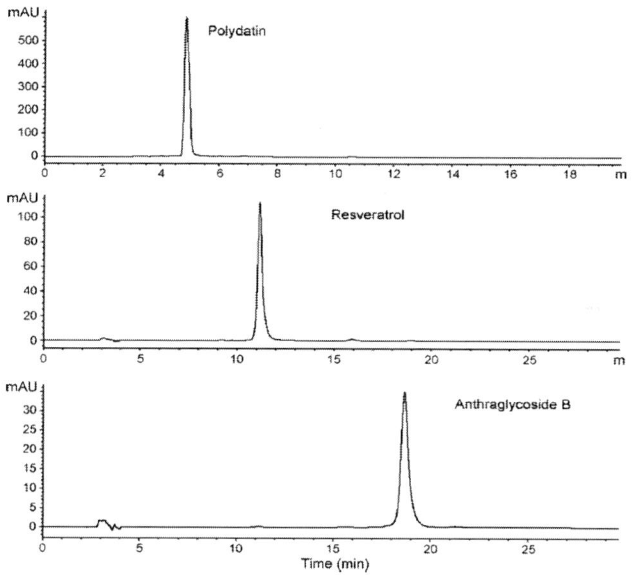

Figure 5.7 Profile of HLPC purification of polydatin, resveratrol, and anthraglycoside by PDVB-SPG column. Reprinted from Ref. 21, with permission from Pergamon.

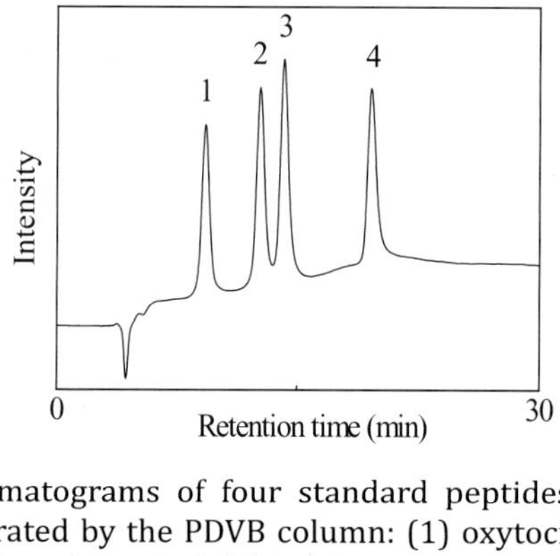

Figure 5.8 Chromatograms of four standard peptides (1, 2, 3, and 4) separated by the PDVB column: (1) oxytocin, (2) angiotensin II, (3) angiotensin I, (4) insulin.

From the above results, uniform porous hydrophobic polymer microspheres several tens micrometers in size could be demonstrated to be a promising chromatographic media for purification of peptides, bioactive natural products, and the as-mentioned separating system, with high separation efficiency, lower work pressure, and high flow rate.

5.3.2 Hydrophilic Microspheres Used for BSM Separation

Some BSM show moderate or high hydrophilicity. The hydrophobic microspheres are not suitable to separate these hydrophilic BSM since they do not provide adequate retention and separation for hydrophilic BSM. To circumvent this problem, recently, the demand for hydrophilic media has rapidly increased with the development of biotechnology and synthetic polymers. With abundant hydroxyl groups on its surface, porous poly(2-hydroxyethyl methacrylate) (PHEMA) microspheres have proved to be an excellent hydrophilic media for applications in chromatographic separation and immobilization of bioactive molecules (24,25). Qu *et al.* (26,27) successfully prepared uniform porous poly(2-hydroxyethyl methacrylate-ethylene dimethacrylate) [P(HEMA-EDMA)] microspheres with control

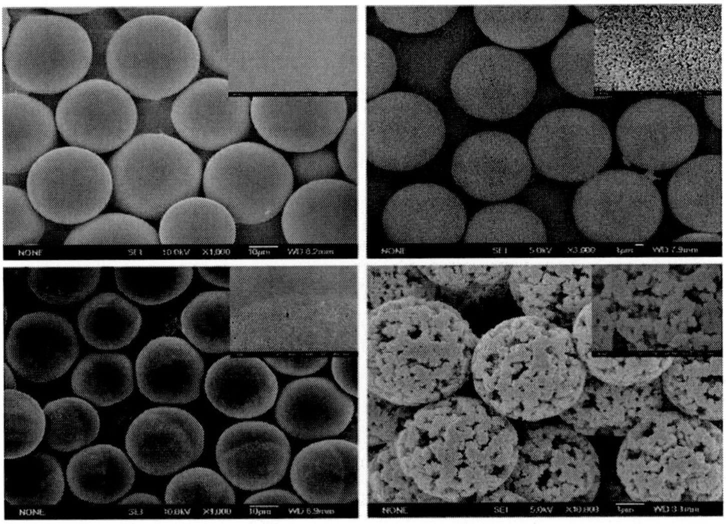

Figure 5.9 SEM micrographs of P(HEMA-EDMA) microspheres with different microstructures. Reprinted with permission from Ref. 27. Copyright 2008 American Chemical Society.

lable particle and pore size by combining the membrane-emulsification technique and the swelling process of uniform droplets, as shown in Fig. 5.9. The porous P(HEMA-EDMA) microspheres with narrow size distribution showed many advantages, such as low performance pressure, high flow rate, and high separation resolution. Wang *et al.* (28) successfully prepared another hydrophilic uniform polymer microspheres, poly(glycidyl methacrylate-divinylbenzene) [P(GMA-DVB)] and poly(glycidyl methacrylate-ethylene dimethacrylate) [P(GMA-EGDMA)] microspheres by membrane emulsification–polymerization technology. Covered with many epoxy side groups, PGMA microspheres have important potential applications in BSM separation (29,30).

Although some hydrophilic microspheres have been prepared to be used as reversed-phase separation media to purify BSM with moderate hydrophilicity, still some highly hydrophilic BSM cannot be sufficiently retained in the reversed-phase columns, resulting in the target compounds getting eluted at or near the void volume (31). HILIC has become another version of normal phase liquid chromatography (NPLC, polarity based). In the HILIC mode separation, aqueous organic solvents are used as mobile phases on more polar stationary phases, causing solutes to elute in the order of increasing hydrophilicity. Recently, literature and research on HILIC has increased drastically, along with various stationary phases developed for HILIC (32,33). Many highly hydrophilic peptides, such as phosphopeptides and glycopeptides in tryptic digests, were successfully separated with high resolution using HILIC (34). Theoretically, any polar chromatographic surface can be used for HILIC separations. Commonly used HILIC phases can be grouped into five categories of neutral polar or ionic surfaces.

(1) simple unbonded silica silanol or diol bonded phases
(2) amino or anionic bonded phases
(3) amide-bonded phases
(4) cationic bonded phases
(5) zwitterionic bonded phases

Various silica-based polar stationary phases have been used in HILIC, but the main drawbacks of silica, such as its slight dissolution and low thermal stability, make researcher seek more stable polar stationary phases, such as polymer-based microspheres. Hydrophobic microspheres with cross-linked polymer as support and hydrophilic polymer as coating layer were reported, which are promising for high-speed chromatographic separation of peptides or proteins. Agarose-coated polystyrene (35), glycerol monomethacrylate/dimethacrylate-coated polystyrene-divinylbenzene (36), and P(*N*-isopropylacrylamide-*co*-*N*-*tert*-butylacrylamide) [P(NIPAAm-*co*-tBAAm)]-coated poly(hydroxy methacrylate) (PHMA) (37) have been developed and some have been successfully applied for HILIC (Table 5.3).

Table 5.3 Polymer-based stationary phase for hydrophilic chromatography

Support	Coated layer	Reference
Polystyrene	Agarose	35
Polystyrene-divinylbenzene	Glycerol monomethacrylate and glycerol dimethacrylate	36
Poly(hydroxy methacrylate)	P(NIPAAm-*co*-tBAAm)	37

5.3.3 Thermo-Responsive Microspheres Used for BSM Separation

Classical chromatographic separations are often carried out by gradient elution, which is achieved by changing the content of organic solvents in mobile phase. The presence of organic solvents is often destructive to the bioactivities of peptides or other BSM. Separation under purely aqueous mobile phases might prevent the denaturation of peptides during liquid chromatographic separation. However, it is not easy to change the interaction between the analytes and the surface of stationary phase by just using purely aqueous mobile phases. An alternative method for changing the interaction between the analytes and the stationary phase is to adjust the surface properties of stationary phase with thermo-responsive materials.

Recently, Mizutani *et al.* reported a new method to prepare thermo-responsive microspheres, which may change their surface properties from hydrophilic to hydrophobic as a response to temperature (38–40). The main advantage of the temperature-sensitive microspheres is that they can be used to separate peptide mixtures or proteins in a pure aqueous solution under isocratic elution. The thermo-responsive microsphere was prepared by modification of porous polystyrene microsphere with poly (*N*-isopropylacrylamide) (PNIPAAm) using surface-initiated atom-transfer radical polymerization (ATRP). PNIPAAm has a sharp phase transition at 32°C (lower critical solution temperature, LCST). Below the LCST, the polymer chains are strongly hydrated and show an expanded conformation. Above the LCST, the expanded conformation of PNIPAAm can change to the compact forms caused by the dehydration and the intermolecular hydrogen bond between the carbonyl and NH groups. The application of the thermo-responsive microsphere to separate the solutes without using any organic solvents may show many advantages over classical microsphere since the hydrophobic interactions can be modulated by changing the temperature (38).

Silica beads can be used as the core materials for preparation of polymer-grafted beads since the silica-based stationary phases are widely used in HPLC systems. As shown in Fig. 5.10, copolymer such as P(NIPAAm-*co*-tBAAm) was used as functional groups grafted on silica beads by surface-initiated ATRP (40). However, silica beads are not stable enough due to the hydrolysis of silica, especially under aqueous alkaline conditions. Therefore, polystyrene-based thermo-responsive beads have been designed to improve the bead stability under aqueous mobile phase (39). Using PNIPAAm-grafted polystyrene beads, all-aqueous separation of peptides were achieved by changing the operational temperature (38). Due to the strong hydrophobicity of polystyrene, peak tailing is often observed at evaporation temperature. To improve the resolution of peptide separation and the stability of the stationary phase, hydrophilic beads, PHEMA, were used as the core material and P(NIPAAm-*co*-tBAAm) were grafted by surface-initiated ATRP method (Fig. 5.11) (37). Thermo-responsive microspheres were successfully used as the stationary phase to separate insulin, insulin chain A, and insulin chain B by changing column temperature, and higher resolution was obtained by increasing operating temperature (Fig. 5.12). It was found that the retention time of insulin chain A and B in the chromatogram increased with temperature. The amount of grafted copolymer of PHEMA shows great influence on the retention behaviors and resolution of peptides. The optimal operational temperature is also associated with the amount of grafted copolymer.

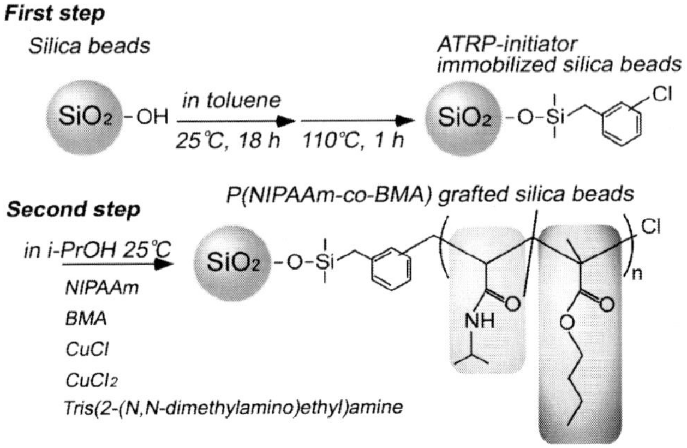

Figure 5.10 Preparation of PNIPAAm-grafted silica beads using surface-initiated ATRP. Reprinted with permission from Ref. 40. Copyright 2010 American Chemical Society.

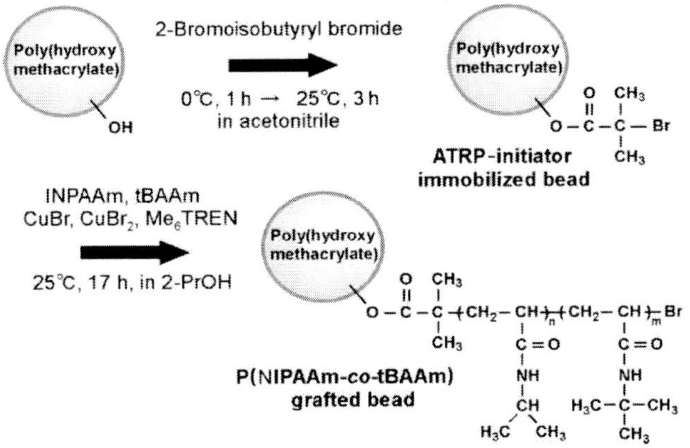

Figure 5.11 Preparation of poly (*N*-isopropylacrylamide-*co*-*N*-*tert*-butylacrylamide) [P(NIPAAm-tBAAm)] brush-grafted PHMA beads using surface-initiated ATRP. Reprinted from Ref. 37, with permission from Elsevier BV.

Lee *et al.* synthesized a series of thermo-responsive hydrogels with *N*-isopropylacrylamide (NIPAAm), glycidyl methacrylate (GMA), and *N,N*-dimethyl acrylamide (DMA). It was found that the ratio of GMA/DMA showed great influence on the swelling behaviors of the synthesized hydrogels. Increase of the content of DMA in the copolymeric hydrogel induced higher swelling ratio of hydrogel in deionized water, but the mechanical strength decreased with increasing DMA content (41).

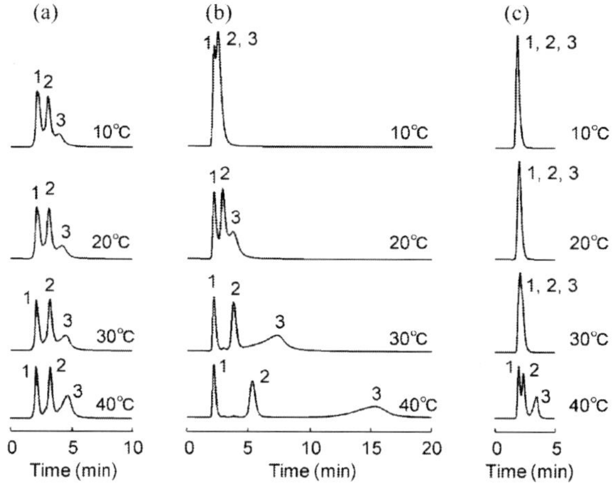

Figure 5.12 Chromatograms of peptides on the control unmodified PHEMA microsphere (a) and P(NIPAAm-*co*-tBAAm)-grafted PHEMA microsphere at different temperatures, the amounts of grafted copolymer is 1.6 mg/m^2 (b) and 5.1 mg/m^2 (c) respectively. Peak 1: Insulin chain A; Peak 2: Insulin chain B; Peak 3: Insulin. Reprinted from Ref. 37, with permission from Elsevier BV.

For temperature-modulated chromatographic separation, the column temperature has to be strictly controlled, normally using the temperature-circulating bath and column oven. Yagi *et al.* developed a novel technique that combined temperature-responsive polymeric beads with inductive heating via an alternating magnetic field (AMF). The beads were prepared from magnetite and silica, which was further modified with poly(*N*-isopropylacrylamide-*co*-2-carboxyisopropylacrylamide) [poly(NIPAAm-*co*-CIPAAm)]. The novel temperature-responsive beads could be used as stationary phase, which influence the elution of the model compounds by AMF-induced heating generated by the surrounding coil (42). The separation efficiency was tested with hydrocortisone, dexamethasone, and testosterone as model compounds. The resolution of the three model compounds were improved by the AMF-induced heating generated by the surrounding coil.

5.3.4 pH-Responsive Microspheres Used for BSM Separation

pH-sensitive or pH-responsive polymers are prepared with materials that are responsive to the pH of the surrounding media by varying their dimensions. The conformational structure of such materials may swell or collapse according to the environmental pH. In recent years, some novel microspheres with pH-responsive surface have been used as stationary phase for chromatographic separation of peptides or other bioactive molecules. The chromatographic mode might be switched between reversed-phase liquid chromatography (RPLC) and HILIC just by changing the environmental pH. In RPLC mode, the retention time of hydrophobic compounds decreased with the increase in the pH of the mobile phase. The opposite result was obtained in HILIC mode, and the separation of some hydrophilic compounds can be achieved by decreasing the pH of mobile phase.

Poly(acrylic acid) (PAAc) is a well-recognized pH-responsive polymer and has been widely used for various industries. When being exposed to different pH conditions, the macromolecule chains of the PAAc swell or shrink, resulting in hydrophilic or hydrophobic properties accordingly. However, the hydrophobic property of PAAc is limited since the alkyl chain is relatively short. To increase the hydrophobicity of the polymer, Ma *et al.* introduced longer alkyl group into the polymer and prepared poly (AAc-*co*-BA) (Fig. 5.13). When the new polymer was grafted onto silica surfaces, the microsphere surface characteristics related highly to the environmental pH. With the pH-responsive microsphere as

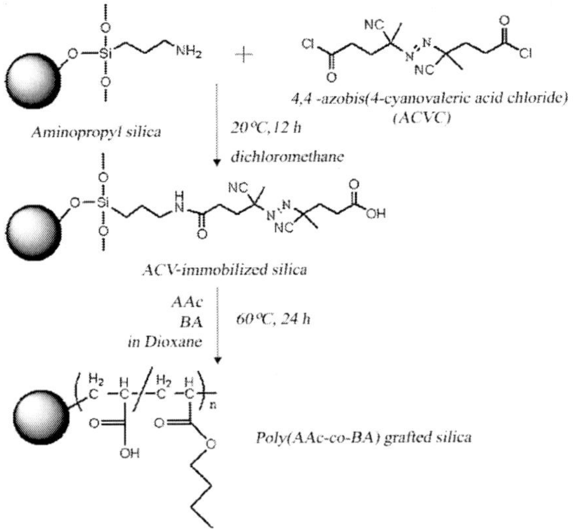

Figure 5.13 Scheme for preparation of pH-responsive poly(acrylic acid-*co*-butyl acrylate) [poly(AAc-*co*-BA)]-grafted silica beads. Reprinted from Ref. 43, with permission from Elsevier BV.

stationary phase, both RPLC- and HILIC-separation mode can be achieved since the hydrophobic/hydrophilic surface can be changed by altering the solution pH (43). With the soybean isoflavones as mode compounds, it was found that the relative elution orders in the two chromatographic modes were opposite.

To prepare polymers with multifunctional properties, other groups, such as ion-exchange group, can be introduced to the copolymer. Sakamoto *et al.* prepared P(NIPAAm-*co*-tBAAm-*co*-AAc) copolymer-coated microsphere that respond to the solution pH and temperature. The ion-exchange groups on the polymer chain surface can be exhibited and hidden according to the surrounding conditions. The interactions between the analyte and the stationary phase could be freely controlled by changing the temperature and the pH. Then, the microsphere might alter its surface properties from hydrophilic to hydrophobic and from charged to non-charged state according to the temperature and pH changes during chromatographic separation (44,45). In a recently published paper, pH- and temperature-responsive microspheres were successfully used to separate phosphorylated peptides with controllable electrostatic and hydrophobic interactions during the chromatographic process (46).

5.3.5 Core–Shell Structure Microspheres Used for BSM Separation

The concept of core–shell particles was first introduced by Horvath *et al.* and Kirkland *et al.* in the late 1960s (47,48). In recent years, microspheres with core–shell structure have attracted considerable attention, and some have been successfully used for chromatographic separation. Normally, the core–shell microspheres to be used as stationary phase have a nonporous core and a porous shell with a well-defined thickness (Fig. 5.14) (49). For traditional totally porous microspheres, the solute molecules are exposed to mass-transfer resistances due to the interaction with the stationary phase, which cause significant band broadening during chromatographic separation. For nonporous microspheres, the mass-transfer resistance may be reduced by decreasing the diameter of microspheres, but the column capacity might also be decreased in comparison with the porous microsphere with the same diameter.

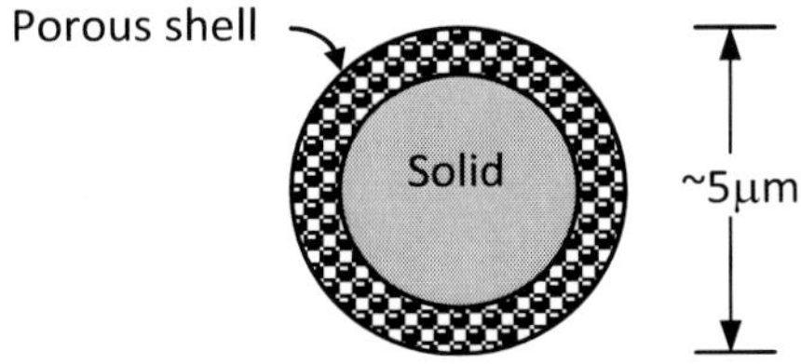

Figure 5.14 The structure of core–shell microsphere. Reprinted from Ref. 49, with permission from Elsevier BV.

Microspheres with core–shell structure possess several advantages over nonporous or totally porous microspheres. The shorter diffusion path due to the presence of nonporous core results in smaller band broadening. The surface area of porous layer is larger than nonporous particles, which is beneficial for increasing sample capacity.

Microspheres with core–shell structure can be classified into several types according to the materials of the porous shell and the non-porous core. Natural polymers, such as agarose, dextran, cellulose, chitoson, synthetic polymers, such as polymethacrylate, polyacrylamide, trisacryl, and polystyrene, can be used for preparation of porous shell. Inorganic material, such as silica, zirconium oxide, glass, hydroxyapatite, and composite materials can be used as non-porous core (49,50).

Now, three new brands of core–shell microspheres are commercially available: Poroshell (Agilent), Halo (Advanced Materials Technology, USA), and Kinetex (Phenomenex, USA) (51–55). Felinger compared the mean diffusion time in fully porous and shell-type packing materials using amicroscopic random-walk model (56). It was found that the relative decrease of the diffusion time is rather significant with the core–shell packing materials when only intra-particle diffusion was considered. As shown in Fig. 5.15, as long as the diffusion time is considered, the 2.7 μm Halo and Poroshell 120 phases are equivalent to a 1.8 μm fully porous packing material. The 2.6 μm Kinetex packing material is equivalent to a 1.34 μm fully porous stationary phase. Kiss *et al.* measured and compared the characteristics and performance of core–shell particles (a 1.7 μm solid core is covered with a 0.5 μm–thick shell of porous silica) and totally porous silica particles (3.5 μm commercial silica) with human insulin as mode analyte (51). It was found that the retention factor of insulin on the column packed with core–shell microspheres was higher than that on the column packed with totally porous microspheres under the same pressure. The plate height for insulin also decreased for core–shell microspheres packed column.

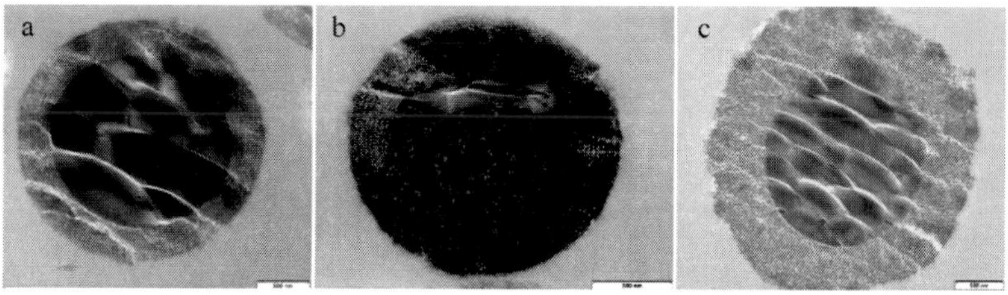

Figure 5.15 Transmission electron microscope (TEM) photographs of the cross-sections of core–shell particles: (a) 2.6 μm Kinetex, (b) 1.7 μm Kinetex, and (c) 2.7 μm Halo. Reprinted from Ref. 55, with permission from Elsevier BV.

5.4 Summary and Prospects

Chromatography, as one of most widely used separation techniques, has been used extensively for many years. Microspheres play a key role in column chromatography. In recent years, great efforts have been made to prepare new microspheres and some have been successfully used as stationary phases. However, there are always new challenges in the development of novel microspheres to meet the requirements for separation of mixtures with extreme complexity. The diffusional mass transport, back pressure, mobile-phase viscosity, and tolerance for temperature are permanent problems to be faced. The ever-faster analysis and ever-broader application of HPLC encourage more chemical engineers and analyzers to overcome these problems. It is possible that more and more microspheres with versatile properties appear and act as new stationary phases in the near future.

References

1. Katz, Elena D. (Ed.) (1996). *High Performance Liquid Chromatography: Principles and Methods in Biotechnology*. Chichester: John Wiley.
2. Zheng, X. F., Chan, T. F., & Zhou, H. H. (2004). Genetic and genomic approaches to identify and study the targets of bioactive small molecules. *Chemistry & Biology,* **11**, 609–618.

3. Challis, G. L., & Hopwood, D. A. (2007). Chemical biotechnology: Bioactive small molecules targets and discovery technologies. *Current Opinion in Biotechnology,* **18**, 475–477.
4. Tang, C. S., & Pu, D. (2005). Emphasize the role of bioactive small molecules in the homeostasis regulation of cardiovascular system. *Acta Academiae Medicinae Sinicae,* **27**(4), 440–442.
5. Schmidt-Traub, H. (Ed) (2005). *Preparative Chromatography of Fine Chemicals and Pharmaceutical Agents.* Weinheim, Germany: Wiley-VCH Verlag GmbH.
6. Ersahin, S., Gunal, H., Kutlu, T., *et al.* (2006). Estimating specific surface area and cation exchange capacity in soils using fractal dimension of particle-size distribution. *Geoderma,* **136**(3–4), 588–597.
7. Sun, M., Qiu, H. D., Wang, L. C., *et al.* (2009). Poly(1-allylimidazole)-grafted silica, a new specific stationary phase for reversed-phase and anion-exchange liquid chromatography. *Journal of Chromatography A,* **1216**, 3904–3909.
8. François, I., Sandra, K., & Sandra, P. (2009). Comprehensive liquid chromatography: Fundamental aspects and practical considerations: A review. *Analytica Chimica Acta,* **641**, 14–31.
9. Michel, M., Baczek, T., Studzinsk, S., *et al.* (2007). Comparative evaluation of high-performance liquid chromatography stationary phases used for the separation of peptides in terms of quantitative structure–retention relationships. *Journal of Chromatography A,* **1175**, 49–54.
10. Soukupova, K., Krafkova, E., Suchankova, J., *et al.* (2005). Comparison of zirconia- and silica-based reversed stationary phases for separation of enkephalins. *Journal of Chromatography A,* **1087**, 104–111.
11. Claessens, H. A., & Van Straten, M. A. (2004). Review on the chemical and thermal stability of stationary phases for reversed-phase liquid chromatography. *Journal of Chromatography A,* **1060**, 23–41.
12. Korhonen, H., & Pihlanto, A. (2006). Bioactive peptides: Production and functionality, *International Dairy Journal,* **16**, 945–960.
13. Sarmadia, B. H., & Ismail, A. (2010). Antioxidative peptides from food proteins: A review. *Peptides,* **31**, 1949–1956.
14. Lloyd, L. L., Millichip, M. I., & Watkins, J. M. (2002). Reversed-phase poly(styrene–divinylbenzene) materials optimised for large scale preparative and process purification of synthetic peptides and recombinant proteins. *Journal of Chromatography A,* **944**(1–2), 169–177.

15. Ma, G. H., Gong, F. L., Hu, G. H., *et al.* (2005). Multi-scale structures in emulsion and microsphere complex systems. *China Particuology,* **6**, 296–303.

16. Sun, H., Li, X. N., Ma, G. H., *et al.* (2005). Polystyrene-type uniform porous microsphere enables high resolution and low-pressure chromatography of natural products-a case study with icariin purification. *Chromatographia,* **61**, 9–15.

17. Sun, H., Li, X. N., Ma, G. H., *et al.* (2005). A facile two-column chromatographic process for efficient purification of paclitaxel from crude extract. *Journal of Liquid Chromatography & Related Technologies,* **28**, 625–637.

18. Qin, L., Lin, H., Gong, F. L., *et al.* (2007). Preparation of uniform-sized porous PST-DVB microspheres by combining membrane emulsification with suspension polymerization. *The Chinese Journal of Process Engineering,* **7**, 526–530.

19. Hao, D. X., Gong, F. L., Wei, W., *et al.* (2008). Porogen effects in synthesis of uniform micron-sized poly(divinylbenzene) microspheres with high surface area. *Journal of Colloid and Interface Science,* **323**, 52–59.

20. Hao, D. X., Gong, F. L., Hu, G. H., *et al.* (2009). The relationship between heterogeneous structures and phase separation in synthesis of uniform microns polyDVB microspheres. *Polymer,* **14**, 3188–3195.

21. Zhang, D. L., Li, X. N., Hao, D. X., Li, G. S., Xu, B. M., Ma, G. H., *et al.* (2009). Systematic purification of polydatin, resveratrol and anthraglycoside B from Polygonum cuspidatum Sieb. et Zucc. *Separation and Purification Technology,* **66**, 329–339.

22. Lloyd, L. L., Millichip, M. I., & Watkins, J. M. (2002). Reversed-phase poly(styrene-divinylbenzene) materials optimised for large scale preparative and process purification of synthetic peptides and recombinant proteins. *Journal of Chromatography A,* **944**, 169–177.

23. Wang, J., Qu, H. H., Hao, D. X., Gong, F. L., & Ma, G. H. (2009). Preparation of uniform polydivinylbenzene microspheres and their application in reversed-phase chromatographic separation of small molecular compounds and peptide. *The Chinese J. Proc. Eng.* **9**, 763–769.

24. Garcý´a, D. M., Escobar, J. L., Noa, Y., Bada, N., Herna´ez, E., & Katime, I. (2004). Timolol maleate release from pH-sensible poly(2-hydroxyethyl methacrylate-*co*-methacrylic acid) hydrogels. *European Polymer Journal,* **40**, 1683–1690.

25. Omi, S., Kaneko, K., Nakayama, A., Katami, K., Taguchi, T., Iso, M., *et al.* (1997). Application of porous microspheres prepared by SPG (Shirasu

porous glass) emulsification as immobilizing carriers of glucoamylase (GluA). *Journal of Applied Polymer Science*, **65**, 2655–2664.

26. Qu, H. H., Gong, F. L., Ma, G. H., & Su, Z. G. (2007). Preparation and characterization of large porous poly(HEMA-*co*-EDMA) microspheres with narrow size distribution by modified membrane emulsification method. *Journal of Applied Polymer Science*, **105**, 1632–1641.
27. Qu, H. H., Gong, F. L., Wang, J., Ma, G. H., & Su, Z. G. (2008). Strategies for the design of hydrophilic microspheres with controlled size and microstructure. *Industrial & Engineering Chemistry Research*, **47**, 6399–6404.
28. Wang, R. W., Zhang, Y., Ma, G. H., & Su, Z. G. (2006). Preparation of uniform poly(glycidyl methacrylate) porous microspheres by membrane emulsification-polymerization technology. *Journal of Applied Polymer Science*, **102**, 5018–5027.
29. Kotha, A., Raman, R. C., Ponrathnam, S., Kumar, K. K., & Shewale, J. G. (1996). Beaded reactive polymers, 2. Immobilisation of penicillin G acylase on glycidyl methacrylate-divinyl benzene copolymers of differing pore size and its distribution. *Reactive & Functional Polymers*, **28**, 235–242.
30. Zhang, S. H., Huang, X., Yao, N. S., & Horvath, C. (2002). Preparation of monodisperse porous polymethacrylate microspheres and their application in the capillary electrochromatography of macrolide antibiotics. *Journal of Chromatography A*, **948**, 193–201.
31. Yoshida, T. (2004). Peptide separation by hydrophilic-interaction chromatography: A review. *Journal of Biochemical and Biophysical Methods*, **60**, 265–280.
32. Pesek, J. J., Matyska, M. T., Fischer, S. M., *et al.* (2008). Analysis of hydrophilic metabolites by high-performance liquid chromatography-mass spectrometry using a silica hydride-based stationary phase. *Journal of Chromatography A*, **1204**, 48–55.
33. Ikegami, T., Tomomatsu, K., Takubo, H., *et al.* (2008). Separation efficiencies in hydrophilic interaction chromatography. *Journal of Chromatography A*, **1184**, 474–503.
34. Boersema, P. J., Mohammed, S., & Heck, A. J. R. (2008). Hydrophilic interaction liquid chromatography (HILIC) in proteomics. *Analytical and Bioanalytical Chemistry*, **391**(1), 151–159.
35. Qu, J. B., Wan, X. Z., Zhai, Y. Q., *et al.* (2009). A novel stationary phase derivatized from hydrophilic gigaporous polystyrene-based microspheres for high-speed protein chromatography. *Journal of Chromatography A*, **1216**, 6511–6516.

36. Hosoya, K., Kishii, Y., & Kimat, K. (1995). Uniform-size hydrophobic polymer-based separation media selectively modified with a hydrophilic external polymeric layer. *Journal of Chromatography A*, **690**(1), 21–28.

37. Mizutani, A., Nagase, K., Kikuchi, A., *et al.* (2010). Preparation of thermo-responsive polymer brushes on hydrophilic polymeric beads by surface-initiated atom transfer radical polymerization for a highly resolutive separation of peptides. *Journal of Chromatography A*, **1217**, 5978–5985.

38. Mizutani, A., Nagase, K., Kikuchi, A., *et al.* (2010). Effective separation of peptides using highly dense thermo-responsive polymer brush-grafted porous polystyrene beads. *Journal of Chromatography B*, **878**, 2191–2198.

39. Mizutani, A., Nagase, K., Kikuchi, A., *et al.* (2010). Thermo-responsive polymer brush-grafted porous polystyrene beads for all-aqueous chromatography. *Journal of Chromatography A*, **1217**(4), 522–529.

40. Nagase, K., Kumazaki, M., Kanazawa, H., *et al.* (2010). Thermo-responsive polymer brush surfaces with hydrophobic groups for all-aqueous chromatography. *ACS Applied Materials & Interfaces*, **2**(4), 1247–1253.

41. Lee, W. F., & Huang, C. T. (2008). Immobilization of trypsin by thermo-responsive hydrogel for the affinity separation of trypsin inhibitor. *Desalination*, **234**, 195–203.

42. Yagi, H., Yamamoto, K., & Aoyagi, T. (2008). New liquid chromatography method combining thermo-responsive material and inductive heating via alternating magnetic field. *Journal of Chromatography B*, **876**, 97–102.

43. Ma, Q., Chen, M., Yin, H. R., *et al.* (2008). Preparation of pH-responsive stationary phase for reversed-phase liquid chromatography and hydrophilic interaction chromatography. *Journal of Chromatography A*, **1212**, 61–67.

44. Sakamoto, C., Okada, Y., Kanazawa, H., *et al.* (2004). Temperature- and pH-responsive aminopropyl-silica ion-exchange columns grafted with copolymers of *N*-isopropylacrylamide. *Journal of Chromatography A*, **1030**, 247–253.

45. Ayano, E., Suzuki, Y., Kanezawa, M., *et al.* (2007). Analysis of melatonin using a pH- and temperature-responsive aqueous chromatography system. *Journal of Chromatography A*, **1156**, 213–219.

46. Nishio, T., Ayano, E., Suzuki, Y., *et al.* (2011). Separation of phosphorylated peptides utilizing dual pH-and temperature-

responsive chromatography. *Journal of Chromatography A*, **1218**, pp. 2079–2084.

47. Horvath, C. G., & Lipsky, S. R. (1969). Rapid analysis of ribonucleosides and bases at the picomole level using pellicular cation exchange resin in narrow bore columns. *Analytical Chemistry*, **41**, 1227–1234.
48. Kirkland, J. J. (1992). Superficially porous silica micro-spheres for the fast high-performance liquid chromatography of macromolecules. *Analytical Chemistry*, **64**, 1239–1245.
49. Jungbauer, A. (2005). Chromatographic media for bioseparation. *Journal of Chromatography A*, **1065**, 3–12.
50. Gritti, F., Leonardis, I., Abia, J., *et al.* (2010). Physical properties and structure of fine core–shell particles used as packing materials for chromatography relationships between particle characteristics and column performance. *Journal of Chromatography A*, **1217**, 3819–3843.
51. Kiss, I., Bacskay, I., & Kilar, F. (2010). Comparison of the mass transfer in totally porous and superficially porous stationary phases in liquid chromatography. *Analytical and Bioanalytical Chemistry*, **397**, 1307–1314.
52. Gritti, F., & Guiochon, G. (2007). Unusual behavior of the height equivalent to a theoretical plate of a new poroshell stationary phase at high temperatures. *Journal of Chromatography A*, **1169**, 125–138.
53. Marchetti, N., Cavazzini, A., Gritti, F., *et al.* (2007). Gradient elution separation and peak capacity of columns packed with porous shell particles. *Journal of Chromatography A*, **1163**, 203–211.
54. Gritti, F., & Guiochon, G. (2010). Performance of columns packed with the new shell Kinetex-C18 particles in gradient elution chromatography. *Journal of Chromatography A*, **1217**, 1604–1615.
55. Gritti, F., Leonardis, I., Shock, D., *et al.* (2010). Performance of columns packed with the new shell particles, Kinetex-C18. *Journal of Chromatography A*, **1217**, 1589–1603.
56. Felinger, A. (2011). Diffusion time in core–shell packing materials. *Journal of Chromatography A*, **1518**, 1939–1941.

Chapter 6

Microspheres for Separation of PEG-Modified Biomolecules

Yongdong Huang, Yanqin Zhai, and Zhiguo Su

State Key Laboratory of Biochemical Engineering, Institute of Process Engineering, Chinese Academy of Sciences, Beijing, P. R. China

zgsu@home.ipe.ac.cn

6.1 Introduction

Considerable interest has been shown in the last 20 years for PEG-modified biomolecules. These products are formed by covalent attachment of polyethylene glycol (PEG) to proteins, peptides, oligonucleotides, and other molecules of pharmaceutical values. The technique, known as PEGylation, was developed as early as 1970s (1) and has achieved commercial success with PEG-protein conjugates. These products distinguish themselves from their non-PEGylated counterparts by possessing more useful clinical properties, such as increased stability, less enzymatic degradation, and reduced renal clearance. The conjugates also show better therapeutic effects due to the improved solubility, reduced immunogenicity and antigenicity, less toxicity, lower peak plasma concentration, less-frequent administration, and so forth (2–3). Most of these positive changes have been attributed to the increased hydrodynamic radius by the PEG attachment and to the protection of the protein surface by the steric action of the PEG chain (4). After more than 30 years' development, PEGylation has become a well-established method used extensively to transform biomolecules into more potent pharmaceuticals than their corresponding unmodified native

Microspheres and Microcapsules in Biotechnology: Design, Preparation, and Applications
Edited by Guanghui Ma and Zhiguo Su

ISBN 978-981-4316-47-7 (Hardcover), 978-981-4364-62-1 (eBook)
www.panstanford.com

molecules. It has been proved to be a versatile modification and delivery technique (5,6).

The first PEG-modified pharmaceutical, approved by US Food and Drug Administration (FDA) is Adagen® (Enzon Pharmaceuticals, USA) (7). Adagen® is a PEGylated form of adenosine deaminase and is used for the treatment of severe combined immunodeficiency (SCID). Other major PEGylated therapeutics approved by FDA are Oncaspar®, PEGintron®, PEGasys®, Neulasta®, PEGvisomant®, Macugen®, Mircera®, etc. Most of these products are PEGylated proteins, except Macugen®, which is a PEG-modified nucleic acid. In addition, PEGylated antibodies, antibody fragments, and single-chain antibodies are also in various stages of clinical development. These PEGylated drugs have achieved great economic and social benefits. For example, the two PEG-interferon alpha products (PEGintron® and PEGasys®) and PEG-G-CSF (Neulasta®) all have about 1 billion $US market and all have reduced dosing frequency. PEGylation will continue to attract global attention due to its notable advantages and potential benefits in providing long circulation and improved efficiency.

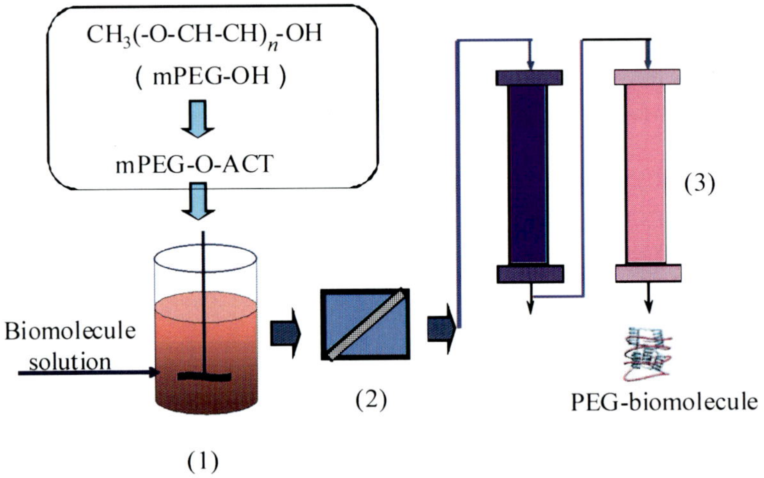

Figure 6.1 A schematic process of PEGylation: (1) PEGylation reactor. (2) Membrane separator. (3) Chromatographic columns.

For protein or other biomolecules, the common PEGylation route is shown as Fig. 6.1. Monomethoxy PEG (mPEG) was activated to form a derivative with active end group. Here we use mPEG-O-ACT to illustrate various derivatives that are formed through different chemistry (6). These can be aldehyde, maleimide, or NHS ether, which can react subsequently with a biomolecule, such as a protein, at mild reaction condition to preserve the biological activity of the biomolecule. The derivative, mPEG-O-ACT, is added to the solution of biomolecule to be PEGylated. The whole process involves two stages: reaction and separation. In the reaction stage, mPEG-O-ACT reacts with the corresponding group of the biomolecule, usually amine, thio, or N terminal of a protein. In most cases of protein PEGylation, the reaction is not very specific because there could be more than one group reacting with the PEG molecule. As a consequence, isomers of different PEGylation sites could be formed. If the reaction is not controlled at the optimum condition, a complex mixture might be produced, which contains mono-, di-, or tri-PEGylated products.

Another important issue is the separation and purification of the reaction mixture. After PEGylation, the reaction mixture is usually composed of unreacted PEGs, PEGylated biomolecules, native biomolecules, and even hydrolyzed small molecules. Therefore, the purification of PEGylated biomolecules from the reaction mixture is vital to the control of the purity and safety of the product. Except for the target product of PEGylated biomolecules, all the other components should be removed during the purification process. In clinical application, most of the target PEGylated products are composed of a specific PEGylated form (4). However, only a few cases of PEGylation reactions are very specific. Therefore, subfractionation of PEG-modified biomolecules with different degrees of PEGylation or different PEGylation sites is also necessary.

6.2 Separation of PEGylated Biomolecules

Similar to other biomolecules, the design of separation strategy for PEGylated biomolecules is mainly based on the differences in physicochemical properties between PEGylated biomolecules and other molecules in the reaction mixture. The physicochemical properties that are widely used to purify biomolecules can also be applied to PEGylated biomolecules with varying effectiveness (8). These properties commonly include molecular weight, size, and shape for size-exclusion chromatography (SEC) and membrane separations, isoelectric point and surface charge distribution for ion-exchange chromatography (IEC), and relative hydrophobicity for hydrophobic interaction chromatography (HIC) and reverse-phase chromatography (RPC), etc. (4,8).

6.2.1 Properties of PEG-Biomolecules

The physicochemical properties of PEG-biomolecules, such as molecular weight, size and shape, electrostatic charge, and relative hydrophobicity may change to a certain extent after PEGylation. Among them, the change of molecular weight, size or shape is obvious, because the viscosity radii of random-coil PEG molecules are much larger than those of proteins with the same relative molecular weight (M_r) (Table 6.1) (8). However, the increment of the size of PEG-biomolecules will lower their diffusivity and access to internal pores in chromatographic microspheres, resulting in worse resolution and lower capacity in chromatography, especially when the microporous chromatographic microspheres are used. The changes of the surface charge and pI are relatively complex, which are mainly determined by the original charge, PEGylated site and extent, PEG's shield effect, the hydrogen bond formed between the biomolecule and PEG, etc. For example, the number of charged residues of bovine serum albumin (BSA) was large, and the changes of its net charge with each PEG chain added were relatively small, while the number of charged residues of α-interferon 2A was small, and the changes of its net charge with each PEG chain added were relatively obvious (8). In addition, the change of relative hydrophobicity is relatively small, since it arises mainly from the attachment of relatively inert PEG chains to the biomolecules.

Table 6.1 Equivalent protein molecular weights for common PEG molecules

PEG M_r (kDa)	Viscosity radius (nm)	Equivalent globular protein M_r (kDa)
1	0.91	1.4
5	2.24	20.4
10	3.29	65.4
20	48.5	209.5
50	80.9	962.1

Source: Adapted from Ref. 8.

6.2.2 Methods for the Separation of PEG-Biomolecules

For the separation of PEGylated biomolecules, all the common separation methods used widely in the purification of biomolecules have also been applied. Among those, IEC and SEC are the most commonly used methods, followed by ultrafiltration, HIC, RPC, affinity chromatography (AFC), etc. In practical application, ultrafiltration and diafiltration are used infrequently and mainly for concentration or buffer exchange rather than separation of PEGylated biomolecules from the reaction mixture, while AFC is only used in the case of on-column PEGylation (4). Typical chromatographic methods for the separation of PEG-biomolecules are shown in Table 6.2.

Table 6.2 Purification of PEGylated proteins

Protein	Purification step	Chromatographic media	References
HSA	IEC	DEAE Sepharose FF	(9)
Staphylokinase	IEC	Q Sepharose FF (big beads)	(9)
Calcitonin	SEC	Superose 12	(10)
rh-GCSH	SEC	SP Sepharose FF	(11)
Epidermal growth factor	SEC RPC	Shodex protein KW-800 Spherisorb C-18	(12)
rhIL-1RA	IEC	SP Sepharose FF or Mono S	(13, 14)
rh-ngEPO	IEC	Resource S or Source 30S	(15, 16)
α-interferon 2b, interleukin10	IEC	Mono-S	(17)
Myelopoietin	IEC	Monolithic media (QA-CIM and SO_3-CIM)	(18)
β-interferon	IEC SEC	SP Sepharose FF Superose 12	(19)
BSA, β-lactoglobulin, lysozyme, caricain	HIC	Fractogel TSK-Butyl 650	(20)
Immunoadhesin	IMAC	IDA Sepharose	(21)

6.2.2.1 Size-exclusion chromatography

SEC has two types: analytical and preparative. Analytical SEC is commonly used in analysis of PEGylation reaction mixture or the purity of the products, while preparative SEC is mainly used to separate or purify PEGylation reaction mixture; both of them are based on the size differences between PEG-biomolecules and unPEGylated biomolecules (native biomolecules) or unreacted PEG (10–12,19). Analytical SEC uses commercially pre-packed columns with high plate number, such as TSK columns. Preparative SEC columns are often packed by the operator according to the large volume of the chromatographic medium, such as Sepharose or Superdex that fulfill the need of productivity.

The efficiency of SEC is mainly dependent on the difference of the molecular sizes and the pore sizes of the chromatographic medium. The sizes of PEG, protein, and PEGylated proteins can be estimated using the following equations (22,23):

$$R_{\mathrm{h,prot}} \approx (0.82 \pm 0.02) M_{\mathrm{r,prot}}^{1/3} \tag{6.1}$$

$$R_{\mathrm{h,PEG}} = 0.1912 M_{\mathrm{r,PEG}}^{0.559} \tag{6.2}$$

$$\begin{aligned} R_{\mathrm{h,PEGprot}} = {} & \frac{1}{6}\left[108R_{\mathrm{h,prot}}^{3} + 8R_{\mathrm{h,PEG}}^{3} + 12(81R_{\mathrm{h,prot}}^{6} + 12R_{\mathrm{h,prot}}^{3}R_{\mathrm{h,PEG}}^{3})^{1/2}\right]^{1/3} \\ & + \frac{2}{3}\frac{R_{\mathrm{h,PEG}}^{2}}{\left[108R_{\mathrm{h,prot}}^{3} + 8R_{\mathrm{h,PEG}}^{3} + 12(818R_{\mathrm{h,prot}}^{6} + 12R_{\mathrm{h,prot}}^{3}R_{\mathrm{h,PEG}}^{3})^{1/2}\right]^{1/3}} \\ & + \frac{1}{3}R_{\mathrm{h,PEG}} \end{aligned} \tag{6.3}$$

Here $R_{\mathrm{h,pro}}$, $R_{\mathrm{h,PEG}}$, and $R_{\mathrm{h,PEGprot}}$ are the viscosity radii of the native protein, PEG molecule, and PEGylated protein, respectively. $M_{\mathrm{r,pro}}$ and $M_{\mathrm{r,PEG}}$ are the relative molecular weights of the native protein and PEG molecule, respectively. In most cases the viscosity radius of PEG is much larger than that of protein with the same relative molecular weight according to Eqs. 6.1 and 6.2, especially when M_{r} is larger than 1000 Da.

To achieve satisfied SEC separation at the preparative scale, the proteins should differ in molecular weight by about 100% (4). According to Eq. 6.1, this requires a size ratio of 1.26 between protein molecules. Figure 6.2 illustrates the size ratio between species with different PEGlation extents when a 15 kDa protein is PEGylated. The lines in the figure are calculation results from

Eq. 6.3. If the number of PEG per protein molecule is less than 3, all the PEGylated species can be separated from their unPEGlyated counterparts (4). The figure indicates that an increase in the PEG size can facilitate SEC separation, while attachment of more PEGs to one protein will decrease the resolution of SEC.

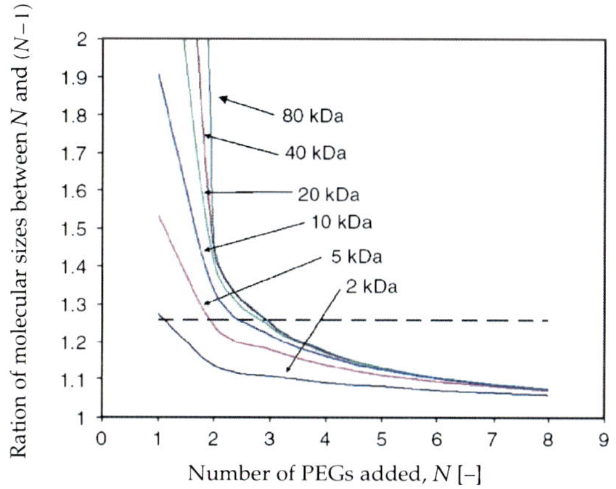

Figure 6.2 Ratio of hydrodynamic radii of PEGylated species differing by a single PEG adduct. The dotted line shows the normal guideline for obtaining sufficient resolution for SEC. Reprinted from Ref. 4, with permission from Elsevier.

Since SEC separates proteins or protein conjugates based on the difference of size, it would be difficult to separate the position isomers of PEGylation. There is usually more than one group on the surface of the protein that can react with PEG, resulting in more than one PEGylation product even if the number of PEG per protein is the same. In this case, we have to consider another type of chromatography to carry out the separation. These are IEC and RPC that distinguish different PEGylation products by their surface properties.

6.2.2.2 Ion-exchange chromatography

IEC is based on the charge differences between PEG-biomolecules and unPEGylated biomolecules. It is one of the most commonly used techniques for the purification of PEGylated biomolecules (9,13–19,24), especially in large-scale production. In addition, high performance IEC is also successfully applied to the subfractionation of PEG-proteins on the basis of their degree of PEGylation and positional isomerism (24–26).

PEG is a neutral polymer, but it may affect the charge properties of biomolecules in three ways: shielding the surface charges of a biomolecule and weakening electrostatic interaction, altering their charge nature, and affecting their charge versus pH behavior (4). Generally, the attachment of PEG will weaken the ionic interaction between PEG-biomolecules and ion-exchange media. As a result, more highly PEGylated species often flow through the IEC column, while mildly PEGylated species are eluted at lower salt concentrations than those of the native proteins (Figs. 6.3 and 6.4) (24).

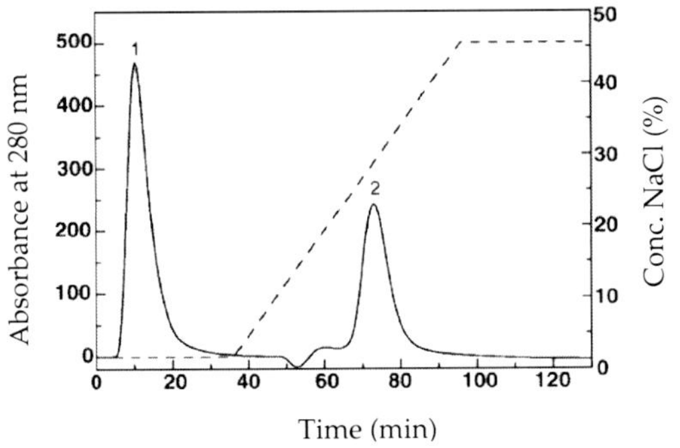

Figure 6.3 Separation of PEGylated rhG-CSF from the intact protein by cation-exchange chromatography using SP Sepharose FF media. (Peak 1 was the mixture of conjugates, while peak 2 was native rhG-CSF.) Reprinted from Ref. 24, with permission from Elsevier.

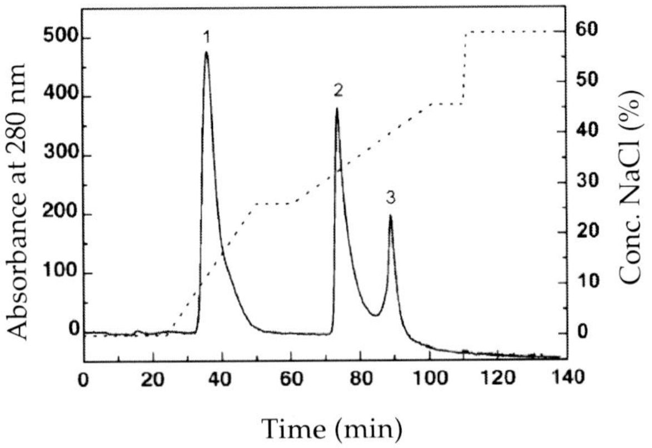

Figure 6.4 Fractionation of the individual PEGylated species by anion-exchange chromatography using Source 30Q media. (Peaks 1, 2, and 3 were tri-, di-, and mono-PEGylated rhG-CSF, respectively.) Reprinted from Ref. 24, with permission from Elsevier.

Except for the change in surface charge, PEGylation dramatically affects molecular size, and this may be reflect in lower capacity of ion-exchange media due to the decrement in diffusivity or access to internal pores in ion-exchange media. Pabst *et al.* studied the purification of PEG-BSA with six different anion-exchange media (27). Results indicated that the purity and recovery yield of PEG-BSA purified with different anion-exchange media were nearly the same. However, the dynamic binding capacities (DBCs) of PEG-BSA were decreased dramatically due to modulation of electrostatic interactions caused by the neutral PEG chain and the increased mass-transfer resistance associated with the large size of the molecule (Table 6.3) (27). To overcome the steric hindrance, the pore size of the media should be increased properly. According to the report of Fee *et al.* (28), a macroporous cation exchange microsphere has been shown to achieve higher DBCs than a conventional microsphere for PEGylated species. We also found that the purification efficiency of PEGylated rhG-CSF was improved by a novel macroporous ion-exchange media (SP-AP). The DBC of SP-AP was higher than 10 mg/mL medium even at a high speed of 1224 cm/h, which was about five times higher than that of traditional SP Sepharose FF media.

Table 6.3 DBCs binding capacities of strong anion-exchangers obtained from breakthrough experiments at different superficial velocities

Media	BSA (mg/mL bed) 250 (cm/h)	12 kDa PEG-BSA (mg/mL bed) 50 (cm/h)	12 kDa PEG-BSA (mg/mL bed) 250 (cm/h)	30 kDa PEG-BSA (mg/mL bed) 50 (cm/h)	30 kDa PEG-BSA (mg/mL bed) 250 (cm/h)
Q Hyper DF	61.4	5.0	4.8	1.3	1.3
Macroprep High Q	15.6	7.6	6.1	5.6	3.2
Capto Q	144.2	50.8	37.4	8.1	6.8
Fractoprep TMAE	113.5	13.8	13.1	0.4	0.6
Toyopearl Super Q 650C	35.9	3.7	2.2	0.6	0.4
Q Sepharose FF	52.8	29.5	27.4	18.2	13.9

Source: Reprinted from Ref. 27, with permission from Elsevier.

6.2.2.3 Hydrophobic interaction chromatography and reverse-phase chromatography

HIC and RPC are based on the differences of relative hydrophobicity between PEG-biomolecules and native biomolecules. Until now, only a little work has been done for the separation of PEG-proteins by HIC. For example, the separation of growth hormone modified with 5 kDa PEG-NHS was started with HIC using a phenyl TSK 5PW media. It was then further purified by IEC after ultrafiltration and buffer exchange on a G-25 Sephadex column (29). Vincentelli *et al.* found that a PEGylated form of β-lactoglobulin bound more strongly to a Fractogel TSK-Butyl 650 hydrophobic interaction media than either the free PEG or the native protein, indicating that conformational changes to the PEG-protein may improve its surface hydrophobicity (20). Compared with SEC and IEC, HIC was less used in the purification of PEG-biomolecules. The reason is that the change of hydrophobicity is insignificant. Therefore, HIC is usually not effective enough to separate various PEGylated forms of a biomolecule, either on the basis of PEGylation extent or PEGylation site.

RPC is usually used for the analysis of PEG-biomolecules. It is an effective method for the analysis of PEG-biomolecules because of its high resolution, even for the analysis of positional isomers (30,31). It also can be used for the purification of PEG-biomolecules only if the biomolecules are stable during the process of RPC. For example, mono-PEGylated salmon calcitonin was first separated by SEC, and then its positional isomers were further separated by RPC (32). RPC has also been used for stability analysis of a PEGylated rhIL-1RA using a Vydac 214TP54 column (33).

6.3 Microspheres for the Separation of PEG-Modified Biomolecules

Although various separation techniques are used to purify PEG-modified biomolecules, SEC and IEC are dominant ones. Therefore, most of the media for PEG-modified biomolecules purification are SEC and IEC media. A list of the primary media used for PEG-modified biomolecules separation is shown in Table 6.4.

Table 6.4 Properties of chromatographic microspheres considered for the purification of PEG-modified biomolecules

Media	Type	Mean particle size (μm)	Maximum recommended pressure (bar)	Base matrix	Mean pore diameter (nm)	Exclusion limit (Dalton, globular protein)
Superdex 75	SEC	34 (24–44)	3~4	Dextran and agarose	—[a]	3×10^3–7×10^4
Superdex 200	SEC	34 (24–44)	3~4	Dextran and agarose	—	1×10^4–5×10^6
DEAE, Q, CM and SP Sepharose FF	IEC	90 (45–165)	3	Agarose	~50	1×10^4–4×10^6
Mono S	IEC	10	30	P(ST-DVB)	—	—
Mono Q	IEC	10	30	P(ST-DVB)	—	—
Resoure S	IEC	15	15	P(ST-DVB)	—	—
Source 30S	IEC	30	30	P(ST-DVB)	—	—
Macroprep High Q	IEC	50	—[a]	Methacrylate copolymer	100	—
Macrocap SP	IEC	50	3	Dextran and bisacrylamide	—	$<15 \times 10^4$
Toyopearl Super Q 650	IEC	50–150	—	Methacrylate copolymer	100	—
SP-AP	IEC	55 (40–70)	30	P(ST-DVB) and agarose	300	—
Phenyl TSK 5PW	HIC	30 (20–40)	20	Methacrylate	100	$<5 \times 10^6$

[a]Not reported.

6.3.1 Microspheres of SEC

The SEC studies of PEGylated proteins by Fee and Van Alstine (23) suggest that existing SEC media are suitable for separation of PEG-proteins from native proteins and moderate fractionation of PEG-proteins of differing N. One of the potential disadvantages of SEC media used may be the relative soft matrix. Among those SEC separation media, Superdex 75 and Superdex 200 are the most popularly used for PEGylation extent analysis and industrial scale separation of the PEGylated proteins. Both media are prepared with a unique composite matrix of dextran and agarose, which combines the excellent gel-filtration properties of dextran with the physical and chemical stability of highly cross-linked agarose. With their high physical and chemical stability, they are ideal for all stages from research and process development through scale-up and into production.

6.3.2 Microspheres of IEC

For IEC, many commercial media have been developed for biomolecular separation. Theoretically, most of the PEGylated biomolecules could also be separated by the media used for the separation of their counterparts (native biomolecules). However, the DBC of PEGylated proteins decreased to about 1–55% of those of the native proteins (Table 6.3). A comparison of six strong anion exchangers for the purification of a PEGylated BSA (27) showed that agarose based Q sepharose FF and Capto Q achieved the higher DBC of PEG-BSA than the other four media (Q Hyper DF, Macroprep High Q, Fractoprep TMAE, and Toyopearl Super Q 650C) under evaluated flow rate. The lower DBC of the other four media could be attributed to their relatively small pore sizes and the intense diffusion hindrance that excluded the PEG-BSA conjugate. Interestingly, Capto Q showed higher DBCs of native BSA (144.2 mg/mL media for Capto Q, 52.8 mg/mL media for Q Sepharose FF) and 12 kDa PEG-BSA compared with Q Sepharose FF (50.8 mg/mL media for Capto Q, 29.5 mg/mL media for Q sepharose FF). However, for 30 kDa PEG-BSA, things are inverse (5.6 mg/mL media for Capto Q, 18.2 mg/mL media for Q Sepharose FF). Such a phenomenon was attributed to their difference in base matrix. The base matrix of Q Sepharose FF is 6% cross-linked agarose, while the base matrix of Capto Q is 6% cross-linked agarose fabricated by grafted dextran. Compared with Q Sepharose FF, the DBC of Capto Q for relatively small biomolecules (native BSA and 12 kDa PEG-BSA) were improved by the stacking of adsorbed protein in grafted dextran layer (34), while its DBC for 30 kDa PEG-BSA was less because the presence of dextran extending into the pore appeared to cause more diffusional hindrance (27).

The mean pore radius of Q sepharose FF determined by inverse size-exclusion chromatography (ISEC) is about 28.8 nm (35), which is about 4.2 times larger than the hydrodynamic radius

(6.87 nm) of 30 kDa PEG-BSA (27). However, the DBC of 30 kDa PEG-BSA is still only about 25% of that of BSA, indicating that the hindered diffusion effect still exists. To overcome the hindered diffusion effect of PEG-biomolecules, the pore size of the media should increase greatly. Macroporous media, also denominated as perfusion media, can be operated at a high speed while maintaining high column efficiency and DBC for the purification of proteins (36). It is expected that macroporous media also can improve the DBC of PEG-biomolecules and their chromatographic efficiency. Macrocap SP, developed by GE Healthcare, was a cation media designed for the purification of PEGylated proteins and other large biomolecules at a high sample load. Mono-PEGylated proteins can be separated to high purity from oligo-PEGylated and non-PEGylated proteins, most often in a single run. Moreover, the hydrophilic nature of the base matrix reduces nonspecific binding and consequent fouling, thus allowing for a long media lifetime. The pore size of Macrocap SP is about 100 nm evaluated by SEM analysis (Fig. 6.5). Therefore, the capacity of Macrocap SP is much higher than that of traditional SP Sepharose HP in the purification of mono-PEG40 avidin according to the Application Note (28-4094-65 AA) supplied by GE Healthcare.

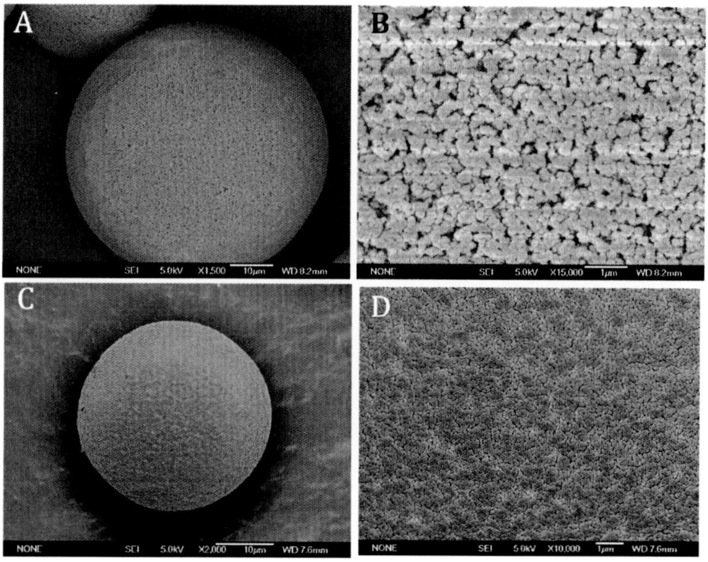

Figure 6.5 SEM images of Macrocap SP (A: 1500×, B: 15,000×) and SP Sepharose FF (C: 2000×, D: 10,000×). Images C and D were reprinted from Ref. 37, with permission from Elsevier.

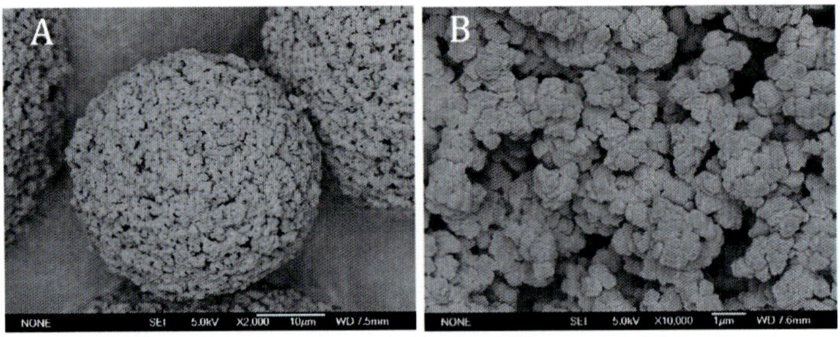

Figure 6.6 SEM images of SP-AP (A: 2000×, B: 10,000×). Reprinted from Ref. 37, with permission from Elsevier.

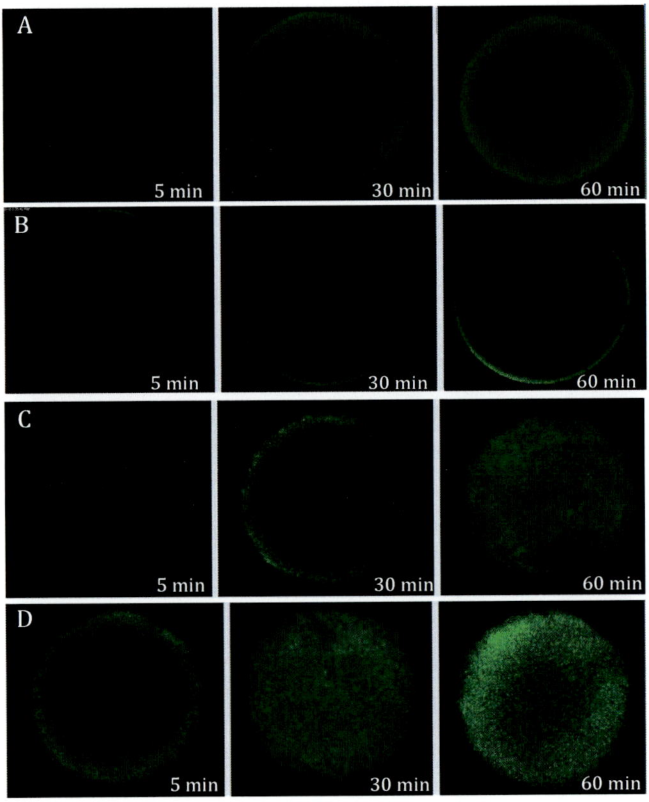

Figure 6.7 Laser scanning confocal microscopy (LSCM) images showing the mass transfer of native-GCSF (A, C) and PEG-GCSF (B, D) on SP-6FF microsphere (A, B) and on SP-AP microsphere (C, D). Microscopy was performed through chromatographic particle midsections taken under dynamic passive (bulk mass adsorption) conditions. Reprinted from Ref. 37, with permission from Elsevier.

SP-AP strong cation-exchange medium is another macroporous medium developed by our lab for the purification of PEG-GCSF (37). The preparation procedure of SP-AP was as follows: (i) macroporous poly(styrene-divinyl benzene) [P(ST-DVB)] microspheres were prepared by surfactant reverse micelles swelling method (38); (ii) P(ST-DVB) microspheres were hydrophilically modified with agarose (AP), which was easily incorporated in the inner pores by utilizing the macropores; (iii) sulphonyl group was coupled to AP microspheres. SP-AP has a rigid matrix of P(ST-DVB) and a macroporous pore structure as shown in Fig. 6.6. The pore size of SP-AP is about 300 nm, which is larger than those of Macrocap SP and traditional SP Sepharose FF. Such a pore size is large enough to eliminate the steric hindrance of PEG-GSCF as shown in Fig. 6.7.

Except for pore size, particle size and size distribution also have important roles in the separation of PEG-proteins and other biomolecules. For example, microspheres with uniform particle size may result in less backpressure and higher resolution. Mono S, Mono Q, Resource S, and SOURCE 30S are composed of monodispersed P(ST-DVB) particles with a particle size of 10, 15, or 30 μm, which are designed for fast, high-resolution cation-exchange separations of proteins, peptides, and PEG-proteins as well. These are also effective in the analysis of PEG-biomolecules (14,24). In addition, these P(ST-DVB)-based media are suitable for large-scale production due to their good physical stability at maximum recommended pressure as high as 30 bars.

6.3.3 Microspheres of Hydrophobic Interaction Chromatography

Since HIC was rarely used in the separation of PEG-biomolecules, only few reports are available on HIC microspheres. Among them, Phenyl TSK 5PW is relatively popularly used. This chromatographic microsphere is composed of a methacrylic polymer incorporating high mechanical and chemical stability. Its maximum recommended pressure is as high as 20 bars. In addition, it is a macroporous microsphere with a mean pore size of 100 nm, which can weaken the steric hindrance of PEGylated proteins.

6.4 Conclusion and Prospects

In this chapter, the PEGylation process of biomolecules, the properties of PEGylated biomolecules, the chromatographic methods, and related microspheres for the separation of PEGylated biomolecules were discussed. In the PEGylation process, site-specific modification is beneficial for further purification and improvement in the recovery of target product. In the separation process, SEC and IEC are the most commonly used methods because the changes in molecular size and surface charge are relatively obvious after PEGylation. According to the process of PEGylation and the properties of PEG-protein, an ideal chromatographic media for the separation of PEG-modified biomolecules should possess the following characters: (i) high hydrophilicity, (ii) no non-specific adsorbability, (iii) chemical and physical stability, (iv) macroporosity (the pore size should match with the size of PEG-proteins), (v) high capacity and high resolution, (vi) uniformity of particle size. In a word, the chromatographic media should have the properties of high speed, high resolution, high capacity, and high recovery yield. However, none of the commercial chromatographic media fulfills all the criteria mentioned above, and they are all compromises. Other than the common criteria for the native biomolecules' purification, a large pore size is generally needed for the purification of PEG-biomolecules. Additional effort should be focused on the regulation of pore size and the improvement of particle uniformity, especially for the preparation of macroporous media.

References

1. Abuchowski, A., Es, T. V., Palczuk, N. C., & Davis, F. (1977). Alteration of immunological properties of bovine serum albumin by covalent attachment of polyethylene glycol. *Journal of Biological Chemistry*, **252**(11), 3578–3581.
2. Molineux, G. (2002). PEGylation: Engineering improved pharmaceuticals for enhanced therapy. *Cancer Treatment Reviews*, **28**(Suppl. A), 13–16.
3. Bailon, P., & Won, C. Y. (2009). PEG-modified biopharmaceuticals. *Expert Opinion on Drug Delivery*, **6**(1), 1–16.

4. Fee, C. J., & Van Alstine, J. M. (2006). PEG-proteins: Reaction engineering and separation issues. *Chemical Engineering Science,* **61**(3), 924–939.
5. Roberts, M. J., Bentley, M. D., & Harris, J. M. (2002). Chemistry for peptide and protein PEGylation. *Advanced Drug Delivery Reviews,* **54**(4), 459–476.
6. Veronese, F. M., & Harris, J. M. (2008). Peptide and protein PEGylation III: Advances in chemistry and clinical applications. *Advanced Drug Delivery Reviews,* **60**(1), 1–2.
7. Davis, S., Abuchowski, A., Park, Y. K., & Davis, F. F. (1981). Alteration of the circulating life and antigenic properties of bovine adenosine deaminase in mice by attachment of polyethylene glycol. *Clinical and Experimental Immunology,* **46**, 649–652.
8. Conan, J. (2009). Protein conjugates purification and characterization. In F. M. Veronese (Ed.), *PEGylated Protein Drugs: Basic Science and Clinical Applications* (pp. 113–125), Switzerland: Birkhäuser Verlag.
9. Suo, X. Y., Lu, X. L., Hu, T., Ma, G. H., & Su, Z. G. (2009). A solid-phase adsorption method for PEGylation of human serum albumin and staphylokinase: Preparation, purification and biochemical characterization. *Biotechnology Letters,* **31**, 1191–1196.
10. Lee, L. S., Conover, C., Shi, C., Whitlow, M., & Filpula, D. (1999). Prolonged circulating lives of single-chain Fv proteins conjugated with polyethylene glycol: A comparison of conjugation chemistries and compounds. *Bioconjugate Chemistry,* **10**(6), 973–981.
11. Zhai, Y. Q., Zhao, Y. J., Lei, J. D., Su, Z. G., & Ma, G. H. (2009). Enhanced circulation half-life of site-specific PEGylated rhG-CSF: Optimization of PEG molecular weight. *Journal of Biotechnology,* **142**, 259–266.
12. Lee, H., & Park, T. G. (2002). Preparation and characterization of mono-PEGylated epidermal growth factor: Evaluation of *in vitro* biologic activity. *Pharmaceutical Research,* **19**(6), 845–851.
13. Yu, P. Z., Zheng, C. Y., Chen, J., Zhang, G. F., Liu, Y. D., Suo, X. Y., *et al.* (2007). Investigation on PEGylation strategy of recombinant human interleukin-1 receptor antagonist. *Bioorganic & Medicinal Chemistry,* **15**, 5396–5405.
14. Yu, P. Z., Zhang, G. C., Bi, J. X., Lu, X. L., Wang, Y. S., & Su, Z. G. (2007). Facile purification of mono-PEGylated interleukin-1 receptor antagonist and its characterization with multi-angle laser light scattering. *Process Biochemistry,* **42**, 971–977.

15. Wang, Y. J., Hao, S. J., Liu, Y. D., Hu, T., Zhang, G. F., Zhang, X., *et al.* (2010). PEGylation markedly enhances the *in vivo* potency of recombinant human non-glycosylated erythropoietin: A comparison with glycosylated erythropoietin. *Journal of Controlled Release*, **145**, 306–313.

16. Wang, Y. J., Liu, Y. D., Chen, J., Hao, S. J., Hu, T., Ma, G. H., *et al.* (2010). Efficient preparation and PEGylation of recombinant human non-glycosylated erythropoietin expressed as inclusion body in *E. coli*. *International Journal of Pharmaceutics*, **386**, 156–164.

17. Wylie, D. C., Voloch, M., Lee, S., Liu, Y. H., Cannon-Carlson, S., Cutler, C., *et al.* (2001). Carboxyalkylated histidine is a pH-dependent product of PEGylation with SC-PEG. *Pharmaceutical Research*, **18**(9), 1354–1360.

18. Hall, T., Wood, D. C., & Smith, C. E. (2004). Preparative and analytical chromatography of PEGylated myelopoietin using monolithic media. *Journal of Chromatography A*, **1041**(2), 87–93.

19. Arduini, R. M., Li, Z., Rapoza, A., Gronke, R., Hess, D. M., Wen, D., *et al.* (2004). Expression, purification, and characterization of rat interferon-[beta], and preparation of an *N*-terminally PEGylated form with improved pharmacokinetic parameters. *Protein Expression and Purification*, **34**(2), 229–242.

20. Vincentelli, J., Paul, C., Azarkan, M., Guermant, C., El Moussaoui, A., & Looze, Y. (1999). Evaluation of the polyethylene glycol-kf-water system in the context of purifying PEG-protein adducts. *International Journal of Pharmaceutics*, **176**(2), 241–249.

21. Chamow, S. C., Kogan, T. P., Venuti, M., Gadek, T., Harris, R. J., Peers, D. H., *et al.* (1994). Modification of CD_4 immunoadhesin with monomethoxypoly (ethylene glycol) aldehyde via reductive alkylation. *Bioconjugate Chemistry*, **5**, 133–140.

22. Hagel, L. (1998). Gel filtration. In J. C. Janson & L. Rydén (Eds.), *Protein Purification* (pp. 79–143). New York: Wiley.

23. Fee, C. J., & Van Alstine, J. A. (2004). Prediction of viscosity radius and size exclusion chromatography behavior of PEGylated proteins. *Bioconjugate Chemistry*, **15**(6), 1304–1313.

24. Yun, Q., Yang, R. E., Chen, T., Bi, J. X., Ma, G. H., & Su, Z. G. (2005). Reproducible preparation and effective separation of PEGylated recombinant human granulocyte colony-stimulating factor with novel "PEG-pellet" PEGylation mode and ion-exchange chromatography. *Journal of Biotechnology*, **118**, 67–74.

25. Monkarsh, S. P., Ma, Y., Aglione, A., Bailon, P., Ciolek, D., Debarbieri, B., *et al.* (1997). Positional isomers of monoPEGylated interferon-a2a: Isolation, characterization, and biological activity. *Analytical Biochemistry*, **247**, 434–440.

26. Foser, S., Schacher, A., Weyer, K. A., Brugger, D., Dietel, E., Marti, S., *et al.* (2003). Isolation, structural characterization, and antiviral activity of positional isomers of monoPEGylated interferon [alpha]-2a (PEGASYS). *Protein Expression and Purification*, **30**(1), 78–87.

27. Pabst, T. M., Buckley, J. J., Ramasubramanyan. N., & Hunter., A. K. (2007). Comparison of strong anion-exchangers for the purification of a PEGylated protein. *Journal of Chromatography A*, **1147**, 172–182.

28. Fee, C. J., Bergstrom, J., Stadler, J., Magnusson, R., & Van Alstine, J. M. (2005). Challenges related to the processing of PEG-modified proteins, In *International Conference on Biopartitioning and Purification (BPP 2005)*. Netherlands: Delft.

29. Clark, R., Olson, K., Fuh, G., Marian, M., Mortensen, D., Teshima, G., *et al.* (1996). Long-acting growth hormones produced by conjugation with poly(ethylene glycol). *Journal of Biological Chemistry*, **271**(21), 969–977.

30. Youn, Y. S., Na, D. H., Yoo, S. D., Song, S. C., & Lee, K. C. (2004). Chromatographic separation and mass spectrometric identification of positional isomers of polyethylene glycol-modified growth hormone-releasing factor (1–29). *Journal of Chromatography A*, **1061**(1), 45–49.

31. Park, E. J., Lee, K. C., & Na, D. H. (2009). Separation of positional isomers of mono-poly(ethylene glycol)-modified octreotides by reversed-phase high-performance liquid chromatography. *Journal of Chromatography A*, **1216**(45), 7793–7797.

32. Lee, K. C., Tak, K. K., Park, M. O., Lee, J. T., Woo, B. H., Yoo, S. D., *et al.* (1999). Preparation and characterization of polyethylene-glycol-modified salmon calcitonins. *Pharmaceutical Development and Technology*, **4**(2), 269–275.

33. Yu, P. Z., Qin, D., Qin, G. H., Fan, B., Ma, G. H., & Su, Z. G. (2009). PEGylation of rhIL-1RA increased its solution stability at room temperature. *Process Biochemistry*, **44**, 1340–1345.

34. Yao, Y., & Lenhoff, A. M. (2006). Pore size distributions of ion exchangers and relation to protein binding capacity. *Journal of Chromatography A*, **1126**(1–2), 107–119.

35. Shi, Q. H., Jia., G. D., & Sun, Y. (2010). Dextran-grafted cation exchanger based on superorous agarose gel: Adsorption isotherms, uptake kinetics and dynamic protein adsorption performance. *Journal of Chromatography A*, **1217**, 5084–5091.

36. Qu, J. B., Wan, X. Z., Zhai, Y. Q., Zhou, W. Q., Su, Z. G., & Ma, G. H. (2009). A novel stationary phase derivatized from hydrophilic gigaporous polystyrene-based microspheres for high-speed protein chromatography. *Journal of Chromatography A*, **1216**, 6511–6516.

37. Zhai, Y. Q., Zhou, W. Q., Wei, W., Qu, J. B., Lei, J. D., Su, Z. G., Ma, G. H. (2012). Functional gigaporous polystyrene microspheres facilitating separation of poly(ethylene glycol)-protein conjugate. *Analytica Chimica Acta*, **712**, 152–161.

38. Zhou, W. Q., Gu, T. Y., Su, Z. G., & Ma, G. H. (2007). Synthesis of macroporous poly (styrene-divinyl benzene) microspheres by surfactant reverse micelles swelling method. *Polymer*, **48**, 1981–1988.

Chapter 7

Microspheres for Solid-Phase Organic Synthesis

Jing Zhang and Zhiguo Su

State Key Laboratory of Biochemical Engineering, Institute of Process Engineering, Chinese Academy of Sciences, Beijing, P. R. China

zgsu@home.ipe.ac.cn

7.1 Solid-Phase Organic Synthesis

7.1.1 Introduction

In 1963, Robert B. Merrifield first described his method of "solid-phase" peptide synthesis (1). In this method, an amino acid was bound to a polymer through the esterification of the carboxyl group. This revolutionary approach resulted in Merrifield being awarded the 1984 Nobel Prize for chemistry. Also in 1963, Letsinger and Kornet (2) described an alternative solid-phase synthesis procedure. In this method, an amino acid was bound to microspheres or bulk materials through the amino group. This procedure has been developed into the basis of solid-phase oligonucleotide synthesis and has been proved to be invaluable for synthesis of oligonucletides.

Solid-phase organic synthesis (SPOS) is a chemical cross-linking reaction which is accomplished on insoluble polymer microspheres or other materials. In SPOS, polymer microspheres are the most used medium because of their economical availability and variable operation. Generally we called these microspheres "supports" or "beads" in SPOS. The general reaction mechanism of SPOS is shown in Fig. 7.1. First, the initial reactant is conjugated to the supporting

Microspheres and Microcapsules in Biotechnology: Design, Preparation, and Applications
Edited by Guanghui Ma and Zhiguo Su

ISBN 978-981-4316-47-7 (Hardcover), 978-981-4364-62-1 (eBook)
www.panstanford.com

beads through some cleavable linkage; then, other reactants react with the components on the beads sequentially. Finally, the product is cut off by using a certain cleavage reagent.

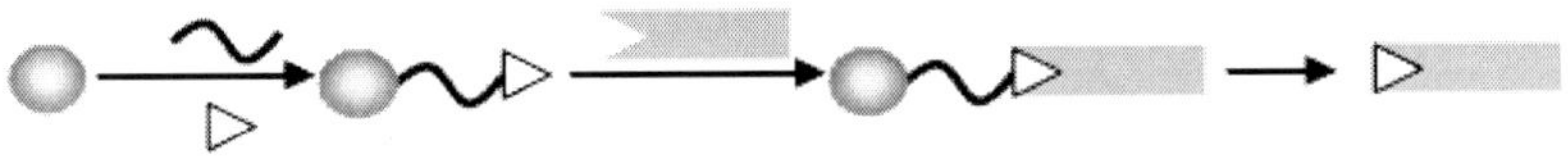

Figure 7.1 General reaction mechanism of solid-phase organic synthesis.

The advantages of SPOS over traditional solution phase organic synthesis are as follows:

- Ease of product isolation: During the entire process, supported intermediates could be separated from excess reactants cleanly and easily by filtration and solvent washing.
- High product purity: The reaction could almost reach the completion stoichiometrically by using excess reagents.
- Low yield of by-product: Active sites of the reaction were scattered over the surface of supporting beads. The density was neither too high to trigger side reactions nor too low to induce excessively low yield of the final product.
- Feasible automated process: For simple and convenient operation of SPOS, automated machines have been invented, which developed quickly in the past several decades.

As far as outstanding advantages are concerned, SPOS has greatly contributed to efficiency, speed, and flexibility in organic chemistry and biochemistry. It has been widely used in many fields, for example, peptide synthesis, oligonucleotide synthesis, and combinatorial chemistry. Over the last decades, thousands of papers have been published on these topics (3–6).

7.1.2 Solid-Phase Peptide Synthesis

There is no doubt that SPOS is one of the most sufficient methodologies for the synthesis of almost all peptides. Small peptides are synthesized by solid-phase strategy alone. Larger peptides are synthesized by the combination of solid-phase synthesis of protected segments and solution phase conjugation of these segments.

In the solid-phase peptide synthesis (Fig. 7.2), first, the amino acid corresponding to the carboxyl end of the desired peptide chain is attached to the beads through a certain linker. After the amino group of this amino acid has been deprotected, the chemical cross-linking reaction proceeds effectively between it and the excess carboxyl group of the next amino acid. The chain extends toward the amino end of the peptide by stepwise coupling of activated amino acid derivatives. Filtration and washing remove the soluble reactants and by-products. Finally, after the synthesis of the desired peptide, the chain is cleaved from the beads and purified by column chromatography (7).

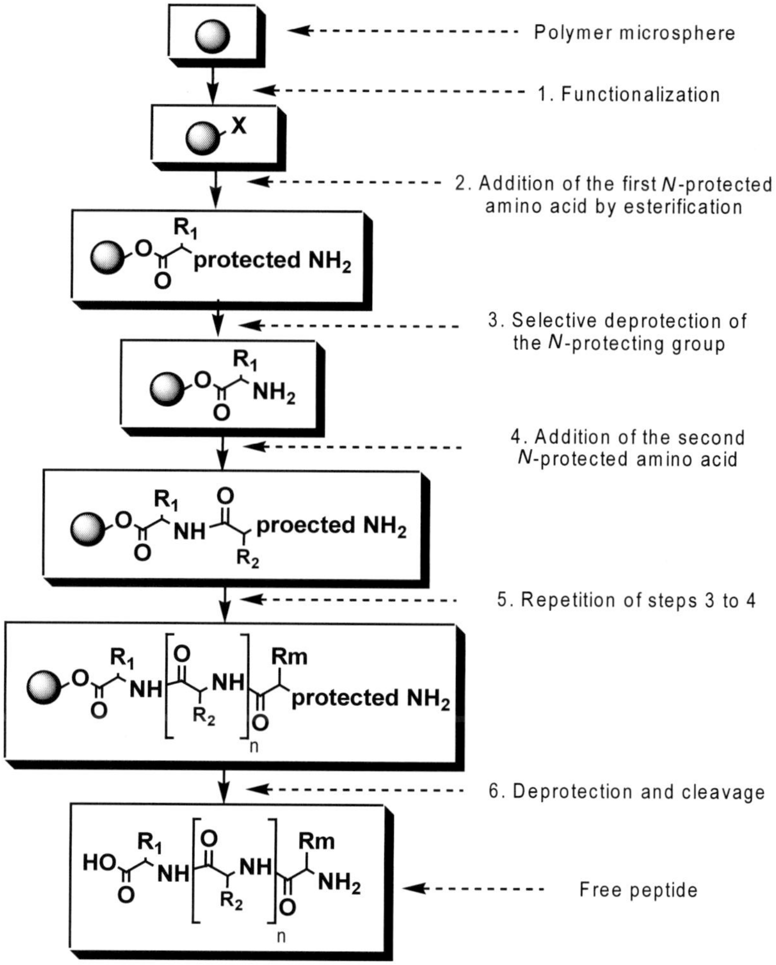

Figure 7.2 Scheme of solid-phase peptide synthesis.

Generally, the mentioned peptide synthesis can be performed on polystyrene (PS) beads using standard tert-butoxycarbonyl (*t*-Boc) or 9-fluorenylmethyl-oxycarbonyl (Fmoc) (8,9).

The *t*-Boc solid-phase peptide synthesis relies on *t*-Boc group to protect α-amino group of the amino acid temporarily. *T*-Boc group can be removed by trifluoroacetic acid (TFA) to expose a positive-charged amino group, which can be neutralized through the following coupling by using the next excess activated amino acid. During the process, resins have to withstand the repeating use of TFA in the deprotection of Boc group of each amino acid, which often leads to some undesired cleavage of intermediates. The final cleavage of peptide from resins should be done with a strong acid, such as hydrofluoric acid (HF) or trifluoromethanesufonic acid (TFMSA), which would aggravate the progresses of side-reactions besides those undesired cleavage of intermediates. In 2010, microwave-assisted Boc synthesis was reported by Kluczyk and his coworkers (10). Under microwave irradiation, a peptide could be cleaved from Merrifield resins much more efficiently because the shortened cleavage time relatively reduced the loss of target product.

Considering the above-mentioned evident limitations, Boc strategy is generally used to synthesize some base-sensitive peptide analogs. Researchers prefer much milder method, Fmoc protecting, which is based on base deprotection

The Fmoc solid-phase peptide synthesis utilized Fmoc group as the protecting group of α-amino. The deprotection can be applied under mild base condition. For example, Fmoc group can be removed entirely by using piperidine in *N*,*N*-dimethylformamide (DMF) (20–50%). The base neutralizes the exposed amino group, and the neutralization of the peptide resin is not required. At the same time, the final cleavage of the peptide from supports and the removal of side chain–protecting groups are performed with TFA solution which greatly takes the place of HF or TFMSA used in Boc method. Independent mechanisms occur during the deprotection of amino group and the cleavage of peptide. As expected, undesired cleavage of intermediates and side reactions decreases and the yield of peptide increases obviously under much milder reaction conditions. As the neutralization of the peptide resin is not needed in the presence of a base, therefore the lack of electrostatic repulsions between the peptide chains often leads to peptide aggregations, which is an innegligible disadvantage of Fmoc strategy. Besides, Fmoc amino acids are more expensive than their corresponding Boc amino acids. Therefore, several factors, such as the secondary structure of the target peptide, should be considered when choosing a suitable method to synthesize peptides.

7.1.3 Solid-Phase Synthesis of Oligonucleotides

In the past 30 years or so, almost all syntheses of oligonucletides has been carried out on solid supports. It is undoubtedly the most rapid and effective method of synthesizing small amounts of oligonucleotides required for biological research or multigram quantities of them for clinical trials of potential oligonucleotide drugs (11). It took more than 10 years to turn the aforementioned research of Letsinger and Kornet (2) to a feasible approach. By the late 1970s, the phosphotriester strategy had been established and applied by many people (12). In the mid of 1980s, this method was superseded by phosphoramidite base synthesis (13).

Solid-phase synthesis of oligonucleotides processed by using phosphoramidite method in a 3′ to 5′ direction is shown in Fig. 7.3.

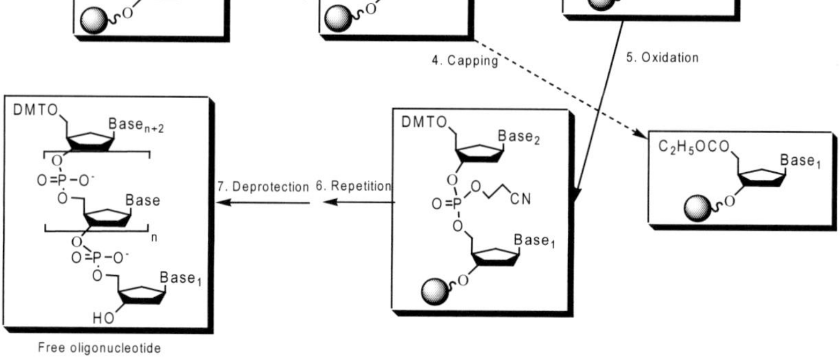

Figure 7.3 Scheme of solid-phase DNA synthesis.

In the synthesis of DNA (14,15), the most common nucleotide phosphoramidite building blocks are acid-labile dimethoxytrityl (DMT) protected 5′-hydroxyl group and base-labile groups protected exocyclic amino groups. In a typical synthesis cycle, first, the solid support possessing the first protected 5′-hydroxyl group is loaded into the reaction column. After removing the DMT protected 5′-hydroxyl group by adding a base and repeatedly washing the column by acetonitrile, the appropriate nucleoside phosphoramidite solution, premixed with activator (tetrazole), is sent to the reaction column to form an unstable phosphite linkage with the preceding hydroxyl group. The coupling reaction is allowed to take place for few minutes, and then excess reagents are flushed out of the column. Capping reagent (acetic anhydride) is then sent to the column to prohibit the uncoupling hydroxyl group from growing to an undesired chain. Then the solution of iodine in water, pyridine, or tetrafuran is added to oxidize the unstable phosphate linkage to a more stable phosphate linkage. The cycle is then repeated for the desired number of times. Each cycle has a routine step yield of about 98–99.5%, which can be estimated by spectrophotometrically determining the released DMT cation. After the nucleotides have been added, the protecting groups are removed and the chain is cut off simultaneously by using concentrated ammonia at high temperature. The obtained mixture is then purified by polyacrylamide gel electrophoresis (PAGE) or reversed-phase high performance liquid chromatography (RP-HPLC). While controlled-pore glass (CPG) (16,17) is used as the generally used supports, some other beads can also be used in the synthesis of DNA. For example, Farre *et al.* synthesized a 23-mer hepatitis B virus (HBV) probe by using nanosized silica particles, which were conjugated to microsized silica through reversible immobilization as supports (18). The obtained nano-on-micro (NOM) system can be well adapted to oligodeoxyribonucleotide (ODN)-synthesis in an automated instrument and can finally provide ODN-functionalized nanosized particles that can be used to target DNA in biology samples.

The synthesis of RNA is regarded far more difficult than the synthesis of DNA because it is difficult to find a compatible 2′-protecting group that is stable throughout the chain assembly and can be removed selectively without isomerization or degradation. Based on the efficient solid-phase DNA synthesis method, great efforts are currently being directed toward the development of reliable methods for solid-phase synthesis of RNA, especially by the solid-phase phosphoramidite method (19). The key success in these endeavors lies in finding suitable and mutually compatible protecting groups for both the 2′- and 5′-hydroxy groups. Generally, the protection of 2′-hydroxyl group by bis(2-acetoxyethoxy)-methyl orthoester (ACE) group (20), *t*-butyldimethylsilyl (TBDMS) group, or *t*-butyldimethylsilyl-oxymethyl (TOM) group (21) begins only after the protection of the 5′-hydroxy group has finished. However, the acid-labile protecting group in the ACE strategy is not stable enough under the deprotected conditions of 5′-hydroxy group. The TBDMS and TOM approaches (Fig. 7.4) have proved as much greater improvements in solid-phase RNA synthesis, as both of them can be removed by treatment with fluoride ion and the latter also offers slightly higher yields of RNA (19). Besides the above-mentioned protecting strategies, there are also other protecting groups which could be used in solid-phase synthesis of RNA. For example, Lackey *et al.* (22) carried on 2′-*O*-Lv (Levulinyl group) strategy to produce oligoribonucleotides. In this method, the monomers produced were harder than TBDMS monomers. They were synthesized by sequential *N*-levulination of the exocyclic amines, 5′-dimethoxytritylation, 2′/3′-*O*-levulination, and finally, phosphitylation. The separated 2′-*O*-phosphoramidite regio-isomers were then obtained through flash silica gel column chromatography in low yield. However, in this strategy, the coupling reaction is 10 times faster than that in the TBDMS strategy and the on-column unblocking step at the end of the synthesis greatly simplifies and speeds up the post-synthesis process.

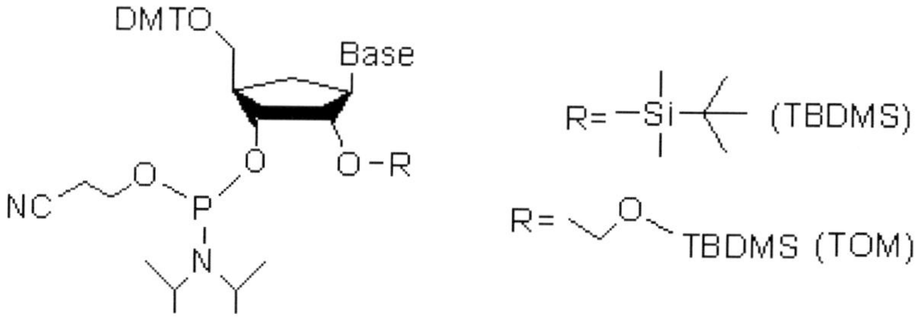

Figure 7.4 2′-*O*-protected ribonucleoside phosphoramidites.

7.1.4 Solid-Phase Synthesis of Peptide–Oligonucleotide Conjugates

Synthetic oligonuceotide is a class of therapeutic agents that is used for the treatment of viral infections and genetic diseases by therapeutic regulation of gene expression. However, chemical modification such as peptide–oligonucleotide conjugation has been used to solve the problem of the poor cell uptake of these agents and their inability to reach the nucleus of cells. Stepwise solid-phase synthesis without intermediate purification is an efficient method to produce the covalently linked conjugates. The starting functional groups on the beads such as the amino group and the hydroxyl group are similar to those used in the synthesis of peptides and oligonucletides. However, the incompatibility of the usual protecting strategies for them has resulted in a different way of synthesizing them, even when the traditional methods of solid-phase synthesis of both oligonucleotides and peptides are well established (23). The existing protocols of oligonucleotide synthesis are based on acid-labile temporary protection and base-labile global protection or both protections are acid-labile, whereas for peptide synthesis, the temporary protection is base-labile and global protection is acid-labile. Therefore, the existing strategy should be modified to meet the needs of the chain conjugation. So far, the stepwise strategies have been carried out based on about the following two ways: (i) Using a linker which bears an amino group for peptide synthesis and a hydroxyl group for oligonucleotide synthesis as the starting material in solid-phase synthesis and (ii) After the supported bounded peptide was prepared, the linker was conjugated to the terminal amino group of the peptide, which possesses hydroxyl group as the starting point of the oligonucleotide. Very recently, the review on these methods has been reported by Lonnberg (24). In this report, most of the synthesized peptide oligonucleotide conjugates (POCs) contain peptides at the 3′-end of the oligonucleotides. Generally in this strategy, the peptide is first synthesized by using Fmoc *N*-terminal-protected amino acids containing acid-labile protecting group (Boc), which can be removed by treating it with TFA-1,2-ethanedithiol (25,26).

Though PS beads and CPG supports are used in the synthesis of peptides and oligonucletides, respectively, CPG and its derivatives are far more commonly used in the synthesis of POC.

During the past 20 years, a number of POCs were synthesized. It has been shown that the conjugation of peptides and oligonucleotides improved some of the desired properties, such as cell penetration, stability toward exonuleases, and rate of hybridization (23). So far, the synthesis of POC through SPOS is the methodology used. In this methodology, the conjugation yield is very high and the purification is very easy, but the choice of suitable protecting groups, which should be compatible with peptides and oligonucleotides both, is restricted and still deserves great considerations.

7.1.5 Solid-Phase Synthesis in Combinatorial Chemistry

Combinatorial chemistry is a relative new field of chemistry which developed in the last decade of the twentieth century. It has significantly contributed to the drug discovery process because of some of its distinctive advantages, including the entire completion of the reaction, easy separation, and highly developed automation. Since the aim of this field is the preparation of rationally designed libraries, the use of solid supports for the generation of libraries has been a great advance in enhancing productivity. With a group of resins and simple operations, thousands of numerous and different but related beaded-tethered compounds could be generated under similar reaction conditions. The elegant and powerful "split-and-pool" synthesis strategy for combinatorial chemistry (Fig. 7.5) (27,28) is one of the most commonly used method to produce compounds libraries. In the beginning, beads are split into corresponding vessels, and each portion reacts with only a single reactant. Once the first reaction step is accomplished, all the bead-bounded products are mixed and washed together. These beads are then split again into requisite number of reaction vessels. These split-and-pool steps are repeated until the desired compound libraries are obtained. In this approach, each compound is prepared in equal quality or equal concentration. Moreover, besides increasing the synthesizing efficiency over the traditional way, the automated procedure provides great accessibility to purification and characterization of high-throughput parallel reactions in combinatorial chemistry. Progress in this field is quick and the developments will be indispensable to almost all combinatorial chemistry efforts. For example, Willoughby *et al.* (29) synthesized the 3-(aminoalkyl)-2-arylindole series, which was targeted for a variety of G-protein-coupled receptors. A 128,000-member library, organized in 320 pools containing 400 compounds each, was prepared and several of these compounds were revealed to be highly potent for certain receptors.

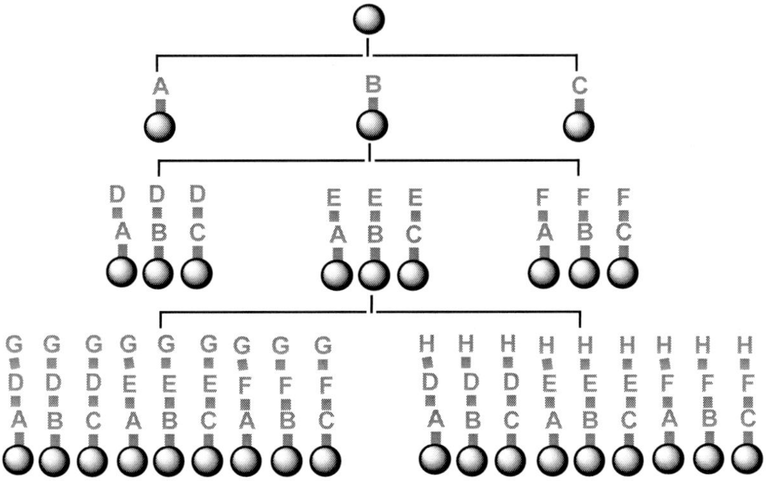

Figure 7.5 Scheme of "split-and-pool" synthesis strategy in combinatorial chemistry.

7.1.6 Solid-Phase Synthesis of Other Chemicals and Biomolecules

Inspired by the success of solid-phase synthesis of peptides and nucleotides, the study on the solid-phase synthesis of oligosaccharide has been carried on in 1970s. As we know, solid-phase synthesis of oligosaccharides also includes repeating activations and deprotection steps. The reaction can be driven to completion, and the separation proceeds thoroughly by using excess dissolved reagents and washing solvent. The central feature of the reaction is the formation of a glycosidic bond through the coupling of two glycosyls in solid-phase synthesis of oligosaccharides. Because of the complex structure of glycoside (Fig. 7.6), more factors should be considered in solid-phase synthesis of oligosaccharide compared to that of peptides or oligonucleotides: (i) choice of synthetic strategies with reducing or nonreducing end of the chain; (ii) proper selection of suitable supports and linkers which are stable under all reaction conditions but easy to be cleaved smoothly at the end of the routine; (iii) stable protecting-group strategy compatible with the desired oligosaccharide; and (iv) stereospecific and high-yielding glycosylation reactions. These aspects have been reviewed by Seeberger and Haase in 2000 (30). In recent years, various new methods in automated solid-phase synthesis of oligosaccharides have been exploited (31). Since most of them are methodologies, more work is still needed. On the whole, SPOS is a well-developed strategy to produce some biomolecules and a series of chemicals. However, there are some disadvantages and problems that deserve more consideration: (i) Heterogeneous reactions carried out on solid supports restrain the rate of reactions to a certain extent. This limitation makes the solid-phase synthesis method applicable only to some high-speed organic synthesis reaction but not to popular solution reaction; (ii) The use of excess of solvents and reactants solutions greatly reduces the rate of transformation stoichiometrically; (iii) The analysis on the reactions proceeded on solid beads are not as feasible and reliable as those on solution reactions; and (iv) The conjugations between supports and initial reactants are not stable to nucleophilic agents, reducing agents, and so on. Series of researches are being carried out to solve these problems (32,33).

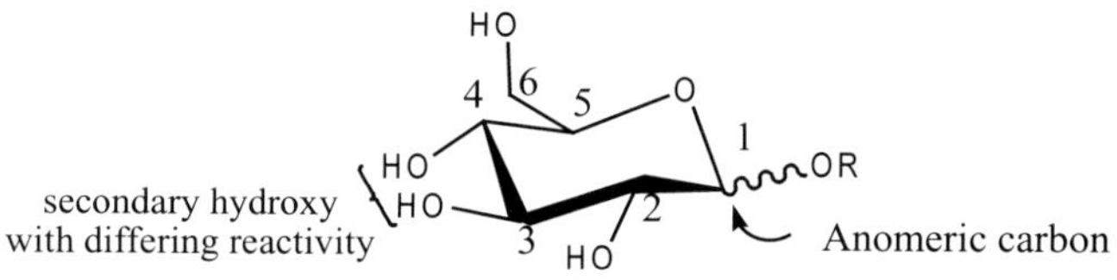

Figure 7.6 Monosaccharide contains multiple reactive sites and an anomeric carbon.

7.2 Polymer Microsphere Supports in Solid-Phase Synthesis

In SPOS, the polymer microsphere supports play the most pivotal role because the kinetics of SPOS is not always similar to that of solution phase reaction. It has been known that the supported reaction system is complex. The polymer–solvent–reagent system possesses various physical and chemical parameters which determine the procedure of the reaction. As a consequence, these limitations can be partly overcome by certain carrier materials, which were tailored with alternative properties suited for defined purposes. For the completion of the reaction carried out on the carriers, these polymer microsphere supports should possess the following characteristics:

- The polymer microspheres should be swollen but should not be dissolved in the reaction solvents. This helps the reactants in the solution to access the reactive site on the surface of those beads.
- The flexibility and the mechanical strength of polymer supports should be at a certain level to make the entire procedure feasible, including conjugation, reactions, filtrations, and deprotection.
- To avoid side reactions or the decomposition of the beads, the polymer supports should be inert to all reagents and solvents except the initial reactant.
- The beads should be capable of functionalization through some specific reactions. The conjugation of the initial reactant and the cleavage of the final product should be processed thoroughly under mild but special conditions.

A wide variety of polymer microspheres have been investigated considering the above criteria. In the following sections, we will introduce those most used polymer microsphere supports briefly.

7.2.1 Polystyrene Microspheres

The most extensively used and one of the earliest developed polymer supports in SPOS are cross-linked PS beads. They are available in various sizes, with different loading capacities (amount of functionalization) and different degrees of cross-linking.

PS beads can be prepared by suspension polymerization (34,35) or membrane emulsification combined with suspension polymerization (36). The basis of these processes is the dispersion of an organic phase consisting of monomer, radical initiator, and cross-linking agent into an aqueous phase. The size of the polymer beads is adjusted by the emulsifying process. In order to obtain insoluble beads, a certain amount of cross-linking agent should be used in the polymerization. The cross-linking agent is generally a multifunctional molecule that can conjugate two or more growing chains to interconnected networks. The most popular cross-linking agent is divinylbenzene (DVB) whose content varies approximately between 0% and 100%.

Low-cross-linked (1–2% DVB) PS-DVB microspheres are the most commonly used supports in SPOS, and they process a homogenous network with a glassy and transparent appearance. Obviously, cross-linked PS microspheres need further functionalization for loading the initial reactant in solid-phase synthesis, such as post-reaction with the aromatic rings of PS via eletrophilic substitution or Friedel–Crafts reaction. For example, Merrifield resin (chloromethyl PS resin) and Wang resin (hydroxymethylphenoxy resin) are the most traditionally used and commercially available supports in solid-phase peptides synthesis. Because of their unfriendly production conditions and rigorous coupling reactions with protected amino acids, some new resins have been developed based on traditional resins (Table 7.1). In Boc solid-phase peptide synthesis, Merrifield resin, oxymethylphenylacetamidomethyl (PAM) resin, and paramethylbenzhydrylamide (MBHA) resin are separately used in C-teminal carboxylic acid and C-terminal amide peptides synthesis. In Fmoc strategy, Wang resin and 2-chlorotrityl chloride (2-CTC) resin are used in C-terminal carboxylic acid peptides synthesis and the Rink amide MBHA resin is used in C-terminal amide peptides synthesis.

Table 7.1 Commonly used polymer microsphere supports in solid-phase peptide synthesis

Boc strategy		Fmoc strategy	
C-teminal acid	**C-teminal amide**	**C-teminal acid**	**C-teminal amide**
Merrifield resin 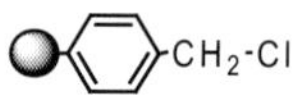	MBHA resin 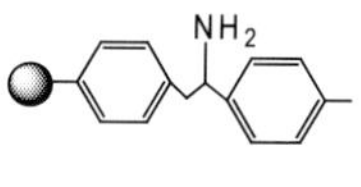	Wang resin 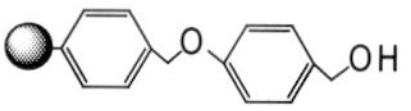	Rink amid MBHA resin 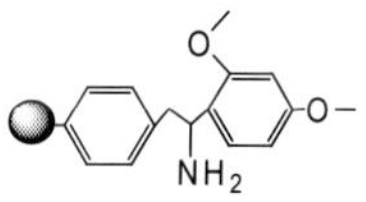
PAM resin 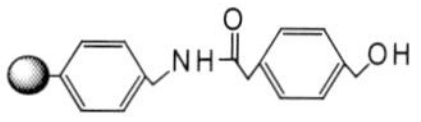		2-CTC resin 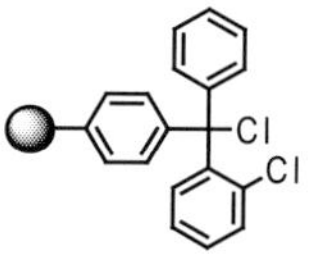	

Besides low-cross-linked PS beads, highly cross-linked PS-DVB microspheres, such as POLYHYPE (polystyrene-polyacrylamide composite), have been gradually developed in recent years. They are generally macroporous PS-based microsphere supports. In SPOS, reagents transport through the pore structure of the polymer rather than on the surfaces alone, which makes their washing easier and quicker than classical PS resins. It is worth mentioning that because of their extremely rigid structure compared with the soft low–cross-linked PS microspheres, macroporous PS resins are more suitable choices in continuous-flow automated solid-phase synthesis but not in usual mixing and bubbling solid-phase synthesis (37,38).

Compared with other supports, PS-based microspheres have high mechanical and chemical stability in various reagents and solvents. However, because of the hydrophobic property of the PS polymer, the formation of a truncated chain of the targeted molecule by incomplete coupling and deprotection pose a major problem in the overall yield of pure products. In some researches, these problems could be partly overcome by appropriate choice of cross-linker or linker. For example, Roice and Pillai (39) prepared poly(styrene-co-glycerol dimethacrylate) as a resin for gel-phase peptide synthesis. Sasikumar *et al.* (40) synthesized PS support with glycerol-based cross-linker for peptide synthesis. Generally these derivatized PS beads could show higher efficiency in solid-phase synthesis than pure PS polymer. While linkers modified the local chemical environment around the reactive sites of these resins, they did not have enough significant effect on the intrinsic hydrophobicity of the PS resin. Many efforts have been made for the development of other polymer supports.

7.2.2 PEG-Grafted Microspheres

Since the reaction kinetics may be restrained by the hydrophobic microenviroment and the steric-hindrance effect of PS supports in polar solvent, which has been proven to be undesirable for biochemical reactions, solutions of PEG-grafted PS polymers (41), such as Tentagel (monofunctional PEG-grafted PS) and ArgoGel (bifunctional PEG-grafted PS) (42,43) have been developed as another class of most-used resins for SPOS. Generally, the PEG-grafted PS polymers are prepared by grafting short-chain PEG to PS resin, which enhances the pressure stability of PS alone. The content of grafted PEG chain can be as high as 70%. These PEG chains determine the extended solvent compatibility of the resin and higher degrees of swelling in polar solvents than those of PS-DVB alone. For example, Tentagel resins exhibit almost equally good swelling property in several different solvents, which makes them suitable for packing columns and carrying out continuous solid-phase synthesis (44). Moreover, the high degree of flexibility of the PEG chains results in a good solution-like microenviroment, which, in turn, makes the on-bead reaction monitoring, such as gel-phase NMR, possible. However, there are some disadvantages of PEG chains, which restrict their use in SPOS. First, the loading of active site is not as high as that of normal PS derivatives for those active functional groups that lie at the end of the PEG chain and not on the aromatic ring of PS. Second, leakage of PEG chains often occurs when some aggressive reagents are used in solid-phase synthesis or a PEG complex is formed when some Lewis acids are used. Moreover, PEG-grafted PS polymers are much more expensive than PS resins.

Under some circumstances, solid-phase reactions are still affected by the proximity of the attached reactants to the polymer matrix even when these beads are highly solvated (Fig. 7.7). To overcome these problems, Cho and his group (45) synthesized core–shell-type PEG-grafted PS microspheres (CutiCore), which showed much higher efficiency in the solid-phase peptide synthesis. Compared to traditional PEG–PS supports, these microspheres could be free from the matrix effect and diffusion problems, and the functional group spread on the surfaces of beads allowing the coupling reaction to proceed in homogeneous status successfully (Fig. 7.7B). Cho *et al* (46) also synthesized highly cross-linked (>4%) core–shell-type PS–PEG microspheres that performed well in peptide synthesis. Furthermore, a highly PEG-based support was made by Roice *et al.* (47) that presented high efficiency in peptide synthesis.

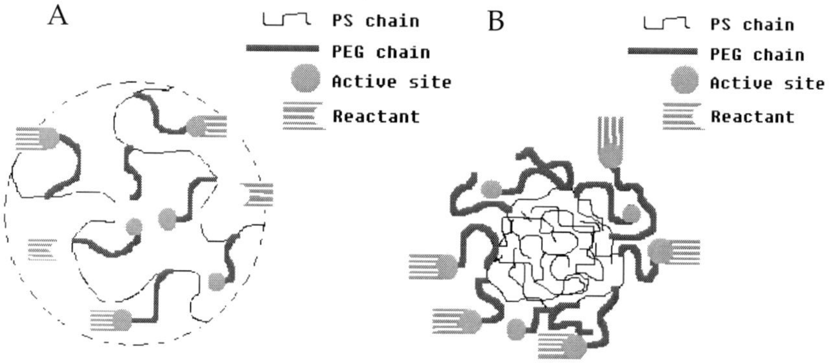

Figure 7.7 Reactions proceeding on different PEG–PS polymer microspheres: (A) Traditionally grafted, (B) Core–shell-type grafted.

Considering the advantages of undissolved cross-linked PEG polymer in polar solvents, other copolymers of PEG besides PS–PEG were synthesized to meet the growing demands of solid-phase synthesis, such as PEG derived with oxetane, SPOCC^{TM} (48); polyoxyethylenepolyoxypropylene, POEPOP (49); PEG-polyacryl–amide, PEGA^{TM} (50); PEG-MA (51); and poly(vinyl alcohol)-graft-poly(ethylene glycol) resins (52–54). Furthermore, totally PEG polymer, ChemMatrix (49,55), has also been developed as support for solid-phase peptide synthesis (56).

However, the PEG–PS polymer microspheres still have some non-negligible disadvantages. First, PEG fragments that drop from the polymer during the synthesis process often contaminate the final products. Second, since PEG is slightly soluble in THF, which is the most commonly used solvent in metal–organic chemistry, the application of the polymer is greatly restrained in solid-phase synthesis in these fields. Moreover, low loading of PEG–PS polymer microspheres compared with that of PS beads, poor mechanical stability, and their high cost also greatly limits their application in SPOS.

7.2.3 Polyacrylamides Microspheres

Sometimes the PS-based polymer beads may not be best suited for certain reactions because of their hydrophobic backbones. For example, a peptide chain may become incompatible with the polymer matrix when the sequence grows longer in peptide synthesis. To solve this problem, a series of microspheres have been explored as solid supports to be used in SPOS. Generally, such polymer microspheres are produced by the copolymerization of (i) *N,N′*-dimethylacrylamide, (ii) ethylenebisacrylamide, and (iii) acryloylsarcosine methyl ester, in which (i) is the monomer, (ii) is the cross-linker, and (iii) serves as a functionalizing substituent (57). The resulting polyacryamide supports have great mechanical characteristics and are structurally more like the peptide chains than PS. These characteristics greatly improve supports' performances in peptide synthesis. First, the presence of a high proportion of tertiary amide groups gives good swelling characteristic in suitable polar organic solvents. Second, the cross-linkers are more stable in the repeated treatment with an acid than the usual methylenebisacryamide. Finally, the preparation of the polymer microspheres is easy enough to minimize other chemical operations. The beads, on the other hand, are so greatly compatible with hydrophilic solvent that they can even partly dissolve in polar solvents and may not swell well in non-polar solvents, restricting their application in SPOS in certain fields.

On the basis of polyacryamide, Wang and his coworkers (58) prepared a new kind of hydrophilic and thermosensitive copolymer, P(DEA-co-HEMA), that had a macroporous network structure and showed high loading and good compatibility in polar solvents. The copolymerization of HEMA increased the mechanical strength of the microsphere. The thermosensitivity of the polymer, which was determined by the content of DEA, makes the polymer swell better under lower critical solution temperature (LCST) and shrink above LCST. This transformation facilitates filtration on heating during the multiple steps of SPOS. However, the coupling yield of this polymer was not as high as traditional PS beads and, therefore, it needs more efforts to optimize its synthesis reaction conditions. Furthermore, Kempe and Barany (59) produced cross-linked ethoxylate acrylate resin (CLEAR™) by radical copolymerization. As expected, the resin showed excellent swelling properties in an unusually broad range of solvents. However, during the synthesis of peptides, the supports in some cases are labile to certain strong bases except when they are entirely stable under standard peptide synthesis conditions.

7.2.4 Controlled-Pore Glass (CPG) Beads

In contrast to other SPOS, the synthesis of oligonucleotide proceeds better on non-swelling solid supports than on other beads. CPG beads have been proven as the most-accepted support in the solid-phase synthesis of nucleotide (60). Compared to traditional polymer supports, CPG beads are more stable and do not swell in almost all solvents, which allows the introduced functional groups to only spread on their surface. Therefore, the loading of active sites is relatively lower than other polymer microspheres and the reactions proceed only on their surfaces. On the other hand, the low loading of active sites may lead to relatively higher reaction feasibility, and the non-swellable glass beads are the most suitable medium for column package in continuous-flow SPOS. Furthermore, on the basis of non-swellable CPG, amino silica colloid/polymer hybrid microspheres (61) were also prepared for oligonucleotide synthesis for high yield and purity. These stable aggregates were prepared by the covalent coupling of bifunctional PEG and colloidal aminated silica particles. The new supports were suitable not only for solid-phase oligonucleotide synthesis but also for *in vitro* diagnostic assay as the accessibility and the number of DNA probes immobilized on support surface and specific surface being offered for the capture are both greatly enhanced.

7.2.5 Other Supports

Besides the above-mentioned generally used polymer microspheres, new beads are being synthesized increasingly by researchers for SPOS. For example, Dai *et al.* (62) have synthesized a fluorescent DNA probe, which can be used as a DNA biosensor, by using poly(divinylbenzene) core/poly(4-vinylpyridine) shell beads in solid-phase synthesis. Barth and Rademann (63) have synthesized a series of microspheres based on polyethyleneimine cross-linking, possessing both high loading and good chemical reactivity, and have named them ultragels or ultraresins. Greenland *et al.* (64) have prepared a series of new gel-type cross-linked poly(vinyl ether) networks called superior liquid-uptake resin for polymer-supported synthesis (SLURPS) (65) to overcome the shortcomings of commercial beads mentioned earlier in this chapter.

7.3 Conclusion and Prospects

As a whole, great advances been made by numerous research groups to improve polymer microspheres, coupling strategies, protecting groups, and automation have enable the development of SPOS. Combined with its ease of separation, purification, and high yield of products, SPOS can now be applied in more and more solution phase synthesis effectively.

Since solid support is one of the key factors in solid-phase synthesis, lightly cross-linked PS microspheres and PEG-grafted PS polymer continuously act as the main forces in solid-phase synthesis methodology. However, considering some of their disadvantages, such as unconstant swelling property in different solvents of hydrophobic PS beads and the leakage of PEG chains of the PEG–PS polymer, more efforts should be made to optimize the structure of polymer microspheres. We believe that more and more new supports can be developed to meet the growing demands of solid-phase synthesis in chemistry, biochemistry, combinatorial chemistry, and so on.

References

1. Merrifield, R. B. (1963). Solid-phase peptide synthesis. I. The synthesis of a tetrapeptide. *Journal of the American Chemical Society*, **85**, 2149–2154.
2. Letsinger, R. L., & Kornet, M. J. (1963). Popcorn polymer as a support in multistep syntheses. *Journal of the American Chemical Society*, **85**, 3045–3046.
3. Li, H. D., Fedorova, O. S., Trumble, W. R., *et al.* (1997). Site-specific photomodification of DNA by porphyrin–oligonucleotide conjugates synthesized via a solid-phase H-phosphonate approach. *Bioconjugate Chemistry*, **8**(1), 49–56.
4. Mitchell, A. R. (2008). Bruce Merrifield and solid-phase peptide synthesis: A historical assessment. *Biopolymer (Peptide Science)*, **90**(3), 175–184.
5. Venkatesan, N., & Kim, B. H. (2006). Peptide conjugates of oligonucleotides: Synthesis and applications. *Chemical Reviews*, **106**, 3712–3761.
6. Plante, O. J., Palmacci, E. R., & Seeberger, P. H. (2001). Automated solid-phase synthesis of oligosaccharides. *Science*, **291**(5508), 1523–1527.
7. Sewald, N., & Jakubke, H. D. (2009). *Peptides: Chemistry and Biology* (2nd ed.). Weinheim: Wiley-VCH.
8. Schnölzer, M., Alewood, P., Jones, A., *et al.* (1992). *In situ* neutralization in Boc-chemistry solid-phase peptide synthesis: Rapid, high yield assembly of difficult sequences. *International Journal of Peptide and Protein Research*, **40**(3–4), 180–193.
9. Wurtz, N. R., Turner, J. M., Baird, E. E., *et al.* (2001). Fmoc solid-phase synthesis of polyamides containing pyrrole and imidazole amino acids. *Organic Letters*, **3**(8), 1201–1203.
10. Kluczyk, A., Rudowska, M., Stefanowicz, P., & Szewczu. Z. (2010). Microwave-assisted TFA cleavage of peptides from merrifield resin. *Journal of Peptide Science*, **16**(1), 31–39.
11. Reese, C. B. (2005). Oligo- and Poly-nucleotides: 50 years of chemical synthesis. *Organic & Biomolecular Chemistry*, **3**(21), 3851–3868.

12. Sproat, B. S., & Gait, M. J. (1984). In M. J. Gait (Ed.), *Oligonucleotide Synthesis. A Practical Approach*. USA: Oxford University Press.

13. Caruthers, M. H. (1987). In S. A. Narang (Ed.), *Synthesis and Applications of DNA and RNA*. Orlando: Academic Press.

14. Haraguchi, K., Delaney, M. O., Wiederholt, C. J., *et al.* (2002). Synthesis and characterization of oligodeoxynucleotides containing formamidopyrimidine lesions and nonhydrolyzable analogues. *Journal of the American Chemical Society*, **124**(13), 3263–3269.

15. Eritja, R., Robles, J., Fernandez-Forner, D., *et al.* (1991). NPE-Resin, a new approach to the solid-phase synthesis of protected peptides and oligonucleotides I: Synthesis of the supports and their application to oligonucleotide synthesis. *Tetrahedron Letters*, **32**(11), 1511–1514.

16. Gough, G. R., Brunden, M. J., & Gilham, P. T. (1981). Recovery and recycling of synthetic units in the construction of oligodeoxyribonucleotides on solid supports. *Tetrahedron Letters*, **22**, 4177–4180.

17. Perez-Rentero, S., Garibotti, A. V., & Eritja, R. (2010). Solid-phase synthesis of oligodeoxynucleotides containing *N*-4-(2-(*t*-butyldisulfanyl)ethyl)-5-methylcytosine moieties. *Molecules*, **15**(8), 5692–5707.

18. Farre, C., Lansalot, M., Bazzi, R., *et al.* (2010). Automated oligonucleotide solid-phase synthesis on nanosized silica particles using nano-on-micro assembled particle supports. *Langmuir*, **26**(7), 4941–4950.

19. Marshall, W. S., & Kaiser, R. J. (2004). Recent advances in the high-speed solid-phase synthesis of RNA. *Current Opinion in Chemical Biology*, **8**, 222–229.

20. Scaringe, S. A., Wincott, F. E., & Caruthers, M. H. (1998). Novel RNA synthesis method using 5′-*O*-silyl-2′-*O*-orthoester protecting groups. *Journal of the American Chemical Society*, **120**, 11820–11821.

21. Usman, N., Ogilvie, K. K., Jiang, M. Y., & Cedergren, R. J. (1987). The automated chemical synthesis of long oligoribonucleotides using 2′-*O*-silylated ribonucleoside 3′-*O*-phosphoramidites on a controlled-pore glass support: Synthesis of a 43-nucleotide sequence similar to the 3′-half molecule of an *Escherichia coli* formylmethionine tRNA. *Journal of the American Chemical Society*, **109**, 7845–7854.

22. Lackey, J. G., Sabatino, D., & Damha, M. J. (2007). Solid-phase synthesis and on-column deprotection of RNA from 2′- (and 3′-) *O*-Levulinated (Lv) ribonucleoside monomers. *Organic Letters*, **9**(5), 789–792.

23. Liu, K., Duan, Q. P., Ma, L., *et al.* (2010). Chemical strategies for the synthesis of peptide- oligonucleotide conjugates. *Bioconjugate Chemistry*, **21**, 187–202.

24. Lonnberg, H. (2009). Solid-phase synthesis of oligonucleotide conjugates useful for delivery and targeting of potential nucleic acid therapeutics. *Bioconjugate Chemistry*, **20**(6), 1065–1094.

25. Tong, G., Lawlor, J. M., Tregear, G. W., & Haralambidis., J. (1995). Oligonucleotide-polyamide hybrid molecules containing multiple pyrene residues exhibit significant excimer fluorescence. *Journal of the American Chemical Society*, **117**, 12151–12158.

26. Haralambidis, J., Angus, K., Pownall, S., *et al.* (1990). The preparation of polyamideoligonucleotide probes containing multiple non-radioactive labels. *Nucleic Acids Research*, **18**, 501–505.

27. Nicolaou, K. C., Hanko, R., & Hartwig, W. (2002). *Handbook of Combinatorial Chemistry. Drugs, Catalysts, Materials.* Weinheim: Wiley-VCH.

28. Shin, D. S., Kim, D. H., Chung, W. J., *et al.* (2005). Combinatorial solid-phase peptide synthesis and bioassays. *Journal of Biochemistry and Molecular Biology*, **38**(5), 517–525.

29. Willoughby, C. A., Hutchins, S. M., Rosauer, K. G., *et al.* (2002). Combinatorial synthesis of 3-(Amidoalkyl) and 3-(Aminoalkyl)-2-arylindole derivatives: Discovery of potent ligands for a variety of G-protein coupled receptors. *Bioorganic & Medicinal Chemistry Letters*, **12**, 93–96.

30. Seeberger, P. H., & Haase, W. C. (2000). Solid-phase oligosaccharide synthesis and combinatorial carbohydrate libraries. *Chemical Reviews*, **100**, 4349–4393.

31. Dhanawat, M., & Shrivastava, S. K. (2009). Solid-phase synthesis of oligosaccharide drugs: A review. *Mini-Reviews in Medicinal Chemistry*, **9**, 169–185.

32. Hodge, P. (1997). Polymer-supported organic reactions: What takes place in the beads? *Chemical Society Reviews*, **26**, 417–424.

33. Poschalko, A., Rohr, T., Gruber, H., *et al.* (2003). SUBPOL: A novel sucrose-based polymer support for solid-phase peptide synthesis and affinity chromatography applications. *Journal of the American Chemical Society*, **125**, 13415–13426.

34. Zhou, W. Q., Gu, T. Y., Su, Z. G., *et al.* (2007). Synthesis of macroporous poly(styrene-divinyl benzene) microspheres by surfactant reverse micelles swelling method. *Polymer*, **48**(7), 1981–1988.

35. Zhu, X. X., Zhang, J. H., Gauthier, M., *et al.* (2006). Large uniform-sized polymer beads for use as solid-phase supports prepared by ascension polymerization. *Journal of Combinatorial Chemistry*, **8**(1), 79–84.

36. Hao, D. X., Gong, F. L., Hu, G. H., *et al.* (2008). Controlling factors on droplets uniformity in membrane emulsification: Experiment and modeling analysis. *Industrial & Engineering Chemistry Research*, **47**, 6418–6425.

37. Ley, S. V., Baxendale, I. R., Bream, R. N., *et al.* (2000). Multi-step organic synthesis using solid-supported reagents and scavengers: A new paradigm in chemical library generation. *Journal of the Chemical Society, Perkin Transactions I*, 3815–4195.

38. Kirschning, A., Monenschein, H., & Wittenberg, R. (2001). Functionalized polymers-emerging versatile tools for solution-phase chemistry and automated parallel synthesis. *Angewandte Chemie, International Edition in English*, **40**(4), 650–679.

39. Roice, M., & Pillai, V. N. R. (2005). Poly(Styrene-co-Glycerol Dimethacrylate): Synthesis, characterization, and application as a resin for gel-phase peptide synthesis. *Journal of Polymer Science, Part A: Polymer Chemistry*, **43**(19), 4382–4392.

40. Sasikumar, P. G., Kumar, K. S., & Pillai, V. N. R. (2004). A novel chemically and mechanically stable glycerol-based crosslinked polystyrene support for polypeptide synthesis: A comparative study with merri-field resin. *Journal of Applied Polymer Science*, **92**(1), 288–294.

41. Kates, S. A., McGuinness, B. F., Blackbum, C., *et al.* (1998). "High-load" polyethylene glycol–polystyrene (PEG–PS) graft supports for solid-phase synthesis. *Biopolymers (Peptide Science)*, **47**, 365–380.

42. Maeda, M., Kamada, H., Hojo, K., *et al.* (2001). Amino acids and peptides. XXXVIII. Facile synthesis of laminin-related Peptide-Poly(ethylene glycol) hybrids by the solid-phase method. *Chemical & Pharmaceutical Bulletin*, **49**(4), 488–491.

43. Gooding, O. W., Baudart, S., Deegan, T. L., & Heisler, K. (1999). On the development of new Poly(styrene–oxyethylene) graft copolymer resin supports for solid-phase organic synthesis. *Journal of Combinatorial Chemistry*, **1**(1), 113–122.

44. Bayer, E. (1991). Towards the chemical synthesis of proteins. *Angewandte Chemie International Edition*, **30**(2), 113–216.

45. Cho, J. K., Park, B. D., Park, K. B., & Lee, Y. S. (2002). Preparation of core–shell-type Poly(ethylene glycol) grafted polystyrene resins and their characteristics in solid-phase peptide synthesis. *Macromolecular Chemistry and Physics*, **203**(15), 2211–2217.

46. Cho, J. K., Park, B. D., & Lee, Y. S. (2000). A novel core–shell type polymer support for solid-phase peptide synthesis. *Tetrahedron Letters*, **41**(39), 7481–7485.

47. Roice, M., Johannsen, I., & Meldal, M. (2004). High capacity Poly (ethylene glycol) based amino polymers for peptide and organic synthesis. *QSAR & Combinatorial Science*, **23**(8), 662–673.

48. Rademann, J., Grøtli, M., Meldal, M., & Bock, K. (1999). SPOCC: A resin for solid-phase organic chemistry and enzymatic reactions on solid-phase. *Journal of the American Chemical Society*, **121**, 5459–5466.

49. Renil, M., & Meldal, M. (1996). POEPOP and POEPS: Inert polyethylene glycol crosslinked polymeric supports for solid synthesis. *Tetrahedron Letters*, **37**(34), 6185–6188.

50. Meldal, M. (1992). PEGA: A flow stable polyethylene glycol dimethyl acrylamide copolymer for solid-phase synthesis. *Tetrahedron Letters*, **33**(21), 3077–3080.

51. Kita, R., Svec, F., & Frechet, J. M. J. (2001). Hydrophilic polymer supports for solid-phase synthesis: Preparation of Poly(ethylene glycol) methacrylate polymer beads using "Classical" suspension polymerization in aqueous medium and their application in the solid-phase synthesis of hydantoins. *Journal of Combinatorial Chemistry*, **3**(6), 564–571.

52. Luo, J. T., Pardin, C., Lubell, W. D., & Zhu, X. X. (2001). Poly(vinyl alcohol)-Graft-Poly(ethylene glycol) resins and their use in solid-phase synthesis and supported TEMPO catalysis. *Chemical Communications*, **21**, 2136–2138.

53. Luo, J. T., Pardin, C., Zhu, X. X., & Lubell, W. D. (2007). Preparation, characterization, and application of Poly(Vinyl Alcohol)-Graft-Poly(ethylene glycol) resins: Novel polymer matrices for solid-phase synthesis. *Journal of Combinatorial Chemistry*, **9**(4), 582–591.

54. Wang, Z., Luo, J. T., Zhu, X. X., Jin, S. J., & Tomaszewski, M. J. (2004). Functionalized cross-linked Poly(Vinyl Alcohol) resins as reaction scavengers and as supports for solid-phase organic synthesis. *Journal of Combinatorial Chemistry*, **6**(6), 961–966.

55. Cote, S. New polyether based monomers and highly cross-linked amphiphile resins. WO2005012277, 2005-02-10.

56. Garcia-Martin, F., Quintanar-Audelo, M., Garcia-Ramos, Y., *et al.* (2006). ChemMatrix, a Poly(ethylene glycol)-based support for the solid-phase synthesis of complex peptides. *Journal of Combinatorial Chemistry*, **8**(2), 213–220.

57. Arshady, R., Atherton, E., Gait, M. J., *et al.* (1979). Easily prepared polar support for solid-phase peptide and oligonucleotide synthesis. Preparation of substance P and a nonadeoxyribonucleotide. *Journal of the Chemical Society, Chemical Communications*, **9**, 423–424.

58. Wang, Z., Song, Z. J., & Zhu, X. X. (2003). Preparation and characterization of crosslinked poly(*N*,*N*-dethylacrylamide-*co*-2-hydroxyethyl methacrylate) resins and their application as support in solid-phase peptide synthesis. *Journal of Polymer Science, Part A: Polymer Chemistry*, **41**(11), 1681–1690.

59. Kempe, M., & Barany, G. (1996). CLEAR: A novel family of highly cross-linked polymeric supports for solid-phase peptide synthesis. *Journal of the American Chemical Society*, **118**(30), 7083–7093.

60. Damha, M. J., Gianmaris, P. A., & Zabarylo, S. V. (1990). An improved procedure for derivatization of controlled-pore glass beads for solid-phase oligonucleotide synthesis. *Nucleic Acids Research*, **18**(13), 3813–3821.

61. Pacard, E., Brook, M. A., Ragheb, A. M., *et al.* (2006). Elaboration of silica colloid/polymer hybrid support for oligonucleotide synthesis. *Colloids and Surfaces, B: Biointerfaces*, **47**(2), 176–188.

62. Dai, Z., Li, P., Sun, X. X., *et al.* (2009). Preparation of fluorescent DNA probe by solid-phase organic synthesis. *Express Polymer Letters*, **3**(8), 483–491.

63. Barth, M., & Rademann, J. (2004). Tailoring ultraresins based on the cross-linking of polyethylene imines. Comparative investigation of the chemical composition, the swelling, the mobility, the chemical accessibility, and the performance in solid-phase synthesis of very high loaded resins. *Journal of Combinatorial Chemistry*, **6**(3), 340–349.

64. Greenland, B. W., Liu, S. Y., Cavalli. G., Alpay, E., & Steinke, J. H. G. (2010). Synthesis of beaded Poly(Vinyl Ether) solid supports with unique solvent compatibility. *Polymer*, **51**(14), 2984–2992.

65. Cavalli, G., Shooter, A. G., Pears, D. A., *et al.* (2007). Novel non-PEG derived polyethers as solid supports. 2. Solid-phase synthesis studies. *Journal of Combinatorial Chemistry*, **9**(6), 1012–1027.

Chapter 8

Microspheres for Solid-Phase Modification of Proteins

Tao Hu and Zhiguo Su

State Key Laboratory of Biochemical Engineering, Institute of Process Engineering, Chinese Academy of Sciences, Beijing, P. R. China

thu@home.ipe.ac.cn

8.1 Introduction

Protein-based pharmaceuticals (1,2) are limited in their clinical applications because of their poor solubility, proteolytic degradation, short *in vivo* half-life, immunogenicity, and toxicity (3–5). PEGylation, that is, conjugation with polyethylene glycol (PEG) has been widely used to overcome these disadvantages (6–9). Conjugation with the serum protein albumin (albuminylation) is another strategy to achieve this objective (10,11). However, traditional PEGylation and albuminylation methods may lead to low biological activity and heterogeneity of the modified proteins. To circumvent these disadvantages, the microsphere-based solid-phase method has been developed for chemical modification of proteins in recent years (12). This method has the potential to avoid multi-modification of proteins, maintain the biological activity of proteins, and minimize post-modification purification of the modified proteins.

The microspheres used for solid-phase modification of proteins are usually the liquid chromatography media. Liquid chromatography includes ion-exchange chromatography (IEC), size-exclusion chromatography (SEC), and affinity chromatography. These contain the liquid mobile phase and the stationary phase (13). Typically,

Microspheres and Microcapsules in Biotechnology: Design, Preparation, and Applications
Edited by Guanghui Ma and Zhiguo Su

ISBN 978-981-4316-47-7 (Hardcover), 978-981-4364-62-1 (eBook)
www.panstanford.com

the stationary phase is mechanically stable and chemically inert under chromatographic conditions used. It is also desirable that the microspheres act as a support for chemical groups with desirable properties, such as hydrophobicity, ion exchange, or affinity ligands (14). Therefore, microspheres are most commonly used as the stationary phase (i.e., chromatography media) in chromatography of samples.

The microspheres can prevent the adsorbed proteins from reacting with each other, thus favoring intramolecular reaction of proteins with reagents (15). In addition, the microsphere-based solid-phase method for modification of proteins can achieve at least partial separation of the modified protein from the unreacted modification reagent, unmodified proteins, and byproducts (16). Therefore, this method helps to control the modification process and avoid the heterogeneity of the modified proteins. Moreover, partial active sites of proteins may be covered by the interface of the microspheres. The proteins can then be protected from chemical modification. Therefore, the bioactivity of the modified proteins prepared by the solid-phase modification method is higher than that by random modification in solution.

In the following section, the solid-phase modification methods will be introduced for PEGylation and albuminylation of proteins. The solid-phase used includes the IEC media (i.e., microspheres), the SEC media, and the affinity chromatography media.

8.2 Modification Based on Ion-Exchange Chromatography

IEC can be used to separate proteins, which is based on attraction of oppositely charged molecules (17). The ion exchanger is composed of a base microsphere usually in the form of porous microsphere. This can provide enough surface area for adsorption of charged molecule (e.g., proteins) (18). Cation exchangers are negatively charged and can be used for adsorption of positively charged molecule. In contrast, anion exchangers are positively charged and can be used for adsorption of negatively charged molecule. In an IEC experiment, the sample is applied to a stationary phase which has been charged with the ion to be exchanged (Fig. 8.1). Since the protein carries surface charges depending on pI and pH of the environment, the type of the ion exchanger and the elution buffer should be carefully chosen to adsorb the protein. The adsorbed protein can be eluted from the ion exchanger using pH gradient or salt gradient.

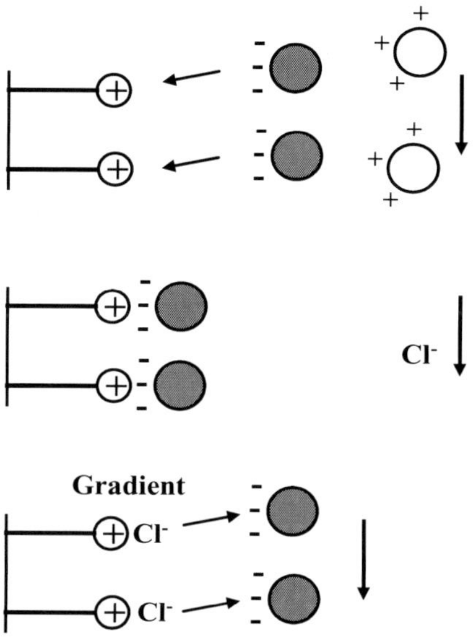

Figure 8.1 A typical ion-exchange experiment. (a) Cl^- is attached to a positively charged ion exchanger. (b) Negatively charged protein exchanges itself with the Cl^- and binds to the ion exchanger. (c) A competing Cl^- (from a NaCl gradient) elutes the protein.

Polysaccharides (cellulose, dextran, and agarose), synthetic organic polymers (polyacrylamide, polymethacrylate, polystyrene), and inorganic materials (silica, hydroxyapatite) have been used as microspheres in the design of chromatography media (19). To produce a mechanically stable and functional media, the materials are chemically cross-linked and provided with a functionally charged ligand. In many cases, the chromatographic media exhibited a typical size distribution of both particles and pores. Based on the properties of the IEC media, the IEC media were used to adsorb the charged proteins for solid-phase modification of proteins.

8.2.1 PEGylation of Proteins Using Ion-Exchange Chromatography

Succinimidyl carbonate PEG (SC-PEG) is a major commercial PEGylation reagent that has been widely used (20). SC-PEG is frequently targeted at ε-amino groups of lysine residues and α-amino groups of *N*-terminal amino acid residues of proteins (21,22) (Fig. 8.2). Such PEGylation reaction usually leads to complex reaction mixtures of the PEG-protein conjugates (e.g., mono-, di-, tri-, and higher PEGylated conjugates), differing in the number and the sites of the attached PEG chains (23,24). As therapeutic drugs, the PEGylated proteins or peptides should ideally be homogeneous with well-defined biological activities. Solid-phase PEGylation using IEC media is a good choice to control the SC-PEG-mediated PEGylation process and achieves the purpose.

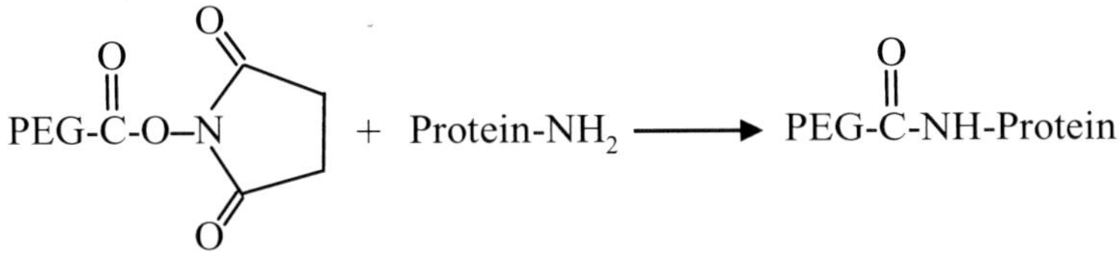

Figure 8.2 The reaction between SC-PEG and proteins.

Suo's research group has established a solid-phase modification method for PEGylation of human serum albumin (HSA, 66 kDa) and staphylokinase (SAK, 15 kDa) (25). Q Sepharose Big Bead is a cross-linked agarose coated with quaternary ammonium ligand. It is also a strong anion-exchange chromatography media. In contrast, DEAE Sepharose Fast Flow is a cross-linked agarose coated with diethylaminoethyl ligand and a weak anion-exchange chromatography media. HSA and SAK have different pI values of 4.7 and 6.8, respectively. To ensure full adsorption of proteins, Q Sepharose Big Bead and DEAE Sepharose Fast Flow were used as the solid-phase for modification of HSA and SAK, respectively. Briefly, HSA at pH 7.0 and SAK at pH 9.0 were adsorbed on Q Sepharose Big Beads and DEAE Sepharose Fast Flow, respectively. Then, the two adsorbed proteins were incubated with SC-PEG for a certain time. Finally, the PEGylated proteins and the proteins were eluted from the solid-phase using a linear salt gradient (Fig. 8.3).

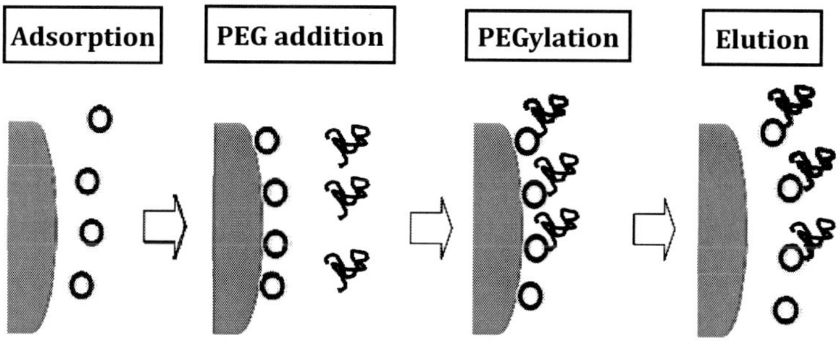

Figure 8.3 Solid-phase PEGylation of proteins based on IEC.

SEC studies demonstrated that solid-phase PEGylation generated the monoPEGylated HSA and the monoPEGylated SAK with yields of ~40% and ~35%, respectively. In contrast, PEGylation of proteins in aqueous solution led to multiPEGylation, in spite of the higher yield of monoPEGylation (>50%). This is presumably due to the steric hindrance of the microsphere to the PEG reagent and the interaction between the amino groups of proteins and the microspheres. The higher yield of the monoPEGylated HSA is presumably due to the high ligand density of Q Sepharose Big Bead. The higher M_w of HSA also decreases the steric hindrance of Q Sepharose Big Bead. Moreover, the solid-phase PEGylation could maintain 65% of the biological activity of SAK, which is higher than PEGylation of SAK in aqueous solution (30%). Presumably, partial active sites of SAK were adsorbed by the solid-phase interface and were thus protected for PEGylation at these active sites.

Lee *et al.* (16) reported a solid-phase modification method, using CM Sepharose as the solid-phase. CM Sepharose, a weak cation exchanger, is a cross-linked agarose coated with carboxymethyl ligand. This method is used for site-specific PEGylation of interferon (IFN) at the *N*-terminus using reductive alkylation (aldehyde chemistry) under acidic condition (Fig. 8.4). In brief, IFN was adsorbed on the solid-phase and PEGylated by PEG aldehydes. A salt-gradient elution efficiently purified the PEGylated IFN from the solid-phase with a yield of 50–60%. Although site-specific PEGylation at the *N*-terminus can be achieved in solution reaction, the solid-phase PEGylation can integrate the site-specific PEGylation of IFN and primary isolation of the PEGylated product in a single unit operation.

$$\text{IFN-NH}_2 + \text{PEG-}\overset{\overset{\displaystyle O}{\|}}{C}\text{-H} \xrightarrow{\text{NaCNBH}_3} \text{G-CSF-NH-CH}_2\text{-PEG}$$

Figure 8.4 The reaction between the PEG aldehyde and the protein.

8.2.2 Albuminylation Using Ion-Exchange Chromatography

Albumin is emerging as a versatile protein carrier for drug targeting and for improving the pharmacokinetic profile of protein-based drugs (10). Typically, albuminylation of proteins using bifunctional cross-linker (e.g., glutaraldehyde) may suffer from losing control of the conjugation process. IEC media can be used as a solid-phase to solve this disadvantage for albuminylation of proteins.

A well-defined bovine serum albumin (BSA)–hemoglobin (Hb) conjugate was prepared by a solid-phase modification method based on the IEC media (15). Since BSA has a low pI value of 4.7, Q Sepharose Fast Flow was used for its wide work range of pH. In brief, BSA at pH 7.0 was first adsorbed by the solid-phase. The absorbed BSA molecules were then allowed to react with the cross-linker glutaraldehyde. The spacing out of BSA molecules on the solid-phase was assumed to limit glutaraldehyde mediated polymerization of BSA molecules. However, some molecules bound closely on the solid-phase resulting in minor BSA dimer formation (Fig. 8.5). The monomeric BSA was separated from the BSA dimers by SEC based on a Superdex 200 column. The purified monomeric BSA reacted with the amino groups of bovine Hb at 4–8°C and pH 9.5. A well-defined BSA–Hb conjugate (i.e., BSA:Hb = 1:1) was thus purified from the resultant reaction mixture using the Superdex 200 column. To further minimize the formation of HSA dimers, low ligand density on the media and uniform size of the microsphere are required for the solid-phase.

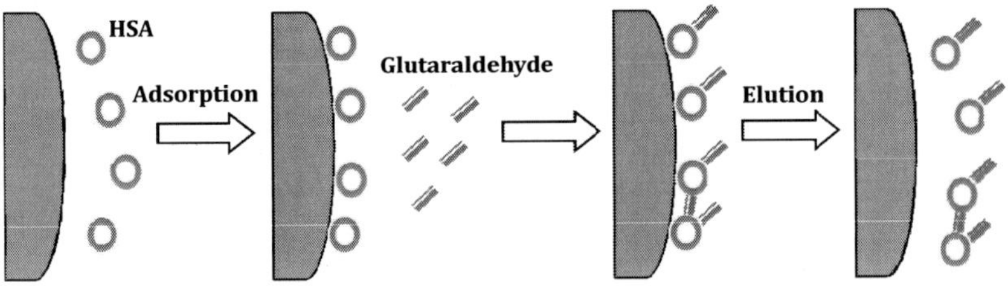

Figure 8.5 Solid-phase modification of HSA using glutaraldehyde.

The purity of the cross-linker (e.g., glutaraldehyde) is important to functionalize the proteins adsorbed on the solid-phase (26). Glutaraldehyde, a five-carbon dialdehyde, may form oligomers and polymers that can also cross-link with proteins (27). This means that various distances may exist between the two aldehyde residues. Thus, bovine Hb was adsorbed by Q Sepharose Fast Flow, followed by reaction with glutaraldehyde. When glutaraldehyde (grade II) containing these oligomers and polymers was used, undesired oligmeric Hb were formed (Fig. 8.6). Obviously, the long distance between the two aldehyde residues could diminish the spacing out of Hb molecules and result in the formation of Hb oligmers. In contrast, the bimodal distribution of Hb was observed using glutaraldehyde (grade I), a specifically purified form without oligomers and polymers, as the cross-linker (Fig. 8.6). Therefore, the purity of the cross-linker is important to control the reaction occurring in the solid-phase.

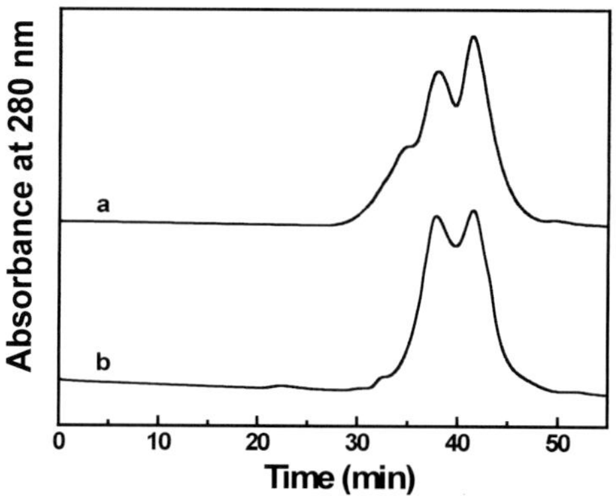

Figure 8.6 Comparison between the oligomeric Hb using glutaraldehyde (grade II) (a) and glutaraldehyde (grade I) (b). The samples were loaded on the Superdex 200.

As mentioned above, the ligand density of the microsphere and the steric hindrance of the microsphere to PEG may determine the process of PEGylation and albuminylation. The efficacy of the solid-phase modification also depends on the properties of the adsorbed protein (e.g., pI and the molecular size). In order to design IEC microsphere suitable for solid-phase modification, large size and low ligand density of the microsphere are preferable. In addition, the size uniformity of the microsphere is also desirable to control the batch-to-batch reproducibility of the solid-phase modification.

8.3 Modification Based on Size-Exclusion Chromatography

SEC is performed using porous, microspheres (beads) as the chromatographic support (28). A column constructed from such microspheres will have two measurable liquid volumes (29). The external volume consists of the liquid between the microspheres. The internal volume consists of the liquid within the pores of the microspheres. Large molecules can equilibrate only with the external volume, due to that they cannot enter into the pores of the microspheres. In contrast, small molecules can equilibrate both with the external and internal volumes, due to that they cannot enter into the pores (Fig. 8.7). This is referred to as the cutoff value or exclusion limit of the stationary phase.

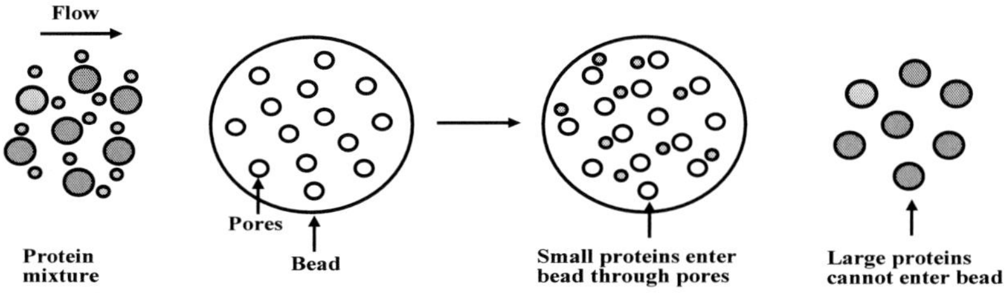

Figure 8.7 SEC of protein mixtures. Proteins smaller than the bead-exclusion limit can enter the beads. Larger proteins are excluded from the beads.

The important dimensions of microspheres are the diameter of the pores that access the internal volume and the hydrodynamic diameter of the protein molecules (30). Accordingly, Fee (31) proposed a reaction method, size-exclusion reaction chromatography (SERC) to control the modification process. The SERC method combines control of the PEGylation process with simultaneous purification of the products. Superdex 75, a cross-linked agarose with a diameter of 22–44 μm was used as the solid-phase. SERC utilizes a moving reaction zone within a SEC column. SERC can control the time of contact between reactants, selectively remove products from the reaction zone, and inhibit reactions based on molecular size. A short pulse of PEG reagent is injected onto the SEC column running at a constant flow rate. A subsequent pulse of native protein will eventually catch up to the slower-moving PEG pulse, forming a moving reaction zone (Fig. 8.8). Thus, the PEGylated product is partitioned differently from either of the two reactants.

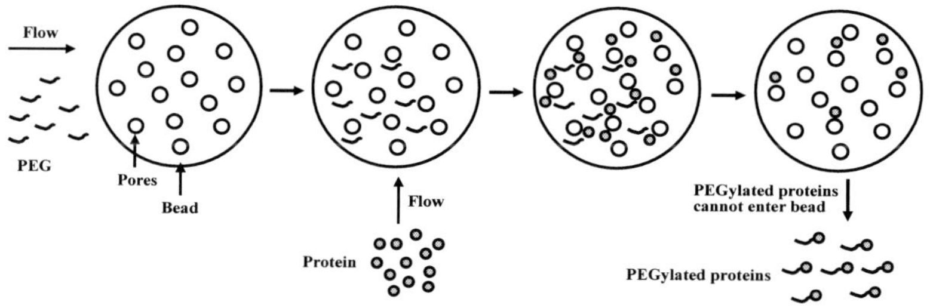

Figure 8.8 Solid-phase PEGylation of proteins based on SEC.

For example, α-lactalbumin and β-lactoglobulin were PEGylated using the SERC method (31). The large M_W products of PEGylation could be partitioned into the void volume of the column and be selectively removed from the slower-moving reaction zone, thereby minimizing the multiple PEGylation reactions. In addition, the SEC media may prevent the formation of a product larger than the pore size. Therefore, the PEGylated proteins can be directly separated from the reaction mixtures using the SERC method.

The operating conditions may alter the PEGylation kinetics. Also, the pore size of the microspheres and its distribution may influence the extent of reaction. By design and preparation of the microspheres with proper pore size, the PEGylated product will be partitioned into the void space and move relatively quickly out of the reaction zone. Additional column length will allow good separation of the PEG reagent and products.

Compared to the solid-phase PEGylation based on IEC media, the SERC PEGylation is milder and can avoid the tedious gradient salt elution. On the other hand, since the loading capacity of SEC media is much lower than that of IEC media, the amount of proteins loaded for the SERC PEGylation is much less than that for the IEC media–based method.

In order to design the microspheres suitable for the SERC method, the diameter of the pores are very important. In addition, PEGylation may significantly increase the hydrodynamic volume of the proteins (32,33). The mono-, di-, and multi-PEGylated proteins may show great difference in the hydrodynamic volume. Therefore, the proper diameter of the pores should be designed to ensure that the mono-PEGylated protein can enter into the pore whereas the di- and multi-PEGylated proteins cannot. Accordingly, the PEGylation process may be controlled and the multi-PEGylation may be minimized using SERC.

8.4 Modification Based on Affinity Chromatography

Affinity chromatography is a method for selective purification of a molecule from complex mixtures based on highly specific biological interaction between two molecules (34,35). The interaction is typically reversible. The purification is achieved through a biphasic interaction with the ligand immobilized to a surface while the target molecule is in a mobile phase as part of a complex mixture. Affinity chromatography media should ideally be macroporous with high chemical and physical stability (36). It also should selectively capture the target molecule while simultaneously avoid non-specific adsorption of other molecules.

However, large pores can reduce the surface area and the ligand density. In most cases, a compromise is made between the pore size and the surface area. The selection of the microsphere size will largely depend on the purpose and method of the separation (37). The size distribution should be as uniform as possible so that the smaller microspheres do not fill the void volume. Common microspheres include beaded agarose and cellulose. The microspheres should be specific for a protein of interest and inert to all other compounds in the sample (38). Accordingly, the solid-phase modification method based on affinity chromatography media has been developed to control the modification process.

8.4.1 PEGylation Using Affinity Chromatography

Affinity chromatography media can be used to orientate a protein so that its active site is held toward the microspheres interface. Thus, the active sites of the protein can be avoided from PEGylation through reversible adsorption on the affinity chromatography media (39). For example, glycoproteins were adsorbed on the microspheres, and their carbohydrate residues were located on the outside of the active sites. Thus, PEGylation of the adsorbed glycoproteins resulted in high stability and high bioactivity.

Baran *et al.* (40) used PEG to modify L-asparaginase (ASNase) and catalase. Two ligand-immobilized affinity column systems, L-asparagine-2 Sepharose CL-4B and Procion red-Sepharose CL-4B, were used for solid-phase PEGylation of ASNase and catalase, respectively (Fig. 8.9). In brief, catalase and ASNase were adsorbed on the microspheres and the PEG reagent was added. The active sites of the proteins were protected by orienting the enzyme-active sites toward the microspheres. The microspheres offer a convenient way to remove free reactive PEG reagent. Thus, this approach is suitable to control attachment of various polymers (e.g., PEG). Intravenous administration of the PEG-ASNase showed prolonged *in vivo* retention time. No adverse effects were observed in presensitized mice.

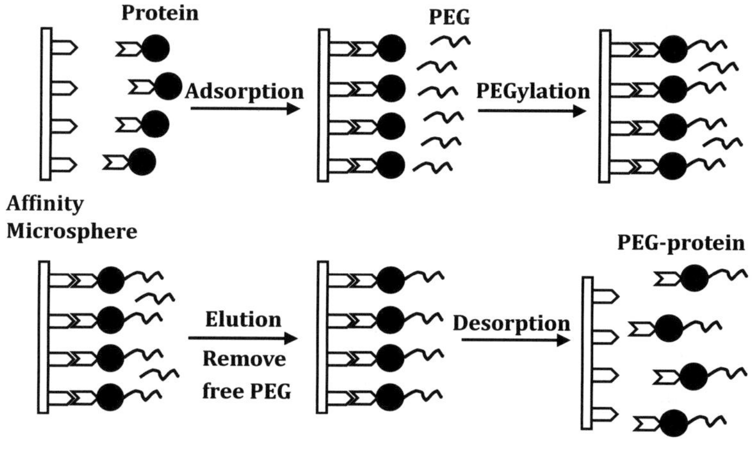

Figure 8.9 Solid-phase PEGylation of proteins based on affinity chromatography.

8.4.2 Preparation of Protein–Protein Conjugate Using Affinity Chromatography

Historically, albuminylation of proteins is derived from covalent conjugation between two dissimilar proteins. Both the two proteins should be in stoichiometric equivalence, preferably in the form of monoconjugates. Most methods of protein–protein conjugation employing bifunctional reagents (e.g., glutaraldehyde) are of low efficiency and result in mixed large conjugates of indeterminate stoichiometry.

Pillai and Bauihawat (41) proposed a solid-phase conjugation method based on affinity chromatography media to prepare the monoconjugates. Succinyl concanavalin A Sepharose is a cross-linked agarose coated with concanavalin A and was used as the solid-phase to specifically adsorb the glycoprotein enzyme, invertase. The adsorbed invertase was allowed to react with glutaraldehyde and the excess glutaraldehyde was eluted from the solid-phase. A non-glycoprotein, urease, was reacted with the resultant invertase and the conjugate was eluted from the media (Fig. 8.10). After elution with α-methyl glucoside, enzymes were detected in the eluate and the conjugate was a monoconjugate. Immunodiffusion analysis indicated that invertase could retain its immunological reactivity in the conjugate.

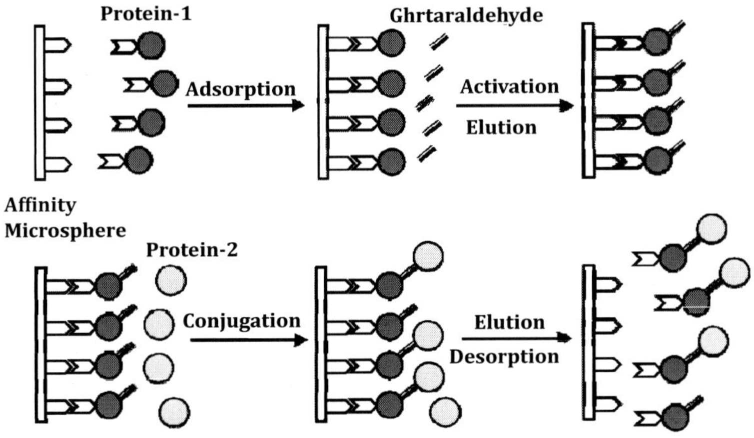

Figure 8.10 Preparation of protein–protein conjugate using affinity chromatography.

Similarly, an ^{125}I-labeled insulin was conjugated to invertase using affinity chromatography media as the solid-phase (42). Insulin and invertase are approximately equimolar (i.e., ~1:1) in the conjugate. Thus, the solid-phase method based on affinity chromatography media has significantly facilitated synthesis of protein–protein monoconjugates.

To obtain protein–protein monoconjugates, the minimum concentration of the adsorbed protein should be determined. This can ensure that the inter-protein distances are sufficient to avoid intermolecular bridge formation. For example, a polysaccharide media was used for affinity adsorption of the galactose-binding toxin from *Ricinus communis* at various densities (43). Inter-molecular cross-links were not formed below the critical density and the galactose-binding toxin was monoconjugated to lysozyme. This critical density may vary from protein to protein, which depends on the type of microspheres and parameters such as size and shape of protein.

In order to design the microspheres that are suitable for PEGylation and protein–protein monoconjugation, the surface area, the ligand density and the spacer of the microspheres should be carefully considered. For example, large surface area and low ligand density can increase the minimum concentration of the adsorbed protein due to their efficiency to avoid intermolecular bridge formation of adsorbed protein. The short spacer of the microspheres can efficiently decrease the formation of the multiple protein–protein conjugate. In addition, the microspheres should be carefully screened for selectivity, loading capacity, and low non-specific binding.

8.5 Conclusion and Prospects

Solid-phase modification of proteins based on the microspheres has been successfully developed in recent years. The microspheres used usually are the liquid chromatography media, including IEC media, SEC media, and affinity chromatography media. The modified proteins prepared by the methods based on the IEC and affinity chromatography media are homogeneous and have higher biological activity than that prepared in aqueous solution. The SERC method can inhibit the formation of di- and multi-PEGylated proteins. In addition, the methods can achieve primary isolation of the modified protein from the unmodified protein.

However, these methods need to be further optimized to control the modification process and obtain a well-defined modified product. For example, the yield of the modified product is relatively low in the method based on IEC media (25). The mono-PEGylated protein was not well-separated from the multi-PEGylated protein in the SERC method (31). The amount of protein loaded on the affinity chromatography media is low in order to avoid the intermolecular bridge formation (43). However, the presently used commercial microspheres cannot fully solve these problems during the solid-phase modification process. Therefore, it is necessary to design novel microspheres or select proper microspheres for modification of proteins. Accordingly, the diameter of the pore, the size and uniformity of the microspheres, the ligand chemistry, and the ligand density on the microspheres should be carefully considered and designed. In addition, the purity of the cross-linker is an important parameter to prepare a well-defined modified product.

Chemically modified proteins have been used in clinical therapy of many patho-physiological diseases (44–49). Some related products have been approved by FDA and launched in the market for their long-acting properties (50,51). Since the solid-phase modification method can solve the disadvantage of the modification occurring in aqueous solution, it is of interest to develop well-defined, long-acting proteins as therapeutics using this proposed method. It is fully anticipated that the microspheres will continue to play key roles in solid-phase modification of proteins in the future.

References

1. Nuttall, S. D., & Walsh, R. B. (2008). Display scaffolds: Protein engineering for novel therapeutics. *Current Opinion in Pharmacology*, **8**, 609–615.
2. Gebauer, M., & Skerra, A. (2009). Engineered protein scaffolds as next-generation antibody therapeutics. *Current Opinion in Chemical Biology*, **13**, 245–255.
3. Malik, D. K., Baboota, S., Ahuja, A., Hasan, S., & Ali, J. (2007). Recent advances in protein and peptide drug delivery systems. *Current Drug Delivery*, **4**, 141–151
4. Kompella, U. B., & Lee, V. H. L. (1991). Pharmacokinetics of peptide and protein drugs. In V. H. L. Lee (Ed.), *Peptide and Protein Drug Delivery* (pp. 391). New York: Marcel Dekker.
5. Rabkin, R., & Dahl, D. C. (1993). Renal uptake and disposal of proteins and peptides. In K. L. Audus & T. J. Raub (Eds.), *Biological Barriers in Protein Delivery* (pp. 299). New York: Plenum Press.
6. Abuchowski, A., McCoy, J. R., Palczuk, N. C., van Es, T., & Davis, F. F. (1977). Effect of covalent attachment of polyethylene glycol on immunogenicity and circulating life of bovine liver catalase. *Journal of Biological Chemistry*, **252**, 3582–3586.
7. Wong, K., Cleland, L. G., & Poznanski, M. J. (1980). Enhanced anti-inflammatory effects and reduced immunogenicity of bovine liver

superoxide dismutase by conjugation with homologous albumin. *Agent Action*, **10**, 231–239.

8. Veronese, F. M. (2001). Peptide and protein PEGylation: A review of problems and solutions. *Biomaterials*, **22**, 405–417.
9. Roberts, M. J., Bentley, M. D., & Harris, J. M. (2002). Chemistry for peptide and protein PEGylation. *Advanced Drug Delivery Reviews*, **54**, 459–476.
10. Bertucci, C., & Domenici, E. (2002). Reversible and covalent binding of drugs to human serum albumin: Methodological approaches and physiological relevance. *Current Medicinal Chemistry*, **9**, 1463–1481.
11. Pristoupil, T. I., Kramlova, M., Fricova, V., & Kraml, J. (1983). The reproducibility of fast protein liquid chromatography of pyridoxalated haemoglobin copolymerized with serum albumin. *Journal of Chromatographic*, **280**, 165–166.
12. Fee, C. J., & Van Alstine, J. M. (2006). PEG-proteins: Reaction engineering and separation issues. *Chemical Engineering Science*, **61**, 924–939.
13. Jungbauer, A. (2005). Chromatographic media for bioseparation. *Journal of Chromatography, A*, **1065**, 3–12.
14. Mueller, E. (2005). Properties and characterization of high capacity resins for biochromatography. *Chemical Engineering & Technology*, **28**, 1295–1305.
15. Hu, T., & Su, Z. (2003). A solid-phase adsorption method for preparation of bovine serum albumin-bovine hemoglobin conjugate. *Journal of Biotechnology*, **100**, 267–275.
16. Lee, B. K., Kwon, J. S., Kim, H. J., Yamamoto, S., & Lee E. K. (2007). Solid-phase PEGylation of recombinant interferon α-2a for site-specific modification: Process performance, characterization, and *in vitro* bioactivity. *Bioconjugate Chemistry*, **18**, 1728–1734.
17. Pabst, T. M., & Carta, G. (2007). pH transitions in cation exchange chromatographic columns containing weak acid groups. *Journal of Chromatography, A*, **1142**, 19–31.
18. Kaltenbrunner, O., Tauer, C., Brunner, J., & Jungbauer, A. (1993). Isoprotein analysis by ion exchange chromatography using a linear pH gradient combined with a salt gradient. *Journal of Chromatography*, **639**, 41–49.
19. Necina, R., Amatschek, K., & Jungbauer, A. (1998). Capture of human monoclonal antibodies from cell culture supernatant by ion exchange media exhibiting high charge density. *Biotechnology and Bioengineering*, **60**, 689–698.

20. Miron, T., & Wilchek, M. (1993). A simplified method for the preparation of succinimidyl carbonate polyethylene glycol for coupling to proteins. *Bioconjugate Chemistry*, **4**, 568–569.

21. Hooftman, G., Herman, S., & Schacht, E. (1996). Poly(ethylene glycol)s with reactive endgroups. II. Practical consideration for the preparation of protein-PEG conjugates. *Journal of Bioactive and Compatible Polymers*, **11**, 135–159.

22. Delgado, C., Francis, G. E., & Derek, F. F. (1992). The uses and properties of PEG-linked proteins. *Critical Reviews in Therapeutic Drug Carrier System*, **9**, 249–304.

23. Harris, J. M., & Chess, R. (2003). Effect of PEGylation on pharmaceuticals. *Nature Reviews Drug Discovery*, **2**, 214–221.

24. Hamidi, M., Azadi, A., & Rafiei, P. (2006). Pharmacokinetic consequences of PEGylation. *Drug Delivery*, **13**, 399–409.

25. Suo, X., Lu, X., Hu, T., Ma, G., & Su, Z. (2009). A solid-phase adsorption method for PEGylation of human serum albumin and staphylokinase: Preparation, purification and biochemical characterization. *Biotechnology Letters*, **31**, 1191–1196.

26. Hu, T., Li, D., & Su, Z. (2003). Preparation and characterization of dimeric bovine hemoglobin tetramers. *Journal of Protein Chemistry*, **22**, 411–416.

27. Beauchamp, R. O. Jr., St Clair, M. B., Fennell, T. R., Clarko, D. O., Morgan, K. T., & Kari, F. W. (1992). A critical review of the toxicology of glutaraldehyde. *Critical Reviews in Toxicology*, **22**, 143–174.

28. Hagel, L. (1998). Gel filtration. In J. C. Janson, & L. Ryden (Eds.), *Protein Purification* (pp. 145). New York: Wiley.

29. Hagel, L., Ostberg, M., & Andersson, T. (1996). Apparent pore size distributions of chromatography media. *Journal of Chromatography, A*, **743**, 33–42.

30. Hagel, L., Hartmann, A., & Lund, K. (1993). Analytical gel filtration of dextran for study of the glomerular barrier function. *Journal of Chromatography*, **641**, 63–70.

31. Fee, C. J. (2003). Size-exclusion reaction chromatography (SERC): A new technique for protein PEGylation. *Biotechnology and Bioengineering*, **82**, 200–206.

32. Hu, T., Li, D., Manjula, B. N., Brenowitz, M., Prabhakaran, M., & Acharya, S. A. (2009). PEGylation of Val-1(α) destabilizes the tetrameric structure of hemoglobin. *Biochemistry*, **48**, 608–616.

33. Wang, J., Hu, T., Liu, Y., Zhang, G., Ma, G., & Su, Z. (2011). Kinetic and stoichiometric analysis of the modification process for *N*-terminal PEGylation of staphylokinase. *Analytical Biochemistry*, **412**, 114–116.
34. Ostrove, S. (1990). Affinity chromatography: General methods. *Methods in Enzymology*, **182**, 357–371.
35. Wilchek, M., Miron, T., & Kohn, J. (1984). Affinity chromatography. *Methods in Enzymology*, **104**, 3–55.
36. Lowe, C. R., Lowe, A. R., & Gupta, G. (2001). New developments in affinity chromatography with potential application in the production of biopharmaceuticals. *Journal of Biochemical and Biophysical Methods*, **49**, 561–574.
37. Labrou, N. E. (2003). Design and selection of ligands for affinity chromatography. *Journal of Chromatography, B: Analytical Technologies in the Biomedical and Life Sciences*, **790**, 67–78.
38. Labrou, N. E. (2000). Dye-ligand affinity chromatography for protein separation and purification. *Methods in Molecular Biology*, **147**, 129–139.
39. Turková, J. (1999). Oriented immobilization of biologically active proteins as a tool for revealing protein interactions and function. *Journal of Chromatography, B*, **722**, 11–31.
40. Baran, E. T., Ozer, N., & Hasirci, V. (2003). Solid-phase enzyme modification via affinity chromatography. *Journal of Chromatography, B*, **794**, 311–322.
41. Pillai, S., & Bauihawat, B. K. (1977). Protein–protein conjugation on a lectin matrix. *Biochemical and Biophysical Research Communications*, **75**, 240–245.
42. Pillai, S., & Bauihawat, B. K. (1979). Stoichiometric enzyme-peptide conjugation of an affinity matrix. *FEBS Letters*, **101**, 301–304.
43. Pillai, S., Surolia, A., & Bachhawat, B. K. (1979). Critical density for protein–protein monoconjugation on a polysaccharide matrix. *Biochimica et Biophysica Acta*, **580**, 216–219.
44. Greenwald, R. B. (2001). PEG drugs: An overview. *Journal of Controlled Release*, **74**, 159–171.
45. Duncan, R., & Kopecek, J. (1984). Soluble synthetic polymers as protein drug carriers. *Advances in Polymer Science*, **57**, 53–101.
46. Ramon, J., Saez, V., Baez, R., Aldana, R., & Hardy, E. (2005). PEGylated interferon-alpha2b: A branched 40 K polyethylene glycol derivative. *Pharmaceutical Research*, **22**, 1374–1386.

47. Harris, J. M., Martin, N. E., & Modi, M. (2000). PEGylation: A novel process for modifying pharmacokinetics. *Clinical Pharmacokinetics*, **40**, 539–551.

48. Conover, C. D., Gilbert, C. W., Shum, K. L., & Shorr, R. L. (1997). The impact of polyethylene glycol conjugation on bovine hemoglobin's circulatory half-life and renal effects in a rabbit top-loaded transfusion model. *Artificial Organs*, **21**, 907–915.

49. Ashihara, Y., Kono, T., Yamazaki, S., & Inada, Y. (1978). Modification of E. coli L-asparaginase with polyethylene glycol: Disappearance of binding ability to anti-L-asparaginase serum. *Biochemical and Biophysical Research Communications*, **83**, 385–391.

50. Yamaoka, T., Tabata, Y., & Ikada, Y. (1995). Fate of water-soluble polymers administered via different routes. *European Journal of Pharmaceutical Sciences*, **84**, 349–354.

51. Kodera, Y., Matsushima, A., Hiroto, M., Nishimura, H., Ishii, A., Ueno, T., *et al.* (1998). PEGylation of proteins and bioactive substances for medical and technical applications. *Progress in Polymer Science*, **23**, 1233–1271.

Chapter 9

Microspheres and Microcapsules for Protein Drug Delivery

Lianyan Wang, Tingyuan Yang, and Guanghui Ma

State Key Laboratory of Biochemical Engineering, Institute of Process Engineering, Chinese Academy of Sciences, Beijing, P. R. China

ghma@home.ipe.ac.cn

9.1 Introduction

With the recent progress in biotechnology and genetic engineering, a variety of protein drugs have formed a very important class of therapeutic agents. For example, insulin is used for diabetic patients, tumor necrosis factor (TNF) and interferon for cancer, and vaccines and mono-antibodies for therapy. However, most protein drugs have short half-lives *in vivo* requiring multiple treatments to provide efficacy. Therapeutic proteins are presently delivered by injections. In some cases, in order to have a better control, multiple daily injections are necessary, which result in poor compliance of patients. At the same time, proteins are also unstable in biological fluids and can be easily digested by enzymes, and may not fully absorbed from the gastrointestinal tract (GIT) because of their relatively high molecular weights, which can lead to their low bioavailability *in vivo*. For these reasons, the need for more a suitable drug-delivery system has been growing. These systems can not only effectively prevent degradation of protein in the body, but can also improve the stability and achieve sustained-release or controlled-release (1–3).

Sustained release systems for therapeutic proteins or peptides have received extensive attention in recent years, due to

Microspheres and Microcapsules in Biotechnology: Design, Preparation, and Applications
Edited by Guanghui Ma and Zhiguo Su

ISBN 978-981-4316-47-7 (Hardcover), 978-981-4364-62-1 (eBook)
www.panstanford.com

their great clinical potential. Among lots of delivery systems for protein and peptide (PP), microspheres and microcapsules attract more interests of scientists due to their excellent properties and potential for clinic applications. They are able to protect these agents against rapid degradation and clearance, and control their release from the site of administration (4).

With proteins and peptides (PPs) as drug carriers, the size of microparticles, drug-loading efficiency, drug activity, and release profile have important influence on bio-distribution, adsorption, and drug efficacy *in vivo*. In this chapter, we will describe the design, preparation, and application of microspheres and microcapsule for loading PP drugs based on their characteristics.

9.2 Properties of Protein and Peptide

Though the initial problems related to obtaining non-immunogenic PP drugs in purer form at commercial scales have been overcome to quite some extent, their formulation and optimum delivery still remain the biggest challenges to pharmaceutical scientists. Use of PPs as therapeutic agents is limited due to lack of an effective route and method of delivery. Various critical issues associated with PP delivery that have drawn the attention of formulation scientists include the following aspects. (i) PPs are high-molecular weight biopolymers which serve as enzymes, structural elements, hormones, or immunoglobulins and are involved in several biological activities. However, due to their large molecular weight and size, they show poor permeability characteristics through various mucosal surfaces and biological membranes. (ii) Many PP drugs are efficacious, in large part because of their tertiary structure. The tertiary structure can be lost under various physical and chemical environments, resulting in their denaturation or degradation with consequent loss in biological activity, hence, making these molecules inherently unstable. (iii) Many PPs have very short biological half-lives *in vivo* due to their rapid clearance in liver and other body tissues by proteolytic enzymes. (iv) PP drugs have very specific actions, and precise clinical dosing is of utmost importance.

The most important consideration when designing an effective delivery system for any drug is that of achieving a predictable and reproducible absorption into systemic circulation with high bioavailability. In the case of PP drugs, an interplay of poor permeability characteristics; luminal, brush border, and cytosolic metabolism; and hepatic clearance mechanisms results in their poor bioavailability from oral and non-oral mucosal routes. Hence, at present these drugs are usually administered by parenteral route. However, inherent short half-lives of PPs and almost warranted long-term therapy requirements in a majority of cases make their repetitive dosing necessary. Frequent injections, oscillating blood drug concentrations, and low patient acceptability make even the simple parenteral administration of these drugs problematic (5).

During designing and preparing microspheres and microcapsule for PPs carriers, the preparation process, drug-loading ways, inner structure, and surface characteristics of microparticles have important influence on drug-loading efficiency, drug activity, and release profile, and these would further affect the delivery, adsorption, and bioavailability of PPs. Therefore, it is necessary to consider all these factors during design and preparation of drug carriers for PPs.

9.3 Characteristics and Advantages of Microsphere-Based Drug-Delivery Systems

A controlled release drug-delivery system should be able to achieve the following benefits: (i) maintenance of optimum therapeutic drug concentration in the blood with minimum fluctuation; (ii) predictable and reproducible release rates for extended duration; (iii) enhancement of activity duration for short half-life drugs; (iv) elimination of side effects, frequent dosing, and wastage of drug; and (v) optimized therapy and improved patient compliance.

To achieve these benefits, the design of a controlled release system requires simultaneous considerations of several factors, such as the chemical and physical properties of the drug, the route of administration, the nature of the delivery vehicle, the mechanism of drug release, the potential for targeting, and biocompatibility. Low encapsulation efficiency (EE), and protein inactivation during the encapsulation process and difficulties of controlling the release of active proteins are some difficulties associated with encapsulation of PP. Therefore, it is necessary to design suitable drug-delivery systems for PP, which is required not only to maintain adequate plasma levels of drug, but also to allow the preservation of PP integrity (6,7).

Until now, many delivery systems for PP are developed, such as liposome, micelle, degradable polymer microspheres, and microcapsules. Among these, polymer microspheres and microcapsules have great potential in delivery of PP drugs, because they own many advantages as follows: (i) Protect PP from enzyme degradation after administration. (ii) Achieve the desired therapeutic effect over a long period of time, while with traditional injections, the drug levels in the blood rise after administration and then decrease until the next one as Fig. 9.1. (iii) Suitable release profile could be obtained by designing the size and inner structure of microspheres and microcapsules. (iv) The target molecule could be conjugated onto the surface of microspheres, and the drug could be delivered into definite location.

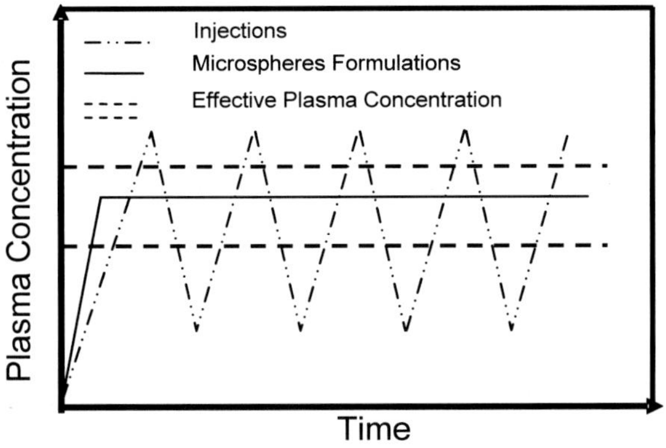

Figure 9.1 Drug levels in blood with traditional repeated dosing and microspheres/microcapsules formulations.

9.4 Drug-Loading Methods

In general, drug loading in micro/nanoparticulate systems can be done at one of the following two stages: (i) during the preparation of particles (incorporation/ encapsulation) or (ii) after the formation of particles (adsorption by physical interaction). When drug is loaded by adsorption, microspheres should be prepared at first, and then put the microspheres into drug solutions to load drug into microspheres. When drug is loaded by encapsulation, it should be added or dispersed into dispersed phase and then embedded into microspheres or microcapsules during preparation of microparticles. Therefore, design and preparation strategies of microspheres and microcapsules vary with drug-loading methods. As PP delivery systems, different strategies will have significant influence on drug activity, loading efficiency, release profile *in vitro*, and drug efficacy *in vivo*. In addition, the size and size control of microspheres and microcapsules are also important because administration methods, microparticles distribution, and drug efficacy *in vivo* are highly dependent on particle size. Furthermore, design of drug formulations and delivery systems is also influenced by other several factors. For example, a protein's physicochemical properties often affect its pharmacokinetics and toxicity and may have an impact on its clinical use, and upon administration, the clearance and metabolism of the protein depends upon *in vitro* and *in vivo* stability. Therefore, the development of delivery systems, such as microspheres and nanoparticles (NPs), for therapeutic proteins and their evaluation should systematically consider biophysical, biochemical, and physiological characteristics of protein molecules, including their molecular size, biological half life, immunogenicity, conformational stability, dose requirement, site and rate of administration, pharmacokinetics, pharmacodynamics, and so on (8). All of the factors shown in Fig. 9.2 have interdependencies that must be considered when designing a formulation or delivery system. The design strategy, the size, property and structure of microspheres, and their effects on loading efficiency, release profile, and stability of active drug will be individually described.

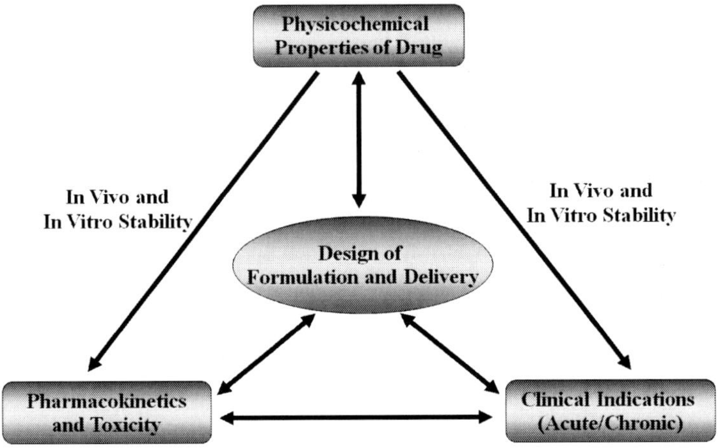

Figure 9.2 Factors involving design of protein-delivery system.

9.4.1 Encapsulation

(i) Factors for protein encapsulation

The most important aspect affecting delivery and biological effectiveness is the complexity of protein structure. Protein must maintain specific, folded, three-dimensional structure (conformation) during encapsulation process to allow delivery of biologically active native protein upon administration. Loss of biological activity and change in immunogenicity as a result of protein aggregation or denaturation are the most important challenges faced in protein encapsulation and delivery (9). Studies have shown that several types of delivery systems including poly(lactic-co-glycolic acid) (PLGA) microspheres can cause physical or chemical degradation of protein. If degraded protein is released from microspheres, it may cause unwanted immunogenicity when administered. In addition, degradation of the protein may affect its release rate and bioactivity. It is, therefore, important to devise microsphere-based formulation strategies to preserve protein stability. These approaches include addition of stabilizing agents and development of fabrication processes for delivery systems that are benign to proteins. Stabilizing additives used in the formulation of proteins are diverse and include proteins, sugars, polyols, amino acids, chelating agents, and inorganic salts. These additives can stabilize the protein in solution and also in the frozen and dried states (8).

However, many manufacturing methods for microspheres can damage complex structures of proteins by exposing them to potentially damaging conditions, such as aqueous/organic interfaces, elevated temperatures, vigorous agitation, hydrophobic surfaces, detergents, and so on (10). Moreover, upon administration, the solid protein becomes hydrated and exposed to physiological temperature within the delivery device for long periods of time. At this stage, the polymer matrix micro-environment and the products of polymer degradation can also reduce protein stability. Thus, proteins are exposed to a variety of stresses over the entire life of a controlled release system, including formulation, long-term storage, and release *in vivo* (10). In addition, the size, surface property, uniformity, and structure of microspheres also affect the EE, protein activity, and final drug efficacy.

Therefore, all of these factors must be considered during designing, preparing, and delivering the microspheres for PP encapsulation.

(ii) Requirements for different routes of administration

Microspheres-based formulations for each route of administration may be unique and have a dramatic impact on the *in vivo* stability of the protein as well as the pharmacokinetics and pharmaco-dynamics. Therefore, microspheres designing must be suitable for and according to the special requirements.

The oral administration of peptide and protein drugs faces three big challenges. The first is the metabolic barrier of four major types of proteolytic enzymes (proteases): aspartic proteases (pepsin, rennin), cystinyl proteases (papain, endopeptidase), metallo proteases (carboxypeptidase-A, ACE), and serinyl proteases (thrombin, trypsin) in GIT, which tend to metabolize PPs into smaller fragments of 2–10 amino acids (11). The second is the extreme pH conditions in stomach, which denature proteins after delivery. The third is the epithelial cell barrier which led to lower transmembrane adsorption of hydrophilic drug with large molecular mass as PP (11,12). The microspheres-based formulation for oral delivery must overcome proteases degradation, epithelial cell barrier, and extreme pH conditions to improve oral bioavailability of PP drug. The size, surface properties, and structure of microspheres will have significant influence on protein protection, adsorption efficiency, and drug efficacy by oral delivery, and therefore these factors should be considered while designing and preparing the microspheres.

There are also routes such as intranasal, transdermal, buccal, intraocular, and pulmonary, which deliver the drug to the systemic circulatory system while avoiding transit through the digestive system. A major factor that limits the usefulness of these substances for their intended therapeutic application is that they are easily metabolized by plasma proteases when they reach the peripheral circulation. The design of protein- or peptide-loaded microspheres should take into consideration that the microspheres could effectively protect proteins from proteases degradation, and the major factors including matrix materials, preparation process, protein loading ways, and so on. In addition, the PP drug should be targeted to specific location, which would greatly improve the bioavailability and usefulness of the drug. Therefore, suitable targeting ligand could be conjugated on the surface of microspheres to realize accurate targeting.

The nasal cavity as a site for the systemic absorption of drugs has some advantages, which include relatively large surface area, porous endothelial basement membrane, highly vascularized epithelial layer, high total blood flow per cm^3, avoiding the first pass metabolism, and easy access. Nasal mucociliary clearance is one of the most important limiting factors for nasal drug delivery. It severely limits the time allowed for drug absorption to occur and effectively rules out sustained nasal drug administration. Therefore, mucoadhesive microspheres have been developed to increase the contact time between the dosage form and mucosal layers of nasal cavities, thus enhancing drug absorption.

The challenges in delivery vary in administration routes as shown in Table 9.1. The design and preparation of microspheres should be suitable for different requirements of administration.

Table 9.1 Challenges for noninvasive protein delivery

Routes of delivery	Challenges in delivery
Oral	Epithelial cell barrier, protease degradation, extreme pH conditions, low bioavailability
Transdermal	Stratum corneum barrier, protease degradation, macrophages, limitation in surface area and rate of administration
Pulmonary	Epithelial cell barrier, protease degradation, alveolar macrophages, limits on amount of protein delivered per dose, fixed doses, rapid rate of absorption, permeability challenges in smokers
Ocular	Tear turnover, protease degradation, protein binding, corneal irritation possibility, low patient acceptance
Rectal	Elimination of drug during bowel movements, low patient compliance, limited surface area
Nasal	Low bioavailability, protease degradation, variable absorption, possibility of nasal mucosal irritation, and pathological changes (on prolonged administration)
Buccal	Lack of dosage form retention, presence of drug-metabolizing enzymes (oxidases, reductases, cyclooxygenases, peptidases)

Source: Reprinted from Ref. 13, with permission from Bentham Science.

(iii) Polymers used for preparation of microspheres

A variety of synthetic and naturally occurring biodegradable polymers have been studied in preparation of microspheres including polyesters, polyanhydrides, polyorthoesters, polyphosphazenes, and pseudopolyamino acids, out of which polyesters have found more widespread use. A majority of investigation of natural polymers in the microsphere-based drug-delivery system has centered on proteins (e.g., collagen, gelatin, and albumin) and polysaccharides (e.g., starch, dextran, inulin, cellulose and hyaluronic acid). Biodegradable polymers are natural or synthetic in origin and are degraded *in vivo*, either enzymatically, nonenzymatically, or both to produce biocompatible, toxicologically safe by-products which are further eliminated by the normal metabolic pathways (14).

Despite of the many advantages, polyesters like PLGA or polylactide (PLA) also have some inherent shortcomings. They are quite hydrophobic compared with most of the proteins to be microencapsulated. Lack of protein polymer compatibility leads to stability problems of the protein during storage or under *in vivo* release conditions. Since hydration and degradation of PLGA are prerequisites for the release of protein during bioerosion phase, this results in an acidic microenvironment (due to formation of lactic and glycolic acids), leading to sufficiently low pH, which can denature active drugs such as PP. Both hydrolytic degradation and formation of aggregates occur at low pH. One approach to improve protein polymer compatibility is by coencapsulating buffer salts and stabilizers for proteins, which are thought to modify internal pH of the microspheres. Another way can be realized by the modification of PLA or PLGA itself (9).

There is a sort of polymers called smart polymers. Smart polymers are materials composed of polymers that respond in a dramatic way to very slight changes in their environment. Different types of smart polymer-based delivery systems were developed until now, such as those sensitive to temperature, phase, pH, electric charge, light, magnetic field, and biochemicals. For example, the polymer of poly(*N*-isopropylacrylamide) is sensitive to temperature, and polysaccharide of chitosan and alginate sensitive to pH, and so on. Smart polymers, however, face challenges with regard to high burst release, unpredictable behavior in later part of biphasic release profile, overall drug release kinetics, conformational stability during processing, and preserving biological activity after getting released.

In the preparation of mucoadhesive microspheres, some semi-synthetic polymers, such as cellulose derivatives can be used. Hydroxypropyl methylcellulose (HPMC), hydroxypropyl cellulose (HPC), and carboxymethylcellulose sodium (CMC Na) have all been investigated for nasal drug administration.

As PP-delivery system, suitable polymers could be chosen according to the property of PP, administration route, release profile, and so on.

(iv) Release-related consideration

Release properties of microspheres are the key considerations to design microsphere-delivery systems. Since the release kinetics of protein dominantly depend on polymer nature, morphology and drug distribution within microspheres, fundamental understanding of the relationship among these key characteristics and release mechanisms is essential to yield useful products.

Biodegradable microspheres may be produced from a variety of synthetic and natural polymers and may provide a continuous or pulsatile release of the protein over a period of days, weeks, or months. The polymeric microspheres loading PP with suitable release profile could be designed and prepared according to actual requirements.

It is believed that proteins are released from solid microspheres by means of various routes as shown in Fig. 9.3 (15): (i) desorption of drugs adsorbed onto surfaces; (ii) diffusion through the polymer matrix; and (iii) physical leakage through porous cavities or surfaces along with biodegradation of the polymer matrix.

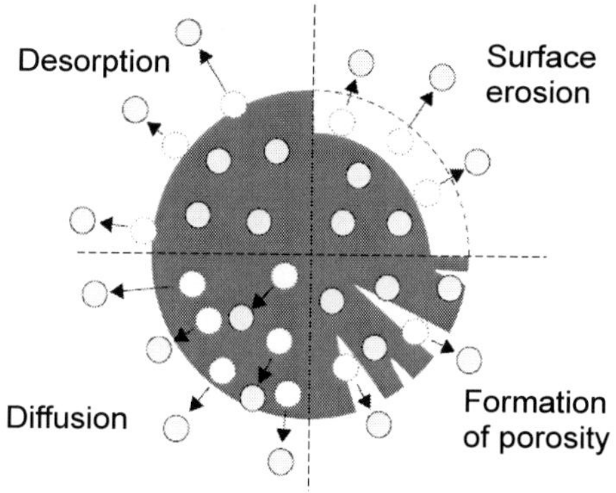

Figure 9.3 Various routes of drug release from microspheres. Reprinted from Ref. 15, with permission from Ashley Publications.

The parameters that affect the release behaviors of drugs from microspheres can be classified as follows. First are the characteristics of the particles, such as water content, size, shape, viscosity, porosity, and rigidity. The second group is the physicochemical properties of the proteins, such as solubility, acid-dissociation constant (pKa), molecular mass/volume, crystallinity, and polarity. These characteristics affect both the distribution (loci) of the drug in the particle and the compatibility/affinity between the drug and the particle. Third, the release rate depends on environmental conditions, such as ionic strength and pH of surrounding aqueous phase, temperature, and presence of biological components, such as proteins, lipids, and cells.

Among these, the factors of particle size, loading efficiency, protein solubility and molecular weight, polymer composition, and the dimensions and shape of the matrix are very important during microspheres designing.

(v) Preparation methods

Although, a number of microencapsulation techniques have been developed and reported to date, each of them has their own advantages and disadvantages. The choice of the technique depends on the nature of the polymer, protein property, the intended use, and the duration of the therapy. The microencapsulation method employed must be suitable for the following requirements (14):

(a) The stability and biological activity of the proteins should not be adversely affected during the encapsulation process or in the final microsphere product.
(b) The drug EE should be high enough.
(c) The microsphere quality and the drug-release profile should be reproducible within specified limits.
(d) The microspheres should be produced as a free-flowing powder and should not exhibit aggregation or adherence.

Up to now, the preparation methods developed include phase separation–coacervation, emulsification–solidification technique, spray drying, interfacial deposition, phase inversion microencapsulation, *in situ* polymerization, emulsion chemical and thermal cross-linking, but the most-widely used techniques for microsphere preparation of proteins are spray drying, emulsification–solidification, and phase separation–coacervation. We will introduce these methods and their influencing factors, respectively.

(1) Spray drying

In this method, the biodegradable polymer is dissolved in a volatile organic solvent, such as dichloromethane or acetone, the drug in solid form is dispersed in polymer solution by high-speed homogenization, and this dispersion is atomized in a stream of heated air. From the droplets formed, the solvent evaporates instantaneously, yielding microspheres in typical size ranges from 1 μm to 100 μm depending upon atomizing conditions. The microspheres are collected from airstream by a cyclone separator. Residual solvents are removed by vacuum drying. The process can be operated under aseptic condition. Important advantages of this technique over other encapsulation techniques are very rapid, convenient, easy to scale-up, less dependent on the solubility parameter of the drug, and the process is quite tolerant to small changes of polymer specifications. The disadvantages include high capital investment, encapsulation of proteins using this technique requires lyophilization of protein before dispersion and homogenization in organic polymer solution. These processing conditions are likely to induce aggregation and denaturation to sensitive proteins and antigens, so stability of microencapsulated proteins during processing, release, and storage becomes a major concern.

Gavini *et al.* prepared vancomycin (VA)-loaded PLGA microspheres as carriers for the topical ocular delivery using an emulsification/spray drying method, which is proposed as an alternative to double emulsification. A single W/O emulsion, where VA dissolved in the aqueous dispersed phase and PLGA dissolved in the organic continuous phase, was spray dried. The drug encapsulation efficiencies were close to the theoretical values (84.2–99.5%), and the average particle size, expressed as mean volume–surface diameter (d_{vs}), was about 11 μm (16).

There may be a significant loss of the product during spray drying due to adhesion of the microparticles to the inside wall of the spray-drier apparatus and can also produce agglomeration of the microparticles. In order to overcome these problems, a novel double-nozzle spray-drying technique was developed, which involved use of mannitol as an anti-adherent. A solution or a dispersion (W/O emulsion) of the drug in PLGA solution was sprayed from one nozzle and an aqueous mannitol solution from another nozzle was sprayed simultaneously. The surface of the spray-dried microspheres was coated with mannitol and the extent of agglomeration was decreased. Also, this method produced microspheres with higher yield and EE as compared to those prepared from the emulsification–solidification method (14).

Actually the spray dryer is typically operated at relatively high temperature to achieve a fast convective drying rate. Exposure to high temperatures in the spray dryer, where the outlet spray-dryer temperatures vary in the range of 60–70°C, however, is prohibitive when low melting point or thermally sensitive biodegradable polymers, such as poly(caprolactone) (PCL) and poly(propylene succinate), are employed as therapeutic carriers

(17,18). A novel low-temperature spraying method for preparing microspheres has been reported. Owing to its nonelevated temperature operation, spray-freeze-drying (SFD) represents a viable alternative to spray drying to produce dry powder. In terms of its applications, SFD has primarily been employed to produce dry powder–loading proteins at sub-zero temperature, and unlike the loading at high temperature, SFD does not adversely affect protein structures and functions. The detailed physical mechanisms of the process are shown in Fig. 9.4 (17).

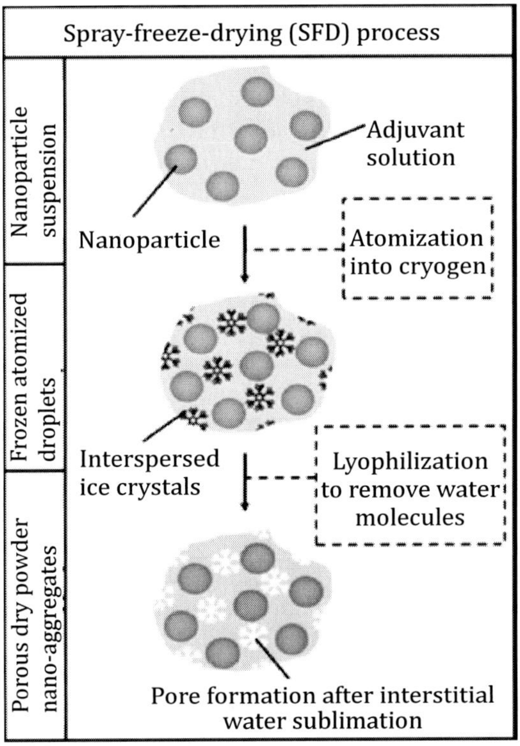

Figure 9.4 Physical mechanisms of the SFD process. Reprinted from Ref. 17, with permission from Elsevier.

(2) Emulsification–solidification method

(a) Single-emulsion process

These processes involve oil-in-water (O/W), oil-in-oil (O/O), and water-in-oil (W/O) emulsification. The process of O/W is widely used to encapsulate lipid-soluble drugs. The polymer is first dissolved in a water-immiscible volatile organic solvent. The protein is then added to the polymer solution to produce a solution as dispersed phase (19). This polymer–solvent–drug solution is then emulsified (with appropriate stirring and temperature conditions) in a larger volume of water in presence of an emulsifier as polyvinyl alcohol (PVA) to yield an O/W emulsion. The emulsion is then subjected to solvent removal by either evaporation or extraction process to harden the oil droplets. In the former case, the emulsion is maintained at reduced pressure or at atmospheric pressure to enable the volatile solvent to evaporate (19). In the latter case, the emulsion is transferred to a large quantity of water (with or without surfactant) or other quench medium, into which the solvent associated with the oil droplets is diffused out (20). The obtained solid microspheres are then washed and collected by filtration, sieving, or centrifugation. These are then dried under appropriate conditions or are lyophilized to give the final free-flowing microsphere product.

The O/O solvent evaporation method has also been used extensively to prepare microparticles, which were composed of PLA, PLGA, and synthetic biodegradable polymers (21). A water-miscible organic solvent like acetonitrile is employed to solubilize the drug in which PLGA or PLA are also soluble. This solution is then dispersed into another oil phase such as light mineral oil in presence of an oil-soluble surfactant like Span to yield the O/O emulsion. Microspheres are finally obtained by evaporation or extraction of the organic solvent from the dispersed oil droplets and the oil is washed off by solvents like *n*-hexane. This process is also used to encapsulate lipid-soluble drugs.

Although this procedure is simple, halogenated solvents (e.g., methylene chloride) that are commonly used for the preparation of the dispersed phase are considered harmful for the body and the environment. Therefore, their use is being regulated. Preparation methods for microspheres that do not use halogenated solvents have been previously investigated. Ethyl acetate (22), ethyl formate (23), and methyl ethyl ketone (24) were substituted for the halogenated solvents for the preparation of microspheres.

The processes of O/W and O/O are suitable for encapsulation of hydrophobic proteins. During the prepared process, several key process parameters would impact the microencapsulation process, activity of protein drugs, and the final microsphere product, such as the solubility of protein drug in inner oil phase, the polymer phase properties as concentration, composition, and molecular weight, the drug/polymer ratio, the organic solvent used, the temperature, and stirring/agitation speed of the emulsification process (19,20, 25,26). For example, the organic solvent not only affected the activity and structure of protein drugs during encapsulation but also had important influence on solvent evaporation, which would further have significant impact on microsphere formation, EE, and release profile of microspheres. Very rapid solvent evaporation may cause local explosion inside the droplets and lead to formation of porous structures on the microsphere surface or lack of microsphere formation; very slow solvent evaporation may cause lower EE and higher burst (17). Therefore, it is necessary to optimize these parameters in order to obtain ideal preparation conditions of microspheres for hydrophobic protein encapsulation.

Hydrophilic PPs could be entrapped into water-soluble polymer microspheres like chitosan, collagen, and gelatin by emulsification combined with definite solidification method such as cross-linking. The detailed preparation processes are shown as Fig. 9.5 (27). Hydrophilic drug is dissolved into polymer solution at first, and then the solution containing drug is emulsified by mechanical stirring, ultra-sonification technique, or homogenization into W/O emulsion, and finally the emulsion is solidified into microspheres by cross-linking. The critical parameters affecting EE, protein drug activity, and release profile include environment of water phase and cross-linking process. The environment of water phase as pH values and additives would not only exert important influence on drug activity, but also have impact on microsphere formation, morphology, and structure, which would further affect the EE and release profile of drugs. Cross-linking process includes physical and chemical cross-linking methods. Physical cross-linking can usually maintain high activity of protein drugs, while the lower EE and fast release can be obtained due to weak solidification for microparticles. However, the higher EE and lower release profile could be achieved when the microparticle was solidified by chemical cross-linking. Therefore, the suitable water phase and cross-linking process should be chosen during preparation of hydrophilic protein loaded microspheres by W/O emulsion method.

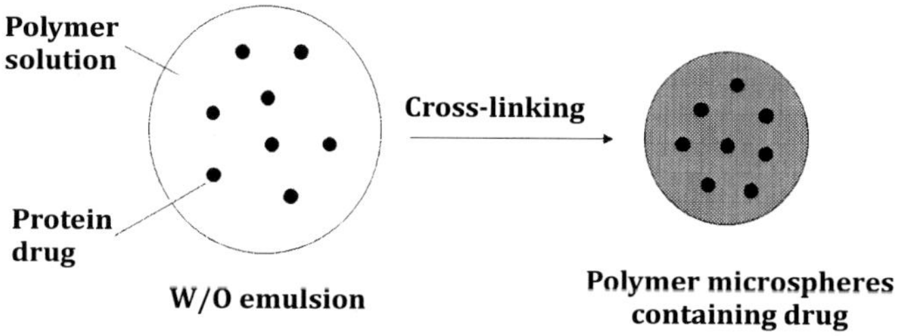

Figure 9.5 The schematic process of water-soluble polymeric microspheres containing protein.

(b) Double-emulsion method

Double-emulsion solvent extraction/evaporation technique is the most commonly used method to encapsulate hydrophilic drugs, especially PP drugs, into hydrophobic polymeric microspheres.

The detail process is shown in Fig. 9.6. Protein in aqueous solvent (sometimes containing a viscosity building and/or stabilizing protein-like gelatin) is added to a nonmiscible organic solution of polymer with vigorous stirring to form a primary W/O emulsion. The organic solvent dichloromethane is mainly used, and the homogenization step is carried out using either high-speed homogenizer or sonicators. This primary emulsion is then rapidly transferred to an excess of an aqueous medium, containing a stabilizer, such as PVA. Again homogenization or intensive stirring is necessary to initially form a double emulsion of W/O/W. The emulsion is then subjected to solvent removal of organic solvent by heat, vacuum, or both to produce microspheres.

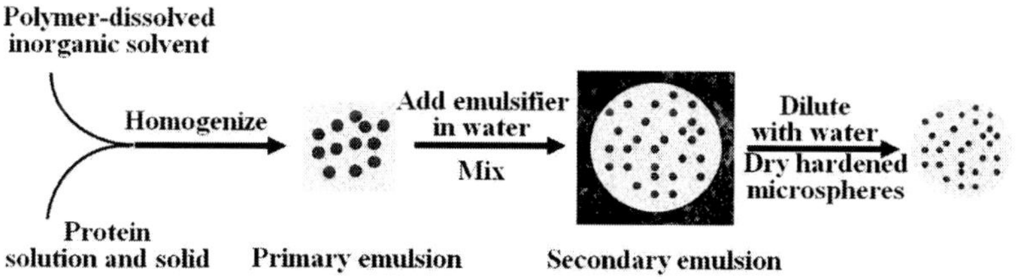

Figure 9.6 The process of double emulsion solvent extraction/evaporation technique.

The advantages of this method include that the proteins can be used for encapsulation as an aqueous solution, scaling up is possible, high yields and encapsulation efficiencies are obtained. The disadvantages include complex process, sensitive to polymer properties and difficulties in modifying release profiles of drug from microspheres.

Although a water-in-oil-in-water (W/O/W) emulsion method has been frequently used for the fabrication of protein-loaded microspheres, inherent protein instability problems, such as unfolding, covalent/noncovalent aggregation, and nonspecific adsorption, occurred during this formulation process. It is because the presence of the harsh preparation or formulation conditions, such as the aqueous/organic interface produced by a W/O microemulsion and the intense shearing strength brought by a high-speed homogenizer or a sonicater, caused protein denaturation and aggregation. Therefore, some improvements were carried out to avoid the aqueous/organic interface, such as S/O/W emulsion method which may be suitable for protein drugs, because it can reduce the contact area of protein molecules and the organic solvent (28). In order to improve protein-activity retention during encapsulation, PEG is commonly used in this method, which includes two processes. The first process is the micronization of protein by freeze drying with PEG, which is based on the aqueous phase–separation phenomenon induced by freezing-condensation. The second process is the encapsulation by a S/O/W emulsion method. PEG is not only a phase-separation inducing agent, but also an amphiphilic polymer. PEG can play an important role in the modification of protein release from microspheres. By this technology, protein can be efficiently encapsulated into microspheres without any loss of activity (29).

Another improvement for the retaining activity of protein was to use anhydrous method of solid-in-oil-in-oil (S/O/O), in which the outer phase is washed with organic volatile solvents and dried with air. The hydrophilic proteins would not diffuse out of the microspheres before they harden. This method has shown to be advantageous for the preservation of the native protein structure. The S/O/O method was developed by Emami *et al.* (29). They modified the original O/O solvent evaporation method into S/O/O method, and it not only maintained the high activity of protein drug, but also improved the EE and release pattern of the protein via PLGA microparticles. Chromatograms of insulin released in phosphate buffer solution (pH 7.4, 20 mM), extracted from microspheres (in 0.01 N HCl, pH = 2), and standard insulin are presented in Fig. 9.7. There was no difference in peak shape and retention time between the chromatograms of released, loaded, and standard insulin. This implies that the manufacturing process did not affect the chemical integrity of insulin.

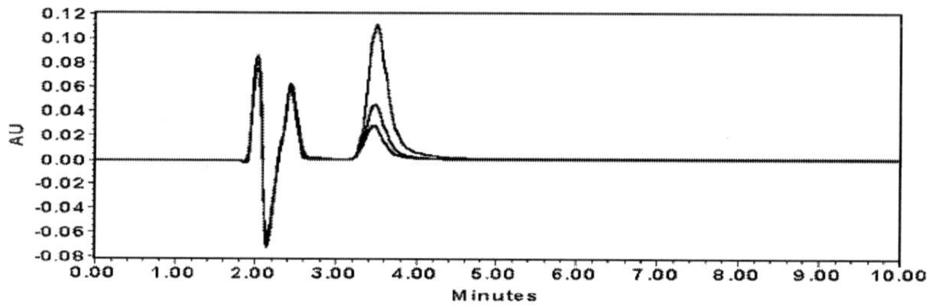

Figure 9.7 HPLC chromatograms of standard (unprocessed) insulin (top), insulin extracted from microspheres (middle), and insulin release from microspheres after 12 h *in vitro* condition (bottom). Reprinted from Ref. 29, with permission from John Wiley & Sons.

(c) Novel emulsification technique — membrane emulsification

The size of droplets or microspheres prepared by the above traditional methods as mechanical stirring, homogenization, and ultrosonication is difficult to be controlled, and the size distribution is very broad. This disadvantage will bring some limitations in applications: (i) the reproducibility of microspheres is poor among batches, which will result in poor repeatability of the release behavior and curative efficacy of drug among doses; it is also difficult to investigate the relationship between doses and treatment effects; (ii) because the accumulated locations of the microspheres containing the drug depend on the size of the particles, the bioavailability of drug will be low and the side-effects of the drug will be increased, especially for anticancer agents if the size distribution of microspheres is broad; (iii) the drug (protein or peptide) bioactivity will be greatly lost due to higher shear force from traditional methods. Therefore, it is necessary to prepare uniform-sized microspheres and control the size of microspheres for their application in drug-delivery system (30).

In order to obtain uniform-sized microspheres and to control the size easily, a novel membrane-emulsification technique was developed. This technique is divided into two types, one is direct membrane emulsification and the other is premix membrane emulsification, and both are suitable for different systems as O/W, W/O and W/O/W (31–33).

The mechanism and equipment of schematic diagram for direct membrane emulsification are shown in Fig. 9.8. A tubular porous glass membrane of size 2 cm (L) × 1 cm (B) was installed into a module. The dispersed phase or primary emulsion was stored in Teflon tank (10 mL), which was connected to a nitrogen gas inlet. The dispersed phase was pressed by the nitrogen gas through the pores of the membrane wall into the continuous phase to form the droplets, and the continuous phase was stirred gently with a magnet bar in a beaker to prevent the creaming of the droplets. Under adequate pressure of nitrogen gas, uniform droplets were formed, which were then stabilized by emulsifier dissolved in the continuous phase. The pressure varied according to the pore size. There are two key points for preparing uniform-sized emulsions by direct membrane emulsification, one is that the pressure should reach to critical pressure and maintain it during preparation process, the other is that the interfacial tension between dispersed phase and membrane should be high enough (contact angle between membrane and dispersed phase should be more than 60°). Direct membrane emulsification is mainly employed to prepare mircoparticles with 1–100 μm. We had prepared uniform-sized chitosan, alginate/chitosan microspheres from W/O emulsion, PLA, and PLGA microspheres or microcapsules from O/W or W/O/W emulsions with this method. For example, the

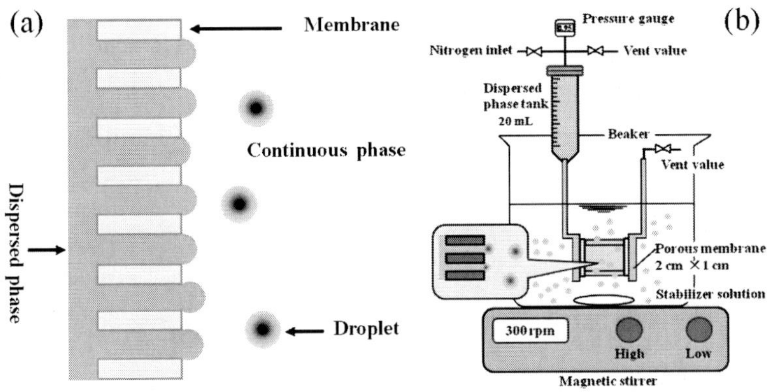

Figure 9.8 (a) The mechanism and (b) equipment of schematic diagram for direct membrane emulsification.

uniform-sized alginate/chitosan microspheres were shown in Fig. 9.9. We employed them to load insulin, and the results showed that the insulin released from microspheres retained higher bioactivity as shown in Fig. 9.10 (34).

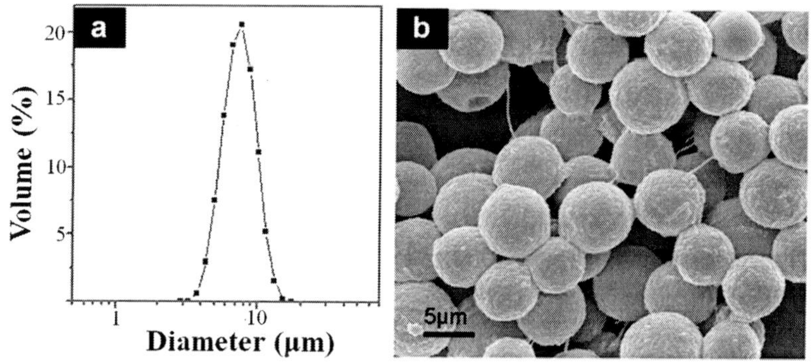

Figure 9.9 Uniform-sized alginate/chitosan microspheres by direct membrane emulsification: (a) size distribution; (b) scanning electron microscope (SEM) micrograph. Reprinted from Ref. 34 with permission from Elsevier.

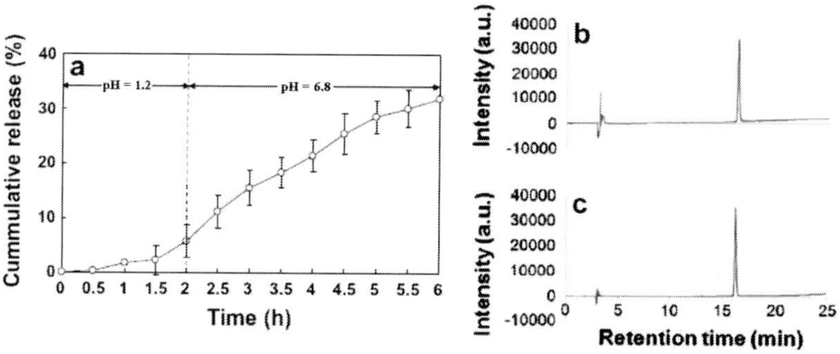

Figure 9.10 The *in vitro* release profile of insulin in simulated gastric fluid and intestinal fluid and HPLC chromatograms of (a) insulin solutions, (b) standard insulin, and (c) insulin released from the microsphere in pH 6.8 after being treated with the simulated gastric fluid containing pepsin for 2 h. Reprinted from Ref. 34, with permission from Elsevier.

Premix membrane emulsification was developed on the basis of direct membrane emulsification to prepare uniformly sized and size-controllable polymer nanospheres with small size. The schematic diagram of preparation equipment is shown in Fig. 9.11, where the coarse emulsion with larger droplet size was pressed through the uniform pores of the porous membrane quickly under higher nitrogen pressure to obtain uniform smaller droplets, which was further solidified into microspheres. The size of microspheres prepared by premix membrane emulsification is from 100 nm to 5 μm. We have successfully prepared uniform-sized chitosan microspheres, PLA microspheres, and PLA microcapsules by premix membrane emulsification. The SEM and size distribution of chitosan microsphere are shown in Fig. 9.12 (35).

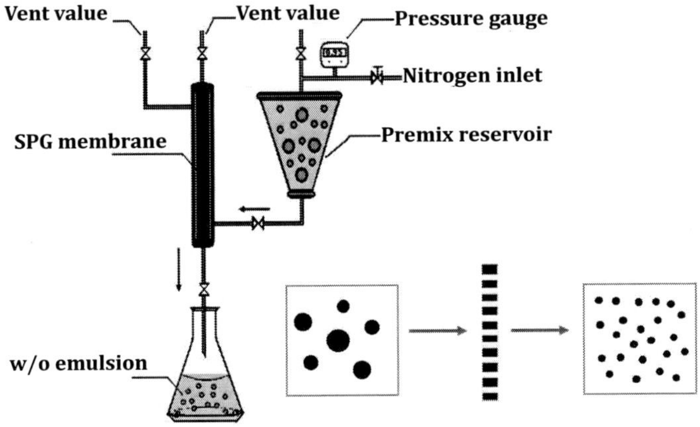

Figure 9.11 The schematic diagram of the mechanism and equipment for premix membrane emulsification.

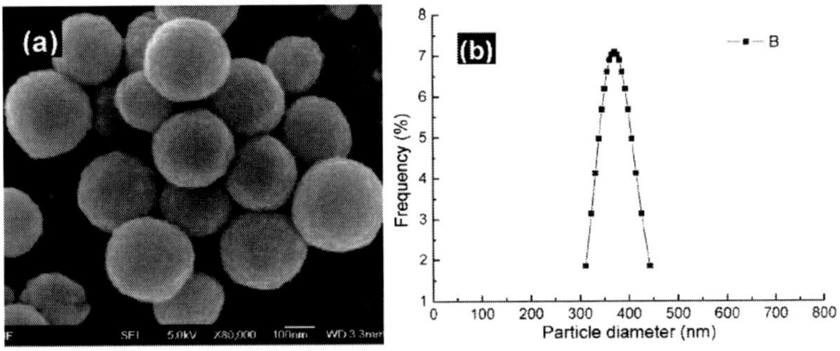

Figure 9.12 (a) SEM micrograph and (b) size distribution of uniform-sized chitosan microspheres by premix membrane emulsification. Reprinted from permission from Ref. 35. Copyright 2009 American Chemical Society.

(3) Phase separation (coacervation)

This process consists of dispersing the proteins into solution of polymer and decreasing the solubility of the encapsulating polymer by addition of a third component to the polymer solution (14). At a particular point, the process yields two phases (phase separation): the polymer containing coacervate phase and the supernatant phase depleted in polymer. The drug which is dispersed or dissolved in the polymer solution is coated by the coacervate. The polymer-rich liquid phase (coacervate) encapsulates the dispersed drug particles and the microspheres are subjected to hardening and washing step. Thus, the coacervation process includes the following three steps: (i) phase separation of the coating polymer solution, (ii) adsorption of the coacervate around the drug particles, and (iii) solidification of the microspheres. The process is quite sensitive to polymer properties, and the residual solvent is also an important issue.

The detailed preparation process varies according to the polymer type. If it is oil-soluble polymer, the process is as follows: The polymer is first dissolved in an organic solution. Peptides and proteins are dissolved in water and dispersed in the polymer solution (W/O emulsion). An organic nonsolvent is then added to the polymer–drug–solvent system with stirring, which gradually extracts the polymer solvent. As a result the polymer is subjected to phase separation, and it forms very soft coacervate droplets (size controlled by stirring) in which the drug is entrapped. This system is then transferred to a large quantity of another organic nonsolvent to harden the microdroplets and form the final microspheres, which are collected by washing, sieving, filtration, or centrifugation and are finally dried.

This process, unlike the O/W emulsification method, is suitable to encapsulate both water-soluble as well as water-insoluble drugs, since it is a nonaqueous method. However, the coacervation process is mainly used to encapsulate water-soluble drugs like peptides, proteins, and vaccines. The addition rate of first nonsolvent should be such that the polymer solvent is extracted slowly so that the polymer has sufficient time to deposit and coat evenly on the drug particle surface during the coacervation process (19). The concentration of the polymer used is important as well, since too high concentrations would result in rapid phase separation and nonuniform coating of the polymer on the drug particles. Due to the absence of any emulsion stabilizer in the coacervation process, agglomeration is a frequent problem in this method (19). The coacervate droplets are extremely sticky and adhere to each other before the complete phase separation or the hardening stages of this method. Adjusting the stirring rate, temperature, or the addition of an additive is known to overcome this problem (19).

Compared to the solvent evaporation/extraction process, the requirement of solvents for the polymer are lower since the solvent need not be immiscible with water, and the boiling point can be higher than that of water (19). Dichloromethane (DCM), acetonitrile, ethyl acetate, and toluene have been used in this process. The nonsolvents affect both the phase separation and the hardening stages of the coacervation process. The nonsolvents should not dissolve the polymer or the drug and should be miscible with the polymer solvent. The second nonsolvent should be relatively volatile and should easily remove the first viscous nonsolvent by washing. Some of the oils used as the first nonsolvent are silicone oil, vegetable oils, light liquid paraffin, low molecular weight liquid polybutadiene, and low–molecular weight liquid methacrylic polymers. Examples of the second nonsolvent include aliphatic hydrocarbons like hexane, heptane, and petroleum ether.

In the phase separation method, the phase equilibrium is never reached and, hence, the system is constantly out of equilibrium. Therefore, the formulation and process variables significantly ffect the kinetics of the entire process and ultimately the characteristics of the final microspheres.

In a classic article, Nihant *et al.* have investigated the effect of several process factors on the coacervation process (36). In another paper, the same group has reported the effects of weight, volume, composition, and viscosity of the coacervate and the supernatant phases on size distribution, surface morphology, and internal porosity of the final microparticles (37,38).

(4) Template-assembly method

In this process, a suitable template was first chosen, then the protein drugs were adsorbed on the surface of template microparticles, and next the polyelectrolyte were assembled on the core template by the layer-by-layer technology, and finally the template was removed by certain method. The electrostatic force between two polyelectrolytes varies in different pH environment, so the nanocapsule prepared by this method has pH sensitivity, which could well control drug release. At the same time, the release of drug also depends on layers of nanocapsules. The drug release could be controlled by altering the thickness of the nanocapsules membrane, and extra layers could be used to slow down the release of drug *in vitro*. In addition, the activity of protein drug could be highly maintained due to the mild preparation conditions, and the whole process could be conducted in aqueous phase and the oil–water interface and intense shear could be avoided.

For example, Shu *et al.* (39) developed a template-assembly method via layer-by-layer assembly of water-soluble chitosan (WSC) and dextran sulfate (DS) by protein-entrapping amino-functionalized silica particles and the subsequent removal of the silica. The detailed process is described as follows: Bovine serum albumin (BSA), a protein which was used as model, was entrapped in the nanocapsules by both physical interaction and electrostatic attraction, and after the required numbers of polyelectrolyte layers like WSC and DS were deposited, the silica core was removed by exposure to an aqueous solution of HF/NH_4F at pH 5.4 as shown in Fig. 9.13.

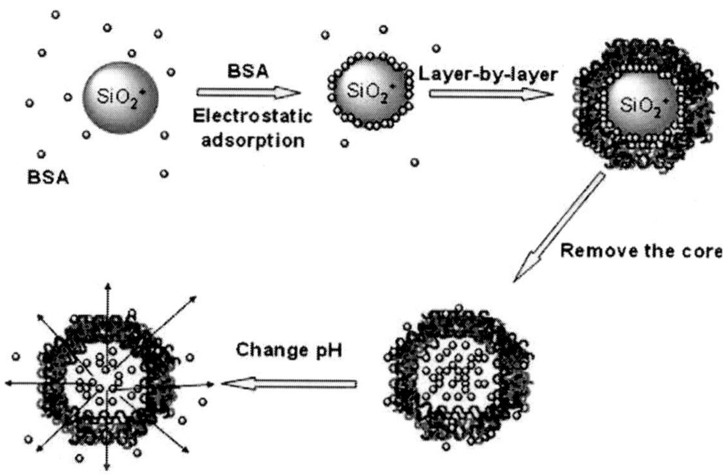

Figure 9.13 Schematic representation of the encapsulation of BSA within polyelectrolyte capsules. Reprinted from Ref. 39, with permission from Elsevier.

The results of EE and loading capacity (LC) of nanocapsules clearly indicated that EE of these particles was affected by the initial concentration of the BSA used. As the initial concentration of BSA was increased, EE decreased slowly while LC increased. Amino-functionalized silica showed higher EE than bare silica, because amino-functionalized silica adsorbed the BSA by electrostatic interaction, while the bare silica mainly by physical adsorption through hydrogen bonds. The isoelectric point of BSA is 4.6, and the BSA molecules are negatively charged in pH 5.4 solution. As a result, strong electrostatic interactions are established between BSA and amino-functionalized silica particles, resulting in enhanced EE.

The results of BSA release from nanocapsules suggested that the nanocapsules exhibit excellent pH sensitivity. At pH 1.4 (simulating the pH in the stomach), the initial burst release was decreased to lower than 10%. Similarly, at pH 7.4 (simulating the pH at the bloodstream), in the first 2 h, more than 60% of BSA was released. At pH 1.4, the amino groups in WSC are protonated and the sulfate groups in DS are ionized ($pKa < 1$). The strong electrostatic attractions between WSC and DS restrain the swelling of nanocapsules, so the rate of drug release is slow. At pH 7.4, most of the amino groups of WSC are in the $-NH_2$ form and most of the sulfate groups of DS are in the $-SO_3-$ form. Therefore, the electrostatic attractions were weakened in nanocapsules.

In conclusion, this method not only improves the EE of proteins, but also controls the release rate of protein by the layers of the nanocapsule, and furthermore, this method greatly maintains protein activity during the encapsulation process due to mild preparation conditions.

(vi) Strategies for maintaining protein activity during encapsulation

Protein inactivation during the encapsulation process is one of the most difficult problems. Therefore, studies on the delivery systems which can not only maintain adequate plasma levels of proteins, but also allow the preservation of protein integrity have been the object of great attention. There are many factors to affect protein activity during encapsulation, such as oil–water interface, shear force, polymeric materials, and release media. Until now, there appeared many strategies for maintaining protein activity and here some examples have been given to elaborate these strategies during microspheres designing (40).

(1) Polymer selection

Several oil-soluble biodegradable polymers such as PLA, PLGA, and PELA have been explored so far in the development of encapsulation of active molecules as proteins, among which the use of amphiphilic polymers, such as PLGA and PLA-PEG offer great promise due to their higher activity retention. We have prepared uniform-sized PLA, PLGA, and PELA microspheres containing recombinant human growth hormone (rhGH), and compared the bioactivity of rhGH after encapsulating them into microspheres made of different materials. SEC-HPLC result for the integrity of rhGH is shown in Fig. 9.14. It was evident that the rhGH released from PELA basically maintained its monomeric and dimeric form ranging from 90% to 100%, while PLA and PLGA microcapsules released the monomeric and dimeric rhGH floating around 70% and 80%, respectively. The result indicated that the aggregates released from PLA and PLGA microcapsules increased more quickly during the release, compared with that of PELA microcapsules. Further physical characterization of the released rhGH was carried out to assess the protein's secondary structure. Circular dichroism (CD) spectra of the control rhGH and released rhGH from PELA microcapsules during the release period are shown in Fig. 9.15, and the result indicated the protein had a high α-helical content, and all the released rhGH samples maintained their structural integrity as judged from the shape of far-UV CD spectra. Therefore, the polymer composition is an important factor for protein encapsulation.

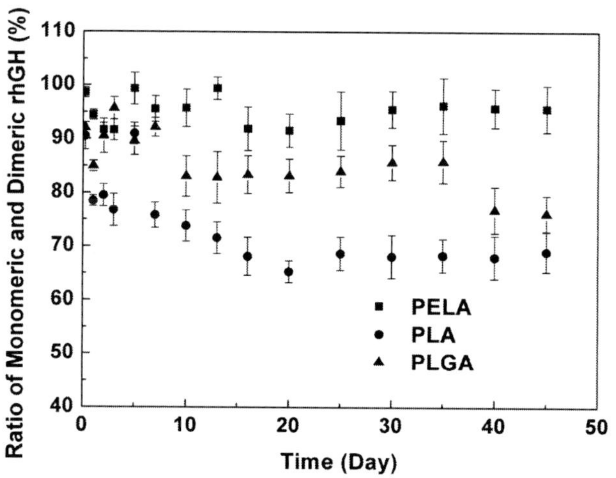

Figure 9.14 Ratio of monomeric and dimeric rhGH from three microcapsules. Reproduced from Ref. 40, with permission from the Royal Society of Chemistry.

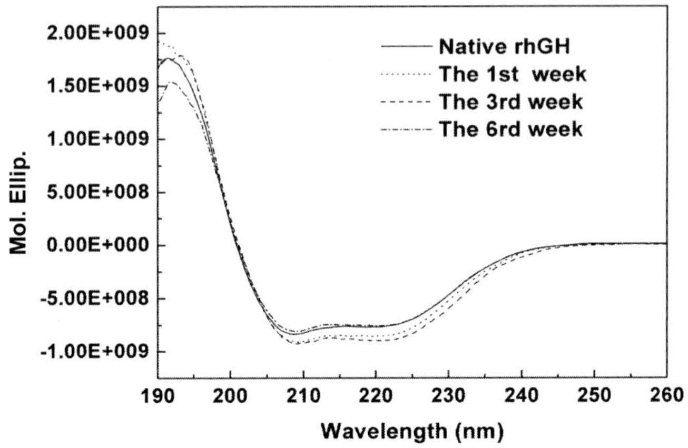

Figure 9.15 CD spectra of rhGH released from PELA microcapsules at different times. Reproduced from Ref. 40, with permission from the Royal Society of Chemistry.

(2) Preparation process selection

Among all the methods for preparing microspheres, the emulsion method is most commonly used. There are W/O interfaces existing while preparing microspheres by this method, which lead to protein aggregation and inactivation. Though solid-in-oil-in-water (S/O/W) may decrease protein aggregates, some proteins still have aggregates due to protein in the surface of microspheres that are in contact with W/O interfaces. S/O/O can almost protect proteins from forming aggregates, but burst release is severe and residue of oil phase may lead to immunity problem. Here, Yuan *et al.* (41) developed a novel multi-emulsion method to prepare microspheres, that is, solid-in-oil phase1-in-oil phase2-in-water (S/O1/O2/W), in which the solid-phase is protein-loaded microparticles, and their size is about 1–5 μm when prepared by low temperature–induced phase separation; O1 is the oil phase of PLGA dichloromethane solution and O2 is a hydrophilic oil (hO) phase and can dissolve in water, so the oil phase is easily removed; W is aqueous phase containing 5% (W/W) sodium chloride aqueous solution. This method could effectively protect proteins activity during the whole preparation process.

Figures 9.16 and 9.17 show the SEC-HPLC results for aggregation of myoglobin (MGB) after microencapsulation by this new process. Figure 9.16a, b shows the SEC-HPLC charts of the original MGB solution and the MGB recovered from PLGA

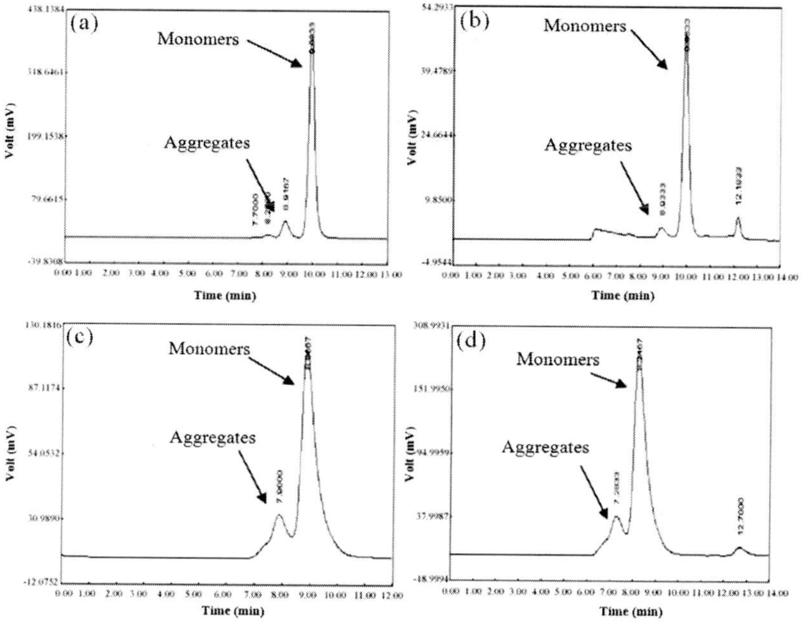

Figure 9.16 Spectra of recovery proteins from PLGA microspheres by SEC-HPLC: (a) original myoglobin; (b) myoglobin-loaded PLGA microspheres by S/O1/O2/W process; (c) original BSA; and (d) BSA-loaded PLGA microspheres by S/O1/O2/W process. Reprinted from Ref. 41, with permission from Elsevier.

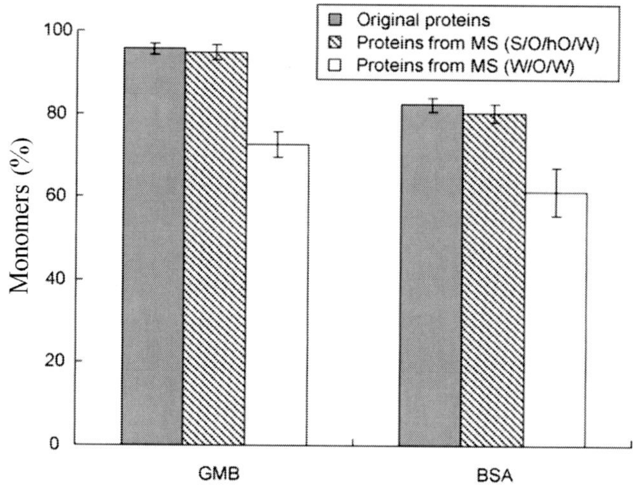

Figure 9.17 Percentage of monomers of recovery proteins from PLGA microspheres by SEC-HPLC (n = 3); MS means microspheres. Reprinted from Ref. 41, with permission from Elsevier.

microspheres. The fractions of dimers and oligomers did not increase during the microencapsulation process as compared with the original sample. Similarly, there was no significant change in the contents of dimers for BSA recovered from PLGA microspheres as compared with original BSA sample (Fig. 9.16c,d). As control, they detected aggregates of BSA and MGB from microspheres using multi-emulsion W/O/W as shown in Fig. 9.17. The results suggested that there were more aggregates of proteins being released from microspheres by W/O/W method. This is possible because the interfaces of multi-emulsion W/O/W exist more than in S/O/O/W.

(3) Solidification method selection

Different solidification method of microspheres containing proteins will lead to different drug activity retention. We (42) employed two methods to solidify chitosan microspheres loading model protein of insulin as shown in Fig. 9.18. The integrity of insulin released from chitosan microspheres by different solidification methods are shown in Fig. 9.19. The results exhibited that there caused a lot of active loss of insulin in solidification method of directly being cross-linked by glutaraldehyde, while less loss of insulin activity was found in solidification method cross-linked stepwise by tripolyphosphate (TPP) and glutaraldehyde. This example suggested that solidification method also has important influence on protein activity, which should be considered during microspheres loading protein drug designing.

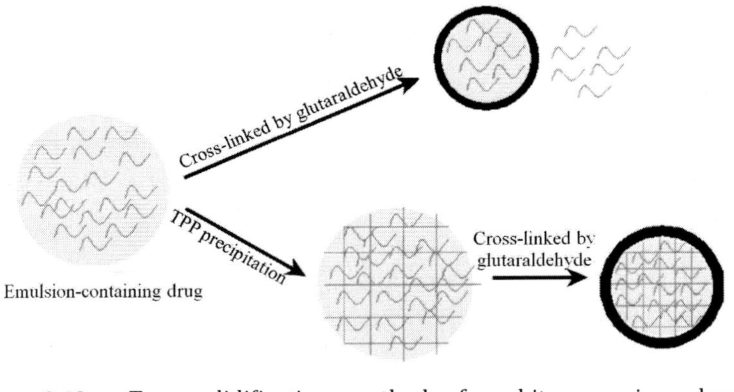

Figure 9.18 Two solidification methods for chitosan microspheres loading insulin. Reprinted from Ref. 42, with permission from Elsevier.

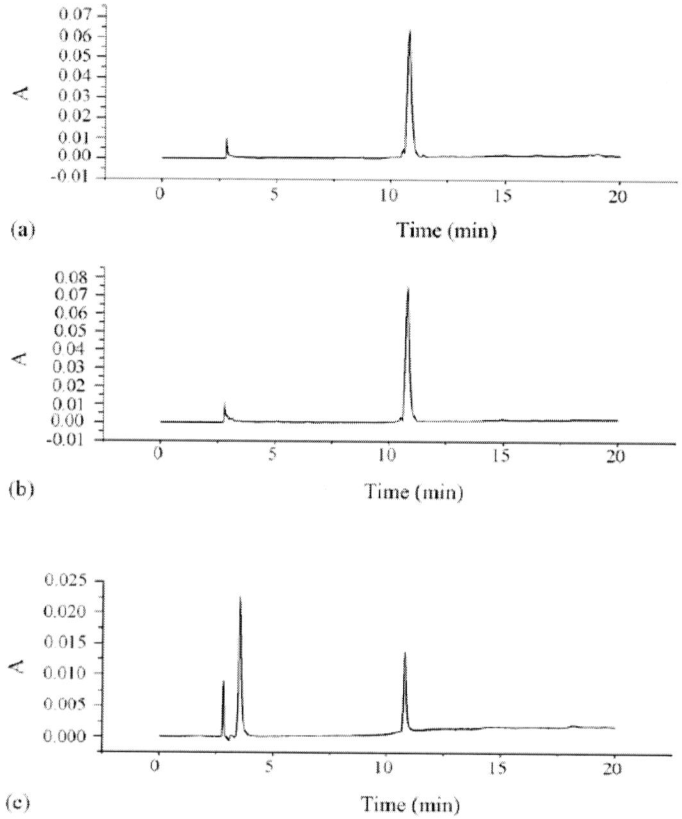

Figure 9.19 The integrity of insulin released from chitosan microspheres by different solidification methods: (a) standard insulin, (b) microspheres solidified by cross-linking them stepwise by TPP and glutaraldehyde, and (c) microspheres solidified by cross-linking them directly by glutaraldehyde. Reprinted from Ref. 42, with permission from Elsevier.

(4) Additives selection

Additives, such as sugars, surfactants, and proteins, could effectively reduce protein enrichment in oil-water interface, and further avoid denaturation of proteins. For example, Emami *et al.* (29) employed PEG to improve the stability of model protein insulin loaded in PLGA microparticles by S/O/O system. The conformational stability of insulin released *in vitro* from microspheres was assessed by CD, and the result is shown in Fig. 9.20. As it is evident from this figure, two minima around 208 nm and 222 nm, which is typical of predominant α-helix structure of proteins, such as insulin, was unaffected by the formulation/process and superimposed almost entirely to that of standard insulin profile. This suggested that the protein maintained its α-helical structure and hence it's secondary structure and physical stability. Among the various protein inactivation sources, there is great concern that proteins incorporated into PLGA microspheres will undergo an aggregation process. Figure 9.21 showed SDS-PAGE of standard insulin, released samples, and supernatant after the insulin-loading process. No aggregation or degradation band or any extra band was observed between the released, encapsulated, and standard insulin. Therefore, they assumed that the loading and release process did not affect the structural integrity of insulin.

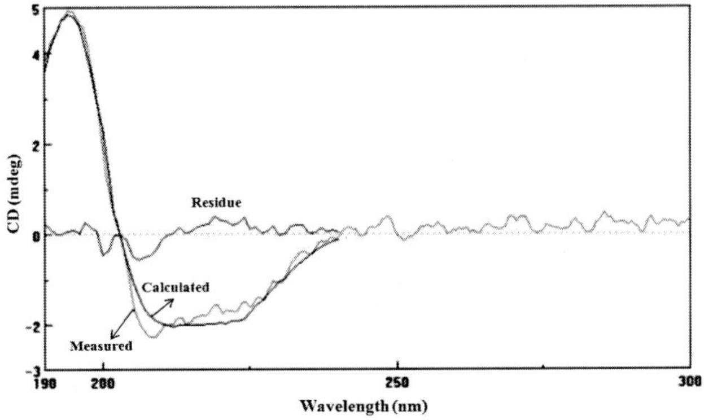

Figure 9.20 Far-UV CD spectra of standard (calculated) and *in vitro* released (measured) insulin from PLGA. Reprinted from Ref. 29 with permission from John Wiley & Sons.

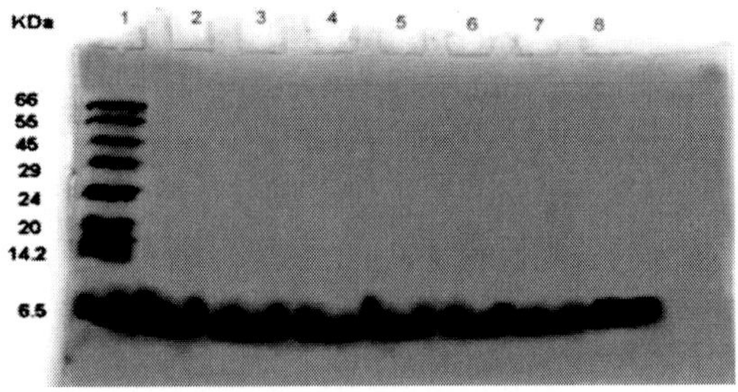

Figure 9.21 SDS-PAGE of insulin: line 1, protein molecular weight standard; line 2, standard insulin; line 3–5, extracted insulin from PLGA microspheres; line 6–8, released insulin during *in vitro* studies. Reprinted from Emami Ref. 29 with permission from John Wiley & Sons.

We (43) studied effects of additives on insulin activity during encapsulation into chitosan microspheres. We chose six types of additives as gelatin, HP-β-CD, PEG2000, PVA, Tween20 and Poloxamer188 to add into aqueous phase, and the effect of different additives on insulin activity is shown in Table 9.2. It was evident that the additives of gelatin and Tween20 could effectively maintain insulin activity.

Table 9.2 Effect of type of additives on encapsulated insulin activity ($n = 3$)

Type of additives	The proportion of area of active insulin peak (%)
Gelatin	78.69 ± 0.68
HP-β-CD	8.69 ± 1.22
PEG2000	5.29 ± 1.08
PVA	3.64 ± 2.26

Source: Reprinted from Ref. 43, with permission from Elsevier.

The above strategy for maintaining protein activity is to add the additives directly to inner water phase containing protein, and another strategy is to first encapsulate protein in the hydrophilic polymer that creates a more "friendly" environment and then embedded into hydrophobic polymer. The hydrogel network protects the molecules from the direct contact with the organic solvent and minimizes the interactions with the hydrophobic surfaces, preventing integrity loss or denaturation. For example, Schoubben *et al.* (44) investigated the possibility to use sodium alginate in the form of microparticles to stabilize peptides and proteins during encapsulation, storage, and release. For instance, alginate, complexing bivalent or trivalent ions (e.g., Ca^{2+}, Ba^{2+}, Fe^{2+}, Fe^{3+}, La^{3+}), can form hydrogels with a characteristic structure called "egg-box." Alginate hydrogels may be able to create the "friendly" environment needed to protect peptides and proteins from the stressing conditions that could denaturate them. Bovine insulin was chosen as a model peptide to study its stability issues. They successfully prepared insulin-loaded alginate/PLGA composite microparticles as shown in Fig. 9.22, and it was observed the microspheres were spherical and had a porous surface.

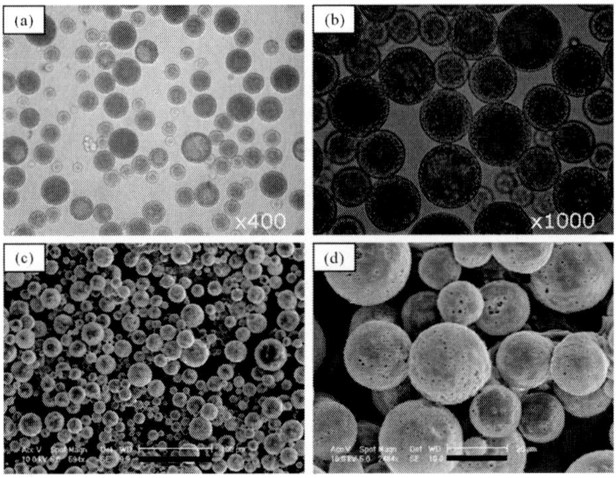

Figure 9.22 Insulin-loaded composite microparticles prepared by the G/O/W (G: hydrogel) method observed by optical microscopy (a and b) and SEM (c and d). Black scale bar corresponds to 100 μm (c) and to 20 μm (d). Reprinted from Ref. 44, with permission from Elsevier.

Insulin-loaded particles were also prepared, using the W/O/W technique, in order to compare the above composite microparticle characteristics with those of conventional microparticles, and the results are shown in Table 9.3. It has been observed that the presence of alginate was able to protect insulin from degradation and denaturation. As shown in Table 9.3, composite microparticles led to higher soluble insulin content with respect to the conventional ones, demonstrating the ability of alginate microparticles in protecting the peptide from denaturation.

Table 9.3 Comparison of the characteristics of insulin-loaded composite and conventional microparticles

Batch name	Insulin (mg)	Theoretical loading (%)	Encapsulation efficiency (%)	Total drug Content (%)[a]	Active drug content (%)[b]
Composite microparticles	60	19.35	38.19 ± 1.83	7.39 ± 0.35	5.65 ± 0.39
Conventional microparticles	60	23.08	17.86 ± 2.41	4.12 ± 0.55	3.72 ± 0.32

[a]The total drug content was determined by microBCA assay.

[b]The active drug content was determined by HPLC method.

Source: Reprinted from Ref. 44, with permission from Elsevier.

In conclusion, the size, polymer types, preparation methods and processes, structure and surface properties of microspheres will have important influence on protein EE, drug-release profile and drug activities, and therefore all these factors should be considered during design of microspheres for encapsulation of PPs.

9.4.2 Adsorption

As described above, there are many methods, such as spray drying, emulsification–solvent removal or cross-linking method, phase separation, and template-assembly method, that are commonly used for encapsulating proteins within microspheres. However, inherent protein instability problems, such as unfolding, covalent/noncovalent aggregation, and nonspecific adsorption, occur during the formulation process. It is because that the presence of the harsh preparation or formulation conditions, such as the aqueous/organic interface produced by a W/O microemulsion and the acute shearing strength brought by a high-speed homogenizer or a sonicater, caused protein denaturation and aggregation. Therefore, some improvements were carried out in order to overcome these problems. Among these, the adsorption of proteins in prepared microspheres is an important and effective method. This technique can be performed in an aqueous solution and at a low temperature, thus improving the prospects for preserved activity of sensitive drug molecules like PP.

The adsorption of PP on the micro- and nanoparticles mainly depends on the affinity of the protein for a particular surface, loading ways, and the extent of structural alteration. For example, most proteins adsorb with higher affinity to hydrophobic surfaces showing lesser native structures, whereas excessive affinity with hard acidic/basic sites on inorganic surfaces may also be accompanied by a reduction in native structures. For the applications using micro- and nanoparticles to deliver proteins, the most important challenge is to fully maintain the conformation and activity of the molecules on a particle surface.

In designing of microspheres or NPs for adsorption of proteins or peptides, the properties of micro- and nanoparticles, such as size, inner structure, and surface chemistry have important influence on the structure, activity, stability, drug-loading efficiency, and release profile *in vitro* of adsorbed PPs. Here, we will give some examples to explain how these factors affect the loading efficiency, release profile, and activity of PPs.

(i) Size of microspheres or NPs

Different size of microparticles will lead to different surface curvature, which will affect the structure of proteins or peptides adsorbed on the microparticles (45). Studies have shown that for specific proteins (lysozyme and human carbonic anhydrase), smaller NPs, possibly due to their higher surface curvature, had higher retention of native-like protein structure and function than their larger counterparts (46–48). For example, Vertegel *et al.* (48) characterized the secondary structure and activity of chicken egg lysozyme adsorbed onto silica NPs of various diameters. They reported a change in protein structure upon adsorption, with a greater loss in α-helix content on particles with larger diameters. The loss of activity of the adsorbed protein correlated well with the decrease in α-helix content. Lundqvist *et al.* (47) studied the adsorption of human carbonic anhydrase variants onto silica NPs of varying sizes, and also observed a larger perturbation of protein secondary structure on particles with larger diameters. These studies suggested that smaller NPs, perhaps as a result of higher surface curvature, promoted the retention of native-like protein structure and function when compared to larger particles. Other studies suggest that the dependence of protein structure retention on nanomaterial curvature is protein-dependent. Roach *et al.* (49) studied the effects of the diameter of silica NPs on the secondary structure of two proteins differing in size and shape, BSA, and fibrinogen. While BSA was increasingly less ordered on larger NPs, consistent with the studies described above, fibrinogen was denatured to a greater extent on smaller NPs. Therefore, the microparticles with different size should be designed according to the adsorbed proteins or peptides in order to maintain structure integrity of drugs.

(ii) Inner structure of microspheres or NPs

Although the protein activity could be improved by adsorbing the drug on the microspheres or NPs, the loading efficiency was usually lower and the burst release was higher than that by encapsulation method. This shortcoming and deficiency could be overcome to a certain extent by designing different inner structure of microspheres or NPs. In addition, the inner structure could further improve the activity of adsorbed proteins.

We have prepared chitosan microspheres with different structures as solid, hollow, hollow-porous, and porous by adjusting the ratio of chitosan and quaternized chitosan and solidification process, and the corresponding confocal images and characteristics are shown in Fig. 9.23 and Table 9.4, respectively. We employ chitosan microspheres with different structure to load model drug of BSA, and the loading efficiency and release profile *in vitro* are investigated. The result of loading efficiency showed that the chitosan microsphere structure had important influence on loading efficiency (Fig. 9.24). It suggested that the hollow-porous microspheres yielded the highest loading efficiency of 400 μg BSA /10^6 microspheres, next being that of porous and hollow microspheres, and the lowest being that of solid microspheres. A possible explanation to the very high loadings is based on the electrostatic effect. The isoelectric point of BSA is 4.7, and thus the BSA molecules are negatively charged in phosphate buffer saline (PBS) buffers. As a result, there are strong electrostatic interactions between BSA and the microspheres containing positively charged quaternized chitosan in the medium, which naturally play an important role in enhancing the loading efficiency. Second, spheres with a hollow structure would provide more space for BSA storage compared with the conventional solid ones. In addition, wall porosity not only facilitates the migration of BSA into the inner space, but also creates a higher surface area and therefore more positions for BSA loading by absorption are possible. The release profile of BSA *in vitro* indicated that different structures of these microspheres resulted in different release profile as shown in Fig. 9.25 (50). The porous microspheres show the highest initial burst due to the large amount of BSA loading in the macroporous outer shell and channels, which released rapidly into the medium. Similarly, the solid microspheres also show a high initial burst due to their poor loading ability on the surface. BSA release in hollow microspheres presents a triphase profile. The hollow structure provides a good explanation for this interesting result. Phase I was a 40% initial burst due to the BSA loading in the shell, and then followed by Phase II, a platform for about 60 h, corresponding to the time taken by the BSA loaded in the cavity to cross the shell, and Phase III was a second burst, which obviously contributed to the outward migration of the mentioned BSA. Compared to the microspheres above, the BSA initial burst has been well controlled within 20% in hollow-porous microspheres, which is followed by a slower linear release. The different release profiles will be suitable for different clinical applications.

Table 9.4 Comparison of the characteristics of chitosan microspheres with different structure

Character	Solid	Hollow	Hollow porous	Porous
Particle size (μm)	7.3	7.6	7.9	7.3
C.V. value (%)	12.4	14.3	13.3	12.9
Zeta potential (mv)	2.4	3.7	20.3	19.7
Surface area (m^2/g)	16	55	142	98
Median pore diameter (nm)	1.8	4.3	16.7	73.2

C.V.: Coefficient variation.
Source: Reprinted from Ref. 50, with permission from Wiley-VCH Verlag GmbH & Co. KGaA.

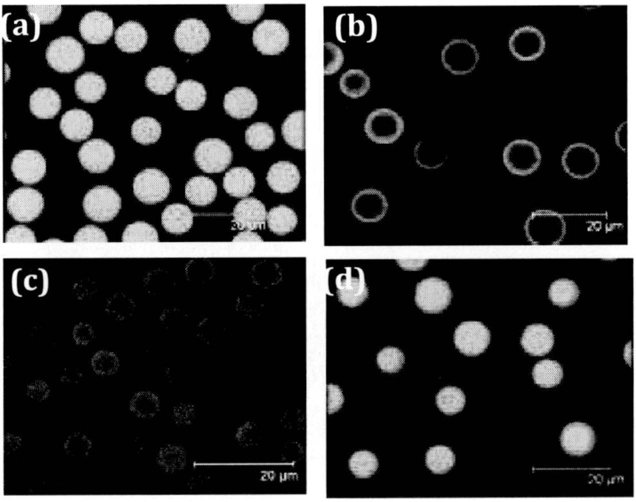

Figure 9.23 The confocal images of chitosan microspheres with different structures: (a) solid, (b) hollow, (c) hollow-porous, and (d) porous.

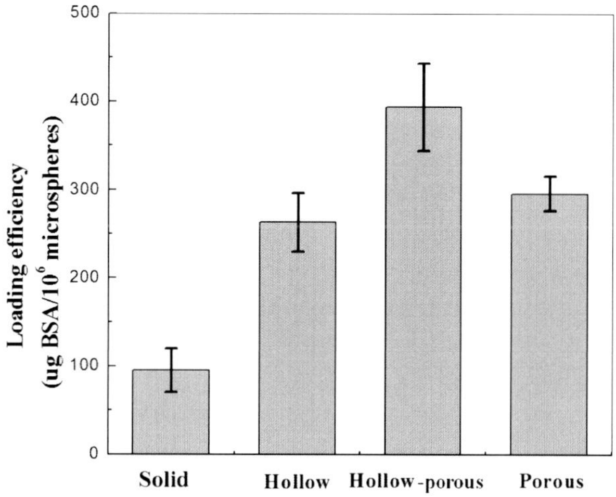

Figure 9.24 BSA loading efficiency for chitosan microspheres with different structure.

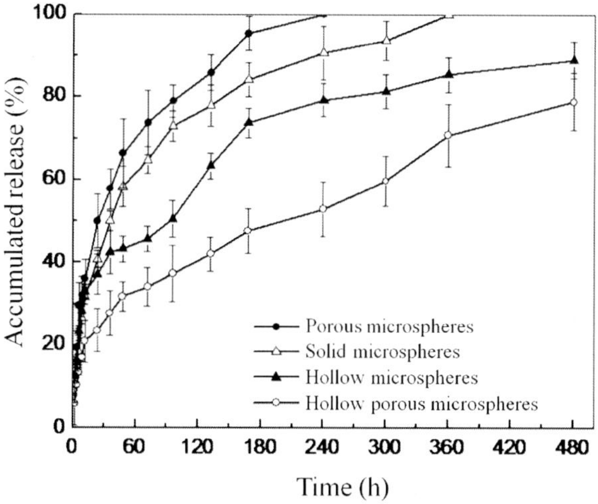

Figure 9.25 BSA release *in vitro* for chitosan micrspheres with different structure. Reprinted from Ref. 50, with permission from Wiley-VCH Verlag GmbH & Co. KGaA.

Here, another example of hollow nanocapsule for protein loading is described. In order to improve loading efficiency and protein activity and reduce burst release, Shu *et al.* (51) prepared gradient shell cross-linked hollow polyelectrolyte nanocapsules composed of cysteamine-conjugated chitosan and DS by layer-by-layer adsorption on β-cyclodextrin (β-CD)-functionalized silica spheres followed by cross-linking thiols and removal of silica core, and the preparation process is shown in Fig. 9.26. They loaded model protein of BSA by adsorption. The designed nanospheres present reduction and pH-dependent release pattern as shown in Fig. 9.27, which can not only protect protein drug from losing in acid environment but also control drug-sustained release in cell environment. The nanospheres showed higher drug EE and drug-LC when BSA was employed as model protein drug. Two possible explanations for the high EE can be proposed based on the role of electrostatic effects and hydrophobic cavities of β-CD. The isoelectric point of BSA is 4.6, and the BSA molecules are negatively charged in pH 5.4 solution. As a result, strong electrostatic interactions occurred between BSA and Si–NH_2–CD. Furthermore, many hydrophobic section of BSA can enter the cavities of β-CD, so these two interactions made it have high EE and LC. CD spectroscopy was used to examine the conformation and self-association of BSA. The result indicated that no significant conformation change was noted for the released BSA as compared with the native BSA.

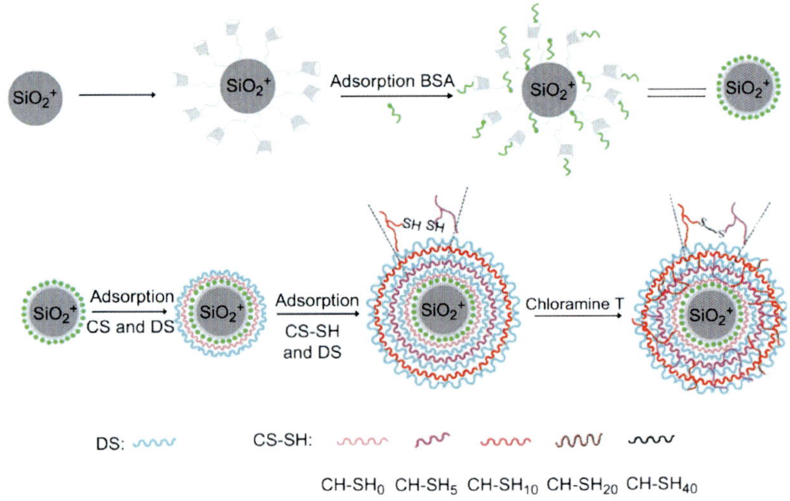

Figure 9.26 Schematic representation of loading BSA of gradient cross-linked NPs. (The degree of thiol modification increased as the color became darker.) Reprinted from Ref. 51, with permission from Pergamon.

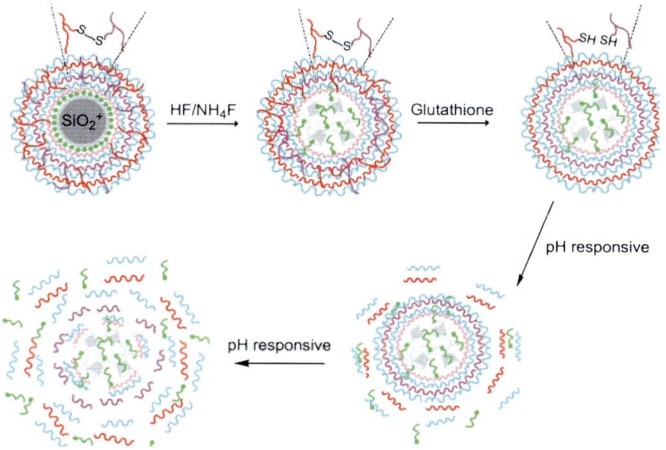

Figure 9.27 Schematic representation of the release of BSA from gradient cross-linked nanocapsules. Reprinted from Ref. 51, with permission from Pergamon.

In order to improve protein-loading efficiency, porous PLGA microspheres with highly interconnecting pores and large porous surface were prepared by Zhou *et al.* as shown in Fig. 9.28 (28). They use these porous microspheres to adsorb model protein of HSA (human serum albumin), and the results (Fig. 9.29) showed that there were higher adsorption efficiency in both pH 4.5 and pH 7.4. However, the porous PLGA microspheres loading HSA had higher burst release (about 90%) *in vitro* as shown in Fig. 9.30. In order to reduce burst release, the authors coated the porous PLGA microspheres loading HSA with calcium alginate gel, and stable and slower release profile *in vitro* was obtained as shown in Fig. 9.30.

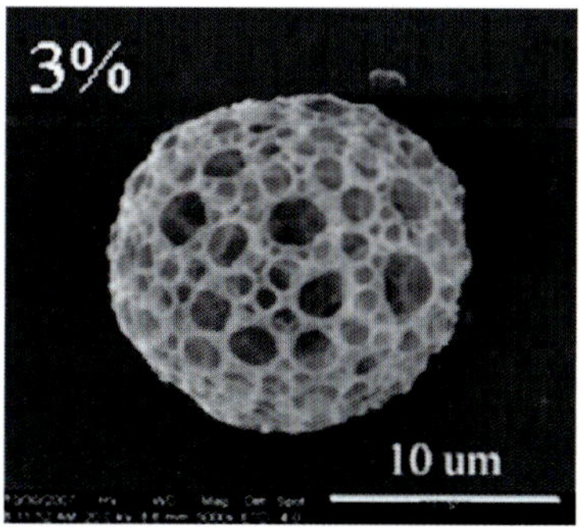

Figure 9.28 SEM micrograph of porous PLGA microsphere. Reprinted from Ref. 28, with permission from Elsevier.

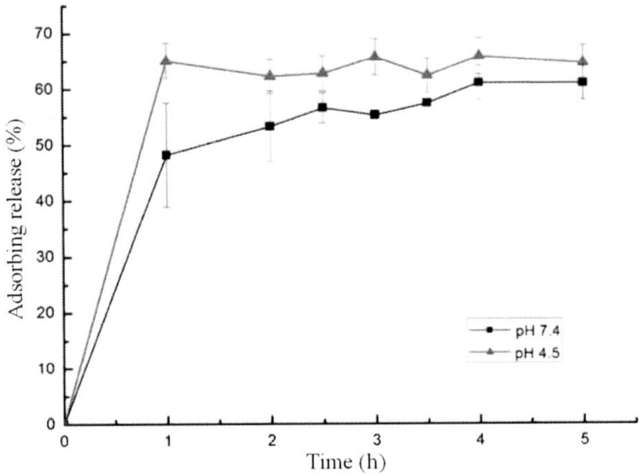

Figure 9.29 Effects of time and pH value on adsorbing efficiency at 4°C. Reprinted from Ref. 28, with permission from Elsevier.

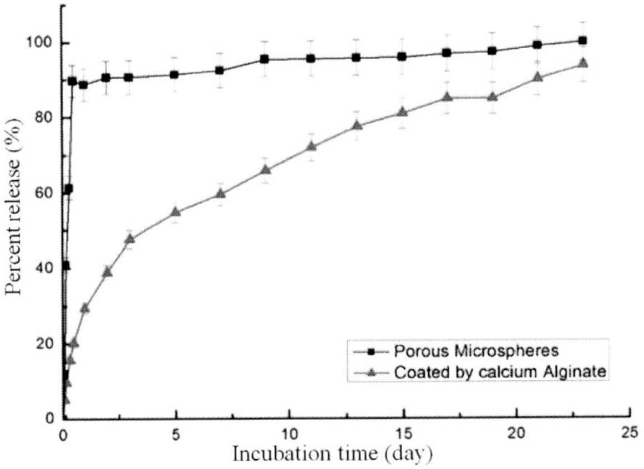

Figure 9.30 Percent release of HSA from microspheres incubation in PBS at 37°C. Reprinted from Ref. 28, with permission from Elsevier.

(iii) Surface property of microspheres or NPs

In terms of surface chemistry, a study comparing hydrophobic and hydrophilic silica spheres showed a greater change in the secondary structure of fibrinogen and BSA (6,49). Figure 9.31 showed that surface property of NPs had important influence on adsorption of protein on microspheres and NPs. The hydrophobic surface could adsorb many proteins in plasma. After PEGylation of the particle, protein binding on the surface of the NP was reduced. When the surface of NP was coated with polysorbate, apolipoprotein E (ApoE) could specifically bind to the NPs (2).

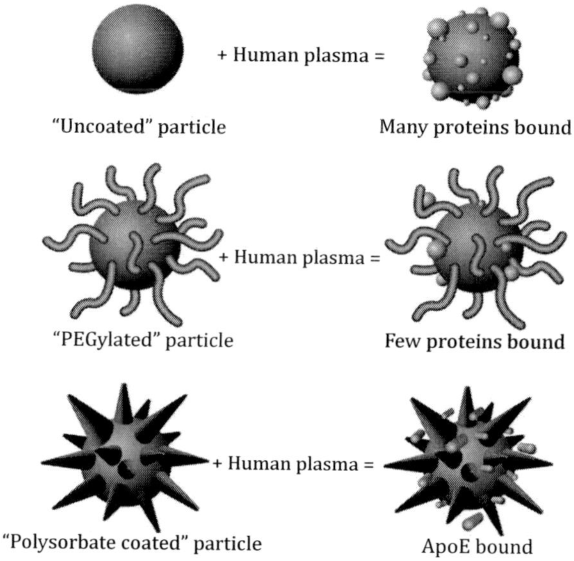

Figure 9.31 Schematic representation of the influence of surface property of microspheres and NPs on protein adsorption. Reprinted from Ref. 52, with permission from Elsevier.

As a usual, the negatively charged microspheres and NPs were suitable for adsorption of positively charged proteins, which could attain a high level of protein loading with full bioactivity. Cai *et al.* (53) described the feasibility of employing a model negatively charged nanocarrier, consisting of a polymer blend of PLGA and poly(styrene-co-4-styrene-sulfonate) (PSS), for increasing the LC of the model positively charged protein lysozyme. Investigations of pH and ionic strength suggest that electrostatic interactions primarily govern the adsorption process and hydrophobic interactions also play a minor role. Moreover, the adsorption behavior of cytochrome C showed different adsorption mechanism that a fine balance of hydrophobic and electrostatic interaction is required for efficient adsorption. The study of BSA adsorption highlighted the importance of hydrophobic interactions between protein and NPs for efficient loading. The lower adsorption capacity of BSA compared to cytochrome C and lysozyme indicated the importance of electrostatic interaction for adsorption process. The release rate of lysozyme from NPs was dependent on drug loading, and the preservation of full protein bioactivity was demonstrated. Taken together, this study suggests a feasible method to improve the LC of proteins based on the electrostatic interactions between oppositely charged proteins and microspheres/NPs.

In conclusion, the main factors of size, inner structure, and surface properties of microspheres/NPs should be considered during microspheres/ NPs design for adsorption of PPs.

9.5 Applications of Microsphere-Based Protein Delivery

After successfully encapsulating or loading protein or peptide in microsphere or NPs, applications will be carried out. The applications have mainly been focused on drug delivery and vaccine adjuvant, and drug delivery can be divided into long-acting injection and mucosal administration according to the route of administration. In this chapter, the different applications will be introduced through some examples.

9.5.1 Protein or Peptide Drug Delivery

(i) Injection administration

When microsphere-based protein-delivery system is administrated by injection, the main target is to realize long-term sustained release of drug, and to reduce administration frequency. In this application, the protein or peptide is usually encapsulated into microspheres in order to realize long-term release. The material, preparation process, size, structure, and surface property of microspheres could be chosen and designed according to the release duration.

For example, Woo *et al.* (54) prepared a 120-day PLA microsphere-delivery system for a luteinizing hormone–releasing hormone (LHRH) analogue as leuprolide by a dispersion/ solvent extraction–evaporation method. Leuprolide is mainly used for treatment of prostatic cancer, which can effectively suppress testosterone level *in vivo*. The average particle size of leuprolide loaded microspheres was 51.7 μm, which is suitable for intramuscular or subcutaneous injections and the leuprolide

EE was 81.2%. The release profile *in vitro* showed that the microspheres released 10% of the peptide within 24 h, followed by a linear release for 150 days. After subcutaneous administration using a rat model, serum leuprolide levels increased immediately to 45.6 ng/mL after administration of the microspheres, but then fell to 4.3 ng/mL at 15 days and ~2.0 ng/mL at 30 days where they remained for 120 days. The testosterone levels increased initially to 15 ng/mL and then decreased to below 0.5 ng/mL by day 4 where they remained for 120 days. The prepared 120-day microsphere of leuprolide showed excellent controlled peptide release characteristics and efficacy *in vivo*.

(ii) Mucosal administration

Mucosal administration includes oral, intranasal, ocular, pulmonary delivery, and so on. The requirements for microspheres, such as materials, size, structure, surface property, and drug-loading ways vary in specific route of administration. Here, some examples will be given to demonstrate the applications of microsphere-based protein or peptide delivery by mucosal administration.

Chalasani *et al.* (55) have developed a VB_{12}-NPs system to enhance the uptake capacity of both NPs and VB_{12} transport to deliver orally effective insulin. NPs were prepared using different molecular weight dextrans (10 K, 70 K, and 200 K) and epichlorohydrin as cross-linker by an emulsion method. NPs' surface was modified with succinic anhydride and conjugated with amino VB_{12} derivatives by carbamate linkage. The preformed NPs conjugates with size of 160–250 nm were loaded with 2%, 3%, and 4% W/W of insulin, and the entrapment was found to be 45–70%. NPs conjugates were found to protect 65–83% of entrapped insulin against *in vitro* gut proteases. *In vitro* release studies exhibit an initial burst followed by diffusion-controlled first-order kinetics with 75–95% release within 48 h. After oral administration of these carriers (20 IU/kg), a nadir of 70–75% reduction in plasma glucose was found in 5 h, which reached basal levels in 8–10 h, and a prolonged second phase was found until 54 h. Compared to NPs without VB_{12}, the percentage of pharmacological availability (PA) of 70 K NPs conjugate containing 2%, 3%, and 4% W/W insulin was 1.1, 1.9, and 2.6 fold higher, respectively. This suggested that the VB_{12}-NPs carrier had great potential for oral-delivery of proteins.

A glucose-responsive composite-microparticles system based on chitosan, concanavalin A, and dextran was constructed by Yin *et al.* (56) for self-regulated insulin delivery. The composite microparticles were fabricated by the process of Fig. 9.32. Insulin could be loaded into the microparticles due to electrostatic and intermolecular interaction with high entrapment efficiency of 92.2% and LC of 9.1%. The insulin release profiles of microparticles with different medium are revealed in Fig. 9.33. A slower release rate of insulin was detected in PBS (pH 7.4) with glucose 0 mg/mL when compared with that in PBS (pH 7.4) with glucose 4 mg/mL. Figure 9.33 shows the release concentration at specific time through the microparticles in response to stepwise changes of the glucose concentration. Pulsatile insulin release patterns were displayed as the glucose concentration changed. As the glucose

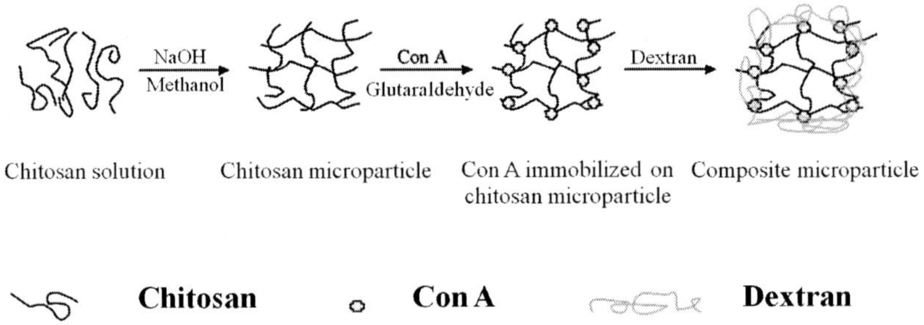

Figure 9.32 Fabrication process of the blank composite microparticles. Reprinted from Ref. 56, with permission from Elsevier.

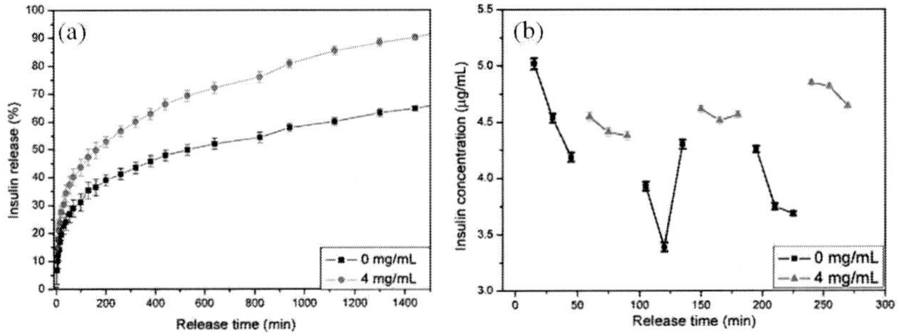

Figure 9.33 Insulin-release profiles from microparticles with different release medium: Glucose 0 and 4 mg/mL in PBS (pH 7.4) and reversible glucose sensitivity of the composite microparticles. Reprinted from Ref. 56, with permission from Elsevier.

concentration of release medium changed between 0 and 4 mg/mL, the released insulin concentration fluctuated at the range from 3 to 5 μg/mL. The release profiles of insulin revealed that the insulin release was in response to the glucose concentration in the medium and the glucose sensitivity was reversible.

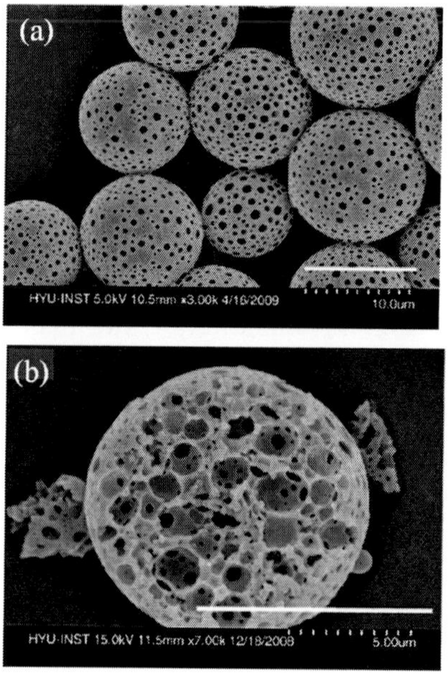

Figure 9.34 (a) Microscopic images of porous PLGA microspheres prepared by osmosis induction methods. (b) Cross-sectional view of porous microspheres (PLGA) = 10%, BSA/PLGA = 0.6, scale bars = 10 μm). Reprinted from Ref. 57, with permission from Elsevier.

Lee *et al.* (57) reported the use of BSA as an osmotic agent in order to control the porous structure of PLGA microspheres prepared by double-emulsion method. BSA was useful to induce osmosis between internal and external water phases during the double emulsion process, resulting in the fabrication of microspheres with controllable, uniform porous structures. The pore size of PLGA microspheres was controlled independently from the particle size by this approach. The use of BSA as an osmotic agent reduced the initial burst of model proteins of vascular endothelial growth factor (VEGF) entrapped in the porous microspheres, and the sustained release of VEGF was achieved for two weeks *in vitro*. This approach to control porous structures of polymer microspheres could be useful to develop novel pulmonary drug-delivery systems. It's shown that porous microspheres fabricated by osmosis induction had a spherical shape with uniform-size distribution (Fig. 9.34a), due to the effective action of BSA as an osmotic agent. Porous structures were uniform on the surface as well as inside the microspheres, and interconnected open-pore structures were observed (Fig. 9.34b). The lung deposit efficiency demonstrated that smaller particle sizes (with similar pore structures) resulted in the higher deposit efficiency of porous microspheres with homogeneous distribution in the lungs as shown in Fig. 9.35. A two-fold increase in the deposit efficiency was observed for porous microspheres with a particle size of 2.9 μm, compared with nonporous microspheres of the same size. Porous PLGA microspheres containing either insulin or VEGF were prepared, and the release of the proteins from the microspheres was monitored *in vitro*. Insulin was released from porous microspheres faster than from nonporous particles, irrespective of the different pore sizes (Fig. 9.36). Surprisingly, the initial burst of VEGF from porous microspheres was remarkably reduced and the sustained release of VEGF was achieved for two weeks, which was dependent on the pore size. This approach to control porous structures of polymer microspheres could be useful to develop novel pulmonary protein-delivery systems.

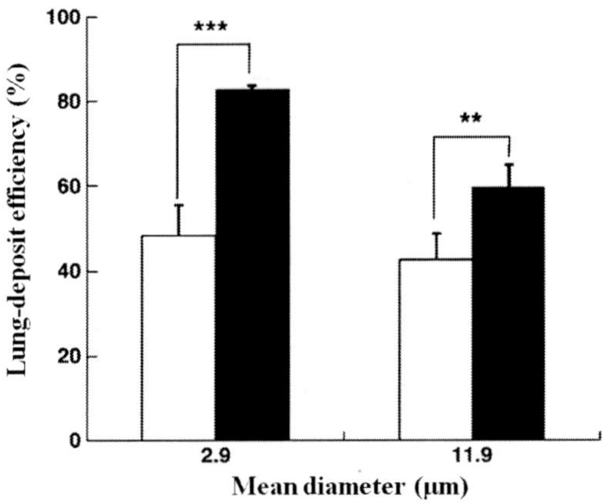

Figure 9.35 Comparison for the lung-deposit efficiency of nonporous (empty bars) and porous microspheres (filled bars) of various particle sizes (**P < 0.01 and ***P < 0.001). Reprinted from Ref. 57, with permission from Elsevier.

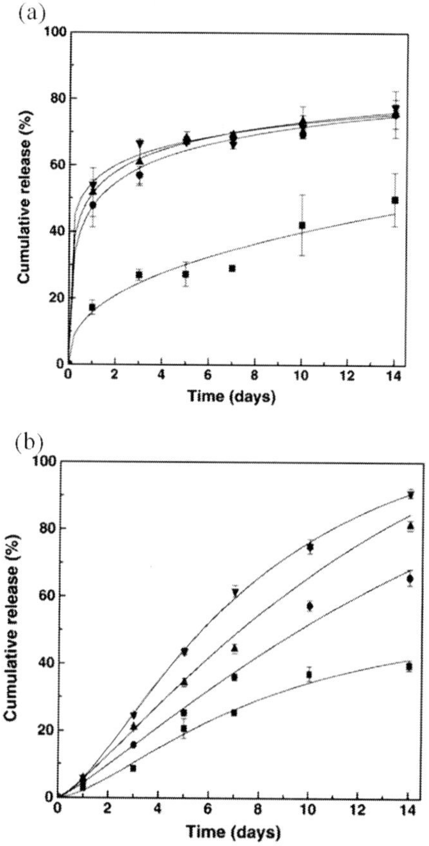

Figure 9.36 *In vitro* release of (a) insulin and (b) VEGF from nonporous microspheres (■) and porous microspheres with various pore sizes (●, 0.57 μm; ▲, 0.88 μm; ▼, 0.92 μm). Mean diameters of particles were kept constant (11 μm). Reprinted from Ref. 57, with permission from Elsevier.

Gavini *et al.* (16) investigated on PLGA microspheres as carriers for the topical ocular delivery of a peptide drug VA. The microspheres were prepared by an emulsification/spray-drying technique with about 11 μm of average particle size. *In vivo* studies were carried out by assessing the pharmacokinetic profile of VA in the aqueous humor of rabbits after topical administration of aqueous suspensions of microspheres (VA_{micr}). High and prolonged VA concentrations and increased Area under curve (AUC) values (2-fold) with respect to an aqueous solution of the drug (VA_{sol}) were observed as shown in Table 9.5. Increasing the viscosity of the microsphere suspension by adding a suspending viscosizing agent (hydroxypropylcellulose, HPC) did not produce an increase of the ocular bioavailability. It showed that PLGA microspheres can be proposed as a system for ocular delivery of peptide drugs.

Table 9.5 Pharmacokinetic parameters in aqueous humor after *in vivo* administration in rabbits of the preparations under study

Preparation	C_{max} (μg/mL ± SE)	T_{max} (min)	AUC (min μg/mL ± SE)	AUC relative
VA_{micro}	2.47 ± 0.49	15	248.2 ± 35.2	2.31
$VA_{micro\text{-}HPC}$	1.80 ± 0.54	30	206.6 ± 52.8	1.92
VA_{sol}	0.94 ± 0.05	30	107.4 ± 17.3	1.00
$VA_{sol\text{-}HPC}$	1.26 ± 0.21	15	122.9 ± 26.7	1.14

Source: Reprinted from Ref. 16, with permission from Elsevier.

9.5.2 Vaccine Adjuvant

The use of polymeric particulate adjuvants as microspheres or NPs could increase antigen uptake by antigen-presenting cells followed by facilitating the induction of potent immune responses. The formulation parameters like materials, particle size, load of antigen, release profile of antigen, surface property of particle, presences of additional adjuvants and route of immunization have major influence on the quality and quantity of immune response.

Panda *et al.* (58) prepared PLA polymer particles to entrap tetanus toxoid (TT) and systemically evaluated the effect of particle size, antigen load, dose, and additional adjuvant on achieving high and sustained anti-TT antibody titer from single-point intramuscular immunization. The results showed that high and long-lasting antibody titer was achieved upon immunization with 2–8 μm size particles, while microparticles within the size range 50–150 μm elicited very low serum antibody response. Immunization with very small particles (<2 μm) and with intermediate size range particles (10–70 μm) elicited comparable antibody response from single-point immunization but lower in comparison to that achieved while immunizing with 2–8 μm size particles. These results indicate the need of optimal particle sizes in micron ranges for improved humoral response from single point immunization. Increasing antigen load on polymer particles was found to have positive influence on generation of antibody titers from particle-based immunization. Maximum peak antibody titer of ~300 µg/mL was achieved on day 50 upon immunization with particles having highest load of antigen (94 µg/mg of polymer) as shown in Fig. 9.37. Increase in dose of polymer-entrapped antigen resulted in concomitant increase in peak antibody titers, indicating the importance of antigen stability, particle size, and load on generating reproducible immune response.

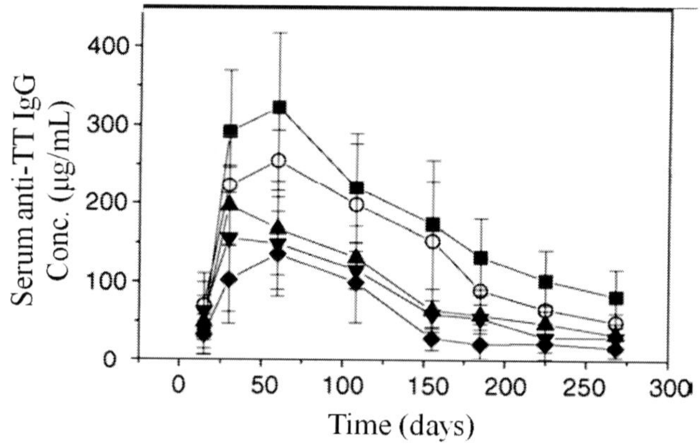

Figure 9.37 Anti-TT IgG concentrations elicited by immunization of TT-loaded PLA microparticles having TT loads of (µg/mg) 94.2 (-■-), 28.2 (-○-), 16.7(-▲-), 5.8 (-▼-), 1.3 (-◆-). Groups of six female wistar rats 2-months old were immunized with 5 Lf microencapsulated TT along with alum intramuscularly. One Lf of TT was equivalent to 2.6 µg of TT. Geometric means of serum antibody titers are presented with standard deviations. Reprinted from Ref. 58, with permission from Elsevier.

Borges *et al.* (59) developed a delivery system that is composed of a chitosan core to which the hepatitis B surface antigen (HBsAg) was adsorbed and subsequently coated with sodium alginate. The adjuvant effect of alginate-coated chitosan NPs for the hepatitis B surface antigen was evaluated after subcutaneous application in mice. Four groups as shown in Table 9.6 were subcutaneously administered to mice and several immunological parameters were evaluated. The adjuvant effect of the alginate-coated chitosan NPs (group I) was clearly demonstrated by the production of high anti-

HBsAg IgG (Immunoglobin G) titers with a clear dominance of Th2-type antibodies (IgG1 > IgG2a) as shown in Fig. 9.38. The co-delivery of antigen-loaded NPs in the presence of the immunopotentiator of cytidine phosphate guanosine oligodeoxynucleotides (CpG ODN) resulted in an increase of anti-HBsAg IgG titers that were not statistically different from group I; however, an increase of the IgG2a/IgG1 ratio from 0.1 to 1.0 was observed. The enhancement of the immune response observed with the antigen-loaded NPs demonstrated that chitosan is a promising platform for parenteral HBsAg delivery and, when co-administered with the CpG ODN, resulted in a mixed Th1/Th2-type immune response.

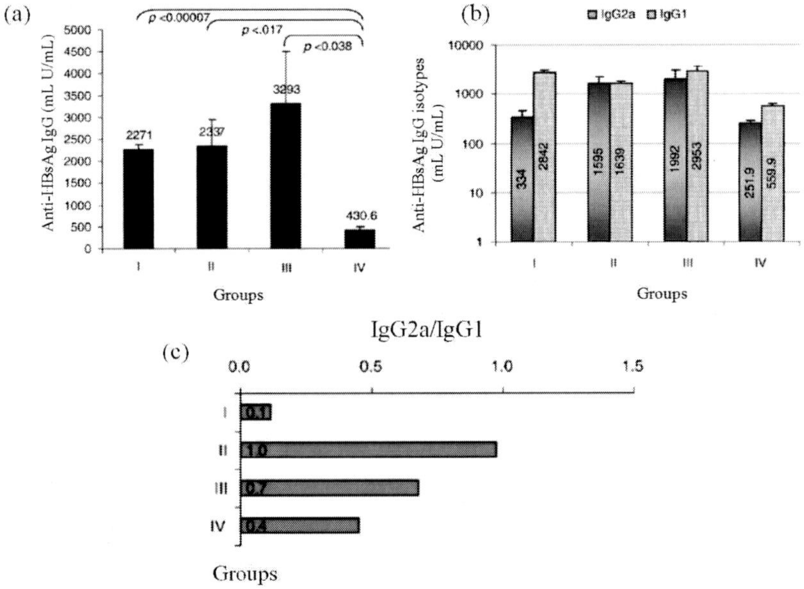

Figure 9.38 (a) Serum anti-HBsAg IgG titers of mice immunized with different formulations of hepatitis B vaccine. Values are expressed as antibody titers of individual mice taken in the end of the experiment. The horizontal bar is the mean of the group. *t*-Test was used for analysis of significance between each group and the group IV. (b) Serum anti-HBsAg IgG1 and IgG2a titers (logarithmic scale) of mice immunized with different formulations of hepatitis B vaccine. The bar corresponds to mean titer in each group. Titers were defined as the highest plasma dilution resulting in an absorbance value twice that of nonimmune plasma (1 mL U/mL = mean + 2 SD of the control group). (c) Ratio IgG2a/IgG1. Reprinted from Ref. 59, with permission from Elsevier.

Table 9.6 Pharmacokinetic parameters in aqueous humor after *in vivo* administration in rabbits of the preparations under study

Group name	Descriptions
Group I	Suspension of alginate-coated chitosan NPs loaded with 10 μg HBsAg
Group II	Suspension of alginate-coated chitosan NPs loaded with 10 μg HBsAg plus 20 μg of CpG ODN in solution
Group III	Suspension of the chitosan NPs loaded with 10 μg HBsAg and 20 μg CpG ODN
Group IV	PBS; pH 7.4 solution with 10 μg HBsAg (reference group)

Source: Reprinted from Ref. 59, with permission from Elsevier.

In summary, therefore, the microspheres or NPs should be chosen or designed according to the requirements in different applications.

9.6 Conclusion

In this chapter, we discussed the design of microspheres/nanopaticles for encapsulation or adsorption of PPs. In encapsulation process, the EE, drug release profile, and protein bioactivity are affected by many factors, such as polymers and preparation method employed, preparation conditions, additives, and so on. Therefore, all these factors should be considered in the design of microspheres/NPs. In adsorption process, the higher loading efficiency, less burst release, and protein activity are very important for applications. The size, inner structure, and surface properties of microspheres/NPs should give more consideration in design of microspheres/NPs for adsorption of PPs. In addition, the requirement for microspheres varies in applications, therefore, the microspheres could be designed according to different applications.

References

1. Morita, T., Sakamura, Y., Horikiri, Y., Suzuki, T., & Yoshino, H. (2000). Protein encapsulation into biodegradable microspheres by a novel S/O/W emulsion method using poly(ethylene glycol) as a protein micronization adjuvant. *Journal of Controlled Release*, **69**, 435–444.

2. Yeo, Y., Baek, N., & Park, K. (2001). Microencapsulation methods for delivery of protein drugs. *Biotechnology and Bioprocess Engineering*, **6**, 213–230.
3. Peppas, N. A., & Carr, D. A. (2009). Impact of absorption and transport on intelligent therapeutics and nanoscale delivery of protein therapeutic agents. *Chemical Engineering Science*, **64**, 4553–4565.
4. Meng, F. T., Ma, G. H., Liu, Y. D., Qiu, W., & Su, Z. G. (2004). Microencapsulation of bovine hemoglobin with high bio-activity and high entrapment efficiency using a W/O/W double emulsion technique. *Colloids and Surfaces, B: Biointerfaces*, **33**, 177–183.
5. Sood, A., & Panchagnula, R. (2001). Peroral route: An opportunity for protein and peptide drug delivery. *Chemical Reviews*, **101**, 3275–3303.
6. Martín del Valle, E. M., Galán, M. A., & Carbonell, R. G. (2009). Drug delivery technologies: The way forward in the new decade. *Industrial & Engineering Chemistry Research*, **48**, 2475–2486.
7. Kokai, L. E., Tan, H., Jhunjhunwala, S., Little, S. R., Frank, J. W., Marra, K. G. (2010). Protein bioactivity and polymer orientation is affected by stabilizer incorporation for double-walled microspheres. *Journal of Controlled Release*, **141**, 168–176
8. Sinha, V. R., & Trehan, A. (2003). Biodegradable microspheres for protein delivery. *Journal of Controlled Release*, **90**, 261–280.
9. Dai, C., Wang, B., & Zhao, H. (2005). Microencapsulation peptide and protein drugs delivery system. *Colloids and Surfaces, B: Biointerfaces*, **41**, 117–120.
10. Fu, K., Klibanov, A. M., & Langer, R. (2000). Protein stability in controlled-release systems. *Nature Biotechnology*, **18**, 24–25.
11. Soltero, R., & Ekwuribe, N. (2001). The oral delivery of protein and peptide drugs. *Innovations in Pharmaceutical Technology*, **1**, 106–110.
12. Hamman, J. H., Enslin, G. M., & Kotzé, A. F. (2005). Oral delivery of peptide drugs: Barriers and developments. *BioDrugs*, **19**, 165–177.
13. Fogueri, L. R., & Singh, S. (2009). Smart polymers for controlled delivery of proteins and peptides: A review of patents. *Recent Patents on Drug Delivery & Formulation*, **3**, 40–48.
14. Jain, R. A. (2000). The manufacturing techniques of various drug loaded biodegradable poly(lactide-co-glycolide) (PLGA) devices. *Biomaterials*, **21**, 2475–2490.
15. Ishihara, T., & Mizushima, T. (2010). Techniques for efficient entrapment of pharmaceuticals in biodegradable solid micro/nanoparticles. *Expert Opinion on Drug Delivery*, **7**, 565–575.

16. Gavini, E., Chetoni, P., Cossu, M., Alvarez, M. G., Saettone, M. F., & Giunchedi, P. (2004). PLGA microspheres for the ocular delivery of a peptide drug, vancomycin using emulsification/spray-drying as the preparation method: *In vitro/in vivo* studies, *European Journal of Pharmaceutics and Biopharmaceutics*, **57**, 207–212.

17. Cheow, W. S., Ng, M. L., Kho, K., & Hadinoto, K. (2010). Spray-freeze-drying production of thermally sensitive polymeric nanoparticle aggregates for inhaled drug delivery: Effect of freeze-drying adjuvants. *International Journal of Pharmaceutics*, **404**, 289–300.

18. Costantino, H. R., Firouzabadian, L., Hogeland, K., Wu, C., Beganski, C., Carrasquillo, K. G., *et al.* (2000). Protein spray-freeze drying. Effect of atomization conditions on particle size and stability. *Pharmaceutical Research*, **17**, 1374–1383.

19. Wu, X. S. (1995). Synthesis and properties of biodegradable lactic/glycolic acid polymers. In Wise D. L. (Ed.) *Encyclopedic Handbook of Biomaterials and Bioengineering* (pp. 1015–1054). New York: Marcel Dekker.

20. Arshady, R. (1991). Preparation of biodegradable microspheres and microcapsules: 2. Polylactides and related polyesters. *Journal of Controlled Release*, **17**, 1–22.

21. Chaw, C. S., Yang, Y. Y., Lim, I. J., & Phan, T. T. (2003) Water-soluble betamethasone-loaded poly (lactide-co-glycolide) hollow microparticles as a sustained release dosage form. *Microencapsul*, **20**, 349–359.

22. Cho, M., & Sah, H. (2005). Formulation and process parameters affecting protein encapsulation into PLGA microspheres during ethyl acetate-based microencapsulation process. *Journal of Microencapsulation*, **22**, 1–12.

23. Sah, H. (2000). Ethyl formate-alternative dispersed solvent useful in preparing PLGA microspheres. *International Journal of Pharmaceutics*, **195**, 103–113.

24. Sah, H., Smith, M. S., & Chern, R. T. (1996). A novel method of preparing PLGA microcapsules utilizing methylethyl ketone. *Pharmaceutical Research*, **13**, 360–367.

25. Woo, B. H., Kostanski, J. W., Gebrekidan, S., Dani, B. A., Thanoo, B. C., & DeLuca, P. P. (2001). Preparation, characterization and *in vivo* evaluation of 120-day poly(DL-lactide) leuprolide microspheres. *Journal of Controlled Release*, **75**, 307–315.

26. Yang, J. F., Qiu, L. Y., Jin, Y., & Zhang, J. X. (2005). Thymosin-loaded enteric microspheres for oral administration: Preparation and *in vitro* release studies. *International Journal of Pharmaceutics*, **301**, 41–47.

27. Wang, L. Y., Wu, J., Wei, W., Zhang, Y. L., Ma, G. H., & Su, Z. G. (2008). Preparation and applications of chitosan microsphere. In R. Jayakumar, & M. Prabaharan (Eds.), *Current Research and Developments on Cchitin and Chitosan in Biomaterial Science* (pp. 143–163). Kerala, India: Research Signpost.

28. Sun, L., Zhou, S. B., Wang, W. J., Li, X. H., Wang, J. X., & Weng, J. (2009). Preparation and characterization of porous biodegradable microspheres used for controlled protein delivery. *Colloids and Surfaces, A: Physicochemical and Engineering Aspects*, **345**, 173–181.

29. Emami, J., Hamishehkar, H., Najafabadi, A. R., Gilani, K., Minaiyan, M., Mahdavi, H., *et al.* (2009). A novel approach to prepare insulin-loaded poly(lactic-co-glycolic acid) microcapsules and the protein stability study. *Journal of Pharmaceutical Sciences*, **98**, 1712–1731.

30. Wang, L. Y., Ma, G. H., & Su, Z. G. (2005). Preparation of uniform sized chitosan microspheres by membrane emulsification technique and application as a carrier of protein drug. *Journal of Controlled Release*, **106**, 62–75.

31. Liu, R., Ma, G. H., Meng, F. T., & Su, Z. G. (2005). Preparation of uniform-sized PLA microcapsules by combining Shirasu porous glass membrane emulsification technique and multiple emulsion-solvent evaporation method. *Journal of Controlled Release*, **103**, 31–43.

32. Zhou, Q. Z., Wang, L. Y., Ma, G. H., & Su, Z. G. (2008). Multi-stage premix membrane emulsification for preparation of agarose microbeads with uniform size. *Journal of Membrane Science*, **322**, 98–104.

33. Wei, Q., Wei, W., Tian, R., Wang, L. Y., Su, Z. G., & Ma, G. H. (2008), Preparation of uniform-sized PELA microspheres with high encapsulation efficiency of antigen by premix membrane emulsification. *Journal of Colloid and Interface Science*, **323**, 267–273.

34. Zhang, Y., Wei, W., Lv, P., Wang, L., & Ma, G. (2011). Preparation and evaluation of alginate-chitosan microspheres for oral delivery of insulin. *European Journal of Pharmaceutics and Biopharmaceutics*, **77**, 11–19.

35. Lv, P. P., Wei, W., Gong, F. L., Zhang, Y. L., Zhao, H. Y., Lei, J. D., *et al.* (2009). Preparation of uniformly sized chitosan nanospheres by a premix membrane emulsification technique. *Industrial & Engineering Chemistry Research*, **48**, 8819–8828.

36. Nihant, N., Granfils, C., Jerome, R., & Teyssie, P. (1995). Microencapsulation by coacervation of poly(lactide-co-glycolide) IV. Effect of the processing parameters on coacervation and encapsulation. *Journal of Controlled Release*, **35**, 117–125.

37. Nihant, N., Stassen, S., Granfils, C., Jerome, R., Teyssie, P., & Goffinet, G. (1994). Microencapsulation by coacervation of poly(lactide-co-glycolide) III. Characterization of the final microspheres. *Polymer International*, **34**, 289–299.

38. Celik, O., & Akbuğa, J. (2007). Preparation of superoxide dismutase loaded chitosan microspheres: Characterization and release studies. *European Journal of Pharmaceutics and Biopharmaceutics*, **66**, 42–47.

39. Shu, S., Sun, C., Zhang, X., Wu, Z., Wang, Z., & Li, C. (2010). Hollow and degradable polyelectrolyte nanocapsules for protein drug delivery. *Acta Biomaterialia*, **6**, 210–217.

40. Wei, Y., Wang, Y. X., Wang, W., Ho, S. V., Wei, W., Ma, G. H. (2011). mPEG-PLA microspheres with narrow size distribution increase the controlled release effect of recombinant human growth hormone, *Journal of Materials Chemistry*, **21**, 12691–12699.

41. Yuan, W., Wu, F., Guo, M., Jin, T. (2009). Development of protein delivery microsphere system by a novel S/O/O/W multi-emulsion. *European Journal of Pharmaceutical Sciences*, **36**, 212–218.

42. Wang, L. Y., Gu, Y. H., Su, Z. G., & Ma, G. H. (2006). Preparation and improvement of release behavior of chitosan microspheres containing insulin. *International Journal of Pharmaceutics*, **311**, 187–195.

43. Wang, L. Y., Gu, Y. H., Zhou, Q. Z., Wan, Y. H., Ma, G. H., & Su, Z. G. (2006). Preparation and characterization of uniform-sized chitosan microspheres containing insulin by membrane emulsification and a two-step solidification process. *Colloids and Surfaces, B: Biointerfaces*, **50**, 126–135.

44. Schoubben, A., Blasi, P., Giovagnoli, S., Perioli, L., Rossi, C., & Ricci, M. (2009). Novel composite microparticles for protein stabilization and delivery. *European Journal of Pharmaceutical Sciences*, **36**, 226–234.

45. Kane, R. S., & Stroock, A. D. (2007). Nanobiotechnology: Protein-nanomaterial interactions. *Biotechnology Progress*, **23**, 316–319.

46. Asuri, P., Bale, S. S., Karajanagi, S. S., & Kane, R. S. (2006). The protein–nanomaterial interface. *Current Opinion in Biotechnology*, **17**, 562–568.

47. Lundqvist, M., Sethson, I., & Jonsson, B. H. (2004). Protein adsorption onto silica nanoparticles: Conformational changes depend on the particles' curvature and the protein stability. *Langmuir*, **20**, 10,639–10,647.

48. Vertegel, A. A., Siegel, R. W., & Dordick, J. S. (2004). Silica nanoparticle size influences the structure and enzymatic activity of adsorbed lysozyme. *Langmuir*, **20**, 6800–6807.

49. Roach, P., Farrar, D., & Perry, C. C. (2006). Surface tailoring for controlled protein adsorption: Effect of topography at the nanometer scale and chemistry. *Journal of the American Chemical Society*, **128**, 3939–3945.
50. Wei, W., Yuan, L., Hu, G., Wang, L. H., Wu, J., Hu, X., et al. (2008). Monodisperse chitosan microspheres with interesting structures for protein drug delivery. *Advanced Materials*, **12**, 2292–2296.
51. Shu, S., Zhang, X., Wu, Z., Wang, Z., & Li, C. (2010). Gradient cross-linked biodegradable polyelectrolyte nanocapsules for intracellular protein drug delivery. *Biomaterials*, **31**, 6039–6049.
52. Aggarwal, P., Hall, J. B., McLeland, C. B., Dobrovolskaia, M. A., & McNeil, S. E. (2009). Nanoparticle interaction with plasma proteins as it relates to particle biodistribution, biocompatibility and therapeutic efficacy. *Advanced Drug Delivery Reviews*, **61**, 428–437.
53. Cai, C., Bakowsky, U., Rytting, E., Schaper, A. K., & Kissel, T. (2008). Charged nanoparticles as protein delivery systems: A feasibility study using lysozyme as model protein. *European Journal of Pharmaceutics and Biopharmaceutics*, **69**, 31–42.
54. Woo, B. H., Kostanski, J. W., Gebrekidan, S., Dani, B. A., Thanoo, B. C., & DeLuca, P. P. (2001). Preparation, characterization and *in vivo* evaluation of 120-day poly(DL-lactide) leuprolide microspheres. *Journal of Controlled Release*, **75**, 307–315.
55. Chalasani, K. B., Russell-Jones, G. J., Yandrapu, S. K., Diwan, P. V., & Jain, S. K. (2007). A novel vitamin B_{12}-nanosphere conjugate carrier system for peroral delivery of insulin. *Journal of Controlled Release*, **117**, 421–429.
56. Yin, R., Han, J., Zhang, J., & Nie, J. (2010). Glucose-responsive composite microparticles based on chitosan, concanavalin A and dextran for insulin delivery. *Colloids and Surfaces, B: Biointerfaces*, **76**, 483–488.
57. Lee, J., Oh, Y. J., Lee, S. K., & Lee, K. Y. (2010). Facile control of porous structures of polymer microspheres using an osmotic agent for pulmonary delivery. *Journal of Controlled Release*, **146**, 61–67.
58. Katare, Y. K., Muthukumaran, T., & Panda, A. K. (2005). Influence of particle size, antigen load, dose and additional adjuvant on the immune response from antigen loaded PLA microparticles. *International Journal of Pharmaceutics*, **301**, 149–160.
59. Borges, O., Silva, M., Sousa, A., Borchard, G., Junginger, H. E., & Cordeiroda Silva, A. (2008). Alginate coated chitosan nanoparticles are an effective subcutaneous adjuvant for hepatitis B surface antigen. *International Immunopharmacology*, **8**, 1773–1780.

Chapter 10

Micro/Nanospheres for Gene Drug Delivery

Jie Wu and Guanghui Ma

State Key Laboratory of Biochemical Engineering, Institute of Process Engineering, Chinese Academy of Sciences, Beijing, P. R. China

ghma@home.ipe.ac.cn

10.1 Introduction

In recent years, micro/nanospheres have not only been used as chemical drug or bio-macromolecule drug-delivery systems but also as gene vectors. In gene therapy, the genetic material need to be introduced into the cytosol or the nucleus of cells, otherwise it does not function effectively and may even cause side effects. However, without the help of other techniques, it would be difficult to introduce genetic material into cells due to its large size and negative charge. One of the most commonly studied techniques is adopting gene vectors to enhance transfection efficiency, such as adenovirus, dendrimer, and micro/nanospheres (1–4).

As gene vectors, the micro/nanospheres are labeled with high biocompatibility, improved stability, and low immunogenicity, which have become a hot topic in gene therapy research. These particles would be especially useful in providing target gene delivery, achieving sustained gene transfer, improving the stability of gene against the enzymatic degradation, and broadening the administration routes. Moreover, since pDNA and siRNA molecules are long, negatively charged polyelectrolyte, the cationic polymers are often used to complex with pDNA and siRNA into nanoscale aggregates for cell endocytosis (5–7).

Microspheres and Microcapsules in Biotechnology: Design, Preparation, and Applications
Edited by Guanghui Ma and Zhiguo Su

ISBN 978-981-4316-47-7 (Hardcover), 978-981-4364-62-1 (eBook)
www.panstanford.com

However, in order to act as ideal gene vectors, there are numerous barriers that need to be conquered from administration to effective transfection *in vivo*. And the choice of polymers, the preparation method, and the characteristics of prepared micro/nanospheres need to be adjusted carefully with these barriers in consideration. The ideal vectors should be able to improve the stability of pDNA or siRNA, help these large molecules into cells, and release them for the nucleus entry. Moreover, the material used should be biodegradable and biocompatible in order to avoid any side effects or immunogenic responses to the vector itself.

In this chapter, the main barriers to gene delivery, characteristics of used polymers, preparation methods, and parts of application fields using polymer micro/nanospheres as gene vectors will be illustrated.

10.2 The Main Barriers for Micro/Nanospheres as Gene Vectors

The transfection process of a gene includes being delivered *in vivo*; being taken by cellular endocytosis; escaping from the endosomal compartments; being transported through the cytosol; and localizing in cell nucleus as shown in Fig. 10.1. During this process, there were many enzymes which tend to degrade genetic material both in extracellular and intercellular circumstances. By adopting micro/nanospheres as gene vector, the encapsulation of DNA could avoid direct contact between the genetic material and the external surroundings. Yu *et al.* prepared poly(ethylene-glycol)-poly(γ-benzyl-L-glutamate) nanospheres with a mean diameter around 80 nm to load the drug. After being incubated with DNase I for 16 h, nearly half of the DNA in the nanospheres remained intact, while almost all of the naked plasmid was digested (8). Adopting polymers with buffering ability was another way to avoid enzymatic degradation, which promotes the escape of the vectors from endosomes prior to fusion with lysosomes (9).

Compared with DNA, RNA is more labile due to ubiquitous RNase activity and chemical degradation. It is hard to protect RNA integrity by the conventional methods used to prevent DNA from degradation (10,11,12). Yagi *et al.* designed a novel nanosphere system called "wrapsome" to deliver siRNA *in vivo*, which had a core–shell structure. siRNA and a cationic lipofection complex were inserted

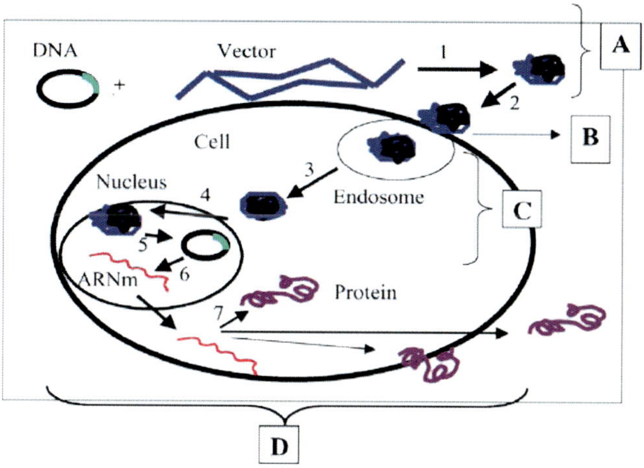

Figure 10.1 Simplified transfection process: (A) Extracellular trafficking; (1) Loading gene on the vector; (B) Cell membrane attachment; (2) Cellular endocytosis; (C) Intracellular trafficking; (3) Escape from the endosomal compartments; (4) Nuclear entry; (D) Gene expression; (5) Dissociation of the vector and release of the gene; (6) ARNm transcription from the gene; (7) Protein translation from ARNm (2). Reprinted from Mansouri Ref. 12 with permission from Elsevier BV.

in a core that was fully enveloped by a neutral lipid bilayer and hydrophilic polymers. The core–shell structure protects siRNA from enzymatic digestion, providing it a long half-life in the systemic circulation (13).

Besides protecting the integrity of genetic material, another important point for gene delivery is to condense pDNA and siRNA into compact structures. Cationic polymers have been reported as efficient polycationic condensing agents for gene delivery owing to the static forces between polymers and genetic material (14,15). Unfolded pDNA had an oval ring structure (about 500 nm long and 250 nm wide) by AFM observation, and after binding with tertiary amine-based cationic polymers, pDNA formed irregular aggregates with largely decreased dimensions (the smallest: 63 nm diameter and 1.5 nm height) (16). It was found that dimension of DNA condensates was sensitive to the characteristics of used polymers, such as nitrogen weight per polymer unit, the molecular weight, and cationic distribution. Moreover, the charge ratio, DNA structure, DNA concentration, polymer length, type of counterion and other reaction conditions also affect the dimension and morphology of complexes, the condensation pathway, and the condensation efficiency (16–23).

For *in vivo* gene delivery, the ability of vectors to achieve long plasma circulation and to distinguish the preserved target cells is equally important. Before arriving near the target cells, the vectors should be stable and must not interact with plasma proteins. Unfortunately, particles composed of cationic polymer and DNA usually do not preserve proper circulatory half-lives. The serum half-life of poly(L-lysine)(PLL)/DNA complexes was shorter than 3 minutes (24). The researchers attributed the rapid removal of these complexes from the blood to protein binding, salt aggregation, and reversible polycation/DNA interactions (25–27). In order to understand how the parameters of cationic polymer and DNA complexes affect the plasma circulation and prolong the circulation term, Ward *et al.* investigated the systemic circulation of PLL/DNA with different molecular weights of PLL and different types of DNA (28). In their study, they showed that complexes formed from high molecular weight (HMW) PLL (211 kD) had longer circulatory half-life than the complexes prepared from low molecular weight PLL (20 kD). PLL211/DNA complexes prolonged the half-life in blood up to 20 times compared to PLL20/DNA. The interactions between PLL and plasma components, hepatic leukocytes, and DNA played an important role and resulted in poor solubility in blood, complement activation, stimulation of leukocytes, and macrophage capture, which affected the final blood circulation of complexes.

After plasma circulation, gene vectors have to leave the bloodstream and reach the targeted organs or cells. They should be able to distinguish target cells from non-target cells *in vivo*. There existed two major targeting modes for micro/nanospheres. One is passive targeting mode, which is usually utilized in tumor-targeting therapy. In most healthy tissues, the intercellular spaces of the capillary are in the order of 5–10 nm (29, 30). However, in solid tumors, the effective pore size of the tumor vessels is between 400 nm and 600 nm in diameter (31). The normal and tumor vasculature are shown in Fig. 10.2. Due to the enhanced vascular permeability, nanospheres can accumulate in tumor tissue effectively, which is known as enhanced permeability and retention effect (the EPR effect) (32). By EPR effect, nanospheres are able to accumulate into tumor tissues and transfer to the appropriate target cells.

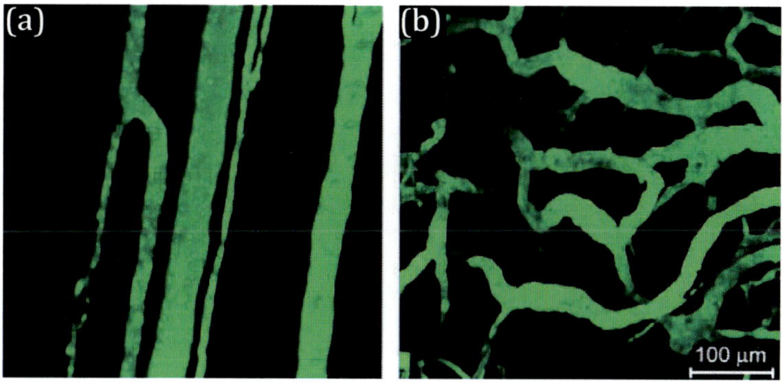

Figure 10.2 (a) Normal and (b) tumor vasculature. Reprinted from Ref. 33, with permission from Oxford University Press.

Another mode is active targeting. In this mode, specific molecules that show affinity toward certain characteristics of target cells, especially those of tumor cells, are grafted on the surface of gene vectors and help selective binding with target cells. Because tumors in the early stages of their development do not show higher permeability than normal vascular vessels, passive targeting mode does not act effectively. The active targeting method is very useful for tumors with immature vasculature. Investigated targeting moieties include antibodies (34,35), peptides (36–38), folate (39–41), transferrin (42,43), saccharide (44,45), and others and are listed in Table 10.1.

Table 10.1 Active targeting moieties used in modifying micro/nanospheres as gene vectors

Targeting tissues	Receptor	Targeting moieties	References
Tumor	Integrin $\alpha_v\beta_3$ (angiogenic endothelial cell)	Integrin $\alpha_v\beta_3$-targeting ligand; RGD peptide	(46–48)
Tumor	Folate receptor	Folate (folic acid)	(39–41,49)
Tumor	Transferrin receptor	Transferrin	(42,43,50)
Tumor	Glucose-regulated protein 78 kDa (GRP-78)	Peptide ligand (WIFPWIQL)	(37)
Tumor	Luteinizing hormone-releasing hormone (LHRH) receptor	LHRH	(51)

(*Continued*)

Table 10.1 (*Continued*)

Targeting tissues	Receptor	Targeting moieties	References
Hepatocellular carcinoma	Epidermal growth factor (EGF) receptor	EGF	(52–54)
Hepatocellular carcinoma	Luciferase; Asialoglycoprotein (ASGP) receptors	Galactose	(44,55)
Hepatocellular carcinoma	ASGP receptor (hepatocyte)	Lactose; asialofetuin	(56–58)
Hepatocellular carcinoma		Linoleic acid	(59)
Ovarian carcinoma	(Erythropoietin-producing hepatocellular) (Eph)2A receptor	YSA peptide (YSAYPDSVPMMSC)	(36)
Metastatic cancer	Nucleolin	Peptide ligand (KAREC)	(60)
Her-2/neu-positive breast cancer	Her-2/neu	Humanized anti-Her-2/neu mAb Trastuzumab	(34,61)
Renal cell carcinoma	G250	G250 monoclonal antibody	(33)
Brain tumors, prostate cancer, sarcoma, and intestinal cancer	Matrix metalloproteinase-2 (MMP-2)-containing lipid raft complex	Chlorotoxin (CTX) peptide	(38)
Bronchial epithelial cells	Lactoferrin (Lf) receptor	Lf	(62)
The epithelium of the intestine, liver, and kidney	Megalin	Bovine serum albumin (BSA)	(63)
Lumen of activated vascular endothelium	E- and P-selectin	Anti E/P-selectin monoclonal antibody	(64)
Islet beta cells	Glutamic acid decarboxylase (GAD)	Anti-GAD antibody	(65)
Human fibroblasts	Human CD90(h Thy-1)	Anti-human IgG1 mAb AS02	(61)
Tumor	Magnetic field	Magnetic particles	(66,67)

After reaching target cells, non-viral vectors need to pass the cell membrane and enter cells. Cellular uptake is considered to be one of the rate-limited steps and major obstacles in the transfection process. In general, vectors composed of small particles are much more easily internalized by cells (68,69). The large polyethylenimine (PEI) particles tended to stick on the cell surface and only be partly internalized. And the small PEI particles were almost completely internalized (observed by laser scanning microscopy) (70). The optimal particle diameter for dendritic cell (DC) uptake was 0.5 μm and below. Besides the size of vectors, the uptake of particles was also related to their hydrophilic–hydrophobic property. Modification with hydrophilic polyethylene glycol (PEG) was frequently performed to prolong half-lives of particles in serum. However, in the presence of PEG, the endocytosis of nanospheres was interrupted by the steric hindrance of PEG (71). To solve the decrease of cellular uptake resulting from PEGlylation, cleavable PEG systems have been reported. In acidic conditions or intracellular reducing environment, PEG moiety was removed from the particles, and then the hydrophobic particles was able to adhere on the cell surface with an increased cellular uptake (72,73).

Intracellularly, the vectors must escape from the endosome to avoid the DNA degradation in the lysosomal pathways. Proton-sponge effect is one of the best known and widely used methods to accomplish the escape of vectors from endocytic vesicles and has been shown in Fig. 10.3 (74). In this process, cationic (usually PEI-composed or -modified) particles first enter into the acidifying lysosomal compartment. Then, the unsaturated amino groups capture protons supplied by the v-ATPase (proton pump). This process leads to the retention of Cl^- ion and water molecules, and subsequently, the lysosome swells and ruptures, which results in the escape of vectors.

As stated before, the cationic polymers have often been used to condense the DNA into nanospheres to protect it and facilitate its entry and passage through target cells. However, once inside the cell, in order to be processed by RNA transcription complexes, the DNA must be dissociate from the particles. For particles based on non-viral vectors, plasmid unpacking is the most important rate-limiting step for gene expression, especially for those polymers that have HMWs or high charge densities.

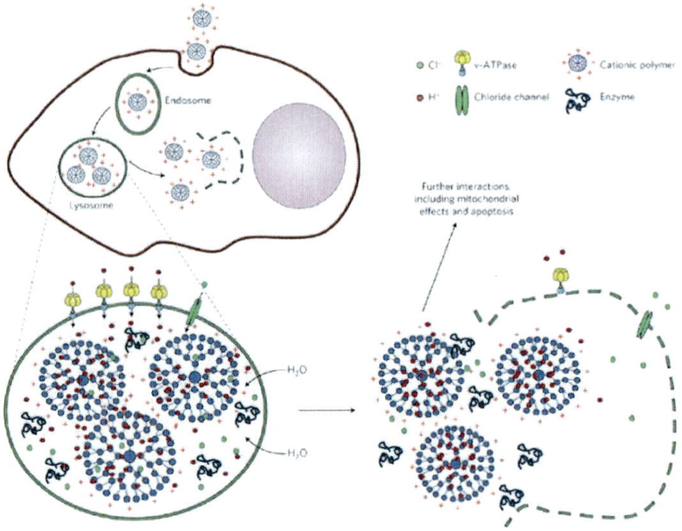

Figure 10.3 Schematic of the proton-sponge effect leading to lysosomal damage and the induction of cytotoxicity by cationic nanospheres. Reprinted from Ref. 74, with permission from Nature Publishing Group.

In order to accomplish effective load of pDNAs and ensure complete drug release in cell, many methods have been investigated. One of them is using reversible or stimuli-sensitive bonds that dissociate in cellular circumstances. For example, thiolated PEI was synthesized and the thiol groups could form cross-linked and caged structure via disulfide bonds. The cross-linking of the shells was favorable to improve the stability of polyplexes in the physiological condition. In cell, the disulfide bonds were cleaved due to the increased glutathione (GSH) concentration, and the particles dissociated, which resulted in the rapid release of pDNA (75). Besides being used to cross-link polymers, disulfide bonds could also be used to cross-link pDNA with polymer. The amino groups of polylysine were substituted with 3-(2-aminoethyldithio)propionyl residues in order to link pDNA onto the polymer backbone via disulfide bond. Compared to the unmodified polylysine, the transfection of 293T7 and HepG2 cells was 10- and 50-fold more efficient with modified polyplexes (76).

The charge density was also adjusted to balance the protection and release of nucleic acid. Liu *et al.* synthesized linear PEI with methyl ester side chains, and these side chains were hydrolyzed gradually under physiologically relevant conditions (77). This resulted in the reduction of the cationic charge density of the polymer and decreased the electrostatic interactions between the polymer and pDNA. By regulating the degree of substitution, it is possible to control the hydrolysis rate and the pDNA release rate. The chain length of polymers also affects the polyplex stability. Increasing the polymer chain length provides more sites to bind with nucleic acids and form more stable polyplex (78).

Thus, in order to act as an ideal gene vectors, there are numerous barriers that need to be conquered from administration to effective transfection *in vivo*. It may be impossible to find or design a gene vector which can conquer all of the stated barriers. However, by taking these problems in consideration before vehicle design and refining existing vectors, it is possible to produce one or several novel vectors to achieve safe and effective gene therapy.

10.3 The Preparation of Micro/Nanospheres as Gene Vectors

To prepare micro/nanospheres as gene vector, there are many factors that need to be investigated and optimized, such as the properties of polymers and the preparation methods. It was found that the transfection efficiency was mainly affected by the characteristics of used vectors and not the characteristics of DNA (79).

10.3.1 Polymers

Since gene drugs are negatively charged, the cationic polymers are generally used to prepare micro/nanospheres, such as polylysine, PEI, and chitosan. By electrostatic interaction, cationic polymers may form a complex with DNA or adsorb DNA on the particle surface.

10.3.1.1 Charge density

The charge density of cationic polymers affects the interaction between the polymers and the nucleic acids. For weak cationic polymers, the complex formed with nucleic acids is easily dissociated when the surrounding pH fluctuates, and the bioactivity of nucleic acids cannot be protected well. However, cationic polymers with too high charge density bind nucleic acids tightly, retard their release, and decrease transfection efficiency. Thus, it is essential to adjust the charge density of cationic polymers carefully.

There are many methods used to adjust the charge density of cationic polymers, such as changing the amount of ionic groups (80), introducing other ionic groups (81), and copolymerizing or grafting with other polymers (82). Furthermore, during the externalization process, the external pH values would change. For example, the pH value of early endosomes, late endosomes, and lysosomes are 7.4–6, 6.0–5.5, and 5.0, respectively (83). It should be noticed that the variation of external conditions would affect the charge density of polymer and affect the DNA–cationic polymer interaction. Rungsardthong *et al.* (84) investigated the effect of solution pH on the degree of ionization of [(dimethylamino)ethyl methacrylate] (DMAEMA) homopolymer, and its interaction with calf thymus DNA. At low pH (4.0 and 6.0), DMAEMA homopolymer had high degree of ionization, strong binding affinity, and high complexation efficiency with DNA. At pH 8.0, DMAEMA homopolymer had a lower binding affinity to DNA due to its low degree of ionization (24%).

Considering the change of external conditions during the transfection process, pH-sensitive polymers were proposed as new non-viral polymers, which could change their charge density according to external pH values. Poly(aspartate-hydrazide)-block-poly(L-lysine) is one of them, which has a PLL segment (pKa 9.4) formed complex with pDNA and a pH-sensitive poly(aspartate-hydrazide) segment (pKa 5.0). The pH-sensitive poly(aspartate-hydrazide) segment could transition from neutral to charged in late endosome and disrupted endosomal membranes. This escape mechanism was different from proton-sponge effect and was minimally toxic to the cell elsewhere (85).

The charge density of particles not only affects the binding efficiency with nucleic acids but also the cellular uptake. For those particles with similar diameter, the cellular uptake of particles relied heavily on the surface charge. Particles with higher positive charge exhibited an enhanced cellular uptake, which resulted from their stronger affinity for the negatively charged cell membrane (86,87). For example, since cells preferred to internalize small particles, and the transfection efficiency of large particles was usually lower than that of small particles. In order to enhance the cellular uptake of large particles, modifying their surface charge was an optional choice. The optimal particle size, which can be internalized quickly and efficiently by DC, was below 0.5 μm. However, after modification by cationic poly-L-lysine and poly-DL-alanine, the cellular uptake of 1 μm particles was greatly enhanced (68).

10.3.1.2 Polymer length

Besides charge density, the molecular weight of polymer also affected the transfection efficiency. As stated in several literatures, HMW polymer tended to achieve better transfection efficiency than low molecular weight polymer (28,88,89). Nemoto *et al.* found that 4-branched poly(*N*,*N*-dimethylaminopropyl acrylamide)polymers with higher Mn (73 kDa) obtained seven times high transfection activity than that for the same polymers with lower Mn (27 kDa) (89). The decreased transfection efficiency of the shorter polycations was attributed to their lower binding ability and the loss of the chain entanglement effect.

However, for chitosan, which is another popular material used in gene delivery, complicated phenomenon were observed. In most studies, the complex composed of chitosan with HMW, higher transfection efficiency was obtained than that with low molecular weight chitosan (90,91). In the study conducted by the team of Zong *et al.*, although HMW chitosan formed tighter association with DNA vaccine and provided better protection against DNase enzyme degradation, the higher transfection efficiency in mucosal gene immunization was achieved by the complex formed with low molecular weight chitosan. The author explained that in this delivery system, the strong affinity of HMW chitosan to DNA hindered DNA release from its carrier, which was regarded as a rate-limiting step for the intracellular release and nuclear entry of DNA, resulting in low transfection efficiency (92).

The molecular weight of polymers directly affected the binding strength with DNA, and thus the DNA protection property and release rate in cells. The choice of polymer molecular weight needs to be considered carefully according to the practical need as one important affecting factor.

10.3.1.3 Toxicity

Concerning cationic polymers, toxicity is their most important obstacle before clinical application. For example, PEI, one of the most widely studied cationic polymers used as gene vector, may induce membrane damage and initiate apoptosis. Moghimi *et al.*

evaluated the toxicity of branch and linear PEI on three human cell lines (Jurkat T cells, umbilical vein endothelial cells, and THLE3 hepatocyte-like cells) (93). They found a two-stage PEI-mediated cytotoxicity: In Phase I cytotoxicity, the membrane destabilization and rapid phosphatidylserine exposure were observed; in Phase II cytotoxicity, apoptosis was observed.

The toxicity of PEIs is related to its molecular weight. PEIs with HMW preserve high toxicity since they are not biodegradable and accumulate *in vivo* (94). In order to decrease the toxicity, PEIs with low molecular weight were investigated as gene vectors. However, with the decrease of molecular weight of PEIs, the transfection efficiency was decreased (70). To achieve high transfection efficiency and reduce toxicity, one method is to synthesize novel PEIs with degradable bonds, such as disulfide bonds (95–98), acetal bonds (99), and ester bonds (100). Another strategy was to graft low molecular weight PEI on the bone of biodegradable polymers, such as chitosan (101,102), alginate (103), cyclodextrin (104), and poly(lactide-co-glycolide) (PLGA) (105). By using these methods, the toxicity will be deduced since these copolymers would be degraded into low molecular weight PEI in cells, and at the same time, these copolymers could achieve high transfection efficiency.

Introducing hydrophilic segments in the vectors is another way to decrease the toxicity of HMW PEI. The hydrophilic polymers would help improve the hydrophilicity, minimize aggregation, and shield the cationic charge. In one study, the application of non-PEGylated DNA/PEI complexes induced high acute toxicity and the lethality of treated mice was up to 50%. After PEGylation, the lethality was decreased to zero with PEGylated complexes at $N/P < 6$ (106,107).

Furthermore, it should be noticed that free PEI showed higher cytotoxicity than PEI-DNA complex (108). And when PEI was used as gene vector, it was important to monitor the concentrations of free and bound PEI during application (109).

Other cationic polymers with better biodegradability were also investigated as non-viral vectors. Chitosan is a biodegradable cationic polysaccharide obtained from deacetylated chitin, and there are many amino groups on its chains according to the deacetylation degree. At acidic solution (pH value lower than its pKa, 6.4), the primary amines were positively charged and formed complex with negatively charged DNA/siRNA via electrostatic interaction (110,111). Most studies on chitosan show that, in general, it is a relatively non-toxic, biocompatible material (112). Apart from chitosan, poly(β-amino ester), dextran, gelatin, alginate, and PLGA-PEG-PLGA have been investigated for their potential as non-viral vectors (113–118). However, compared with PEI, these polymers did not achieve the same high transfection efficiency and more studies need to be promoted.

10.3.2 Preparation Methods

The preparation method of micro/nanospheres greatly influences the transfection efficiency, which is also related to the characteristics of used polymers. For those particles formed with cationic polymers and genes, condensation is one of the most popular and simplest preparation methods. And in order to prepare particles with non-cationic polymers or special structures, more complex preparation methods need to be used.

10.3.2.1 Condensation

The preparation procedures of condensation are very simple. The gene solution is added slowly into polymer solution at stirring state. The static attraction between cationic polymers and genes will compact them into small particles. N/P ratio is one of the most important affecting factors in this method, which is defined as the molar ratio of cationic polymers, amino groups/gene phosphate groups. In certain range, more condense structure and high transfection efficiency was obtained with high N/P ratio (119,120). However, excessive addition of cationic polymers would decrease transfection efficiency and increase cytotoxicity. Besides N/P ratio, the molecular weight (121), structure (122), and charge density (121) also affects the architecture of particles, the cellular uptake, and transfection efficiency.

By condensation, the size of plasmid is reduced efficiently and its activity is protected effectively. However, the condensation method has its own disadvantages. For example, the static attraction between cationic polymers and the gene is the predominant force to form particles. Thus, its application is restricted to cationic polymers. Moreover, by this method, only simple structures can be prepared. In order to prepare particles with non-cationic polymers or special structures, more complex preparation methods were developed and investigated.

10.3.2.2 Emulsification

Although the most popular polymers in gene delivery are cationic polymers, several neutral or negative polymers, such as polylactide (PLA) and PLGA, are also used to protect genes from nuclease degradation and to control the release of drugs after intramuscular administration. (123,124). Emulsification method is frequently adopted to prepare these particles. Typically, in this method, high-speed homogenization or sonication is used to disperse emulsions and form droplets of a size suitable for cellular uptake (125,126). However, the high shear forces may damage plasmid integrity. In order to conquer this problem, some modifications about emulsification method were proposed and developed. For example, Oster *et al.* developed a modified solvent displacement method to prepare polymer nanospheres which did not rely on shear stress to produce nanospheres but on differences in the interfacial tension between water and oil phases. In this method, solution of DNA in water was added to acetone solution containing cationic polyesters [amine-modified poly(vinyl alcohol) backbones grafted with PLGA]. The resulting mixtures were then injected into (2 mL/min) an aqueous Pluronic™ solution, and polymer nanospheres were formed spontaneously. The hydrodynamic diameters of the formed nanospheres ranged from 175 to 285 nm. Agarose gel electrophoresis was used to prove that DNA within nanospheres was protected from shear and enzyme degradation (127).

Alternatively, spray drying technique was combined with emulsification method to reduce the shear forces. While preparing hydrophobic particles of encapsulated plasmid, the emulsion steps need to be performed twice. The first emulsion (w/o) is subjected to shear stress to form droplets with small sizes, then the first emulsion is added into external water phase and subjected to another shear stress to form w/o/w droplets. By combining with spray drying technique, the second emulsion step can be avoided. Better preservation of DNA in PLGA microspheres prepared by spray drying than double-emulsion was observed by Atuah *et al.* (128).

10.3.2.3 Other methods

Besides condensation and emulsification methods, other novel methods, such as spray-drying, phase separation, and supercritical fluid foaming technique have been investigated to prepare particles as gene vectors (129–132). For example, a novel assembly method called the lipid film hydration was developed in recent years, which was used to package condensed DNA with lipid films (133,134). By using this method, multifunctional envelope-type nanodevice (MEND) was prepared, and MEND could enter into cells via macropinocytosis and facilitate transfection activities as efficiently as an adenovirus.

Micro/nanocapsules can also be prepared by layer-by-layer (LbL) technique to obtain encapsulated DNA within tunable, semi-permeable capsules. Zelikin *et al.* used this method and encapsulated single-stranded, linear double-stranded, and plasmid DNA into disulfide cross-linked poly(methacrylic acid) (PMA) capsules successfully (135).

No matter which method is chosen, it has to meet several demands: (i) The structure and integrity of nucleic acid should be protected well during the preparation process and in preservation; (ii) Prepared particles should have proper size distribution and dispersity; (iii) No toxic agent was adopted in preparation, or it should be eliminated completely before use. Moreover, novel methods which can prepare particles with special structure useful in gene delivery have promising prospects in future.

10.3.3 Characteristics

10.3.3.1 Size

Like particles used to load proteins or peptides, the size of particles used as gene carriers also affects drug encapsulation efficiency and release rate. With an increase in particle size, the encapsulation efficiency was enhanced and the release rate was decreased (136). Moreover, particle size also affects the circulation period and biodistribution. In general, particles with small size facilitate the prolongation of circulation period and enhance endocytosis (7,137). But it also needs to be noticed that small particles may be partly excreted by kidney (138) and the treatment efficiency may be decreased. As shown in Fig. 10.4, Jo *et al.* classified the vectors according to their size and illustrated their different characteristics (139).

Xu *et al.* systematically investigated the size-dependent properties of PEI nanogels as gene vectors in cancer cells (140). PEI nanogels with different sizes (38,75,87,121,132, and 167 nm) were selected to load pDNA (pLEGFP-C1) into A549, Bel7402, BGC-823, and Hela cells. The highest transfection efficiency was achieved with samples of 75 nm and 87 nm size in all of the four cell lines. It was assumed that PEI nanogels of 75 and 87 nm were picked up by the cancer cell more easily and released pDNA more quickly than the large ones. The authors attributed the lower transfection efficiency with samples of 38 nm size to their lower quantities of released DNA.

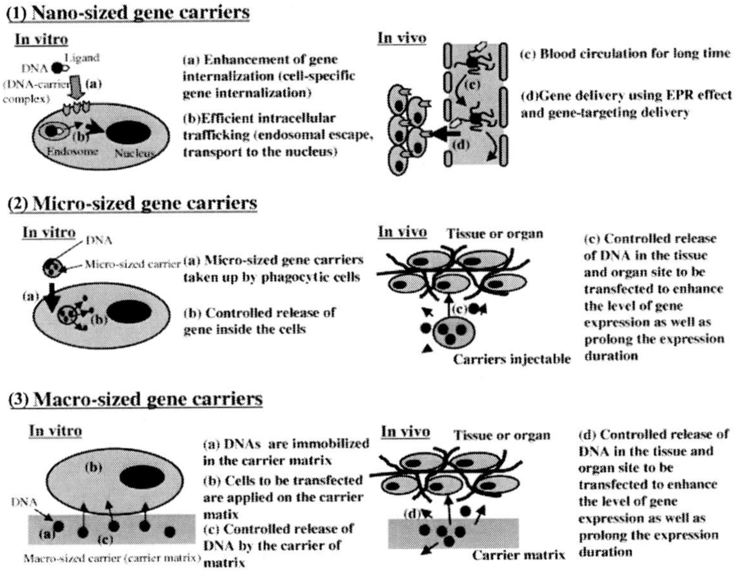

Figure 10.4 Classification of non-viral gene carriers in terms of the carrier size. Reprinted from Ref. 139, with permission from Elsevier BV.

For different cell lines, the size range of particles internalized by cells was different. In Zauner' study, hepatic cell lines (Hepa 1-6, HepG2) internalized only very few particles of 93 nm and 220 nm and hardly any or no microspheres of 560 nm (141). ECV 304, HUVEC, and HNX 14C were able to take up large particles (up to at least 1 μm). The growing cells of mouse squamous cell carcinoma line (KLN 205) were also able to ingest large particles.

10.3.3.2 Morphology

Morphology is another important property of particles, especially at present with the development of preparation methods (142).

Several technologies were developed to prepare particles with novel morphologies, such as "PRINT" technology and microfluidics (143,144). The shape of particles would affect the movement in blood vessels, cellular uptake, and drug release. However, due to the complexity of movement of non-spherical particles and the restricted application range of preparation methods for non-spherical particles, there is still lack of detailed studies about the effect of particle shape on gene delivery. Only few literatures have reported the effect of particle shape on cellular uptake. Compared with spherical particles, non-spherical particles escape phagocytosis, especially those particles that have high aspect ratios (145,146). For example, the internalization is possible only at the ends of the worm-like particles, while the low curvature along the length of the particles inhibits internalization.

Moreover, due to the complexity of interaction between non-spherical particles and cells, there are often confusing and even contradictory results given in different literatures. For example, Alemdaroglu reported that rod-like polymeric particles were take up 12 times more efficiently than their spherical counter parts (147). However, in another report, it was observed that the smaller, spherical nanospheres entered Chinese Hamster Ovary (CHO) cells with higher rate than the larger, cylindrical nanospheres (148). It may be that for different cells and polymer particles, the cellular uptake mechanisms are different and which directly affect the cellular uptake effect and transfection efficiency.

10.3.3.3 Surface characteristics

The surface characteristics, such as charge density and hydrophilic–hydrophobic property, of particles also play an important role in gene transfection. There exists a balance between the hydrophilic–hydrophobic property of particles as gene vectors. Hydrophobic particles enhance physical encapsulation of genetic materials, cell binding, and intracellular pDNA association (149), while hydrophilic particles are useful to prolong the circulation time and improve intracellular pDNA release (150).

In order to achieve high encapsulation efficiency, chitosan was hydrophobically modified with 5β-cholanic acid and formed nanoparticles (NPs) with DNA (151). It has also been observed that with an increase in the modifying degree, the endocytic uptakes and transfection efficiency of nanoparticles is improved. Hydrophobic poly(γ-benzyl-L-glutamate) segment (PBLG) also had been used to modify PEI (152). The introduction of hydrophobic segment on PEI could form smaller particles, lower cytotoxicity, higher bioactivity, and transfection efficiency than unmodified PEI.

For preparing hydrophilic particles, introducing hydrophilic components, especially PEG, on particles is one of the most popular methods (106,153–156). PEG, a linear polymer of ethylene oxide with hydroxyl terminal groups $[H-(O-CH_2-CH_2)_n-OH]$, is a neutral, water-soluble compound, which helps to form hydrophilic shielding around vectors, neutralize the positive surface charge, and hinder the interactions between vector and serum proteins (157). The molecular weight and introducing density of modified PEG affects the circulation time (154,155,158). It is generally assumed that HMW and the modifying density of PEG are beneficial for prolonging the vector's half-life in serum because they provide stronger steric effect (159).

The method used to introduce PEG into particles also affects the circulation time. PEG should be modified to graft on polymer or particle surface in order to prolong the circulation time. Berg *et al.* proved in their study that physical incorporation of PEG into nanospheres did not perform the same function as chemical PEG modification of nanospheres (160). It is possible that the unbound PEG dissolves rapidly and is released from nanospheres into solution.

By altering the modification method, the structure of PEG, or the grafting with other polymers, better circulation time could be obtained. Two different surfactant polymers based upon poly(γ-glutamic acid) (γPGA) and poly(maleic anhydride-alt-1-octadecene) (PMHC18) were bound to nanomaterials and grafted with PEG. The polymer between nanospheres and PEG provided extended hydrophilic groups and remarkably long blood circulation (155).

Besides PEG, there were other hydrophilic polymers or groups being used to modify vectors, such as poly[*N*-(2-hydroxypropyl)m ethacrylamide] (PHPMA) (161–163), ethyleneoxy group (164), and polyvinyl alcohol (PVA) (165,166). Introducing these hydrophilic polymers or groups on the surface of vectors helps to decrease serum protein binding and to increase stability.

It should be noted that besides particle size, morphology, and surface characteristics, other properties of particles also affect transfection efficiency, such as particle stiffness (167). Moreover, none of these factors behave alone, and the final transfection efficiency can be a result of many factors, including the characteristics of particles, administration method, and surroundings *in vivo*. As shown in one literature, besides the size of particles, the molecular weight of polymers and the concentration of particles also affect the EPR effect (168). In addition, the endogenous factors, such as bradykinin, cytokine, and vascular endothelial growth factor, involved in tumor growth also play a role in the EPR effect (169).

10.4 The Applications of Micro/Nanospheres as Gene Vectors

The first gene therapy was performed in 1990 to treat adenosine deaminase deficiency, a genetic disease. In the past twenty years, the application field of gene therapy has been expanding from immunodeficiency diseases to tumor disease, vaccination, diabetes, nervous system diseases, and congenital blindness. Gene therapy attracted much attention and was actively explored especially for diseases, such as tumor and HIV, that could not be treated effectively by other methods.

10.4.1 Cancer Therapy

Many functional classes of genes including suicidal genes, cytokine/chemokines genes, antiangiogenic genes, and tumor-suppressor genes are being evaluated for their anticancer activity (170–175). Before effective treatment effects are accomplished, many barriers related to tumor therapy are still to be crossed: (i) genetic and epigenetic instability, (ii) low transduction efficiency in tumor cells, (iii) inadequate transgene expression level, (iv) short duration of transgene expression, and (v) lack of tumor specificity (176,177). Using nanospheres as gene delivery system is one of the ways to accomplish effective tumor treatment effect, which can target the tumor by EPR effect or by conjugating targeting molecules on the surface of nanospheres, and it has been investigated widely in various cancer therapy (178,179).

For example, in treating ovarian cancer, which is the leading cause of deaths due to gynecologic cancer, there is no effective therapy for patients with advanced ovarian cancer. Gene therapy, including cytotoxic and corrective gene therapy, provides a promising method to conquer ovarian cancer (180). Nanospheres can be used to deliver DNA to tumor cells and to control the expression of DNA inside the cells. Nanospheres composed of poly(β-amino ester) and DNA could enhance the expression of DNA and reduce the size of tumors in mice (181). Vascular targeting of genes to disrupt the growth of new blood vessels is another way to inhibit the growth of tumor. However, without the assistance of vectors, it is hard to deliver genes to vascular endothelial cells in diseased tissues. Hood *et al.* developed cationic nanospheres coupled with an integrin $\alpha_v\beta_3$–targeting ligand, which can deliver genes selectively to angiogenic blood vessels in tumor-bearing mice (46). The apoptosis of the tumor, tumor cell apoptosis, and sustained regression of established primary and metastatic tumors were observed in mice treated by nanospheres.

Furthermore, overexpression of drug-efflux transporters such as P-glycoprotein (P-gp) is a huge obstacle in cancer therapy, which enables cancer cells to develop resistance to multiple anticancer drugs. Using siRNA-mediated RNA interference to silence the expression of the efflux transporter is a promising method to conquer drug resistance. However, in order to achieve ideal therapy effect, the siRNA needs to be localized in the tumor cells. By adopting biotin-modified poly(DL-lactide-co-glycolide) nanospheres as gene vectors to codeliver P-gp-targeted siRNA, anti-tumor drug (paclitaxel) can effectively overcome tumor drug resistance and inhibit tumor growth (182).

In gene therapy for brain tumors, one of the biggest barriers is that the vector cannot pass the blood–brain barrier (BBB). Novel nanospheres were designed to cross BBB, such as co-conjugating brain-specific targetors and PEG. The modification of PEG could prolong the half-life of the vector in plasma, and binding of brain-specific targetor with the vector could enhance the absorptive or receptor-mediated transcytosis. Brain-specific targetors including cationic bovine serum albumin (CBSA) and murine OX26 monoclonal antibody, which have been conjugated on PEG-

PLA nanospheres, have been investigated (183,184). By repeated injections of CBSA-modified PEG-PLA nanospheres, the growth of malignant glioma can be significantly delayed (185). By increasing the number of surface CBSA conjugating per nanosphere enhanced the transcytosis ability across the BBB; however, it decreased the blood concentration curve (AUC) (186).

In order to strengthen the tumor cell transfection *in vivo*, several physical methods were adopted during the administration. Exerting ultrasonication around the tumor helps to induce deformation of cell membranes and enhance particle endocytosis (187,188). Magnetofection is another way to speed vector accumulation and increase vector concentration at tumor tissues by using superparamagnetic nanospheres as gene vectors (67). Moreover, the magnetic fields can be utilized to trigger drug release at tumor site, especially for deep-seated tumors. In one study, dextran-coated iron oxide nanospheres were used as gene vectors, and DNAs were conjugated on particle surfaces by heat-labile linker. By applying electromagnetic field around a tumor, the penetration depth of drug increased nearly six-folds compared to the controls (189).

10.4.2 Vaccine

The study of gene-based vaccination was started in the early 1990s. Plasmid DNA coated with gold microparticles were delivered into the skin of mice by gene gun and antibodies were elicited in response to the delivered antigen (190). This technology offers many attractive advantages compared with traditional vaccine strategies. First, gene vaccine can be simply altered by changing DNA sequencing to develop co-stimulatory molecules, which can be used to stimulate immunity against a range of pathogens or to enhance specific cell-targeting ability and improve immune response. Second, nucleic acid vaccine can result in both neutralizing antibodies and cell-mediated immunity, and the latter is especially important for both prophylactic and therapeutic vaccination strategies (191). Additionally, the adoption of gene vaccine provides higher biosafety than attenuated and inactivated vaccine, which did not contain virus gene and eliminated the need for biocontainment and the risk of exposure to live viral agents.

Micro/nanospheres as vaccine carriers can not only act as delivery system but also as adjuvants (192–194). Several studies showed that the use of micro/nanospheres could efficiently deliver DNA to antigen-presenting cells (APCs) and enhance the cellular uptake (195). Furthermore, with carefully designed structure, micro/nanospheres could realize long, effective, and controlled release of DNA. Thus, adopting micro/nanospheres as vaccine adjuvants, a single injection may be able to induce similar immune response as can be achieved only after several injections by traditional immune methods (196). At present, gene vaccines with micro/nanospheres are being investigated for prevention and therapy of many diseases, such as HIV (197,198), Japanese encephalitis (199), respiratory syncytial virus infection (200), and malaria (201).

As far as the application of DNA vaccines is concerned, many factors affecting immunization, such as the loading method, exist. Compared with encapsulation, adsorption of DNA on the surface of cationic particles achieved high drug loading and avoided negative preparation conditions, such as high shear force, organic solvent, and chemical cross-linking (194). Moreover, after administration, the release of DNA adsorbed on the particle surface was easier and quicker than those entrapped in the interior of particles, which facilitated the transfection of cells. Modified particles with cationic polymers or molecules help to increase the adsorbing capacity of particles and deliver simultaneously other immunomodularory agents within the same vehicle. Chitosan, PEI, and cetyltrimethylammonium bromide (CTAB) have been used to modify the negative PLGA particle surface and enhance the binding capacity with DNA (202–204). Size was another important parameter for particles to be used as vaccine adjuvants, and smaller particles with higher cellular uptake were significantly more immunogenic than larger ones (205). The micro/nanospheres to be used for injections should be smaller than 5 μm (in diameter) in order to be internalized by the phagocytic APCs. Micro/nanospheres that are to be used for oral delivery should be smaller than 10 μm (in diameter), so that they can pass the gut-associated lymphoid tissues (GALT) (195).

10.4.3 Ocular Disease Therapy

Another important application in gene therapy is ocular gene therapy, since many of the eye diseases are due to gene defects and the eye is easily accessible and immune-privileged (206,207). As reported in many literatures, gene therapy has been approved as effective in treating Leber's congenital amaurosis (LCA), a group of inherited blinding diseases that are symptomatic in infancy and lead to total blindness in adulthood (208,209). However, the small surface of eyes, tear flow, blinking, vitreous, and the blood–retinal barrier prevent the permeation of gene drug. Use of mucoadhesive nanospheres as gene vectors enhances the interaction with the ocular mucosa. Two bioadhesive polysaccharides, hyaluronic acid and chitosan were combined to prepare nanospheres and deliver genes to the ocular surface. These nanospheres were internalized into the corneal epithelial cells to deliver pDNA into both the corneal and the conjunctival cells (210). By varying the site of injection, nanospheres carrying genes can be targeted to different tissues within the eyes (211). The nanospheres can transfect almost all cell types of the eye. Most impressively, subretinal delivery of these nanospheres induced robust expression in the outer nuclear layer, retinal pigment epithelium, and the optic nerve of the eyes.

As stated before, the shielding of the surface of cationic particles by PEG chains enhanced the particle stability in serum and neutralized the positive surface charge. Furthermore, PEGylation of polymers or particles helped to prevent the particles from aggregating in vitreous (212). A cell-penetrating peptide for ocular delivery (POD)[CGGG(ARKKAAKA)4], which could bind the plasma membrane of human embryonic retinoblasts and enter the cytoplasm, was PEGylated and used to compact DNA into nanospheres for ocular gene delivery (213). The PEG-POD significantly increased the expression of plasmid DNA by 215-fold compared to the DNA injected alone.

Besides the mentioned diseases, gene therapy has also been used in myocardial repair (214), wound repair (215,216), inhibition of the proliferation of human Tenon's capsule fibroblasts (217), and treatment of diabetes (218,219). With the rapid development of genetic engineering and novel vectors, the application fields of gene therapy would broaden largely in the future.

10.5 The Key Points of Micro/Nanospheres as Gene Vectors

Micro/nanospheres are important non-viral vectors, which provided many advantages in gene delivery, such as cellular or nuclear target ability, structure protection, and controlled release. At the same time, the application of polymer vectors also introduce several key points that need to be considered carefully before any clinical application.

10.5.1 Optimization of the Particles' Design According to Different Administration Routes

The administration route of naked DNA is restricted in injection due to the instability of the plasmid. By using non-viral vectors, the administration route is broadened including oral (220,221), ocular (214,222), nasal (223,224), lung (225), and transdermal (160,226) delivery as shown in Fig. 10.5. Non-injection administration can help improve therapy effect or satisfy specific treatment needs. For example, the nasal delivery of plasmid DNA loaded in chitosan nanospheres triggered systemic, mucosal, and cellular immune responses (227).

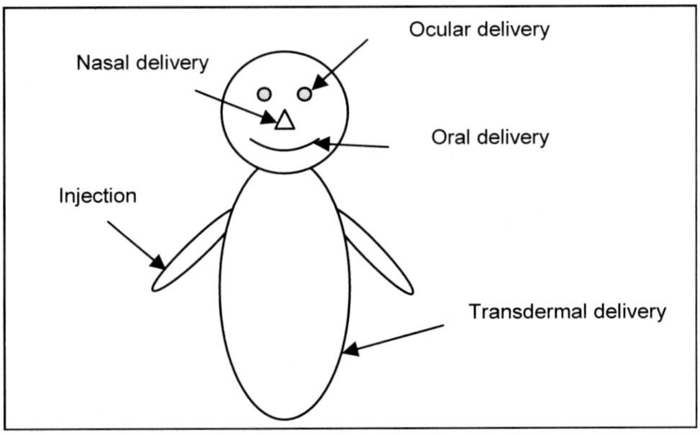

Figure 10.5 Different delivery routes of gene therapy.

For those non-injection administration routes, the gene vector should not only provide protection against enzyme degradation in plasma and intracellular surroundings but also protect the drug during the administration process and enhance the drug absorption at the administration sites. For example, in oral delivery, the intestinal enzymatic attack may decrease the bioactivity of gene drug heavily. Encapsulated DNA in micro/nanospheres can improve the resistance against intestinal enzyme degradation (228). Besides encapsulation, adding enzyme inhibitor in the micro/nanospheres is one of the methods to improve the stability of orally administered genes. In one study, chitosan was covalently bound with enzyme inhibitor (aurintricarboxylic acid, ATA) and complexed with pDNA into nanospheres. The formed nanospheres provided high protection capability against salt solution, lysozyme, DNase, and porcine intestinal fluid (229). Moreover, novel structure design of a vector can help avoid the interaction between plasmid and enzyme. For example, plasmid DNA was encapsulated in gelatin nanospheres, and these nanospheres were further loaded in poly(ε-caprolactone) (PCL) microspheres. The microspheres provided protection against the proteolytic enzymes in stomach and intestine, and the nanospheres traversed through the GI tract and afforded protection during cellular transport for efficient transfection (230).

In mucosal drug delivery, besides enzyme attack, the drug drainage from the site of application and rapid mucociliary clearance also decreases the administration efficiency. Adopting materials with mucoadhesive and penetration promotion ability helps enhance transfection efficiency. For example, in oral delivery, compared with chitosan, the particles prepared with quaternized chitosan and fungal chitosan had stronger attachment to mucus cell membranes and showed higher transfection efficiency (231, 232). In order to enhance permeability across nasal mucosa, PLGA nanospheres were modified on their surface by deslorelin and transferrin. The transport and transfection efficiency show similar tendency: transferrin-PLGA-NPs > deslorelin-PLGA-NPs > PLGA-NPs >> plasmid (233).

10.5.2 Protection of Bioactivity During the Preparation and Degradation Process of Micro/Nanospheres

Although one of the aims for adopting micro/nanospheres as gene vectors is protecting the structure integrity of genes in application, however, without carefully consideration, the structure of genes may be changed in the preparation, lyophilization, and degradation process. The damage of high shear was stated before, and Bozkir *et al.* found that the solvent evaporation method destroyed the supercoiled forms of DNA into open circular forms and fragmented linear forms in the preparation of chitosan-DNA nanospheres. The use of organic solvent may also damage DNA integrity. An alternative method for protecting DNA during encapsulation is using solid plasmid as internal aqueous phase. Because the adoption of solid DNA form decreased the effect of shear stress and interaction between DNA and organic phase, this method could help preserve the supercoiled plasmid DNA (234). Adopting the complex coacervation method also helped to protect the supercoiled form of pDNA effectively (235).

In addition, the bioactivity of DNA is related to pH values in the preparation process. It was observed that exposure of pDNA to strong acidic or basic surroundings would accelerate the hydrolysis of pDNA (234,236). Addition of buffers was approved as an effective method to avoid pH-related degradation, which could preserve the conformation of the original pDNA during the preparation process (237). After preparation, the degradation of particles may also affect the pH values and decrease the activity of DNA. For example, PLGA is one of the most popular biodegradable materials used in biomedical fields. However, the acidic microenvironment during the degradation of the polymer lowers the pH values and damages the bioactivity of DNA (238). Co-encapsulating antacids may help buffer the acidic microenvironment and enhance the stability of plasmid DNA (239).

10.5.3 The Toxicity of Nanospheres

Although nanospheres as gene vectors attracted much attention in the past decades, their toxicity on living cells and organs have raised serious concerns with the development of nanotoxicology (136). Nanospheres administered by intravenous injection were distributed to the lung, bone marrow, liver, spleen, and the lymphatics (240). After systemic circulation, nanospheres were cleared by liver and splenic macrophages. The accumulation of particles in liver and spleen may bring degradation burden and damage these organs. Furthermore, the respiratory system may also be harmed by the inhalation of particles or by dermal and gastrointestinal absorption and injection of particles (241).

The enhanced endocytosis of gene vectors may also increase their toxicity, such as free radicals generated by the interaction of particles with cells, which may result in oxidative stress and cell death (242). Moreover, nanospheres may pass the BBB and the interactions between nanospheres with neurons may affect the central nervous system negatively (243). The most important adverse effects of therapeutically used nanomaterials are shown in Fig. 10.6.

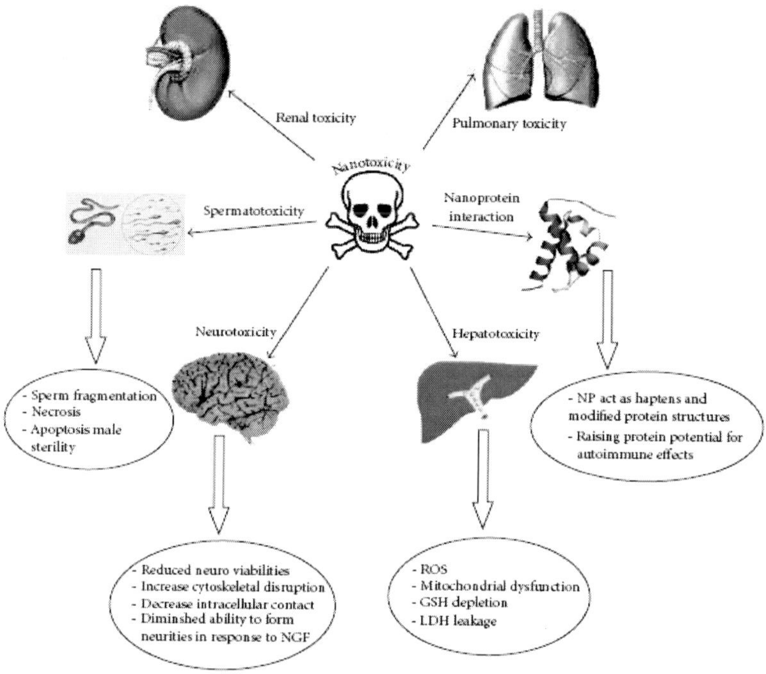

Figure 10.6 A summary of the most important recorded toxic effects of therapeutically used nanospheres reviewed in the present paper. ROS: reactive oxygen species, LDH: lactate dehydrogenase. Reprinted from Ref. 244.

10.6 Conclusion and Prospects

In this chapter, the barriers, the characteristics of polymers, and micro/nanospheres, the preparation methods, and the application fields related to micro/nanospheres as gene vectors were stated and discussed. As stated before, the transfection efficiency depended on several factors such as molecular weight and charge density of polymers, size and morphology of particles, the drug-loading method, administration route, and the cell type. These factors affect condensation of polymers with nucleic acid, the plasma circulation, the cellular uptake, and the vector unpacking. One factor may facilitate one process but damage another one. For example, increasing positive charge of polymers may strengthen the binding with nucleic acid, but may decrease the plasma circulation time.

Thus, in order to obtain optimal transfection efficiency of the factors involved in the preparation of particles as gene vectors, it is essential to consider the effect of these factors comprehensively and choose them carefully. In addition, with the development of materials and preparation techniques, it is in prospect that novel micro/nanospheres with special structures, characteristics, and functions will be developed in the future.

References

1. Selvam, S., Thomas, S. P., Hamm-Alvarez, S. F., Schechter, J. E., Stevenson, D., Mircheff, A. K., *et al.* (2006). Current status of gene delivery and gene therapy in lacrimal gland using viral vectors. *Advanced Drug Delivery Reviews*, **58**, 1243–1257.
2. Park, T. G., Jeong, J. H., & Kim, S. W. (2006). Current status of polymeric gene delivery systems. *Advanced Drug Delivery Reviews*, **58**, 467–486.
3. Dufès, C., Uchegbu, I. F., & Schätzlein, A. G. (2005). Dendrimers in gene delivery. *Advanced Drug Delivery Reviews*, **57**, 2177–2202.
4. Baker, M. (2010). Homing in on delivery. *Nature*, **464**, 1225–1228.
5. Sokolova, V., Kovtun, A., Heumann, R., & Epple, M. (2007). Tracking the pathway of calcium phosphate/DNA nanospheres during cell transfection by incorporation of red-fluorescing tetramethylrhodamine isothiocyanate-bovine serum albumin into these nanospheres. *Advanced Drug Delivery Reviews*, 2007, **12**, 174–179.
6. Saxena, A., Mozumdar, S., & Johri, A. K. (2006). Ultra-low sized cross-linked polyvinylpyrrolidone nanospheres as non-viral vectors for *in vivo* gene delivery. *Biomaterials*, **27**, 5596–5602.
7. Pedraza, C. E., Bassett, D. C., McKee, M. D., Nelea, V., Gbureck, U., & Barralet, J. E. (2008). The importance of particle size and DNA condensation salt for calcium phosphate nanosphere transfection. *Biomaterials*, **29**, 3384–3392.
8. Yu, D., Wang, A., Huang, H., & Chen, Y. (2008). PEG-PBLG nanoparticle-mediated HSV-TK/GCV gene therapy for oral squamous cell carcinoma. *Nanomedicine*, **3**, 813–821.
9. Akinc, A., & Langer, R. (2002). Measuring the pH environment of DNA delivered using nonviral vectors: Implications for lysosomal trafficking. *Biotechnology and Bioengineering*, **78**, 503–508.
10. Orgel, L. E. (2004). Prebiotic chemistry and the origin of the RNA world. *Critical Reviews in Biochemistry and Molecular Biology*, **39**, 99–123.

11. Di Giulio, M. (1997). On the RNA world: Evidence in favor of an early ribonucleopeptide world. *Journal of Molecular Evolution*, **45**, 571–578.

12. Mansouri, S., Lavigne, P., Corsi, K., Benderdour, M., Beaumont, E., & Fernandes, J. C. (2004). Chitosan-DNA nanoparticles as non-viral vectors in gene therapy: Strategies to improve transfection efficacy. *European Journal of Pharmaceutics and Biopharmaceutics*, **57**, 1–8.

13. Yagi, N., Manabe, I., Tottori, T., Ishihara, A., Ogata, F., Kim, J. H., *et al.* (2009). A nanosphere system specifically designed to deliver short interfering RNA inhibits tumor growth *in vivo*. *Cancer Research*, **69**, 6531–6538.

14. Ikonen, M., Murtomäki, L., & Kontturi, K. (2008). Controlled complexation of plasmid DNA with cationic polymers: Effect of surfactant on the complexation and stability of the complexes. *Colloids and Surfaces, B: Biointerfaces*, **66**, 77–83.

15. Morille, M., Passirani, C., Vonarbourg, A., Clavreul, A., & Benoit, J.-P. (2008). Progress in developing cationic vectors for non-viral systemic gene therapy against cancer. *Biomaterials*, **29**, 3477–3496.

16. Liu, L., Yang, Y.-L., Wang, C., Yao, Y., Ma, Y.-Z., Hou, S., *et al.* (2010). Polymeric effects on DNA condensation by cationic polymers observed by atomic force microscopy. *Colloids and Surfaces, B: Biointerfaces*, **75**, 230–238.

17. Mann, A., Richa, R., & Ganguli, M. (2008). DNA condensation by poly-L-lysine at the single molecule level: Role of DNA concentration and polymer length. *Journal of Controlled Release*, **125**, 252–262.

18. Ma, C., & Bloomfield, V. A. (1994). Condensation of supercoiled DNA induced by $MnCl_2$. *Biophysical Journal*, **67**, 1678–1681.

19. Kowalczyk, T., Pasumarthy, M., Gedeon, C., Hyatt, S., Payne, J., Muhammad, O., *et al.* (2001). Type of polylysine counterion influences morphology and biological function of condensed DNA. *Molecular Therapy*, **3**, S259.

20. Morris, V. B., Neethu, S., Abraham, T. E., Pillai, C. K. S., & Sharma, C. P. (2009). Studies on the condensation of depolymerized chitosans with DNA for preparing chitosan-DNA nanoparticles for gene delivery applications. *Journal of Biomedical Materials Research, Part B: Applied Biomaterials*, **89**, 282–292.

21. Danielsen, S., Vårum, K. M., & Stokke, B. T. (2004). Structural analysis of chitosan mediated DNA condensation by AFM: Influence of chitosan molecular parameters. *Biomacromolecules*, **5**, 928–936.

22. Fukushima, T., Hayakawa, T., Okamura, K., Takeda, S., Inoue, Y., Miyazaki, K., *et al.* (2006). Buffer solution can control the porosity of DNA-chitosan complexes. *Journal of Biomedical Materials Research, Part B: Applied Biomaterials*, **76B**, 121–129.

23. Dehousse, V., Garbacki, N., Jaspart, S., Castagne, D., Piel, G., Colige, A., *et al.* (2010). Comparison of chitosan/siRNA and trimethylchitosan/siRNA complexes behaviour *in vitro*. *International Journal of Biological Macromolecules*, **46**, 342–349.

24. Hashida, M., Takemura, S., Nishikawa, M., & Takakura, Y. (1998). Targeted delivery of plasmid DNA complexes with galactosylated poly(L-lysine). *Journal of Controlled Release*, **53**, 301–310.

25. Dash, P. R., Read, M. L., Barret, L. B., Wolfert, M., & Seymour, L. W. (1999). Factors affecting blood clearance and *in vivo* distribution of polyelectrolyte complexes for gene delivery. *Gene Therapy*, **6**, 643–650.

26. Kwoh, D. Y., Coffin, C. C., Lollo, C. P., Jovenal, J., Banaszczyk, M. G., Mullen, P., *et al.* (1999). Stabilisation of poly-L-lysine/DNA polyplexes for *in vivo* gene delivery to the liver. *Biochimica et Biophysica Acta*, **1444**, 171–190.

27. Ziady, A. G., Ferkol, T., Dawson, D. V., Perlmutter, D. H., & Davis, P. B. (1999). Chain length of the polylysine in receptor-targeted gene transfer complexes affects duration of reporter gene expression both *in vitro* and *in vivo*. *Journal of Biological Chemistry*, **274**, 4908–4916.

28. Ward, C. M., Read, M. L., & Seymour, L. W. (2001). Systemic circulation of poly(L-lysine)/DNA vectors is influenced by polycation molecular weight and type of DNA: Differential circulation in mice and rats and the implications for human gene therapy. *Blood*, **97**, 2221–2229.

29. Seymour, L. W. (1992). Passive tumor targeting of soluble macromolecules and drug conjugates. *Critical Reviews in Therapeutic Drug Carrier Systems*, **9**, 135–187.

30. Sheikpranbabu, S., Kalishwaralal, K., Venkataraman, D., Eom, S. H., Park, J., & Gurunathan, S. (2009). Silver nanoparticles inhibit VEGF- and 1L-1β-induced vascular permeability via Src dependent pathway in porcine retinal endothelial cells. *Journal of Nanobiotechnology*, **7**, 8.

31. Yuan, F., Dellian, M., Fukumura, D., Leunig, M., Berk, D. A., Torchilin, V. P., *et al.* (1995). Vascular permeability in a human tumor xenograft: Molecular size dependence and cutoff size. *Cancer Research*, **55**, 3752–3756.

32. Matsumura, Y., & Maeda, H. (1986). A new concept for macromolecular therapeutics in cancer chemotherapy: Mechanism of tumoritropic

accumulation of proteins and the antitumor agent smancs. *Cancer Research*, **46**, 6387–6392.

33. Dreher, M. R., Liu, W., Michelich, C. R., Dewhirst, M. W., Yuan, F., & Chilkoti, A. (2006). Tumor vascular permeability, accumulation, and penetration of macromolecular drug carriers. *Journal of the National Cancer Institute*, **98**, 335–344.
34. Duan, Y., Zheng, J., Han, S., Wu, Y., Wang, Y., Li, D., *et al.* (2008). A tumor targeted gene vector modified with G250 monoclonal antibody for gene therapy. *Journal of Controlled Release*, **127**, 173–179.
35. Chiu, S.-J., Ueno, N. T., & Lee, R. J. (2004). Tumor-targeted gene delivery via anti-HER2 antibody (trastuzumab, HerceptinR) conjugated polyethylenimine. *Journal of Controlled Release*, **97**, 357–369.
36. Blackburn, W. H., Dickerson, E. B., Smith, M. H., McDonald, J. F., & Lyon, L. A. (2009). Peptide-functionalized nanogels for targeted siRNA delivery. *Bioconjugate Chemistry*, **20**, 960–968.
37. Wood, K. C., Azarin, S. M., Arap, W., Pasqualini, R., Langer, R., & Hammond, P. T. (2008). Tumor-targeted gene delivery using molecularly engineered hybrid polymers functionalized with a tumor-homing peptide. *Bioconjugate Chemistry*, **19**, 403–405.
38. Veiseh, O., Kievit, F., Gunn, J., Ratner, B., & Zhang, M. (2009). A ligand-mediated nanovector for targeted gene delivery and transfection in cancer cells. *Biomaterials*, **30**, 649–657.
39. Nayak, S., Lee, H., Chmielewski, J., & Lyon, L. A. (2004). Folate-mediated cell targeting and cytotoxicity using thermoresponsive microgels. *Journal of the American Chemical Society*, **126**, 10258–10259.
40. Zhang, C., Gao, S., Jiang, W., Lin, S., Du, F., Li, Z., *et al.* (2010). Targeted minicircle DNA delivery using folate-poly(ethylene glycol)-polyethylenimine as non-viral carrier. *Biomaterials*, **31**, 6075–6086.
41. Liang, B., He, M.-L., Xiao, Z.-P., Li, Y., Chan, C.-Y., Kung, H.-F., *et al.* (2008). Synthesis and characterization of folate-PEG-grafted-hyperbranched-PEI for tumor-targeted gene delivery. *Biochemical and Biophysical Research Communications*, **367**, 874–880.
42. Bartlett, D. W., & Davis, M. E. (2007). Physicochemical and biological characterization of targeted, nucleic acid-containing nanoparticles. *Bioconjugate Chemistry*, **18**, 456–468.
43. Kursa, M., Walker, G. F., Roessler, V., Ogris, M., Roedl, W., Kircheis, R., *et al.* (2003). Novel shielded transferrin-polyethylene glycol-polyethylenimine/DNA complexes for systemic tumor-targeted gene transfer. *Bioconjugate Chemistry*, **14**, 222–231.

44. Kunath, K., von Harpe, A., Fischer, D., & Kisse, T. (2003). Galactose-PEI–DNA complexes for targeted gene delivery: Degree of substitution affects complex size and transfection efficiency. *Journal of Controlled Release*, **88**, 159–172.

45. Huh, M. S., Lee, S.-Y., Park, S., Lee, S., Chung, H., Lee, S., *et al.* (2010). Tumor-homing glycol chitosan/polyethylenimine nanoparticles for the systemic delivery of siRNA in tumor-bearing mice. *Journal of Controlled Release*, **144**, 134–143.

46. Hood, J. D., Bednarski, M., Frausto, R., Guccione, S., Reisfeld, R. A., Xiang, R., *et al.* (2002). Tumor regression by targeted gene delivery to the neovasculature. *Science*, **296**, 2404–2407.

47. Kim, W. J., & Kim, S. W. (2007). Targeted polymeric gene delivery for anti-angiogenic tumor therapy. *Macromolecular Research*, **15**, 100–108.

48. Hosseinkhani, H., & Tabata, Y. (2004). PEGylation enhances tumor targeting of plasmid DNA by an artificial cationized protein with repeated RGD sequences, Pronectin®. *Journal of Controlled Release*, **97**, 157–171.

49. Chan, P., Kurisawa, M., Chung, J. E., & Yang, Y. Y. (2007). Synthesis and characterization of chitosan-g-poly(ethylene glycol)-folate as a non-viral carrier for tumor-targeted gene delivery. *Biomaterials*, **28**, 540–549.

50. Koppu, S., Oh, Y. J., Edrada-Ebel, R., Blatchford, D. R., Tetley, L., Tate, R., *et al.* (2010). Tumor regression after systemic administration of a novel tumor-targeted gene delivery system carrying a therapeutic plasmid DNA. *Journal of Controlled Release*, **143**, 215–221.

51. Kim, S. H., Jeong, J. H., Lee, S. H., Kim, S. W., & Park, T. G. (2008). LHRH receptor-mediated delivery of siRNA using polyelectrolyte complex micelles self-assembled from siRNA-PEG-LHRH conjugate and PEI. *Bioconjugate Chemistry*, **19**, 2156–2162.

52. Kloeckner, J., Boeckle, S., Persson, D., Roedl, W., Ogris, M., Berg, K., *et al.* (2006). DNA polyplexes based on degradable oligoethylenimine-derivatives: Combination with EGF receptor targeting and endosomal release functions. *Journal of Controlled Release*, **116**, 115–122.

53. Ogris, M., Walker, G., Blessing, T., Kircheis, R., Wolschek, M., & Wagner, E. (2003). Tumor-targeted gene therapy: Strategies for the preparation of ligand-polyethylene glycol-polyethylenimine/DNA complexes. *Journal of Controlled Release*, **91**, 173–181.

54. Kircheis, R., Blessing, T., Brunner, S., Wightman, L., & Wagner, E. (2001). Tumor targeting with surface-shielded ligand-polycation DNA complexes. *Journal of Controlled Release*, **72**, 165–170

55. Park, I. K., Kim, T. H., Park, Y. H., Shin, B. A., Choi, E. S., Chowdhury, E. H., *et al.* (2001). Galactosylated chitosan-graft-poly(ethylene glycol) as hepatocyte-targeting DNA carrier. *Journal of Controlled Release*, **76**, 349–362.

56. Murthy, N., Campbell, J., Fausto, N., Hoffman, A. S., & Stayton, P. (2003). Design and synthesis of pH-responsive polymeric carriers that target uptake and enhance the intracellular delivery of oligonucleotides. *Journal of Controlled Release*, **89**, 365–374.

57. Horiuchi, S., & Aoyama, Y. (2006). Systematic lactose-functionalization of amphiphilic octaamine macrocycle as a gene carrier. Optimization of the charge, size, toxicity, and receptor factors for hepatocyte targeting. *Journal of Controlled Release*, **116**, 107–114.

58. Díez, S., Navarro, G., & de ILarduya, C. T. (2009). *In vivo* targeted gene delivery by cationic nanoparticles for treatment of hepatocellular carcinoma. *Journal of General Microbiology*, **11**, 38–45.

59. Cheong, S.-J., Lee, C.-M., Kim, S.-L., Jeong, H.-J., Kim, E.-M., Park, E.-H., *et al.* (2009). Superparamagnetic iron oxide nanoparticles-loaded chitosan-linoleic acid nanoparticles as an effective hepatocyte-targeted gene delivery system. *International Journal of Pharmaceutics*, **372**, 169–176.

60. Derfus, A. M., Chen, A. A., Min, D.-H., Ruoslahti, E., & Bhatia, S. N. (2007). Targeted quantum dot conjugates for siRNA delivery. *Bioconjugate Chemistry*, **18**, 1391–1396.

61. Strehblow, C., Schuster, M., Moritz, T., Kirch, H.-C., Opalka, B., & Petri, J. B. (2005). Monoclonal antibody-polyethyleneimine conjugates targeting Her-2/neu or CD90 allow cell type-specific nonviral gene delivery. *Journal of Controlled Release*, **102**, 737–747.

62. Elfinger, M., Maucksch, C., & Rudolph, C. (2007). Characterization of lactoferrin as a targeting ligand for nonviral gene deliver to airway epithelial cells. *Biomaterials*, **28**, 3448–3455.

63. Cu, Y., LeMoëllic, C., Caplan, M. J., & Saltzman, W. M. (2010). Ligand-modified gene carriers increased uptake in target cells but reduced DNA release and transfection efficiency. *Nanomedicine: Nanotechnology, Biology and Medicine*, **6**, 334–343.

64. Theoharis, S., Krueger, U., Tan, P. H., Haskard, D. O., Weber, M., & George, A. J. T. (2009). Targeting gene delivery to activated vascular endothelium using anti E/P-Selectin antibody linked to PAMAM dendrimers. *Journal of Immunological Methods*, **343**, 79–90.

65. Jeong, J. H., Lee, M., Kim, W. J., Yockman, J. W., Park, T. G., Kim, Y. H., *et al.* (2005). Anti-GAD antibody targeted non-viral gene delivery to islet beta cells. *Journal of Controlled Release*, **107**, 562–570.

66. Lee, J.-H., Lee, K., Moon, S. H., Lee, Y., Park, T. G., & Cheon, J. (2009). All-in-one target-cell-specific magnetic nanoparticles for simultaneous molecular imaging and siRNA delivery. *Angewandte Chemie, International Edition*, **48**, 4174–4179.

67. Scherer, F., Anton, M., Schillinger, U., Henke, J., Bergemann, C., Krüger, A., *et al.* (2002). Magnetofection: Enhancing and targeting gene delivery by magnetic force *in vitro* and *in vivo*. *Gene Therapy*, **9**, 102–109.

68. Foged, C., Brodin, B., Frokjaer, S., & Sundblad, A. (2005). Particle size and surface charge affect particle uptake by human dendritic cells in an *in vitro* model. *International Journal of Pharmaceutics*, **298**, 315–322.

69. Qaddoumi, M. G., Ueda, H., Yang, J., Davda, J., Labhasetwar, V., & Lee, V. H. L. (2004). The characteristics and mechanisms of uptake of PLGA nanoparticles in rabbit conjunctival epithelial cell layers. *Pharmaceutical Research*, **21**, 641–648.

70. Godbey, W. T., Wu, K. K., & Mikos, A. G. (1999). Size matters: Molecular weight affects the efficiency of poly(ethylenimine) as a gene delivery vehicle. *Journal of Biomedical Materials Research*, **45**, 268–275.

71. Hatakeyama, H., Ito, E., Akita, H., Oishi, M., Nagasaki, Y., Futaki, S., *et al.* (2009). A pH-sensitive fusogenic peptide facilitates endosomal escape and greatly enhances the gene silencing of siRNA-containing nanoparticles *in vitro* and *in vivo*. *Journal of Controlled Release*, **139**, 127–132.

72. Knorr, V., Ogris, M., & Wagner, E. (2008). An acid sensitive ketal-based polyethylene glycol-oligoethylenimine copolymer mediates improved transfection efficiency at reduced toxicity. *Pharmaceutical Research*, **25**, 2937–2945.

73. Lin, S., Du, F., Wang, Y., Ji, S., Liang, D., Yu, L., *et al.* (2008). An acid-labile block copolymer of PDMAEMA and PEG as potential carrier for intelligent gene delivery systems. *Biomacromolecules*, **9**, 109–115.

74. Nel, A. E., Mädler, L., Velegol, D., Xia, T., Hoek, E. M. V., Somasundaran, P., *et al.* (2009). Understanding biophysicochemical interactions at the nano–bio interface. *Nature Materials*, **8**, 543–557.

75. Wang, Y., Zhu, Y., Hu, Q., & Shen, J. (2008). Construction of caged polyplexes with a reversible intracellular unpacking property to improve stability and transfection. *Acta Biomaterialia*, **4**, 1235–1243.

76. Pichon, C., LeCam, E., Guérin, B., Coulaud, D., Delain, E., & Midoux, P. (2002). Poly[Lys-(AEDTP)]: A cationic polymer that allows dissociation of pDNA/cationic polymer complexes in a reductive medium and enhances polyfection. *Bioconjugate Chemistry*, **13**, 76–82.

77. Liu, X., Yang, J. W., Miller, A. D., Nack, E. A., & Lynn, D. M. (2005). Charge-shifting cationic polymers that promote self-assembly and self-disassembly with DNA. *Macromolecules*, **38**, 7907–7914.

78. Danielsen, S., Strand, S., de Lange Davies, C., & Stokke, B. T. (2005). Glycosaminoglycan destabilization of DNA–chitosan polyplexes for gene delivery depends on chitosan chain length and GAG properties. *Biochimica et Biophysica Acta*, **1721**, 44–54.

79. Hsu, C. Y. M., & Uluda, H. (2008). Effects of size and topology of DNA molecules on intracellular delivery with non-viral gene carriers. *BMC Biotechnology*, **8**, 23–37.

80. Kiang, T., Wen, J., Lim, H. W., & Leong, K. W. (2004). The effect of the degree of chitosan deacetylation on the efficiency of gene transfection. *Biomaterials*, **25**, 5293–5301.

81. Opanasopit, P., Petchsangsai, M., Rojanarata, T., Ngawhirunpat, T., Sajomsang, W., & Ruktanonchai, U. (2009). Methylated *N*-(4-*N*,*N*-dimethylaminobenzyl) chitosan as effective gene carriers: Effect of degree of substitution. *Carbohydrate Polymers*, **75**, 143–149.

82. Wong, K., Sun, G., Zhang, X., Dai, H., Liu, Y., He, C., *et al.* (2006). PEI-g-chitosan, a novel gene delivery system with transfection efficiency comparable to polyethylenimine *in vitro* and after liver administration *in vivo*. *Bioconjugate Chemistry*, **17**, 152–158.

83. Forrest, M. L., & Pack, D. W. (2002). On the kinetics of polyplex endocytic trafficking: Implications for gene delivery vector design. *Molecular Therapy*, **6**, 57–66.

84. Rungsardthong, U., Ehtezazi, T., Bailey, L., Armes, S. P., Garnett, M. C., & Stolnik, S. (2003). Effect of polymer ionization on the interaction with DNA in nonviral gene delivery systems. *Biomacromolecules*, **4**, 683–690.

85. Xiong, M. P., Bae, Y., Fukushima, S., Forrest, M. L., Nishiyama, N., Kataoka, K., *et al.* (2007). pH-responsive multi-PEGylated dual cationic nanoparticles enable charge modulations for safe gene delivery. *ChemMedChem*, **2**, 1321–1327.

86. He, C., Hu, Y., Yin, L., Tang, C., & Yin, C. (2010). Effects of particle size and surface charge on cellular uptake and biodistribution of polymeric nanoparticles. *Biomaterials*, **31**, 3657–3666.

87. Kim, W. J., Bonoiu, A. C., Hayakawa, T., Xia, C., Kakimoto, M. A., Pudavar, H. E., *et al.* (2009). Hyperbranched polysiloxysilane nanoparticles: Surface charge control of nonviral gene delivery vectors and nanoprobes. *International Journal of Pharmaceutics*, **376**, 141–152.

88. Grigsby, C. L., & Leong, K. W. (2010). Balancing protection and release of DNA: Tools to address a bottleneck of non-viral gene delivery. *Journal of the Royal Society Interface*, **7**, S67–S82.

89. Nemoto, Y., Borovkov, A., Zhou, Y.-M., Takewa, Y., Tatsumi, E., & Nakayama, Y. (2009). Impact of molecular weight in four-branched star vectors with narrow molecular weight distribution on gene delivery efficiency. *Bioconjugate Chemistry*, **20**, 2293–2299.

90. Huang, M., Fong, C.-W., Khor, E., & Lim, L. Y. (2005). Transfection efficiency of chitosan vectors: Effect of polymer molecular weight and degree of deacetylation. *Journal of Controlled Release*, **106**, 391–406.

91. Liu, X., Howard, K. A., Dong, M., Anderson, M. Ø., Ulrik, L. R., Johnsen, M. G., *et al.* (2007). The influence of polymeric properties on chitosan/siRNA nanoparticle formulation and gene silencing. *Biomaterials*, **28**, 1280–1288.

92. Yang, X., Yuan, X., Cai, D., Wang, S., & Li, Z. (2009). Low molecular weight chitosan in DNA vaccine delivery via mucosa. *International Journal of Pharmaceutics*, **375**, 123–132.

93. Moghimi, S. M., Symonds, P., Murray, J. C., Hunter, A. C., Debska, G., & Szewczyk, A. (2005). A two-stage poly(ethylenimine)-mediated cytotoxicity: Implications for gene transfer/therapy. *Molecular Therapy*, **11**, 990–995.

94. Fischer, D., Bieber, T., Li, Y., Elsässer, H.-P., & Kissel, T. (1999). A novel non-viral vector for DNA delivery based on low molecular weight, branched polyethylenimine: Effect of molecular weight on transfection efficiency and cytotoxicity. *Pharmaceutical Research*, **16**, 1273–1279.

95. Gosselin, M. A., Guo, W., & Lee, R. J. (2001). Efficient gene transfer using reversible cross-linked low molecular weight polyethylenimine. *Bioconjugate Chemistry*, **12**, 989–994.

96. Choi, S., & Lee, K.-D. (2008). Enhanced gene delivery using disulfide-crosslinked low molecular weight polyethylenimine with listeriolysin o-polyethylenimine disulfide conjugate. *Journal of Controlled Release*, **131**, 70–76.

97. Peng, Q., Zhong, Z., & Zhuo, R. (2008). Disulfide cross-linked polyethylenimines (PEI) prepared via thiolation of low molecular weight PEI as highly efficient gene vectors. *Bioconjugate Chemistry*, **19**, 499–506.

98. Breunig, M., Lungwitz, U., Liebl, R., & Goepferich, A. (2007). Breaking up the correlation between efficacy and toxicity for nonviral gene delivery. *Proceedings of the National Academy of Sciences of the United States of America*, **104**, 14454–14459.

99. Knorr, V., Russ, V., Allmendinger, L., Ogris, M., & Wagner, E. (2008). Acetal linked oligoethylenimines for use as pH-sensitive gene carriers. *Bioconjugate Chemistry*, **19**, 1625–1634.

100. Forrest, M. L., Koerber, J. T., & Pack, D. W. (2003). A degradable polyethylenimine derivative with low toxicity for highly efficient gene delivery. *Bioconjugate Chemistry*, **14**, 934–940.

101. Gao, J.-Q., Zhao, Q.-Q., Lv, T.-F., Shuai, W. P., Zhou, J., Tang, G.-P., *et al.* (2010). Gene-carried chitosan-linked-PEI induced high gene transfection efficiency with low toxicity and significant tumor-suppressive activity. *International Journal of Pharmaceutics*, **387**, 286–294.

102. Lu, B., Xu, X.-D., Zhang, X.-Z., Cheng, S.-X., & Zhuo, R.-X. (2008). Low molecular weight polyethylenimine grafted *N*-maleated chitosan for gene delivery: Properties and *in vitro* transfection studies. *Biomacromolecules*, **9**, 2594–2600.

103. Patnaik, S., Aggarwal, A., Nimesh, S., Goel, A., Ganguli, M., Saini, N., *et al.* (2006). PEI-alginate nanocomposites as efficient *in vitro* gene transfection agents. *Journal of Controlled Release*, **114**, 398–409.

104. Yao, H., Ng, S. S., Tucker, W. O., Tsang, Y.-K. T., Man, K., Wang, X.-M., *et al.* (2009). The gene transfection efficiency of a folate-PEI600-cyclodextrin nanopolymer. *Biomaterials*, **30**, 5793–5803.

105. Pai Kasturi, S., Qin, H., Thomson, K. S., El-Bereir, S., Cha, S. C., Neelapu, S., *et al.* (2006). Prophylactic anti-tumor effects in a B cell lymphoma model with DNA vaccines delivered on polyethylenimine (PEI) functionalized PLGA microparticles. *Journal of Controlled Release*, **113**, 261–270.

106. Ogris, M., Brunner, S., Schüller, S., Kircheis, R., & Wagner, E. (1999). PEGylated DNA/transferrin-PEI complexes: Reduced interaction with blood components, extended circulation in blood and potential for systemic gene delivery. *Gene Therapy*, **6**, 595–605.

107. Kleemann, E., Neu, M., Jekel, N., Fink, L., Schmehl, T., Gessler, T., *et al.* (2005). Nano-carriers for DNA delivery to the lung based upon a TAT-derived peptide covalently coupled to PEG-PEI. *Journal of Controlled Release*, **109**, 299–316.

108. Godbey, W. T., Wu, K. K., & Mikos, A. G. (2001). Poly(ethylenimine)-mediated gene delivery affects endothelial cell function and viability. *Biomaterials*, **22**, 471–480.

109. Clamme, J. P., Azoulay, J., & Mély, Y. (2003). Monitoring of the formation and dissociation of polyethylenimine/DNA complexes by two photon fluorescence correlation spectroscopy. *Biophysical Journal*, **84**, 1960–1968.

110. Lavertu, M., Méthot, S., Tran-Khanh, N., & Buschmann M. D. (2006). High efficiency gene transfer using chitosan/DNA nanoparticles with specific combinations of molecular weight and degree of deacetylation. *Biomaterials*, **27**, 4815–4824.

111. Mao, H. Q., Roy, K., Troung-Le, V. L., Janes, K. A., Lin, K. Y., Wang, Y., *et al.* (2001). Chitosan-DNA nanoparticles as gene carriers: Synthesis, characterization and transfection efficiency. *Journal of Controlled Release*, **70**, 399–421.

112. Kean, T., & Thanou, M. (2010). Biodegradation, biodistribution and toxicity of chitosan. *Advanced Drug Delivery Reviews*, **62**, 3–11.

113. Green, J. J., Shi, J., Chiu, E., Leshchiner, E. S., Langer, R., & Anderson, D. G. (2006). Biodegradable polymeric vectors for gene delivery to human endothelial cells. *Bioconjugate Chemistry*, **17**, 1162–1169.

114. Little, S. R., Lynn, D. M., Puram, S. V., & Langer, R. (2005). Formulation and characterization of poly (h amino ester) microparticles for genetic vaccine delivery. *Journal of Controlled Release*, **107**, 449–462.

115. Raemdonck, K., Naeye, B., Buyens, K., Vandenbroucke, R. E., Høgset, A., Demeester, J., *et al.* (2009). Biodegradable dextran nanogels for RNA interference: Focusing on endosomal escapes and intracellular siRNA delivery. *Advanced Functional Materials*, **19**, 1406–1415.

116. Zwiorek, K., Kloeckner, J., Wagner, E., & Coester, C. (2004). Gelatin nanoparticles as a new and simple gene delivery system. *Journal of Pharmaceutical Sciences*, **7**, 22–28.

117. Mittal, S. K., Aggarwal, N., Sailaja, G., Van Olphen, A., HogenEsch, H., North, A., *et al.* (2001). Immunization with DNA, adenovirus or both in biodegradable alginate microspheres: Effect of route of inoculation on immune response. *Vaccine*, **19**, 253–263.

118. Jeong, J. H., Kim, S. W., & Park, T. G. (2004). Biodegradable triblock copolymer of PLGA-PEG-PLGA enhances gene transfection efficiency. *Pharmaceutical Research*, **21**, 50–54.

119. Oh, Y.-K., Suh, D., Kim, J. M., Choi, H.-G., Shin, K., & Ko, J. J. (2002). Polyethylenimine-mediated cellular uptake, nucleus trafficking and expression of cytokine plasmid DNA. *Gene Therapy*, **9**, 1627–1632.

120. Howard, K. A., Rahbek, U. L., Liu, X., Damgaard, C. K., Glud, S. Z., Andersen, M. Ø, *et al.* (2006). RNA interference *in vitro* and *in vivo* using a chitosan/siRNA nanoparticles system. *Molecular Therapy*, **14**, 476–484.

121. Lee, M.-K., Chun, S.-K., Choi, W.-J., Kim, J.-K., Choi, S.-H., Kim, A., *et al.* (2005). The use of chitosan as a condensing agent to enhance emulsion-mediated gene transfer. *Biomaterials*, **26**, 2147–2156.

122. Germershaus, O., Mao, S., Sitterberg, J., Bakowsky, U., & Kissel, T. (2008). Gene delivery using chitosan, trimethyl chitosan or polyethylenglycol-graft-trimethyl chitosan block copolymers: Establishment of structure-activity relationships *in vitro*. *Journal of Controlled Release*, **125**, 145–154.

123. Jang, J.-H., & Shea, L. D. (2006). Intramuscular delivery of DNA releasing microspheres: Microsphere properties and transgene expression. *Journal of Controlled Release*, **112**, 120–128.

124. Patel, M. M., Zeles, M. G., Manning, M. C., Randolph, T. W., & Anchordoguy, T. J. (2004). Degradation kinetics of high molecular weight poly(L-lactide) microspheres and release mechanism of lipid: DNA complexes. *Journal of Pharmaceutical Sciences*, **93**, 2573–2584.

125. Cohen, H., Levy, R. J., & Gao, J. (2000). Sustained delivery and expression of DNA encapsulated in polymeric nanoparticles. *Gene Therapy*, **7**, 1896–1905.

126. Gvili, K., Benny, O., Danino, D., & Machluf, M. (2007). Poly(DL-lactide-co-glycolide acid) nanoparticles for DNA delivery: Waiving preparation complexity and increasing efficiency. *Biopolymers*, **85**, 379–391.

127. Oster, C. G., Wittmar, M., Bakowsky, U., & Kissel, T. (2006). DNA nanocarriers from biodegradable cationic branched polyesters are formed by a modified solvent displacement method. *Journal of Controlled Release*, **111**, 371–381.

128. Atuah, K. N., Walter, E., Merkle, H. P., & Alpar, H. O. (2003). Encapsulation of plasmid DNA in PLGA-stearylamine microspheres: A comparison of solvent evaporation and spray-drying methods. *Journal of Microencapsulation*, **20**, 387–399.

129. Jensen, D. M. K., Cun, D., Maltesen, M. J., Frokjaer, S., Nielsen, H. M., & Foged, C. (2010). Spray drying of siRNA-containing PLGA nanoparticles intended for inhalation. *Journal of Controlled Release*, **142**, 138–145.

130. Li, H.-Y., Seville, P. C., Williamson, I. J., & Birchall, J. C. (2005). The use of absorption enhancers to enhance the dispersibility of spray-dried powders for pulmonary gene therapy. *Journal of General Microbiology*, **7**, 1035–1043.

131. Nie, H., Lee, L. Y., Tong, H., & Wang, C. H. (2008). PLGA/chitosan composites from a combination of spray drying and supercritical fluid foaming techniques: New carriers for DNA delivery. *Journal of Controlled Release*, **129**, 207–214.

132. Zhao, C., Liu, X., Nomizu, M., & Nishi, N. (2004). Preparation of DNA-loaded polysulfone microspheres by liquid-liquid phase separation and its functional utilization. *Journal of Colloid and Interface Science*, **275**, 470–476.

133. Kogure, K., Akita, H., Yamada, Y., & Harashima, H. (2008). Multi-functional envelope-type nano device (MEND) as a non-viral gene delivery system. *Advanced Drug Delivery Reviews*, **60**, 559–571.

134. Kogure, K., Akita, H., & Harashima, H. (2007). Multifuncitonal envelope-type nano device for non-viral gene delivery: Concept and application of Programmed Packaging. *Journal of Controlled Release*, **122**, 246–251.

135. Zelikin, A. N., Becker, A. L., Johnston, A. P. R., Wark, K. L., Turatti, F., & Caruso, F. (2007). A general approach for DNA encapsulation in degradable polymer microcapsules. *ACS Nano*, **1**, 63–69.

136. Varde, N. K., & Pack, D. W. (2007). Influence of particle size and antacid on release and stability of plasmid DNA from uniform PLGA microspheres. *Journal of Controlled Release*, **124**, 172–180.

137. Awasthi, V. D., Garcia, D., Goins, B. A., & Philips, W. T. (2003). Circulation and biodistribution profiles of long-circulation PEG-liposomes of various sizes in rabbits. *International Journal of Pharmaceutics*, **253**, 121–132.

138. Schipper, M. L., Iyer, G., Koh, A. L., Cheng, Z., Ebenstein, Y., Aharoni, A., *et al.* (2009). Particle size, surface coating, and PEGylation influence the biodistribution of quantum dots in living mice. *Small*, **5**, 126–134.

139. Jo, J.-I., & Tabata, Y. (2008). Non-viral gene transfection technologies for genetic engineering of stem cells. *European Journal of Pharmaceutics and Biopharmaceutics*, **68**, 90–104.

140. Xu, D.-M., Yao, S.-D., Liu, Y.-B., Sheng, K.-L., Hong, J., Gong, P.-J., *et al.* (2007). Size-dependent properties of M-PEIs nanogels for gene delivery in cancer cells. *International Journal of Pharmaceutics*, **338**, 291–296.

141. Zauner, W., Farrow, N. A., & Haines, A. M. R. (2001). *In vitro* uptake of polystyrene microspheres: Effect of particle size, cell line and cell density. *Journal of Controlled Release*, **71**, 39–51.

142. Champion, J. A., Katare, Y. K., & Mitragotri, S. (2007). Particle shape: A new design parameter for micro- and nanoscale drug delivery carriers. *Journal of Controlled Release*, **121**, 3–9.

143. Rolland, J. P., Maynor, B. W., Euliss, L. E., Exner, A. E., Denison, G. M., & DeSimone, J. M. (2005). Direct fabrication and harvesting

of monodisperse, shape-specific nanobiomaterials. *Journal of the American Chemical Society*, **127**, 10096–10100.

144. Xu, S., Nie, Z., Seo, M., Lewis, P., Kumacheva, E., Stone, H. A., *et al.* (2005). Generation of monodisperse particles by using microfluidics: Control over size, shape, and composition. *Angewandte Chemie, International Edition*, **44**, 724–728.

145. Champion, J. A., & Mitragotri, S. (2006). Role of target geometry in phagocytosis. *Proceedings of the National Academy of Sciences of the United States of America*, **103**, 4930–4934.

146. Champion, J. A., & Mitragotri, S. (2009). Shape induced inhibition of phagocytosis of polymer particles. *Pharmaceutical Research*, **26**, 244–249.

147. Alemdaroglu, F. E., Alemdaroglu, N. C., Langguth, P., & Herrmann, A. (2008). Cellular uptake of DNA block copolymer micelles with different shapes. *Macromolecular Rapid Communications*, **29**, 326–329.

148. Zhang, K., Fang, H., Chen, Z., Taylor, J.-S. A., & Wooley, K. L. (2008). Shape effects of nanoparticles conjugated with cell-penetrating peptides (HIV Tat PTD) on CHO cell uptake. *Bioconjugate Chemistry*, **19**, 1880–1887.

149. Liu, Z., Zhang, Z., Zhou, C., & Jiao, Y. (2010). Hydrophobic modifications of cationic polymers for gene delivery. *Progress in Polymer Science*, **35**, 1144–1162.

150. Wang, B., He, C., Tang, C., & Yin, C. (2011). Effects of hydrophobic and hydrophilic modifications on gene delivery of amphiphilic chitosan based nanocarriers. *Biomaterials*, **32**, 4630–4638.

151. Yoo, H. S., Lee, J. E., Chung, H., Kwon, I. C., & Jeong, S. Y. (2005). Self-assembled nanoparticles containing hydrophobically modified glycol chitosan for gene delivery. *Journal of Controlled Release*, **103**, 235–243

152. Tian, H., Xiong, W., Wang, Y., Chen, X., Jing, X., & Zhu, Q., (2007). Gene transfection of hyperbranched PEI grafted by hydrophobic amino acid segment PBLG. *Biomaterials*, **28**, 2899–2907

153. Ambegia, E., Ansell, S., Cullis, P., Heyes, J., Palmer, L., & MacLachlan, I. (2005). Stabilized plasmid–lipid particles containing PEG-diacylglycerols exhibit extended circulation lifetimes and tumor selective gene expression. *Biochimica et Biophysica Acta*, **1669**, 155–163.

154. Morille, M., Montier, T., Legras, P., Carmoy, N., Brodin, P., Pitard, B., *et al.* (2010). Long-circulating DNA lipid nanocapsules as new vector for passive tumor targeting. *Biomaterials*, **31**, 321–329.

155. Prencipe, G., Tabakman, S. M., Welsher, K., Liu, Z., Goodwin, A. P., Zhang, L., *et al.* (2009). PEG branched polymer for functionalization of nanomaterials with ultralong blood circulation. *Journal of the American Chemical Society*, **131**, 4783–4787.

156. Funhoff, A. M., Monge, S., Teeuwen, R., Koning, G. A., Schuurmans-Nieuwenbroek, N. M. E., *et al.* (2005). PEG shielded polymeric double-layered micelles for gene delivery. *Journal of Controlled Release*, **102**, 711–724.

157. Merdan, T., Kunath, K., Petersen, H., Bakowsky, U., Voigt, K. H., Kopecek, J., *et al.* (2005). PEGylation of poly(ethylene imine) affects stability of complexes with plasmid DNA under *in vivo* conditions in a dose-dependent manner after intravenous injection into mice. *Bioconjugate Chemistry*, **16**, 785–792.

158. Petersen, H., Fechner, P. M., Martin, A. L., Kunath, K., Stolnik, S., Roberts, C. J., *et al.* (2002). Polyethylenimine-graft-poly(ethylene glycol) copolymers: Influence of copolymer block structure on DNA complexation and biological activities as gene delivery system. *Bioconjugate Chemistry*, **13**, 845–854.

159. Mosqueira, V. C. F., Legrand, P., Morgat, J.-L., Vert, M., Mysiakine, E., Gref, R., *et al.* (2001). Biodistribution of long-circulating PEG-grafted nanocapsules in mice: Effects of PEG chain length and density. *Pharmaceutical Research*, **18**, 1411–1419.

160. Van den Berg, J. H., Oosterhuis, K., Hennink, W. E., Storm, G., Van der Aa, L. J., Engbersen, J. F., *et al.* (2010). Shielding the cationic charge of nanoparticle-formulated dermal DNA vaccines is essential for antigen expression and immunogenicity. *Journal of Controlled Release*, **141**, 234–240.

161. Dash, P. R., Read, M. L., Fisher, K. D., Howard, K. A., Wolfert, M., Oupicky, D., *et al.* (2000). Decreased binding to proteins and cells of polymeric gene delivery vectors, surface modified with a multivalent hydrophilic polymer and retargeting through attachment of transferring. *Journal of Biological Chemistry*, **275**, 3793–3802.

162. Oupický, D., Howard, K. A., Koňák, Č., Dash, P. R., Ulbrich, K., & Seymour, L. W. (2000). Steric stabilization of poly-L-lysine/DNA complexes by the covalent attachment of semitelechelic poly[*N*-(2-hydroxypropyl)m ethacrylamide]. *Bioconjugate Chemistry*, **11**, 492–501.

163. Howard, K. A., Dash, P. R., Read, M. L., Ward, K., Tomkins, L. M., Nazarova, O., *et al.* (2000). Influence of hydrophilicity of cationic polymers on the biophysical properties of polyelectrolyte complexes formed by self-assembly with DNA. *Biochimica et Biophysica Acta*, **1475**, 245–255.

164. Chew, S. A., Hacker, M. C., Saraf, A., Raphael, R. M., Kasper, F. K., & Mikos, A. G. (2009). Biodegradable branched polycationic polymers with varying hydrophilic spacers for nonviral gene delivery. *Biomacromolecules*, **10**, 2436–2445.

165. Takeuchi, H., Kojima, H., Toyoda, T., Yamamoto, H., Hino, T., & Kawashima, Y. (1999). Prolonged circulation time of doxorubicin-loaded liposomes coated with a modified polyvinyl alcohol after intravenous injection in rats. *European Journal of Pharmaceutics and Biopharmaceutics*, **48**, 123-129

166. Takeuchi, H., Kojima, H., Yamamoto, H., & Kawashima, Y. (2001). Evaluation of circulation profiles of liposomes coated with hydrophilic polymers having different molecular weights in rats. *Journal of Controlled Release*, **75**, 83–91.

167. Beningo, K. A., & Wang, Y. L. (2002). Fc-receptor-mediated phagocytosis is regulated by mechanical properties of the target. *Journal of Cell Science*, **115**, 849–856.

168. Kang, J.-H., Toita, R., & Katayama, Y. (2010). Bio and nanotechnological strategies for tumor-targeted gene therapy. *Biotechnology Advances*, **28**, 757–763.

169. Maeda, H., Bharate, G. Y., & Daruwalla, J. (2009). Polymeric drugs for efficient tumor-targeted drug delivery based on EPR-effect. *European Journal of Pharmaceutics and Biopharmaceutics*, **71**, 409–419.

170. Aoi, A., Watanabe, Y., Mori, S., Takahashi, M., Vassaux, G., & Kodama, T. (2008). Herpes simplex virus thymidine kinase-mediated suicide gene therapy using nano/microbubbles and ultrasound. *Ultrasound in Medicine and Biology*, **34**, 425–434.

171. Ramesh, R., Marrogi, A. J., & Freeman, S. M. (1998). Tumor killing using the HSV-tk suicide gene. *Gene Therapy & Molecular Biology*, **1**, 253–263.

172. Wei, M. X., Bougnoux, P., Sacré-Salem, B., Peyrat, M.-B., Lhuillery, C., Salzmann, J.-L., *et al.* (1998). Suicide gene therapy of chemically induced mammary tumor in rat: Efficacy and distant bystander effect. *Cancer Research*, **58**, 3529–3532.

173. Guo, J., Wang, B., Zhang, M., Chen, T., Yu, Y., Regulier, E., *et al.* (2002). Macrophage-derived chemokine gene transfer results in tumor regression in murine lung carcinoma model through efficient induction of antitumor immunity. *Gene Therapy*, **9**, 793–803.

174. Oba, M., Vachutinsky, Y., Miyata, K., Kano, M. R., Ikeda, S., *et al.* (2010). Antiangiogenic gene therapy of solid tumor by systemic injection

of polyplex micelles loading plasmid DNA encoding soluble Flt-1. *Molecular Pharmaceutics*, **7**, 501–509.

175. Zhu, S., Wu, H., Wu, F., Nie, D., Sheng, S., & Mo, Y.-Y. (2008). MicroRNA-21 targets tumor suppressor genes in invasion and metastasis. *Cell Research*, **18**, 350–359.

176. Zhan, J., Ga, Y., Wang, W., Shen, A., Aspelund, A., Young, M., *et al.* (2005). Tumor-specific intravenous gene delivery using oncolytic adenoviruses. *Cancer Gene Therapy*, **12**, 19–25.

177. Frank, O., Rudolph, C., Heberlein, C., Von Neuhoff, N., Schröck, E., Schambach, A., *et al.* (2004). Tumor cells escape suicide gene therapy by genetic and epigenetic instability. *Blood*, **104**, 3543–3549.

178. Maeda, H., Wu, J., Sawa, T., Matsumura, Y., & Hori, K. (2000). Tumor vascular permeability and the EPR effect in macromolecular therapeutics: A review. *Journal of Controlled Release*, **65**, 271–284.

179. Van Vlerken, L. E., & Amiji, M. M. (2006). Multi-functional polymeric nanoparticles for tumour-targeted drug delivery. *Expert Opinion on Drug Delivery*, **3**, 205–216.

180. Sawicki, J. A., Anderson, D. G., & Langer, R. (2008). Nanoparticle delivery of suicide DNA for epithelial ovarian cancer therapy. *Advances in Experimental Medicine and Biology*, **622**, 209–219.

181. Huang, Y.-H., Zugates, G. T., Peng, W., Holtz, D., Dunton, C., Green, J. J., *et al.* (2009). Nanoparticles-delivered suicide gene therapy effectively reduces ovarian tumor burden in mice. *Cancer Research*, **69**, 6184–6191.

182. Patil, Y. B., Swaminathan, S. K., Sadhukha, T., Ma, L., & Panyam, J. (2010). The use of nanoparticles-mediated targeted gene silencing and drug delivery to overcome tumor drug resistance. *Biomaterials*, **31**, 358–365.

183. Lu, W., Tan, Y.-Z., and Hu, K.-L. (2005). Cationic albumin conjugated PEGylated nanoparticle with its transcytosis ability and little toxicity against blood–brain barrier. *International Journal of Pharmaceutics*, **295**, 247–260.

184. Olivier, J.-C., Huertas, R., Lee, H. J., Calon, F., & Pardridge, W. M. (2002). Synthesis of pegylated immunonanoparticles. *Pharmaceutical Research*, **19**, 1137–1143.

185. Lu, W., Sun, Q., Wan, J., She, Z., & Jiang, X.-G. (2006). Cationic albumin-conjugated pegylated nanoparticles allow gene delivery into brain tumors via intravenous administration. *Cancer Research*, **66**, 11878–11887.

186. Lu, W., Wan, J., She, Z., & Jiang, X. (2007). Brain delivery property and accelerated blood clearance of cationic albumin conjugated pegylated nanoparticle. *Journal of Controlled Release*, **118**, 38–53.

187. Chumakova, O. V., Liopo, A. V., Andreev, V. G., Cicenaite, I., Evers, B. M., Chakrabarty, S., *et al.* (2008). Composition of PLGA and PEI/DNA nanoparticles improves ultrasound-mediated gene delivery in solid tumors *in vivo*. *Cancer Letters*, **261**, 215–225.

188. Hauff, P., Seemann, S., Reszka, R., Schultze-Mosgau, M., Reinhardt, M., Buzasi, T., *et al.* (2005). Evaluation of gas-filled microparticles and sonoporation as gene delivery system: Feasibility study in rodent tumor models. *Radiology*, **236**, 572–578.

189. Derfus, A. M., Von Maltzahn, G., Harris, T. J., Duza, T., Vecchio, K. S., Ruoslahti, E., *et al.* (2007). Remotely triggered release from magnetic nanoparticles. *Advanced Materials*, **19**, 3932–3936.

190. Tang, D. C., DeVit, M., & Johnston, S. A. (1992). Genetic immunization is a simple method for eliciting an immune response. *Nature*, **356**, 152–154.

191. Xiang, S. D., Selomulya, C., & Ho, J. (2010). Delivery of DNA vaccines: An overview on the use of biodegradable polymeric and magnetic nanoparticles. *Wiley Interdisciplinary Reviews. Nanomedicine and Nanobiotechnology*, **2**, 205–218.

192. Slütter, B., Plapied, L., Fievez, V., Sande, M. A., Des Rieux, A., Schneider, Y. J., *et al.* (2009). Mechanistic study of the adjuvant effect of biodegradable nanoparticles in mucosal vaccination. *Journal of Controlled Release*, **138**, 113–121.

193. Waeckerle-Men, Y., & Groettrup, M. (2005). PLGA microspheres for improved antigen delivery to dendritic cells as cellular vaccines. *Advanced Drug Delivery Reviews*, **57**, 475–482.

194. Singh, M., Briones, M., Ott, G., & O'Hagan, D. (2000). Cationic microparticles: A potent delivery system for DNA vaccines. *Proceedings of the National Academy of Sciences of the United States of America*, **97**, 811–816.

195. O'Hagan, D. T., Singh, M., & Ulmer, J. B. (2004). Microparticles for the delivery of DNA vaccines. *Immunological Reviews*, **199**, 191–200.

196. Jilek, S., Merkle, H. P., & Walter, E. (2005). DNA-loaded biodegradable microparticles as vaccine delivery systems and their interaction with dendritic cells. *Advanced Drug Delivery Reviews*, **57**, 377–390.

197. Markel, H. (2005). The search for effective HIV vaccines. *New England Journal of Medicine*, **353**, 753–757.

198. Sharpe, S., Hanke, T., Tinsley-Bown, A., Dennis, M., Dowall, S., McMichael, A., *et al.* (2003). Mucosal immunization with PLGA-microencap-

sulated DNA primes a SIV-specific CTL response revealed by boosting with cognate recombinant modified vaccinia virus Ankara. *Virology*, **313**, 13–21.

199. Huang, H.-N., Li, T.-L., Chan, Y.-L., Chen, C. L., & Wu, C. J. (2009). Transdermal immunization with low-pressure-gene-gun mediated chitsosan-based DNA vaccines against Japanese encephalitis virus. *Biomaterials*, **30**, 6017–6025.

200. Boyoglu, S., Vig, K., Pillai, S., Rangari, V., Dennis, V. A., Khazi, F., & Singh, S. R. (2009). Enhanced delivery and expression of a nanoencapsulated DNA vaccine vector for respiratory syncytial virus. *Nanomedicine*, **5**, 463–472.

201. Liu, S., Danquah, M. K., Ho, J., Ma, C., Wang, L., Coppel, R., *et al.* (2009). Preparation and characterization of poly(lactic-co-glycolic acid) microparticles containing DNA molecules encoding a malaria vaccine candidate. *Journal of Chemical Technology and Biotechnology*, **84**, 782–788.

202. Tahara, K., Sakai, T., Yamamoto, H., Takeuchi, H., & Kawashima, Y. (2008). Establishing chitosan coated PLGA nanosphere platform loaded with wide variety of nucleic acid by complexation with cationic compound for gene delivery. *International Journal of Pharmaceutics*, **354**, 210–216.

203. Kasturi, S. P., Sachaphibulkij, K., & Roy, K. (2005). Covalent conjugation of polyethyleneimine on biodegradable microparticles for delivery of plasmid DNA vaccines. *Biomaterials*, **26**, 6375–6385.

204. O'Hagan, D., Singh, M., Ugozzoli, M., Wild, C., Barnett, S., Chen, M., *et al.* (2001). Induction of potent immune responses by cationic microparticles with adsorbed Human immunodeficiency virus DNA vaccines. *Journal of Virology*, **75**, 9037–9043.

205. Xiang, S. D., Scholzen, A., Minigo, G., David, C., Apostolopoulos, V., Mottram, P. L., *et al.* (2006). Pathogen recognition and development of particulate vaccines: Does size matter? *Methods*, **40**, 1–9.

206. Campochiaro, P. A. (2007). Gene therapy for ocular neovascularization. *Current Gene Therapy*, **7**, 25–33.

207. Colella, P., Cotugno, G., & Auricchio, A. (2008). Ocular gene therapy: Current progress and future prospects. *Trends in Molecular Medicine*, **15**, 23–31.

208. Maguire, A. M., Simonelli, F., Pierce, E. A., Pugh, E. N. Jr, Mingozzi, F., Bennicelli, J., *et al.* (2008). Safety and efficacy of gene transfer for Leber's congenital amaurosis. *New England Journal of Medicine*, **358**, 2240–2248.

209. Amado, D., Mingozzi, F., Hui, D., Bennicelli, J. L., Wei, Z., Chen, Y., *et al.* (2010). Safety and efficacy of subretinal readministration of a viral vector in large animals to treat congenital blindness. *Science Translational Medicine*, **21**, 21ra16.

210. De la Fuente, M., Seijo, B., & Alonso, M. J. (2008). Novel hyaluronic acid-chitosan nanoparticles for ocular gene therapy. *Investigative Ophthalmology & Visual Science*, **49**, 2016–2024.

211. Farjo, R., Skaggs, J., Quiambao, A. B., Cooper, M. J., & Naash, M. I. (2006). Efficient non-viral ocular gene transfer with compacted DNA nanoparticles. *PLoS ONE*, **1**, e38.

212. Peeters, L., Sanders, N. N., Demeester, J., & De Smedt, S. C. (2007). Challenges in non-viral ocular gene transfer. *Biochemical Society Transactions*, **35**, 47–49.

213. Read, S. P., Cashman, S. M., & Kumar-Singh, R. (2010). A poly(ethylene) glycolylated peptide for ocular delivery compacts DNA into nanoparticles for gene delivery to post-mitotic tissues *in vivo*. *Journal of General Microbiology*, **12**, 86–96.

214. Haider, H. K., Elmadbouh, I., Jean-Baptiste, M., & Muhammad, A. (2008). Nonviral vector gene modification of stem cells for myocardial repair. *Molecular Medicine*, **12**, 79–86.

215. Benn, S. I., Whitsitt, J. S., Broadley, K. N., Nanney, L. B., Perkins, D., He, L., *et al.* (1996). Particle-mediated gene transfer with transforming growth Factor-β1 cDNAs enhances wound repair in rat skin. *Journal of Clinical Investigation*, **98**, 2894–2902.

216. Eming, S. A., Whitsitt, J. S., He, L., Krieg, T., Morgan, J. R., & Davidson, S. M. (1999). Particle-mediated gene transfer of PDGF isoforms promotes wound repair. *Journal of Investigative Dermatology*, **112**, 297–302.

217. Duan, Y., Guan, X., Ge, J., Quan, D., Zhuo, Y., Ye, H., *et al.* (2008). Cationic nano-copolymers mediated IKKβ targeting siRNA inhibit the proliferation of human Tenon's capsule fibroblasts *in vitro*. *Molecular Vision*, **14**, 2616–2628.

218. Basarkar, A., & Singh, J. (2009). Poly(lactide-co-glycolide)-polymethacrylate nanoparticles for intramuscular delivery of plasmid encoding interleukin-10 to prevent autoimmune diabetes in mice. *Pharmaceutical Research*, **26**, 72–81.

219. Park, J. H., Lee, M., & Kim, S. W. (2006). Non-viral adiponectin gene therapy into obese type 2 diabetic mice ameliorates insulin resistance. *Journal of Controlled Release*, **114**, 118–125.

220. Howarda, K. A., Li, X. W., Somavarapu, S., Singh, J., Green, N., Atuah, K. N., *et al.* (2004). Formulation of a microparticle carrier for oral polyplex-based DNA vaccines. *Biochimica et Biophysica Acta*, **1674**, 149–157.

221. Guliyeva, Ü., Öner, F., Özsoy, Ş., & Haziroğlu (2006). Chitosan microparticles containing plasmid DNA as potential oral gene delivery system. *European Journal of Pharmaceutics and Biopharmaceutics*, **62**, 17–25.

222. Sanders, N. N., Peeters, L., Lentacker, I., Demeester, J., & De Smedt, S. C. (2007). Wanted and unwanted properties of surface PEGylated nucleic acid nanoparticles in ocular gene transfer. *Journal of Controlled Release*, **122**, 226–235.

223. Csaba, N., Sánchez, A., & Alonso, M. J. (2006). PLGA: Poloxamer and PLGA: Poloxamine blend nanostructures as carriers for nasal gene delivery. *Journal of Controlled Release*, **113**, 164–172.

224. Kichler, A., Chillon, M., Leborgne, C., Danos, O., & Frisch, B. (2002). Intranasal gene delivery with a polyethylenimine-PEG conjugate. *Journal of Controlled Release*, **81**, 379–388.

225. Grosse, S., Thévenot, G., Aron, Y., Duverger, E., Abdelkarim, M., Roche, A. C., *et al.* (2008). *In vivo* gene delivery in the mouse lung with lactosylated polyethylenimine, questioning the relevance of *in vivo* experiments. *Journal of Controlled Release*, **132**, 105–112.

226. Lee, P.-W., Peng, S.-F., Su, C.-J., Mi, F.-L., Chen, H.-L., Wei, M.-C., *et al.* (2008). The use of biodegradable polymeric nanoparticles in combination with a low-pressure gene gun for transdermal DNA delivery. *Biomaterials*, **29**, 742–751.

227. Khatri, K., Goyal, A. K., Gupta, P. N., Mishra, N., & Vyas, S. P. (2008). Plasmid DNA loaded chitosan nanoparticles for nasal mucosal immunization against hepatitis B. *International Journal of Pharmaceutics*, **354**, 235–241.

228. Martien, R., Loretz, B., Thaler, M., Majzoob, S., & Bernkop-Schnürch A. (2007). Chitosan–thioglycolic acid conjugate: An alternative carrier for oral nonviral gene delivery? *Journal of Biomedical Materials Research A*, **82**, 1–9.

229. Martien, R., Loretz, B., & Bernkop-Schnürch, A. (2006). Oral gene delivery: Design of polymeric carrier systems shielding toward intestinal enzymatic attack. *Biopolymers*, **83**, 327–336.

230. Bhavsar, M. D., & Amiji, M. M. (2007). Gastrointestinal distribution and *in vivo* gene transfection studies with nanoparticles-in-microsphere oral system (NiMOS). *Journal of Controlled Release*, **119**, 339–348.

231. Zheng, F., Shi, X.-W., & Yang, G.-F. (2007). Chitosan nanoparticle as gene therapy vector via gastrointestinal mucosa administration: Results of an *in vitro* and *in vivo* study. *Life Sciences*, **80**, 388–396.

232. Plapied, L., Vandermeulen, G., Vroman, B., Préat, V., & Des Rieux, A. (2010). Bioadhesive nanoparticles of fungal chitosan for oral DNA delivery. *International Journal of Pharmaceutics*, **398**, 210–218.

233. Sundaram, S., Roy, S. K., Ambati, B. K., & Kompella, U. B. (2009). Surface functionalized nanoparticles for targeted gene delivery across nasal respiratory epithelium. *FASEB Journal*, **23**, 3752–3765.

234. Ando, S., Putnam, D., Pack, D. W., & Langer, R. (1999). PLGA microspheres containing plasmid DNA: Preservation of supercoiled DNA via cryopreparation and carbohydrate stabilization. *Journal of Pharmaceutical Sciences*, **88**, 126–130.

235. Walter, E., Moelling, K., Pavlovic, J., & Merkle, H. P. (1999). Microencapsulation of DNA using poly(DL-lactide-co-glycolide): Stability issues and release characteristics. *Journal of Controlled Release*, **61**, 361–374.

236. Lentz, Y. K., Worden, L. R., Anchordoquy, T. J., & Lengsfeld, C. S. (2005). Effect of jet nebulization on DNA: Identifying the dominant degradation mechanism and mitigation methods. *Aerosol Science*, **36**, 973–990.

237. Li, Y., Ogris, M., Wagner, E., Pelisek, J., & Ruffer, M. (2003). Nanoparticles bearing polyethyleneglycol-coupled transferrin as gene carriers: Preparation and *in vitro* evaluation. *International Journal of Pharmaceutics*, **259**, 93–101.

238. Tse, M. T., Blatchford, C., & Oya Alpar, H. (2009). Evaluation of different buffers on plasmid DNA encapsulation into PLGA microparticles. *International Journal of Pharmaceutics*, **370**, 33–40.

239. Tinsley-Bown, A. M., Fretwell, R., Dowsett, A. B., Davis, S. L., & Farrar, G. H. (2000). Formulation of poly(DL-lactic-co-glycolic acid) microparticles for rapid plasmid DNA delivery. *Journal of Controlled Release*, **66**, 229–241.

240. Kipen, H. M., & Laskin, D. L. (2005). Smaller is not always better: Nanotechnology yields nanotoxicology. *American Journal of Physiology —Lung Cellular and Molecular Physiology*, **289**, L696–L697.

241. Vega-Villa, K. R., Takemoto, J. K., Yáñez, J. A., Remsberg, C. M., Forrest, M. L., & Davies, N. M. (2008). Clinical toxicities of nanocarrier systems. *Advanced Drug Delivery Reviews*, **60**, 929–938.

242. Card, J. W., Zeldin, D. C., Bonner, J. C., & Nestmann, E. R. (2008). Pulmonary applications and toxicity of engineered nanoparticles. *American Journal of Physiology — Lung Cellular* and *Molecular Physiology*, **295**, L400–L411.

243. Yang, Z., Liu, Z. W., Allaker, R. P., Reip, P., Oxford, J., Ahmad, Z., *et al.* (2010). A review of nanoparticle functionality and toxicity on the central nervous system. *Journal of the Royal Society Interface*, **7**, S411–S422.

244. El-Ansary, A., & Al-Daihan, S. (2009). On the toxicity of therapeutically used nanoparticles: An Overview. *Journal of Toxicology*, **2009**, Article ID 754810, 9 pages. doi:10.1155/2009/754810.

Chapter 11

Microspheres for Targeting Delivery of Anticancer Drugs

Wei Wei, Zhanguo Yue, Hua Yue, and Guanghui Ma

State Key Laboratory of Biochemical Engineering, Institute of Process Engineering, Chinese Academy of Sciences, Beijing, P. R. China

ghma@home.ipe.ac.cn

11.1 Introduction

The malignant cancer has become one of the most common diseases endangering human health in modern society. According to world cancer report 2008, the global burden of cancer has more than doubled during the past 30 years. It was estimated that there were over 12 million new cases of cancer diagnosed, 7 million deaths from cancer, and 25 million persons alive with cancer. The continued growth and ageing of the world's population will further intensify the cancer burden (1).

Cancer treatment mainly includes surgery therapy, radiation therapy, and chemotherapy. Compared with other methods, chemotherapy, as a systemic treatment, can kill cancer cells in patients in the maximal degree and thus remains one of the main methods for cancer treatment. However, this method also has to face some difficulties. A key problem is that most anticancer drugs could not be specifically distributed in the cancer-bearing tissues, causing unwanted toxic effects in normal tissues and lower drug concentration at lesion site (2). Another challenge is drug solubility, as about 40% of newly developed active pharmaceutical ingredients by the pharmaceutical industry are rejected because of poor water

Microspheres and Microcapsules in Biotechnology: Design, Preparation, and Applications
Edited by Guanghui Ma and Zhiguo Su

ISBN 978-981-4316-47-7 (Hardcover), 978-981-4364-62-1 (eBook)
www.panstanford.com

solubility (3). These insoluble drugs could assemble to large aggregates, which may block blood capillaries and cause serious safety problems. Recently, protein, peptide, and gene gradually exhibited their efficacy as potential anticancer drugs. When, being administrated alone, these drugs are easily degraded by catabolic enzymes *in vivo*, leading to a shorter half-life time and a lower bioavailability (4).

To overcome aforementioned difficulties, pharmaceutical scientists developed various drug delivery systems (DDSs). A DDS is founded on a drug carrier. Typically, an anticancer drug is encapsulated into or conjugated to a carrier and then delivered to the tumor site. After being particularly functionalized and designed, the carrier can actively target cancer cells via specific cell surface interactions or exhibit controlled release profiles stimulated by the specific microenvironment of cancer tissue such as low pH, redox, and/or hypoxia. These smart abilities ensure therapeutic drug concentration at the tumor site with reduced adverse side effects. Furthermore, the drug carrier can effectively improve the solubility of insoluble drugs by enclosing them with a hydrophilic shell, making it possible to apply a great many of active anticancer ingredients. In addition, the drug carrier could protect its cargoes, especially protein and gene drugs, from degradation, thus prolong the half-life *in vivo* and increase their bioavailability. Based on these advantages, DDSs opened up new prospects for cancer treatment.

Microspheres and microcapsules, consisting of polymeric particles, liposomes, polymeric micelles, microemulsions and so on, are the vital member of DDSs. All of them have been extensively applied and studied in anticancer drug delivery. In this chapter, we will describe DDSs emphatically based on two aspects: anticancer drug carriers and active targeting schemes.

11.2 The Widely Used Anticancer Drug Carriers

To achieve the therapeutic drug concentration at the tumor site with reduced side effects and to raise the drug bioavailability, a lot of drug carriers have been thoroughly studied and developed. The size of these carriers ranged from about ten nanometers to tens of micrometers. Notably, nanoparticles could target the tumor site by enhanced permeability and retention (EPR) effect (5) as well as boost the cellular uptake, resulting in the increased intracellular concentration of drugs. Therefore, the nanoparticles, with booming

... very promising in anticancer drug delivery, and they are also ...

The anticancer drugs investigated involve almost all the available types, including poorly water-soluble drugs, small organic molecules, macromolecules such as proteins and nucleic acids, and inorganic particles for cancer imaging and diagnosis. No matter which drug is to be delivered, the carriers are not limited to one or a few. The widely used anticancer drug carriers consist of polymeric particles, liposomes, polymeric micelles, implanted systems, dendrimers, emulsions, nanocrystals, and carbon nanotubes (CNTs). Each of them owns particular merits, as well as inevitable demerits. The design strategy, preparation techniques, materials used, and practical examples for each type will be described individually.

11.2.1 Polymeric Particles

Polymeric particles have become very important member of the drug carriers because of their significant advantages, including: (i) an effective protection of the encapsulated active agent against degradation, (ii) the possibility to accurately control the release rate of the incorporated drug over periods of hours to months, and (iii) an easy administration (compared with alternative parenteral controlled release dosage forms, such as macro-sized implants).

Conventionally, polymeric particles could be fabricated by dispersion of the preformed polymers or polymerization of monomers (6). For dispersion of preformed polymers, several methods have been suggested such as emulsification/solvent evaporation, spray drying, ionotropic gelation, coacervation technique, etc. Natural polymers, including albumin, fibrinogen, alginate, chitosan, poly lactic acid (PLA), and collagen, are usually employed. Hence, particles fabricated by these materials could show remarkable biocompatibility and biodegradability. Polymer particles can also be prepared by polymerization of monomers. For clinical usage, it is worthwhile to note that most of the synthetic polymers could not be degraded *in vivo* and could be a cause of potential toxicity.

Two key parameters should be considered in the fabricating procedure: particle size and size distribution. Size determines several important functions of particles (Fig. 11.1), such as circulation time, extravasation, targeting, immunogenicity, internalization, intracellular trafficking, degradation, flow properties, clearance, uptake mechanisms, etc. For example, microparticles can be captured by Kupffer cells in liver or physically trapped in the capillary beds, whereas nanoparticles smaller than 100 nm can preferentially leave the blood through the leaky vasculature of cancer. Particle diameter can often be controlled through physical properties of the materials, such as polymer and surfactant concentration, or through the experimental parameters of the fabrication method, such as the mixing method (vortexing, sonication, stirring), nozzle/capillary diameter, and material flow rate (7).

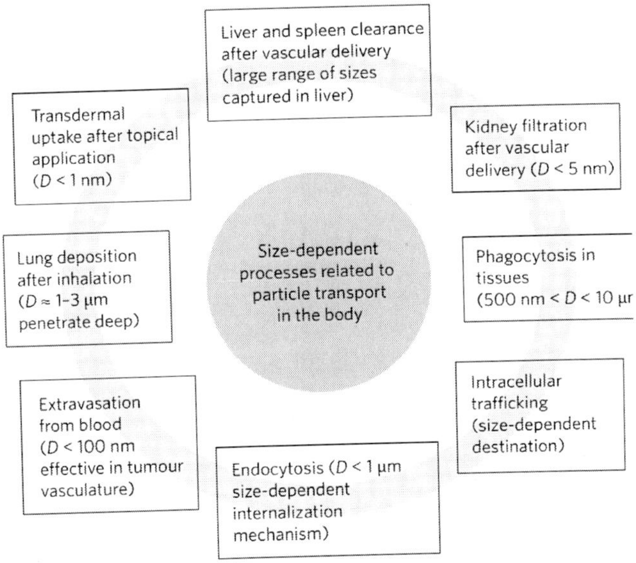

Figure 11.1 Size-dependent processes of particle transport in the human body. Reprinted from Ref. 8, with permission from Nature Publishing Group.

Size distribution is also an important factor that should be taken into account. For traditional fabrication methods, size distribution of particles is very broad, which will bring some limitations for their application: (i) the reproducibility of particles is poor among batches, resulting in poor repeatability of the release behavior. In this case, it is difficult to investigate the relationship between doses and treatment effects; (ii) because the accumulated location of the particles highly depends on the size of the particles, the biodistribution of drug is hardly controlled and side-effects are increased.

Recently, a novel method called membrane emulsification technique has been developed. This method involves using low pressure to force the dispersed phase to permeate through a membrane having a uniform pore-size distribution into the continuous phase. The distinguishing feature is that the resulting droplet size is controlled primarily by the choice of membrane and not by the generation of turbulent droplet break-up. This technique is highly attractive given its simplicity, potentially lower energy demands, need for less surfactant, and the resulting narrow droplet-size distributions (9). So far, a series of uniform-sized micro/nanospheres have been successfully fabricated, such as chitosan micro/nanospheres (9,10), PLA micro/nanospheres (11,12), and albumin nanospheres (13).

Loading drugs into the polymeric particles is achieved by incorporating the drug at the time of particle production or by adsorbing the drug after the formation of particles via incubating them in the drug solution. The former method is more extensively utilized because of its higher loading capacity. For example, biodegradable nanoparticles made of poly(lactic-co-glycolic acid) (PLGA) have been utilized to incorporate paclitaxel (PTX), a microtubule-stabilizing agent that causes cell death by promoting the polymerization of tubuline during cell division. PTX was encapsulated to a very large extent (100% encapsulation efficiency), and this PTX-loaded colloidal system showed a 70% loss of viability of human small-cell lung cancer cells (14). Adsorbing is also a well-studied drug-loading method because it can avoid particle fabrication process, which may reduce the activity of the cargoes. Another characteristic is rapid initial release profile, attributing to the weak binding of the drug on the surface of polymer particles. To achieve optimal release profile, alteration of the structure and surface charge of the particles is effective.

11.2.2 Liposomes

Liposomes, microscopic phospholipid bubbles with a bilayered membrane structure, are derived from Banghman's vesicle, which was discovered in the mid-1960s. Over the past 40 years, liposomes have gained extensive attention as pharmaceutical carriers of great potential (15). It is reported that a number of clinical trials are undergoing worldwide with liposomal formulations of drugs PTX

(Pharmacia corporation), lurtotecan (NX 211, Gilead Sciences), platinum (Aroplatin™, Aronex Pharmaceuticals), vincristine (OncoTCS™, INEX Pharmaceutical Corp.), genes (tgDCC-E1A, Targeted Genetics; AIIovectin™, Vical), antisense molecules, and immunomodulating molecules such as interleukin, interferon gamma, muramyltripeptide (ACT, Jenner Biotherapies), vaccines (Prostatecancer Vaccine™; Novasome®, Novanex Inc.), and proteins (DNA repair enzyme, Dimericine™, AGI Dermatics Inc.) (16).

Liposomes are mostly made up of phospholipids, such as lecithins and cholesterol. Other components can also be applied in the liposome preparation, that is, steroid molecules, charged phospholipids, ganglioside, and polymeric material. In fact, alteration of the composition can modify the biodistribution, the release profile, and the clearance rate of the liposomal DDSs (17), hence particular therapeutic requirement can be achieved. For instance, the inclusion of glucuronate (18) and polyethylene glycol (PEG) (19) in liposomal bilayers has shown to reduce the uptake by reticuloendothelial system (RES) and thus can prolong duration of liposomes in the circulatory system (20). Although, simple bilayer liposome could be prepared by spontaneous self-aggregation of aqueous dispersions of phospholipids, particular procedures must be carried out if a certain type of liposome has to be obtained. For example, thin-layer evaporation (TLE) method allows the formation of multilamellar vesicles, and reverse-phase evaporation vesicles (REVs) method allows the formation of large unilamellar, oligolamellar, and multilamellar vesicles.

Albeit promising, liposome still has its inherent drawback, that is, low drug-loading capacity. In liposomes, the drug contents are generally not greater than 10% to minimize the initial drug release in the blood compartments (premature burst release), while the major component is an inert carrier. A large amount of carriers has to be used to administer a needed dose of a drug, which may cause systemic toxicity such as lipotoxicity and impose an extra burden for the patients to excrete the carriers. This year, Shen *et al.* (21) demonstrated a new concept directly using drug molecules to fabricate nanocarriers in order to minimize use of inert materials, substantially increase the drug loading content, and suppress premature burst release (Fig. 11.2). In the strong hydrophobic anticancer drug camptothecin (CPT), one or two CPT molecule(s) were conjugated to a very short oligomer chain of ethylene glycol (OEG), forming amphiphilic phospholipid-mimicking prodrugs,

OEG-CPT or OEG-diCPT. The resulting prodrugs could form into liposome-like nanocapsules, which achieved higher drug loading (40 or 58 wt%). Using this system, the authors also loaded a water-soluble drug, doxorubicin salts (DOX), resulting in co-delivery and a synergetic cytotoxicity to cancer cells. Using drugs directly as part of a carrier is a novel concept, which facilitates high drug loading.

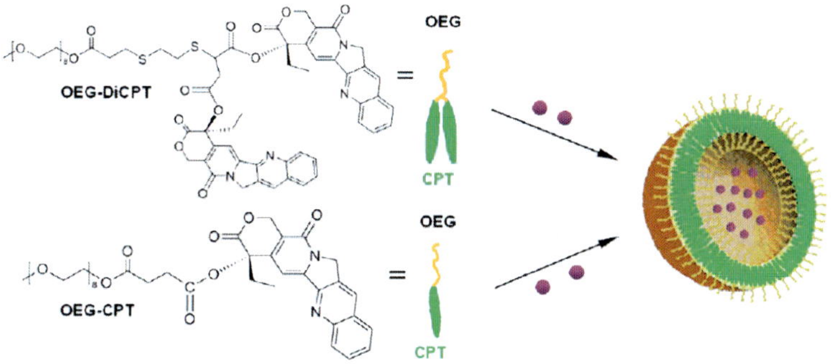

Figure 11.2 Amphiphilic camptothecin (CPT) prodrugs (OEG-CPT and OEG-DiCPT) and their self-assembly into nanoliposomes to load other drugs (●). Hydrophobic membrane is solely made of CPT moieties. Reprinted with permission from Ref. 21. Copyright 2010 American Chemical Society.

11.2.3 Polymeric Micelles

Polymeric micelles, self-assemblies of block copolymers in water above a system-specific critical micelle concentration (CMC), are promising nanocarrier systems for drug delivery. They have a unique core–shell structure, in which an inner core, serving as a nanocontainer of hydrophobic drugs, is surrounded by an outer shell of hydrophilic polymers, such as PEG. Owing to their hydrophilic shell and small size (10–100 nm), the polymeric micelles can exhibit prolonged circulation times *in vivo* and can accumulate in cancer tissues, thus making themselves an appropriate candidate for intravenous (i.v.) administration (22).

Amphiphilicity is the main character of the constituent polymeric monomers, and most studies of polymeric micelles have employed AB or ABA type block copolymers. Critical features of the polymeric micelles as drug carriers, including particle size, stability, loading capacity, and release kinetics of drugs, can also be modulated by the structures and physicochemical properties of the constituent block copolymers (23).

Drugs can be loaded into a polymeric micelle by physical encapsulation (24) or chemical covalent attachment (25). The first polymeric micelle formulation of PTX, Genexol-PM [PEG-poly(DL-lactide)-paclitaxel], is a cremophor-free polymeric micelle-formulated PTX. A phase I and pharmacokinetic study has been conducted in patients with advanced refractory malignancies. Nasongkla and his colleagues (26) have developed a kind of multifunctional polymeric micelle with cancer-targeting capability via $\alpha_v\beta_3$ integrins, controlled drug delivery, and efficient magnetic resonance imaging (MRI) contrast characteristics (Fig. 11.3).

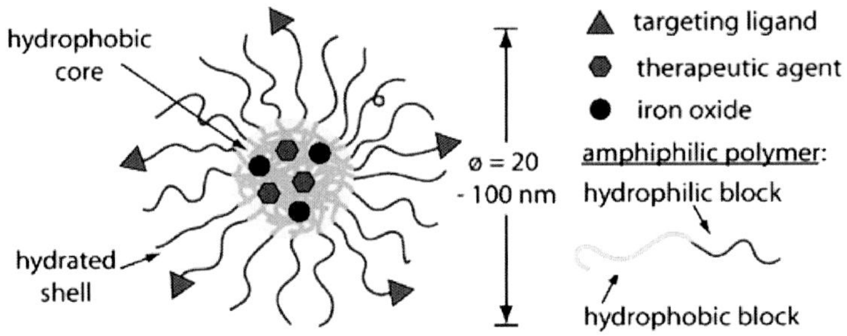

Figure 11.3 Multifunctional nanomedicine platform for targeted drug delivery. Reprinted with permission from Ref. 26. Copyright 2006 American Chemical Society.

Although promising, polymeric micelles still own a defect as a drug carrier, that is, they will disassociate into unimers upon dilution in the body fluids, causing nontargeted drug release and toxicity. To increase stability, polymeric micelles with cross-linked ionic cores (27,28) were designed in recent years. For example, Bronich *et al.* have synthesized a kind of these micelles using a two-step process (27). First, self-assemble of PEO-*b*-PMA copolymers into block ionomer complexes in the presence of divalent metal ions, such as Ca^{2+} (or Ba^{2+}, Sr^{2+}). Then cross-linking of the core of the PEO-*b*-PMA/Ca^{2+} micelles was achieved using a condensation reaction between the carboxylic groups of PMA and the amine groups of 1, 2-ethylenediamine in the presence of a water-soluble carbodiimide.

11.2.4 Implanted Microsystems

Since May in 1958 when the first implantable cardiac pacemaker was placed in an experimental animal, implantable devices have come into our visual field. The development of an implantable pacemaker revolutionized the field of biomedical science and engineering, providing many different implantable biomedical devices to the medical professionals for therapeutic and diagnostic use, among which extensive research has been done on implantable drug delivery devices (29). Implantation of anticancer drugs, in general, can provide prolonged release of drugs from the implants in local area, thus reducing systemic toxicity. Biodegradable polymer implant of PTX (Paclimer™, Guilford Tec.) and Leuprolide (LeuPromaxx™) could provide site-specific and controlled delivery of drugs, and both are at the stage of clinical trials.

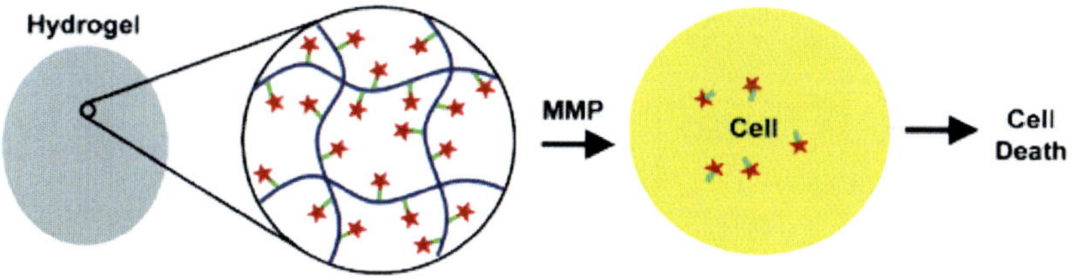

Figure 11.4 Schematic representation of the paradigm proposed for local drug delivery. In this, cisplatin (red stars) is complexed to peptide pendants (light green lines) on the backbone of a hydrogel matrix (dark blue lines). As matrix metalloproteases diffuse through the matrix, the peptides are cleaved, releasing cisplatin–peptide (aqua lines attached to orange stars) complexes. These complexes may dissociate or enter cells as a complex. Cisplatin is then able to cross-link DNA and has a therapeutic effect only when released from the hydrogel. Reprinted with permission from Ref. 30. Copyright 2005 American Chemical Society.

Implant systems are welcome in intratumoral cancer treatment, such as injectable drug depots. Since intratumorally injected liquids may distribute irregularly and be cleared quickly, injectable drug depots have been introduced to prolong the extent of drug release. Examples of intratumoral depots include PLGA (31), alginate (32), and albumin (33) microspheres as well as injectable gels. Gel-type delivery systems are well applied, as they allow the drug to be injected as a liquid, which then solidifies at body temperature, thereby releasing the drug slowly for extended periods. Intracanceral injection of gel formulation of fluorouracil/epinephrine (34) and

PTX (Oncogel™, Macromed) could provide a localized depot which keeps the drug at the tumor site for an extended period. A recent study depicted a novel paradigm for cancer chemotherapy based upon local activation of chemotherapeutic molecules by soluble proteases (Fig. 11.4). The anticancer drug cisplatin was conjugated to a hydrogel through an active peptide, which could be cleaved by matrix metalloproteases (a family of cancer-associated proteases). At lesion site, cisplatin release was doubled owing to the mesh size of the polymer (30).

Another principal application of implanted systems is to address the need for minimally invasive treatment of unresectable cancers. These implants, also termed "polymer millirods," were designed to provide optimal drug release kinetics to improve drug delivery efficiency and anticancer efficacy when treating unresectable cancers (35). In a study, tri-component copolymers of poly(ε-caprolactone)-random-poly(DL-lactide)-block-poly(ethylene glycol)-blockpoly (ε-caprolactone)-random-poly(DL-lactide) (PLCA-PEGPLCA or PLEC) were used as a material to develop cylindrical polymeric rods, which were then applied as the implantable DDS (36). To treat unresectable cancers, another scheme of implanted system is to target the lymphatic system. For example, aclarubicin adsorbed onto activated carbon particles was tested after subcutaneous injection in mice against a murine model (P388 leukemia cells) of lymph node metastases (37). The same system was also used in patients after intracanceral and pericanceral injections as a loco-regional chemotherapy adjuvant for breast cancer (38).

11.2.5 Dendrimers

A dendrimer is a synthetic polymeric macromolecule of nanometer dimensions. A typical dendrimer is built upon an initiator core, with several branching interior layers composed of repeating units and multiple active primary amine groups at its surface (39). Compared with linear polymers, the dendritic architecture can provide several advantages for drug delivery applications. For example, the easily modifiable as well as multivalent surface characteristic of dendrimers enable them to be simultaneously conjugated with several molecules such as imaging contrast agents, targeting ligands, or therapeutic drugs, yielding a multifunctional DDS. In addition, the low polydispersity of dendrimers provides reproducible pharmacokinetic behavior in contrast to that of some linear polymers containing fractions with vastly different molecular weight within a given sample. Furthermore, the more globular shape of dendrimers, as opposed to the random coil structure of most linear polymers, could affect their biological properties, leading to the discovery of interesting effects related to macromolecular architecture (40).

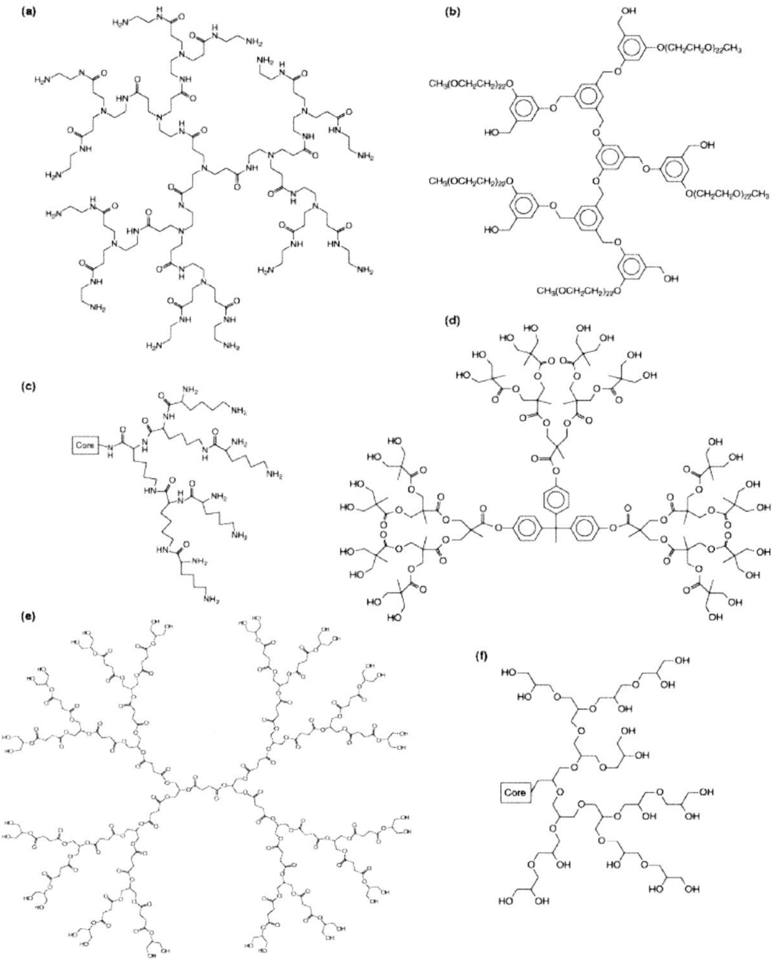

Figure 11.5 Structures of biocompatible dendrimers that have been tested for drug- delivery applications: (a) PAMAM dendrimer, (b) polyaryl ether dendrimer, (c) polylysine dendron, (d) polyester dendrimer based on 2,2-bis(hydroxymethyl)pro pionic acid, (e) polyester dendrimer based on glycerol and succinic acid, and (f) dendritic polyglycerol. Reprinted from Ref. 40, with permission from Elsevier Ltd.

The rapid emergence of dendrimers in biological applications has been accompanied by a growth in the number of dendrimer backbones. Besides the most widely used polyamidoamine (PAMAM) dendrimers (Fig. 11.5), polypropyleneimine, polyaryl ether dendrimers, and peptide-based dendrimers (such as those based on polylysine) have been reported and developed as drug delivery carriers (40). To develop practical synthesis of these highly branched macromolecules, highly efficient and orthogonal reactions have been drawn in, such as "click chemistry" (41). In a recent study, dendrimers were efficiently synthesized by combining kinetic or mechanistic chemo-selectivity with click reactions between the monomers (Fig. 11.6). In the procedure, only separation needed is to remove the little unreacted monomer by simple precipitation or washing. This simple synthesis method could be widely applied to custom-made polyester dendrimers (42).

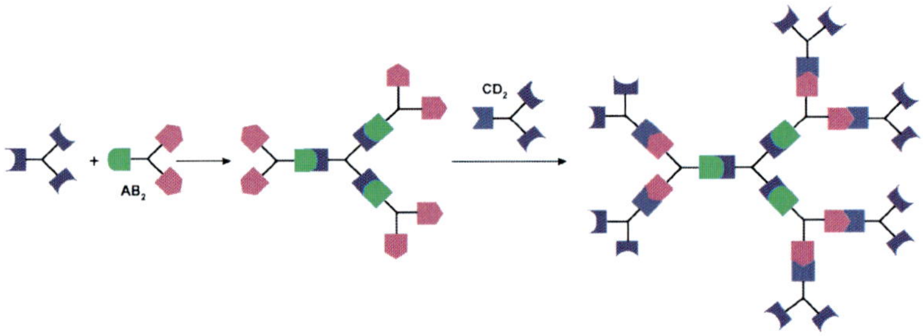

Figure 11.6 Sequential click coupling of asymmetrical monomers for facile polyester dendrimer synthesis. Reprinted with permission from Ref. 42. Copyright 2009 American Chemical Society.

Initial studies of dendrimers as potential delivery systems focused on using them as unimolecular micelles and dendritic boxes for the noncovalent encapsulation of drug molecules. For example, in early studies, DNA was complexed with PAMAM dendrimers for gene delivery applications, and hydrophobic drugs and dye molecules were incorporated into various dendrimer cores (43). An alternative approach to the development of dendrimers as anticancer drug carriers is to exploit their well-defined multivalency for the covalent attachment of drug molecules onto the dendrimer periphery. The drug loading can be tuned by varying the generation number of the dendrimer, and the release of the drug can be controlled by incorporating degradable linkages between the drug and dendrimer. For example, Majoros *et al.* (44) have synthesized PAMAM dendrimer–based multifunctional cancer therapeutic conjugates. The functional molecules fluorescein isothiocyanate (FITC, an imaging agent), folic acid (FA, targets overexpressed folate receptors on specific cancer cells), and PTX (taxol, a chemotherapeutic drug) were conjugated to the modified PAMAM dendrimers, archiving a multifunctional DDS.

11.2.6 Emulsions

The emulsion concept was introduced as early as the 1940s by Hoar and Schulman (45), who generated a clear, single-phase solution by titrating a milky emulsion with hexanol. Today emulsions are defined as "a mixture of water, oil, and amphiphile substances forming a single, optically isotropic, and thermodynamically stable liquid solution". Emulsions present some important characteristics that make them very versatile carriers. For example, the existence of microdomains with different polarity allows the solubilization of both water soluble and insoluble drugs at the same time if this is so desired. Furthermore it is also possible to incorporate amphiphilic drugs into the emulsions (29). Certainly, using emulsions for oral or parenteral drug delivery is complicated by the fact that they are destabilized when diluted by biological aqueous fluids, and the cytotoxicity of the surfactant should also be considered.

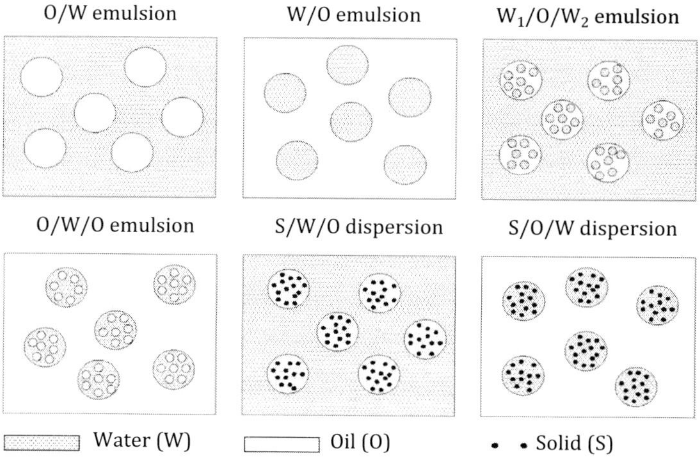

Figure 11.7 Schematic structure of different types of emulsion systems.

To date, various types of emulsions (Fig. 11.7) have been developed, and suitable structure could be selected for specific system and appointed application. Emulsified perfluboron-based contrast agent for imaging to detect solid cancers, known as Imavist™, has been approved by the FDA in United States. Clinical trials are undergoing with emulsion formulation of taxol (S-8184®, Sonus Pharmaceutical). Lipid-coated microbubbles, named Filmix®, could be employed effectively for cancer diagnosis by enhanced ultrasound image and MRI contrast, in cavitation therapy of cancers, and especially for targeted delivery of anticancer drugs to cancers (46). In a review (47), the stabilization of W/O/W emulsion and its application to transcatheter arterial embolization (TAE) therapy were extensively illustrated. It was found that W/O/W emulsion was stabilized by making inner aqueous phase hypertonic, adding chitosan in inner phase, and using techniques of phase-inversion with porous membrane. Using a two-step pumping emulsification procedure, lipiodol W/O/W emulsion could be prepared, and this emulsion was proved as a suitable formulation for the TAE therapy. Recently, nanoemulsion has been widely discussed for the delivery of insoluble drugs such as PTX (48,49) and tributyrin (50).

11.2.7 Nanocrystals

With the development of nanotechnology, various nanocrystals have been successfully fabricated such as quantum dots (QDs) and iron oxide (IO) nanocrystals. These nanocrystals are also applied in cancer therapy, mainly in cancer imaging and diagnosis. Cancer imaging plays a key role in clinical oncology, as it helps to detect solid cancers, determine recurrence, and monitor therapeutic responses. Despite continuous improvements in spatial resolution, imaging modalities using non-targeted contrast agents have limited sensitivity and ability to provide specific and functional information on the disease, which is increasingly being recognized as an obstacle to earlier diagnosis and monitoring treatment responses. Recent advances have stimulated the emergence of the new field of "molecular imaging," which focuses on visualizing or imaging biological events and processes in living systems (51,52). Consequently, selecting and designing an appropriate molecular imaging agent is in high demand. Besides the precisely designed polymer nanoparticles, QD nanocrystals and magnetic IO nanocrystals are also well applied in cancer imaging and diagnosis (53).

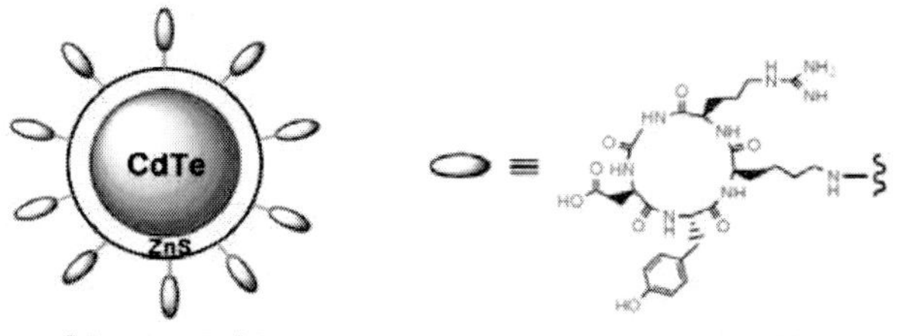

Figure 11.8 The structure diagram of QD705-RGD. Reprinted with permission from Ref. 54. Copyright 2006 American Chemical Society.

QDs are nanometer-scale, light-emitting particles with unique optical and electronic properties such as size-tunable light emission, improved signal brightness, enhanced stability of the fluorescent signal, and the ability to simultaneously excite multiple fluorescent colors (55). These properties are most promising for improving the sensitivity of molecular imaging and quantitative cellular analysis by 1 to 2 orders of magnitude. Nie *et al.* first reported that it is feasible to simultaneously target and image prostatic cancers in living animal models using bioconjugated prostate membrane antigen-targeted QDs (56). Recently, QDs, which can produce near-infrared fluorescence (NIRF) signals, have been developed (54,57). NIRF light can penetrate much more deeply into tissues compared with visible fluorescence and allow for the detection of signals inside animals. Another major advantage of NIRF QDs is that their emission is well beyond the spectral range of the fluorescence signal produced by blood and tissues (autofluorescence), resulting in imaging with a high signal-to-background ratio. A study (54) reported the *in vivo* targeting and imaging of cancer vasculature using arginine–glycine–aspartic acid (RGD) peptide-labeled QDs (Fig. 11.8). Athymic nude mice bearing subcutaneous U87MG human glioblastoma cancers were administered with QD705-RGD intravenously. The cancer fluorescence intensity reached a maximum at 6 h post injection with good contrast.

IO or superparamagnetic iron oxide (SPIO) nanocrystals are becoming increasingly attractive as the precursor for the development of a target-specific MRI contrast agent. IO nanoparticles have unique paramagnetic properties, which generate significant susceptibility effects resulting in strong T2 and T*2 contrast, as well as T1 effects at very low concentrations (58). Recently, significant efforts have been made to develop target-specific MRI contrast agents and ultra-sensitive magnetic nanoprobes. For instance, using magnetism-engineered IO nanoprobes that are conjugated with HER-2 antibodies, Lee *et al.* showed an enhanced sensitivity of MRI for the detection of HER-2 expressing cancer in an animal model compared with that of commonly used SPIO probes (59). This new generation of magnetic nanoparticles (MNs) provided us a powerful contrast agent for cancer detection.

In addition to cancer imaging and diagnosis, nanocrystals could also be used as anticancer drug delivery carrier. Wei *et al.* (60) have proposed a facile method to prepare hierarchical hollow $CaCO_3$ particles, formed by the self-organization of nanocrystallites. This particle was used to carry anticancer drug DOX and exhibited its advantages for localizing drug release by the pH value-sensitive structure and enhancing cytotoxicity by increasing cellular uptake. Besides this, drug nanocrystal has also been reported. Insoluble anticancer drug 10-hydroxycamptothecin (HCPT) was fabricated into nanocrystallites ($\approx$20 nm) with the assistance of galactosylated chitosan (GC) (61). These nanocrystallites owned better water stability and could be directly used for liver cancer therapy (details can be seen in the section of 11.3.1).

11.2.8 Carbon Nanotubes

CNTs also emerged as a highly effective carrier for transporting anticancer drugs across the cell membrane. CNTs are carbon cylinders, composed of benzene rings, and can be classified into single-walled carbon nanotubes (SWNTs) and multi-walled carbon nanotubes (MWNTs) (62). CNTs displayed many peculiar properties such as quasi-one-dimensional nanostructure, unique optical and electronic properties, ultrahigh surface area, and remarkable cell membrane penetrability (63).

However, CNTs are completely insoluble in all solvents, generating some health concerns and toxicity problems. To overcome this problem, CNTs are usually modified with noncovalent supramolecular modifications or covalent functionalization. Modified soluble CNTs display low toxicity and no immunogenicity, hence they hold great potential in the anticancer field. Anticancer drugs could be loaded through noncovalent means (including hydrophobic or stacking interactions) or conjugated to functionalized CNTs. In particular, PEG functionalized SWNTs allow for π-stacking of aromatic cancer drug (DOX, Fig. 11.9) with surprisingly high degree, and the loading capacity could reach to about 60 wt% (64). In another study, insoluble anticancer drug HCPT was covalently combined to a MWNT using hydrophilic diaminotriethylene glycol as the spacer between nanotube and drug moieties. The obtained drug complex achieved relatively long blood-circulation time and higher drug accumulation at lesion site and finally showed an efficient anticancer effect (65).

CNTs have also showed their feasibility in cancer imaging, as the near-infrared (NIR) photoluminescence of semiconducting SWNTs made them promising as NIR fluorescent contrast agents in biological systems. SWNT NIR fluorescent probes with emission mostly in the infrared-A (IR-A, 1–1.4 μm) region are ideal as biological probes because of the inherently low autofluorescence of biological tissues in the NIR range (0.8–1.7 μm) and large Stokes shift between the excitation and emission bands, which allows excitation in the biological transparency window near 800 nm while further reducing the background effects of autofluorescence and scattering. Welsher *et al.* have fabricated biocompatible SWNTs with high quantum yield by surfactant exchanging sodium cholate to form phospholipid–PEG coated nanotubes (66). The exchange procedure was better than directly sonicating the tubes with the phospholipid–PEG because it resulted in less damage to the nanotubes and improved the quantum yield. The whole-animal imaging showed high-resolution intravital microscopy imaging of cancer vessels beneath thick skin.

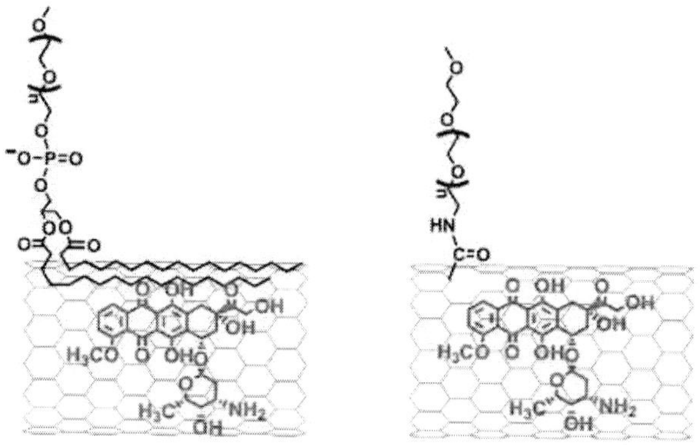

Figure 11.9 Schematic drawings of doxorubicin stacking onto a nanotube prefunctionalized noncovalently by phospholipid (PL)-PEG (left) and covalently by PEGylation of a sidewall -COOH group (right), respectively. Reprinted with permission from Ref. 64. Copyright 2007 American Chemical Society.

11.3 Smart Targeting Strategies in Cancer Therapy

The effectiveness of cancer therapy is directly related to the capacity to target and kill tumor cells while affecting as few healthy cells as possible. In a sense, targeting ability of the treatment determines the degree of change in the patient's quality of life and eventual life expectancy. Recently, cancer targeting approaches have been extensively explored for improved efficacy and reduced toxicity. Among them, active cancer-specific targeting strategies are mostly based on the various discrepancies in molecule expression between the cancer cell and normal cells. Besides, smart targeting device that can respond to pH, and thermal or magnetic stimuli are also being designed for cancer therapy, as the microenvironment around the tumor sites is quite different from that of the normal tissues.

11.3.1 Receptor/Ligand-Mediated Targeting

A typical characteristic for solid tumors is overexpression of specific antigens or receptors on cell surfaces, which help in transmitting signals from the surrounding environment that are essential for cancer growth. In addition, overproduced receptors may cause uncontrolled proliferation, insensitivity to negative growth regulation, angiogenesis, tissue invasion, and evasion of apoptosis. Therefore, drug carriers attached with relevant surface ligands or antibodies is becoming an important delivery mode to selectively and successfully transport drug systems to cancerous tissues (67).

(i) Targeting tumor types or specific organs

With the development of genomics and proteomics, a number of cancer treatments using antibodies for specific cancer types have been approved by the FDA, which help to provide ideal targeting agent for specific tumors. Currently available targeted cancer treatments using antibodies include Anti-CD33 antibody (for relapsed/refractory acute myelogenous leukemia), anti-CD52 antibody (for B-cell chronic lymphocytic leukemia), and so on. Other surface molecules for specific tumor cells are also being explored. In order to shed light on the targeting information of breast cancer, liver cancer, lung cancer, and prostate cancer, we list corresponding instances as follows.

For human breast cancer, a growth factor HER-2 is overexpressed in breast cancer cells or other adenocarcinoma cells. Accordingly, antibodies that bind to HER-2 are mostly exploited for targeting these cells. Chen *et al.* have recently studied the therapeutic efficacy of targeted PLGA nanoparticles loaded with a model toxin against HER-2-positive tumors (68). Carboxylic acid groups on the surface of the nanoparticles were activated and conjugated with primary amine groups on the fragments of humanized anti-HER-2 antibody. Using these particles, a target-specific internalization was demonstrated in HER-positive BT-474 cells compared to control MCF-7 cells. Presentation of the toxin within targeted nanoparticles resulted in significant decrease in the IC50 compared to a toxin-HER-2 conjugate, and more than a 20-fold decrease compared to free toxin or non-targeted particles. Furthermore, *in vivo* intravenous administration of toxin-loaded targeted nanoparticles to mice that previously inoculated with BT-474 cells resulted in significantly decreased tumor growth and systemic toxicity.

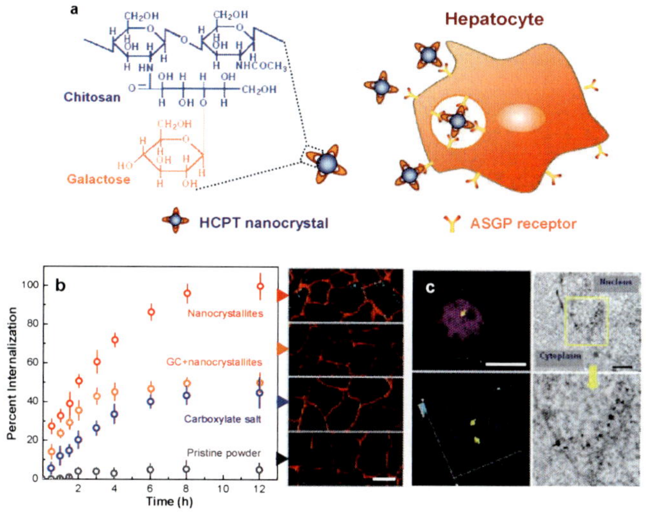

Figure 11.10 (a) Proposed delivery strategy of HCPT nanocrystallites. (b) Internalization profiles of HepG2 cells when treated with different HCPT formulations (10 μg/mL). (c) CLSM images and TEM scans of HepG2 cells treated with HCPT nanocrystallites, which exhibited the advantages of nuclear entry. Scale bars: left panel, 10 μm; right panel, 100 nm. Reprinted from Ref. 61, with permission from Future Medicine.

For liver cancer targeting, a promising receptor is the asialoglycoprotein receptor (ASGP-R, galactose receptor) which can recognize galactose-terminated glycoproteins. As hepatoma cells are known to express large numbers of ASGP-R on the surfaces, the macromolecules with galactose moiety and drug conjugates have been employed as liver-specific drug carriers. In our research group, Wei *et al.* (61) developed a facile method to prepare uniform-sized HCPT nanocrystallites under the soft template effect of GC (Fig. 11.10a). It was revealed that the GC located on the HCPT nanocrystallites not only stabilized the formulation in aqueous medium, but also enhanced the cellular internalization through receptors mediated pathway (Fig. 11.10b). These nanocrystallites formulation also exhibited advantages of nuclear entry and active HCPT delivery (Fig. 11.10c), and consequently, better tumor cytotoxicity was achieved. This work proved the feasibility of GC nanocrystals for liver targeting and represented a very valuable contribution to the field of using nanomedicine for cancer therapy.

With the growing knowledge of molecular deviations involved in the development and progression of lung cancer, targeting approaches are emerging for systemic lung cancer therapies. For example, lung cancer cells often overexpress the sigma receptor and, thus, can be targeted with a specific ligand such as anisamide. Inspired by this mechanism, Li and Huang developed nanocomplexes of antisense RNA or siRNA together with carrier DNA and the cationic polypeptide protamine (69). *In vitro* studies demonstrated that the target ligand anisamide delivery could mediate therapeutic action in H1299 lung cancer cells. In the presence of this nanocomplex–liposome system, the targeted gene at both RNA and protein levels were significantly downregulated. In comparison to the drug cisplatin, this system induced great cell cytotoxicity and susceptibility. Using xenograft models, it was revealed that the targeted complexes preferentially accumulated at the tumor site, exhibiting good potential of this delivery system for cancer therapy. Apart from the sigma receptor, a transmembrane tyrosine kinase receptor c-Met and glycoprotein receptor Erb-B1 are also attracting many interests. As c-Met and Erb-B1 are overexpressed on the surface of lung cancer, using drug carriers with specific antibodies against these receptors is also a promising strategy for lung cancer targeting.

For prostate cancer treatment, prostate-specific membrane antigen (PSMA) is one of the most studied targets. This antigen is a class II transmembrane glycoprotein, which is known to be overexpressed by cancerous prostate tissue. Specifically, Farokhzad and Langer have investigated the targeting potential of aptamers specific to PSMA (70). The nanoparticle systems consisted of a core of biodegradable PLGA polymers surrounded by a shell of PEG chains. The known "stealth" agent PEG not only protects the particles as they travel through the bloodstream and facilitates their escape from vasculature, but also permits covalent conjugation between aptamers and particles. By nanoprecipitation, the drug docetaxel was encapsulated within the nanoparticle core. Promising therapeutic effects for the aptamer-conjugated nanoparticles was demonstrated in xenografts models of PSMA-positive LNCaP prostate cancer cells. The tumor growth was inhibited evidently by using this system, compared to free-drug or drug-loaded nanoparticles without aptamer conjugation.

(ii) Targeting through angiogenesis

Angiogenesis is a process vital to the continued tumor development, which has been intensely researched and has proven to be the result of numerous interactions between regulators, mediators, and stimulatory molecules. Some of the most prominent angiogenesis stimulatory molecules include platelet-derived growth factor (PDGF), vascular endothelial growth factor (VEGF), basic fibroblast growth factor (bFGF), and certain matrix metalloproteinases (MMPs). In contrast to the endogenous angiogenesis stimulators, there are numerous endogenous angiogenesis inhibitors that can block interactions between the tumor cells and the corresponding regulatory molecules. Some molecules and their functional mechanisms are listed in Table 11.1. As so many molecules are involved in angiogenesis, anti-angiogenesis approaches are feasible in limiting tumor growth. Such therapeutic strategies are directed toward inhibited endothelial proliferation or increased expression of angiogenesis inhibitors.

Table 11.1 Endogenous angiogenesis stimulators and inhibitors of stimulators

Endogenous angiogenesis stimulators	Endogenous inhibitors of angiogenesis	Functional mechanisms of inhibitors
VEGF (mitogen, motogen, and survival factor for endothelial cells)	Soluble flt-1	Soluble form of VEGF receptor capable of binding VEGF in the circulation and reducing net function

(*Continued*)

Table 11.1 (*Continued*)

Endogenous angiogenesis stimulators	Endogenous inhibitors of angiogenesis	Functional mechanisms of inhibitors
Angiopoietin-1 [binds VEGF receptor (Tie-2) and stimulates vascular maturation]	Angiopoietin-2	Is involved in maturation of neovessels, blocks effect of angiopoietin-1
PDGF (stimulates proliferation and motility of endothelial cells)	—	—
FGF (one of the most potent mitogens and motogens for endothelial and tumor cells)	Interferon-γ	Blocks induction of angiogenesis by FGF, induces production of IP-10
MMP (secreted by endothelial and tumor cells for wound healing, tumor invasion, and tissue remodeling)	Tissue inhibitor of matrix metalloproteinase	Blocks activity of some MMPs, decreases induction of endothelial cell migration and invasion
Pro-angiogenic molecules (IL-6, tumor necrosis factor-α)	Interferon-α/β	Blocks production of pro-angiogenic molecules
—	Angiostatin	Plasminogen fragment containing kringle domains

Vascular endothelial growth factor receptor (VEGFR) is considered the most relevant inducer of tumor angiogenesis for its induction of the VEGFR signaling cascade. For example, the tumor hypoxia and oncogenes upregulate VEGF levels in the neoplastic cells, resulting in an upregulation of VEGFR-1 (fms-like tyrosine kinase) and VEGFR-2 (fetal liver kinase-1, flk-1) on tumor endothelial cells. As VEGFR-2 is associated with VEGF binding and is highly expressed on endothelial cells in tumor neovasculature, this receptor is of great interest for most active-targeting systems. It was revealed that targeting VEGFR-2 could allow for successful reduction of vessel density in tumors. Using radioisotype-labeled polymerized liposomes, an anti-VEGFR-2 mAb-labeled particle system (anti-VEGFR-2 mAb-NP-90Y) (71) was constructed. This system was more effective in tumor growth delay than anti-VEGF-2 mAb alone or conventional radio-immunotherapy in the mouse melanoma model. Moreover, VEGFR mAb was better at limiting tumor growth compared to an isotype-matched control antibody.

$\alpha_v\beta_3$ integrin is another important targeting molecules, which is highly expressed on neovascular endothelial cells and is important in the calcium-dependent signaling pathway leading to cell migration. The targeting scheme for this integrin has focused on the three–

amino acid sequence RGD, which have shown high efficiencies in targeting specific tumor endothelial cells. After conjugation to the surface of DOX-loaded micelles at different densities, a 30-fold internalization enhancement of micelles was achieved after RGD-functionalization compared with the non-targeted micelles. In addition, an important characteristic of the $\alpha_v\beta_3$ integrin is that it is intrinsically associated with VEGFR-2 signaling. Upon $\alpha_v\beta_3$ integrin binding to the components that harbor the RGD sequence, there is an upregulation of VEGF signaling in cell cultures (72). Thus, targeting the $\alpha_v\beta_3$ integrin with an active targeting system will increase the effectiveness of anti-angiogenic treatments by downregulating the VEGF expression. Similar to $\alpha_v\beta_3$ integrins, MMP receptors and vascular cell adhesion molecule-1 (VCAM-1) are optimal targets that interact with the extracellular matrix (ECM) as well, which are yet readily inducible by angiogenesis and thus can be targeted for cancer therapy.

(iii) Cell proliferation targeting

Cell proliferation markers are a significant target for cancer therapeutics, as many of these markers are highly overexpressed on certain tumor cells. The most established cell proliferation targets used by actively targeting carriers include human endothelial receptors, transferring receptors, and folate receptors. As mentioned above, targeting highly expressed ligands and their receptors is a promising strategy that can ensure the elimination of malignant tumors and metastatic cells that have not become large enough to induce angiogenesis.

The human epidermal receptor (HER) family of receptor tyrosine kinases offers two highly upregulated targets on tumor cell surfaces. Epidermal growth factor receptor (EGFR) and human epidermal receptor-2 (HER-2) are known to mediate a cell signaling pathway for growth and proliferation in response to the binding of the growth factor ligand (73). In this sense, they are the two heavily researched receptors researched in cancer therapy. Particularly, a FDA approved mAb Trastuzumab (Herceptin®), which is designed to antagonize the HER-2 function, is used as a targeting moiety for various nanoparticle systems (74).

Transferrin (Tf) is a serum non-heme iron-binding glycoprotein that helps transport iron to proliferating cells. Upon binding to the Tf receptors distributed on cell surface, the Tf could be endocytosed into acidic compartments where the iron dissociates. Due to the increased requirement of iron in malignant cells, this receptor is often overexpressed. Tf receptor targeting for cancer therapeutics has been used in human clinical trials with several chemotherapy drugs like adriamycin, cisplatin, and diphtheria toxin (75). Davis *et al.* (76) have performed a novel study of small interfering RNA (siRNA) delivery in non-human primates using Tf-conjugated liposomes. Figure 11.11 schematically illustrates the delivery system and its formulation with siRNA to prepare targeted particles. The efficacy of these Tf-conjugated liposomes was demonstrated in metastatic mouse models of Ewing's sarcoma. Moreover, data showed that systemic doses of targeted nanoparticles containing nonchemically modified siRNA could safely be administered to non-human primates. Nevertheless, a firm conclusion could not be drawn regarding the possible influence of anti-Tf Ab formation on nanoparticle kinetics and further experiments that need to be performed by using more znimals to address this issue.

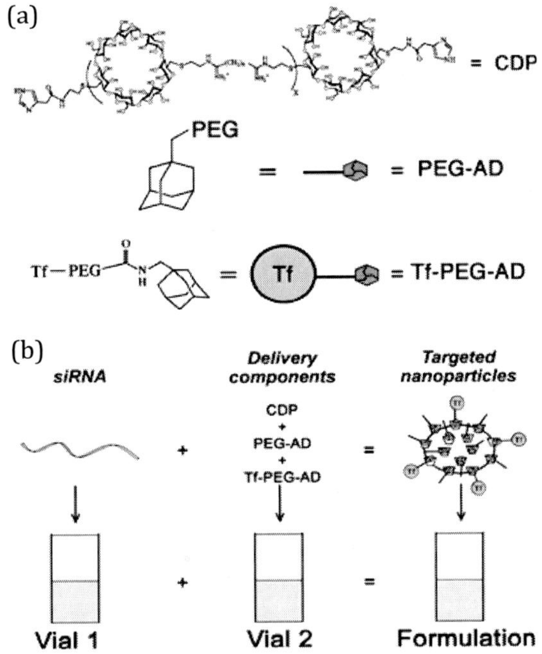

Figure 11.11 Schematic illustration of (a) the delivery system and (b) its formulation with siRNA Reprinted with permission from Ref. 76. Copyright (2007) National Academy of Sciences, U.S.A.

Apart from the above receptors, the folate receptors which are expressed on the surface of most cancer cells is another possible target for cancer treatment. Using FA-coated polymeric nanoparticles, enhanced localization and internalization of nanoparticles within the breast cancer cells was found, where, in a similar example, FA coating induced higher internalization of magnetite nanoparticles intended for tumor imaging. Some researchers attempted to treat folate receptor–positive tumors by developing a vaccine against the folate receptor. This approach in treating folate receptor–positive lung metastases in mice has produced a good therapy effect in tumor-bearing mice (77).

11.3.2 Thermal Targeting

Thermal targeting DDSs, usually used in conjunction with hyperthermia for tumor targeting, are probably the most commonly studied class of environmentally sensitive polymer systems in drug delivery research. They are designed to be systemically soluble when injected *in vivo* but become insoluble and accumulated in a locally heated tumor.

Thermosensitive polymers have gained increasing attention in targeted drug delivery, especially in delivering drugs to tumor sites (78). These polymers show phase transition at lower critical solution temperature (LCST), which can be utilized to deliver therapeutic agents *in vivo*. At LCST, the interaction forces (hydrogen bonding) between water molecules and the polymer becomes unfavorable compared to polymer–polymer and water–water interaction; consequently phase separation occurs as the polymer dehydrates (79).

Examples of thermosensitive polymers are poly (*N*-isopropylacrylamide) (PNIPAAm), PEO-poly(propylene oxide)-PEO triblock copolymers (PEO-PPO-PEO), PEG-poly(lactic acid)-PEG triblocks (PEG-PLA-PEG). These polymers are not only used for the design of thermal targeting DDSs but also to modify the surface properties of other DDSs, including liposomes, polymeric particles, protein nanoparticles, and micelles. Typically, PNIPAAm with a phase transition at LCST (~33°C) has been investigated extensively. This LCST can be adjusted to the desired temperature by copolymerizing the NIPAAm with hydrophilic acryl amide monomers. For example, a LCST of approximately 45°C can be achieved by using 16.6 mol%

acrylamide (AAm) to increase the hydrophilicity of PNIPAAm (80). Shen *et al.* (81) prepared nanospheres with the copolymer chains with the coil to globule transition around 40°C on the surface. The thermosensitive behaviors of copolymers with the incorporation of different comonomers [2-hydroxyethyl methacrylate (HEMA), AAm, *N,N*-dimethylacrylamide (DMAA)] have also been studied. The temperature sensitivity of the conformation change at the transition temperature was found fairly well for PNIPAM-AAm and PNIPAM-DMAA. It was expected that the two kinds of copolymers could be used to encapsulate the anticancer drug for cancer therapy.

Based on the studies that are attempted or proposed to employ the thermosensitive polymers in cancer targeting, several pilot reports are emerging to prepare appropriate thermal targeting delivery systems and evaluate their effect *in vitro* or *in vivo*. For the goal of cancer therapy, the polymers must be designed to be soluble when injected *in vivo* but insoluble to accumulate on a locally heated tumor tissue. This requirement can be accomplished by using polymers whose LCST in water are less than the temperature of a tumor tissue around 42°C under hyperthermia and higher than the physiological body temperature around 37°C. Through the introduction of thermosensitive polymers on nanoparticles, there should be none or a little drug release at the body temperature at normal site and large cumulative release at tumor site.

Li *et al.* have prepared thermosensitive nanoparticles self-assembled from poly(*N,N*-diethylacrylamide-co-acrylamide)-block-poly(γ-benzyl-L-glutamate) for targeted drug delivery in localized hyperthermia (82). The LCST of nanoparticles was adjusted to a level between physiological body temperature and that used in local hyperthermia. He encapsulated PTX in these nanoparticles. The cytotoxicity of PTX-loaded particles increased obviously when hyperthermia was performed, showing an ideal candidate for thermal-triggered antitumor PTX delivery system.

Apart from the above thermosensitive polymer particles, thermosensitive liposomes are also attractive vehicles for the delivery and release of drugs to tumors. These thermosensitive liposomes could release encapsulated molecules near the melting temperature (T_m) of their lipid membrane as it transforms from a gel to a liquid crystalline phase. For instance, Puri *et al.* (83) prepared a thermosensitive small unilamellar vesicle (SUV) liposome consisting of dipalmitoyl phosphatidylcholine (DPPC, $T+$ = 41°C) as the matrix lipid and a maleimide-conjugated PEGylated phospholipid (DSPE-MaL-PEG2000).

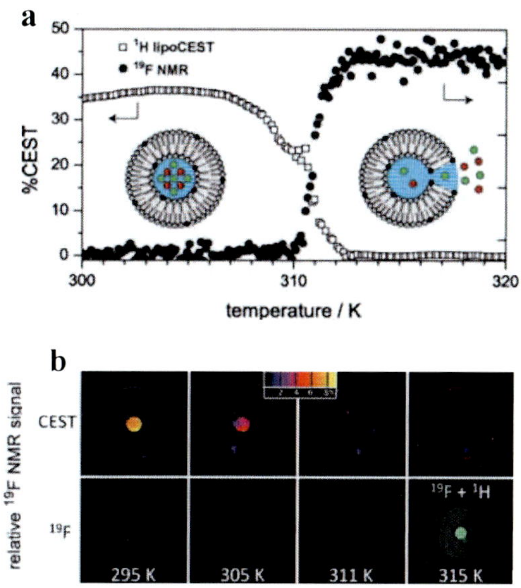

Figure 11.12 (a) The ^{1}H CEST effect ($\mathbf{B}_1$ = 4.5 μT) and the ^{19}F NMR signal intensity of liposomes containing [T_m(hpdo3a)(H_2O)] and NH_4PF_6 as a function of temperature at $\mathbf{B}_0$ = 7 T (lower panel). (b) ^{1}H lipoCEST and ^{19}F MR images of temperature-sensitive liposomes on a clinical 3.0 T MRI scanner. The CEST signal (color scale in percent) vanished at $T \geq 311$ K while the fluorine signal appeared at 315 K (overlay with the ^{1}H image for colocalization and clarity). Reprinted with permission from Ref. 84. Copyright 2009 American Chemical Society.

In another example, Langereis *et al.* (84) employed novel temperature-sensitive liposomal contrast agents. The liposomes contained both a chemical shift agent (for ^{1}H lipoCEST detection) as well as a highly fluorinated compound (for ^{19}F detection). Upon reaching T_m of the agents' lipid membrane, the lipoCEST contrast enhancement vanishes due to the release of the chemical shift agent (Fig. 11.12a). Simultaneously, the ^{19}F MRI probe was freed due to the influence of paramagnetic shift agent, causing the appearance of a ^{19}F MRI signal (Fig. 11.12b). The combined liposomal carrier provided CEST-based contrast enhancement to probe the liposomes before release, while the ^{19}F MR signal could potentially be used to quantify the local release of drugs with MRI (Fig. 11.12). Thus, in response to a local temperature increase, this system allowed drug carrier localization with simultaneous observation and quantification of drug release. Albeit considerable accomplishments are achieved, *in vitro/vivo* evaluation remains to be done in the future work.

11.3.3 pH Targeting

The extracellular pH (pHe) of normal tissues and blood pH are kept constant at pH 7.4. However, pHe in most tumors sites is more acidic (pH 5.8–7.2) than normal tissues. The primary reason for this imbalance in cancer pH is the high rate of glycolysis in cancer cells. As the acidic tumor microclimate is most common in solid tumors (>80%), pH targeting approach is regarded as a more general strategy than conventional targeting approaches.

Recently, the new mode of pH targeting was accelerated by the development of polymers in antitumor drug delivery. In particular, targeting tumor extracellular acidity or targeting the pH in endosomes (85) are the emerging criteria for using pH-sensitive polymers. An important aspect for drug delivery is the possibility to achieve a triggered release of drug entities following certain stimuli. There are two major approaches that have been applied to trigger drug release following a pH change. One is the acidic pH-induced structural change or destabilization undergone by self-assembled particles that formed from polymeric blocks containing ionizable groups. The other is the drug release triggered by the cleavage of pH-labile chemical bonds that link the drug and the constituent polymers.

The use of pH-sensitive systems that exhibit acid-mediated structural changes has been suggested to target tumor pHe. Na *et al.* (86,87) utilized oligomeric sulfadimethoxine (OSDM) and modified hydrophobic pullulan acetate (PA) to prepare pH-sensitive hydrogels. OSDM on the particle surface was fully extended in the water phase at pH 7.4, forming a broad outer hydrophilic shell. As pH decreased from 7.2 to 6.8, the particles shrank and aggregated due to the rapid interior restructuring in this pH range. When DOX was encapsulated in these particles, the amounts of DOX released from the particles at pH 6.8 and 6.2 in 24 h were significantly higher than the amounts released at pH 8.0 and 7.4.

In addition to the structural changes, the structural destabilization of polymers at acidic tumor pHe has also been investigated for cancer therapy. Lee and Bae from the Yonsei University have done many researches on this subject (88). Their approaches to target various solid tumors by pHe include micelle systems with a triggered drug release mechanism. Meanwhile, polyhistidine (His, pKb ≈ 7.0) was selected as an effective pH-buffering agent as its imidazole ring has lone pairs of electrons on the unsaturated nitrogen that endows it pH-dependent amphoteric properties (89,90). They prepared a pH-sensitive micelle (PHSM) system with folate to target the extracellular tumor acidity. This system was created from poly(L-His)-b-PEG and poly(L-lactic acid) (PLLA)-b-PEG-folate, which showed a gradual destabilization below pH 7.0 (91). The drug release using this micelle showed a favorable pH dependency (32 wt% of DOX was released at pH 7.0, 70 wt% of DOX at pH 6.8, and 82 wt% at pH 5.0). Furthermore, the DOX loaded micelle exhibited significant inhibition on the growth of subcutaneous injection (s.c.) (hypodermic injection) MCF-7 xenografts (tumor that developed from implanted MCF-7 cells). After 6 weeks, the tumor volume of mice that were treated with this system was evidently reduced (≈4 times smaller) than those treated with saline solution or free DOX.

Besides the targeting strategy on tumor pHe, acidity in subcellular organelles, especially in endosomes (pH 6.6 to 6.0) and lysosomes (≈pH 5.0), can be utilized for effective cytosolic antitumor DDSs as well. For example, a smart drug-loaded carrier that destabilized at an early endosomal pH of 6.0 was designed (92). This system consisted of poly[His-cophenylalanine (Phe)]-b-PEG and PLLA-b-PEG-folate. The pH sensitivity of the micelle could be controlled by the His/Phe block composition and fine-tuned to target early endosomeby the PLLA-b-PEG composition. The EndoPHSM/f micelle was demonstrated effective in high cytosolic drug accumulation with minimal drug loss during circulation and extracellular environment. *In vitro* studies showed that this micelle was internalized into tumor cells via folate receptor–mediated endocytosis. Subsequently, the drug-loaded micelles destabilized at an early endosomal pH of 6.0 and killed tumor cells through a focal high dose of DOX in the cytosol.

As for the polymeric systems with acidic pH-induced cleavable bonds, acetal, hydrazone, and *N*-ethoxybenzylimidazole (NEBI) bonds have most often been employed to link polymers and drugs together. The cleavage of such chemical bonds by acidic pH can accelerate antitumor drug release from nanovehicles. Hydrazone linkage, which is quite stable at pH 7.4 but hydrolyzes around pH 5–6, was considered an effective approach by Kataoka's group. They designed micelles (size ≈ 65 nm) that formed using folate-conjugated PEGpoly(aspartate hydrazone DOX) copolymers as shown in Fig. 11.13. *In vivo* studies showed that these micelles were stable during blood circulation, had minimal drug leakage, and accumulated in solid tumors through receptor-mediated pathway followed by cytoplasmic DOX release via pH-induced cleavage (pH 6.5). This system provided a promising strategy to achieve significant cytosolic delivery of various antitumor drug molecules, owing to the active internalization, accelerated drug release triggered by endosomal pH, and disruption of endosomal membrane. Moreover, cytosolic drug delivery may be an effective approach for multidrug resistance (MDR) cancers, because drug release that occurs in intracellular regions may avoid the drug efflux activity of P-glycoprotein.

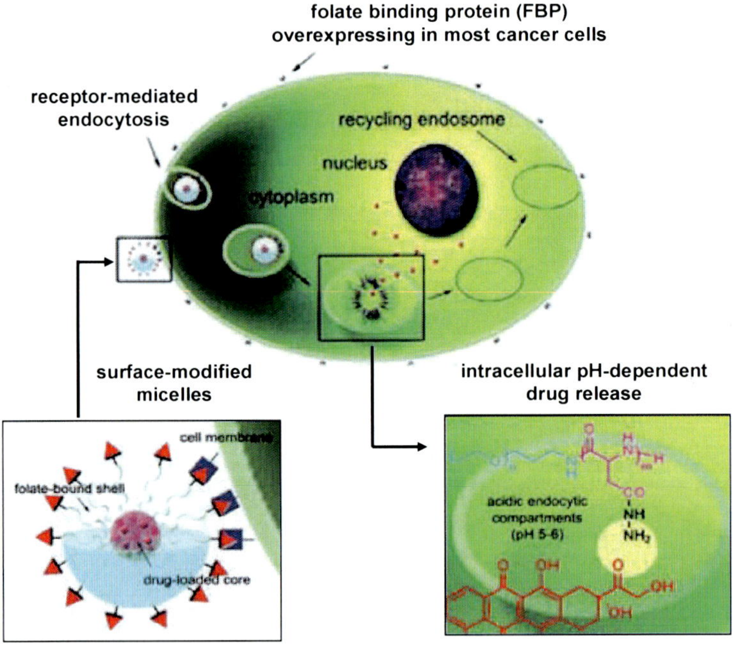

Figure 11.13 Preparation of polymeric micelles with tumor selectivity for active drug targeting and pH sensitivity for intracellular site-specific drug transport. Folic acid with high tumor affinity due to overexpression of its receptors that were conjugated onto the surface of the micelle. Reproduced from Ref. 91, with permission from the Royal Society of Chemistry.

11.3.4 Magnetic Targeting

Owing to the particular magnetic properties, particles with magnetic targeting have been proposed as a promising drug or gene carriers. Under the direction of external magnetic field, therapeutic magnetic particles can be promoted toward selected target area in the human body where they can eventually localize. Therefore, magnetic targeting delivery of therapeutic agents into a tumor mass minimizes toxicity to normal tissues and improves bioavailability of cytotoxic agents. Besides, magnetic particles have also been widely explored for numerous applications in cancer imaging, tissue repair, detoxification of biological fluids, and hyperthermia. To date, IO-based particles have been most extensively researched in magnetic DDS. Thus, we focus on the recent advances that are based on IO particles in this section.

IO particles without any surface coating are not stable in aqueous media and readily aggregate and precipitate. In order to be successfully applied, the IO-based magnetic particles demand combined properties of high magnetic saturation and should be biocompatible and be able to target on certain tissue or organ. Since the nature of the coating, charge state, and functionalization could determine particle biocompatibility, stability, biodistribution, and clearance *in vivo*, numerous strategies are exploited to prepare intelligent magnetic particles. For example, coating of particles with inorganic metals (gold, silica) or with a layer of organic polymer (PEG) allows dispersal into solvents to form homogenous suspensions or ferrofluids, which can interact with an external magnetic field. In addition, the particle surfaces can be modified by attaching bioactive molecules such as antibodies or ligands to allow for specific target aim and functionalization of particles. Moreover, drugs can be linked to the carrier coating, deposited in the surface layer, or trapped within the particles themselves (93). Subsequently, several approaches for cancer therapy have been developed using these well-designed systems.

First, antitumor drugs or agents can be loaded onto the MNs for targeted therapy. Alexiou *et al.* (94) employed the mitoxantrone (MTX)-bounded ferrofluids (FFs) to treat squamous cell carcinoma under a magnetic field. First, the squamous cell carcinoma (VX-2) was implanted in the hind limb of New Zealand White rabbits. After the tumor had reached an appropriate volume, FF-MTX was intra-arterially injected (i.a.; femoral artery) or intravenously injected (i.v.; ear vein); in the meanwhile an external magnetic field was applied on the tumor. Results showed that the FF-MTX i.a. application with magnetic lead to a complete remission of the squamous cell carcinoma with no signs of toxicity, offering a unique opportunity to treat malignant tumors loco-regionally without systemic toxicity. In another recent preclinical study, MNs that were labeled with a near-infrared dye (NIRD) and covalently linked to siRNA (siSuvivin) were injected in tumoral mice (Fig. 11.14). After administration of the contrast agent, there was a significant drop in T2 relativity in images acquired, indicating probe delivery. Then specific tumoral accumulation of MN was demonstrated by *in vivo* MRI and optical imaging. As expected, specific silencing of an apoptosis inhibitor protein was achieved, and *in vivo* studies revealed increased tumor apoptosis and enlarged eosinophilic areas of tumor necrosis (Fig. 11.14c–e) (95).

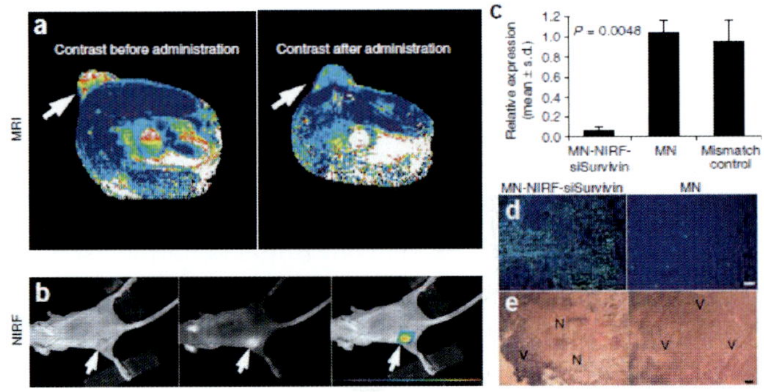

Figure 11.14 Application of MN-NIRF-siSuvivin in a therapeutic tumor model. (a) *In vivo* MRI of mice bearing LS174T human colorectal adenocarcinoma (arrows). (b) NIRF signal on *in vivo* optical images associated with the tumor (left, white light; middle, NIRF; right, color-coded overlay). (c) Quantitative RT-PCR analysis of survivin expression in LS174T tumors. (d) A high density of apoptotic nuclei (green) in tumors treated with MN-NIRF-siSurvivin (left). (e) H&E staining revealed eosinophilic areas of tumor necrosis (N) in tumors treated with MN-NIRF-siSurvivin (left). Reprinted with permission from Ref. 95. Copyright 2007 American Chemical Society.

Second, MNs can be used in hyperthermia for tumor therapy in virtue of their good heat production property under magnetic field and increased specificity through accumulating in target tissues. Recently, Novosad *et al.* developed a novel magnet-sensitive tumor therapy (Fig. 11.15) (96). The as-synthesized gold-coated iron–nickle microdiscs (MDs) possessed spin-vortex ground state. On applying an alternating magnetic field, these MDs' vortices shift, leading to oscillation of the MDs. For specific targeting, the antihuman-IL13 2R antibody was conjugated to MDs to N10 glioma cancer cells. As magnetic field–driven rotation and oscillation of the antibody-conjugated MDs occurred on the surface of the cancer cells, the cellular membrane was destructed and a programmed cell death was initiated. According to their report, approximate 90% cancer-cell destruction can be achieved *in vitro* while a low-frequency field (tens of hertz) was applied.

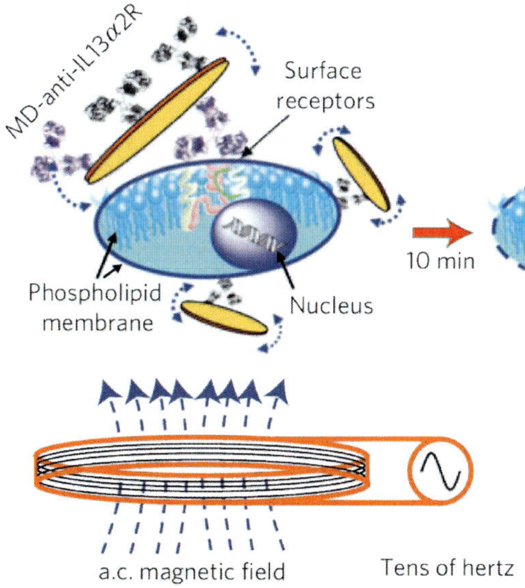

Figure 11.15 The concept of targeted magneto-mechanical cancer-cell destruction using disc-shaped magnetic particles possessing as spin-vortex ground state. The microdiscs are biofunctionalized with antihuman-IL13 2R antibody, specifically targeting human glioblastoma cells. When an alternating magnetic field is applied, the magnetic discs oscillate or swing, compromising membrane integrity and initiating spin vortex–mediated programmed cell death. Reprinted from Ref. 96, with permission from Nature Publishing Group.

Besides the above IO particle-based approaches on cancer therapy, attentions have been attracted to ferromagnetic particles that are ground down to 5–20 nm. These particles exhibit quantum size effects and larger surface areas, leading to superparamagnetism (97). As they are magnetic only in the presence of an external magnetic field and lose this capacity once the external magnetic field is removed, SPIOs have good stability as well as biocompatibility, and the aggregation is also minimized. They are metabolized in the hepato-renal system and enter endogenous iron reserves via hematopoiesis. Based on these properties, several magnet-sensitive tumor delivery systems have been explored to allow early detection and real-time monitoring in a cancer-specific manner for designing and formulating treatments.

11.3.5 Combined Delivery Systems

Successful drug targeting is a very complicated problem. In the search for more effective cancer treatment, current strategies are directed to combined tumor targeting, multi-stimuli response, and multifunctional system. The combined target access allows for even better specificity and selectivity of the tumor, thereby further reducing the needed drug dose as well as the potential harm to the cancer patient.

Since drug carriers can be manipulated with ease to allow for co-encapsulation and efficient delivery of multiple drugs or agents is intended for smart devices. So far, the development of ligands and aptamers provide a way to actively target cancerous cells and then induce receptor-mediated endocytosis for intracellular delivery. Compared to free drug and passive nanomaterial systems, these smart devices have proven to increase therapeutic effects and efficacy in a variety of cellular and animal models. An example for combined tumor targeting is to deliver double ligands (RGD and anginex) using lipsosomes. It demonstrated higher targeting efficiency against tumor microvessels than single-targeting. In another aspect, by fine-tuning the particle size and modifying the surface properties (e.g., PEG shielding), drug carriers can be designed for combined passive–active targeting system to hinder uptake by the reticuloendothelial system (RES) and increase tumor-specific accumulation. Consequently, progression of these techniques will eventually lead to increased accuracy in delivering higher doses and more toxic drugs without harming the healthy tissues.

The microenvironment *in vivo* is very complex, therefore, multi-stimuli targeting is significant to gain a favorable delivery efficacy and therapy effect.

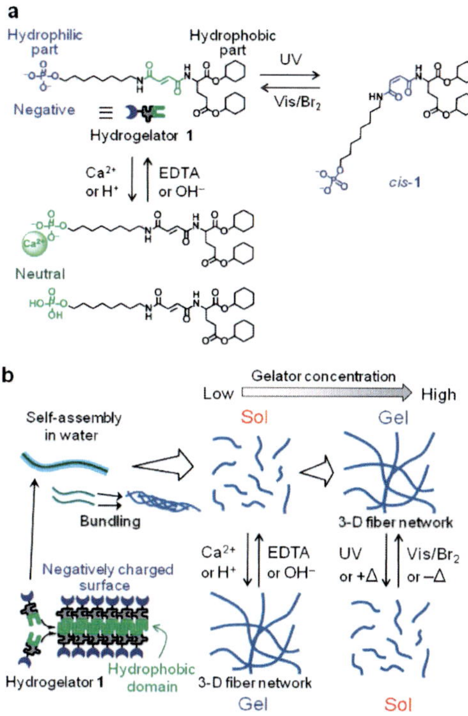

Figure 11.16 (a) Molecular structural changes in hydrogelator **1** in response to a variety of stimuli. (b) Schematic representation of the hierarchical molecular assembly of **1** to form a supramolecular hydrogel and its gel–sol transition triggered by four distinct stimuli (temperature, pH, Ca^{2+}, and light). Reprinted with permission from Ref. 98. Copyright 2009 American Chemical Society.

According to the differences in tumor sites as compared with normal tissues, such as low pH, hypoxia, and thermo or light-sensitive properties, DDSs can be designed to respond to a combination of two or more stimuli. For instance, Komatsu *et al.* (98) prepared a supramolecular hydrogel, comprising the phosphate-type hydrogelator that exhibits gel–sol behavior in response to four distinct input stimuli: temperature, pH, Ca^{2+}, and light (Fig. 11.16). In the presence of various combinations of the four stimuli, several types of stimulus-responsive gel–sol behavior were demonstrated, which could be expressed in four logic gates (AND, OR, NAND, and NOR). In addition, combining the supramolecular gel-based AND logic gate with a photoresponsive supramolecular gel could temporarily modulate the release rate of the bioactive substance. Thus, this multi-stimulus-responsive material is believed to have more promising properties toward the design of controlled drug delivery and release systems.

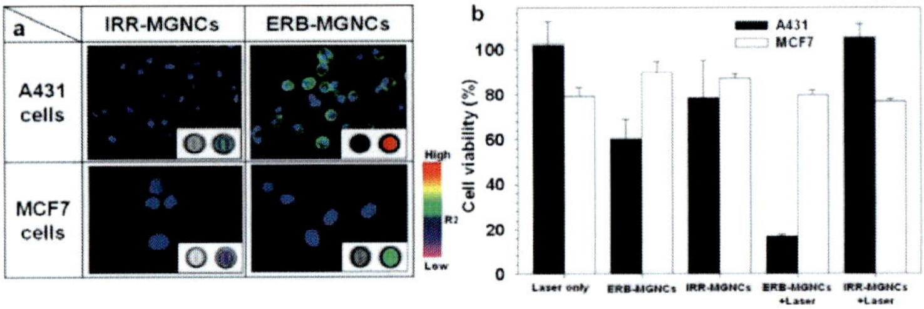

Figure 11.17 (a) Fluorescence microscopy images of A431 and MCF7 cells incubated with ERB-MGNCs and IRR-MGNCs. T2-weighted MR images and their color maps are shown in the left-hand and right-hand insets, respectively. (b) Therapeutic efficacies of ERB-MGNCs and IRR-MGNCs on A431 or MCF7 cells after exposure to an NIR laser. Reprinted from Ref. 101, with permission from Wiley-VCH Verlag GmbH & Co KGaA.

It is highly desirable that drug carriers can not only selectively deliver anticancer drugs to tumor sites but also provide sensitive and specific imaging information in cancer patients (99). Magnetic particles conjugated with targeting ligands is a very potential system that can be used to deliver drugs to treat these tumors and detect disseminated metastatic cells in presence of magnetic field (100). In a recent study, a novel multifunctional magnetic gold nanocomposites (MGNCs) were synthesized for synchronous cancer therapy and diagnosis via MRI (Fig. 11.17) (101). The MGNCs consist of magnetic kernels as MR contrast agents and silica–gold nanocomposites as hyperthermal therapeutic agents. For specific tumor cell targeting, a therapeutic antibody Erbitux (ERB) was conjugated on these composites. Results showed that ERB-conjugated MGNCs could selectively recognize the EGFR overexpression in cancer cells and be effectively internalized. As a result of the therapeutic antibody and NIR laser–induced surface plasmon resonance, ERB-conjugated MGNCs exhibited an excellent synchronous therapeutic efficacy.

Therefore, this study implies the therapeutic potential of MGNCs for simultaneous diagnosis and treatment of cancer.

11.4 Conclusion and Prospects

In this chapter, we discussed the widely applied particles (not merely the microspheres) used in targeting delivery of anticancer drugs based on the anticancer drug carriers and active targeting schemes. Although the results and progress are exciting, studies in this field are still in their infancy. Besides the traditionally used chemical modification, physical properties of the carriers should also be considered, as there is a growing recognition that physical as well as chemical properties of materials can regulate biological responses. Furthermore, new materials are also the hot topic in the anticancer drug carriers. As soon as a novel material appears (e.g., CNT and graphene oxide), its potential in biomedical field begins to be extensively explored. In addition, the development of multifunctional "smart" carriers is currently a growth area. The all-in-one system may be capable of detecting malignant cells (active targeting moiety), visualizing their location in the body (real-time *in vivo* imaging), killing the cancer cells with minimal side effects by sparing normal cells (active targeting and controlled drug release or photothermal ablation), and monitoring treatment effects in real time. Finally, to offset the shortcomings of chemotherapy, immunotherapy and therapeutic cancer vaccines, which use natural function of immune systems and thoroughly decrease the side effects of chemotherapy, can be used for wondrous cancer therapy.

References

1. Boyle, P., & Levin, B. (2008). *World Cancer Report 2008*. France: IARC Press Lyon.
2. Allen, T. M., & Cullis, P. R. (2004). Drug delivery systems: Entering the mainstream. *Science*, **303**, 1818.
3. Lipinski, C. (2002). Poor aqueous solubility–An industry wide problem in drug discovery. *American Pharmaceutical Review*, **5**, 82–85.
4. Ozpolat, B., Sood, A. K., & Lopez-Berestein, G. (2010). Nanomedicine based approaches for the delivery of siRNA in cancer. *Journal of Interferon Research*, **267**, 44–53.

5. Maeda, H., Bharate, G. Y., & Daruwalla, J. (2009). Polymeric drugs for efficient tumor-targeted drug delivery based on EPR-effect. *European Journal of Pharmaceutics and Biopharmaceutics*, **71**, 409–419.
6. Soppimath, K. S., Aminabhavi, T. M., Kulkarni, A. R., & Rudzinski, W. E. (2001). Biodegradable polymeric nanoparticles as drug delivery devices. *Journal of Controlled Release*, **70**, 1–20.
7. Alexis, F., Pridgen, E., Molnar, L. K., & Farokhzad, O. C. (2008). Factors affecting the clearance and biodistribution of polymeric nanoparticles. *Molecular Pharmaceutics*, **5**, 505–515.
8. Mitragotri, S., & Lahann, J. (2009). Physical approaches to biomaterial design. *Nature Materials*, **8**, 15–23.
9. Wang, L. Y., Ma, G. H., & Su, Z. G. (2005). Preparation of uniform sized chitosan microspheres by membrane emulsification technique and application as a carrier of protein drug. *Journal of Controlled Release*, **106**, 62–75.
10. Lv, P. P., Wei, W., Gong, F. L., *et al.* (2009). Preparation of uniformly sized chitosan nanospheres by a premix membrane emulsification technique. *Industrial & Engineering Chemistry Research*, **48**, 8819–8828.
11. Liu, R., Ma, G. H., Meng, F. T., & Su, Z. G. (2005). Preparation of uniform-sized PLA microcapsules by combining Shirasu Porous Glass membrane emulsification technique and multiple emulsion-solvent evaporation method. *Journal of Controlled Release*, **103**, 31–43.
12. Wei, Q., Wei, W., Lai, B., *et al.* (2008). Uniform-sized PLA nanoparticles: Preparation by premix membrane emulsification. *International Journal of Pharmaceutics*, **359**, 294–297.
13. Shen, Z. Y., Ma, G. H., Dobashi, T., Maki, Y., & Su, Z. G. (2008). Preparation and characterization of thermo-responsive albumin nanospheres. *International Journal of Pharmaceutics*, **346**, 133–142.
14. Fonseca, C., Sim Es, S., & Gaspar, R. (2002). Paclitaxel-loaded PLGA nanoparticles: Preparation, physicochemical characterization and *in vitro* anti-tumoral activity. *Journal of Controlled Release*, **83**, 273–286.
15. Torchilin, V. P. (2005). Recent advances with liposomes as pharmaceutical carriers. *Nature Reviews Drug Discovery*, **4**, 145–160.
16. Kim, C. K., & Lim, S. J. (2002). Recent progress in drug delivery systems for anticancer agents. *Archives of Pharmacal Research*, **25**, 229–239.
17. Drummond, D. C., Meyer, O., Hong, K., Kirpotin, D. B., & Papahadjopoulos, D. (1999). Optimizing liposomes for delivery of chemotherapeutic agents to solid tumors. *Pharmacological Reviews*, **51**, 691.

18. Oku, N. (1999). Anticancer therapy using glucuronate modified long-circulating liposomes. *Advanced Drug Delivery Reviews*, **40**, 63–73.

19. Torchilin, V. P. (1998). Polymer-coated long-circulating micro-particulate pharmaceuticals. *Journal of Microencapsulation*, **15**, 1–19.

20. Woodle, M. C. (1998). Controlling liposome blood clearance by surface-grafted polymers. *Advanced Drug Delivery Reviews*, **32**, 139–152.

21. Shen, Y., Jin, E., Zhang, B., *et al.* (2010). Prodrugs forming high drug loading multifunctional nanocapsules for intracellular cancer drug delivery. *Journal of the American Chemical Society*, **132**, 4259–4265.

22. Adams, M. L., Lavasanifar, A., & Kwon, G. S. (2003). Amphiphilic block copolymers for drug delivery. *Journal of Pharmaceutical Sciences-US*, **92**, 1343–1355.

23. Nishiyama, N., & Kataoka, K. (2006). Current state, achievements, and future prospects of polymeric micelles as nanocarriers for drug and gene delivery. *Pharmacology & Therapeutics*, **112**, 630–648.

24. Kataoka, K., Matsumoto, T., Yokoyama, M., *et al.* (2000). Doxorubicin-loaded poly (ethylene glycol)-poly ([beta]-benzyl-aspartate) copolymer micelles: Their pharmaceutical characteristics and biological significance. *Journal of Controlled Release*, **64**, 143–153.

25. Nakanishi, T., Fukushima, S., Okamoto, K., *et al.* (2001). Development of the polymer micelle carrier system for doxorubicin. *Journal of Controlled Release*, **74**, 295–302.

26. Nasongkla, N., Bey, E., Ren, J., *et al.* (2006). Multifunctional polymeric micelles as cancer-targeted, MRI-ultrasensitive drug delivery systems. *Nano Letters*, **6**, 2427–2430.

27. Bronich, T. K., Keifer, P. A., Shlyakhtenko, L. S., & Kabanov, A. V. (2005). Polymer micelle with cross-linked ionic core. *Journal of the American Chemical Society*, **127**, 8236–8237.

28. Gou, P., Zhu, W., & Shen, Z. (2010). Synthesis, self-assembly, and drug-loading capacity of well-defined cyclodextrin-centered drug-conjugated amphiphilic A14B7 Miktoarm star copolymers based on poly(ε-caprolactone) and poly(ethylene glycol). *Biomacromolecules*, **11**, 934–943.

29. Webster, J. G. (Ed.) (2006). *Encyclopedia of Medical Devices and Instrumentation* (Second Edition). New York: John Wiley & Sons.

30. Tauro, J. R., & Gemeinhart, R. A. (2005). Extracellular protease activation of chemotherapeutics from hydrogel matrices: A new paradigm for local chemotherapy. *Molecular Pharmaceutics*, **2**, 435–438.

31. Menei, P., Jadaud, E., Faisant, N., *et al.* (2004). Stereotaxic implantation of 5-fluorouracil-releasing microspheres in malignant glioma. *Cancer*, **100**, 405–410.

32. Arica, B., Calis, S., Kas, H. S., Sargon, M. F., & Hincal, A. A. (2002). 5-Fluorouracil encapsulated alginate beads for the treatment of breast cancer. *International Journal of Pharmaceutics*, **242**, 267–269.

33. Almond, B. A., Hadba, A. R., Freeman, S. T., *et al.* (2003). Efficacy of mitoxantrone-loaded albumin microspheres for intratumoral chemotherapy of breast cancer. *Journal of Controlled Release*, **91**, 147–155.

34. Smith, J. P., Kanekal, S., Patawaran, M. B., *et al.* (1999). Drug retention and distribution after intratumoral chemotherapy with fluorouracil/epinephrine injectable gel in human pancreatic cancer xenografts. *Cancer Chemotherapy and Pharmacology*, **44**, 267–274.

35. Weinberg, B. D., Blanco, E., & Gao, J. (2008). Polymer implants for intratumoral drug delivery and cancer therapy. *Journal of Pharmaceutical Sciences*, **97**, 1681–1702.

36. Nasongkla, N., Akarajiratun, P., & Hongeng, S. (2009). Development of tri-component copolymer rods as implantable drug delivery systems for liver cancer therapy, *IFMBE Proceedings*, **25**, 317–320.

37. Sakakura C., Takahashi T., Sawai K., Hagiwara A., Ito M., Shobayashi S., *et al.* (1992) Enhancement of therapeutic efficacy of aclarubicin against lymph node metastases using a new dosage form: Aclarubicin adsorbed on activated carbon particles. *Gan To Kagaku Ryoho*, **19**, 1560–1563.

38. Hagiwara, A., Takahashi, T., Sawai, K., *et al.* (1997). Selective drug delivery to peri-tumoral region and regional lymphatics by local injection of aclarubicin adsorbed on activated carbon particles in patients with breast cancer–a pilot study. *Anti-Cancer Drugs*, **8**, 666.

39. Majoros, I., & Baker, J. R. (2008). *Dendrimer-Based Nanomedicine*. Singapore: Pan Stanford Publishing.

40. Gillies, E. R., & Frechet, J. (2005). Dendrimers and dendritic polymers in drug delivery. *Drug Discovery Today*, **10**, 35–43.

41. Killops, K. L., Campos, L. M., & Hawker, C. J. (2008). Robust, efficient, and orthogonal synthesis of dendrimers via thiol-ene "click" chemistry. *Journal of the American Chemical Society*, **130**, 5062–5064.

42. Ma, X., Tang, J., Shen, Y., *et al.* (2009). Facile synthesis of polyester dendrimers from sequential click coupling of asymmetrical monomers. *Journal of the American Chemical Society*, **131**, 14795–14803.

43. Meijer, E. W., Jansen, J. F., & Brabander, V. (1994). Encapsulation of guest molecules into a dendritic box. *Science*, **266**, 1226–1229.

44. Majoros, I. J., Myc, A., Thomas, T., Mehta, C. B., & Baker Jr, J. R. (2006). PAMAM dendrimer-based multifunctional conjugate for cancer therapy: Synthesis, characterization, and functionality. *Biomacromolecules*, **7**, 572–579.

45. Hoar, T. P., & Schulman, J. H. (1943). Transparent water-in-oil dispersions: The oleopathic hydro-micelle. *Nature*, **152**, 102–103.

46. Higashi, S., Tabata, N., Kondo, K. H., *et al.* (1999). Size of lipid microdroplets effects results of hepatic arterial chemotherapy with an anticancer agent in water-in-oil-in-water emulsion to hepatocellular carcinoma. *Journal of Pharmacology and Experimental Therapeutics*, **289**, 816.

47. Hino, T., Kawashima, Y., & Shimabayashi, S. (2000). Basic study for stabilization of W/O/W emulsion and its application to transcatheter arterial embolization therapy. *Advanced Drug Delivery Reviews*, **45**, 27–45.

48. Takegami, S., Takara, K., Tanaka, S., *et al.* (2010). Characterization, *in vitro* cytotoxicity and cellular accumulation of paclitaxel-loaded lipid nano-emulsions. *Journal of Microencapsulation*, 206–212.

49. Bagul, M., Kakumanu, S., Wilson, T., & Nicolosi, R. (2010). *In vitro* evaluation of antiproliferative effects of self-assembling nanoemulsion of paclitaxel on various cancer cell lines. *Nano Biomedicine and Engineering*, **2**, 11.

50. Li, Y., Le Maux, S., Xiao, H., & McClements, D. J. (2009). Emulsion-based delivery systems for tributyrin, a potential colon cancer preventative agent. *Journal of Agricultural and Food Chemistry*, **57**, 9243–9249.

51. Atri, M. (2006). New technologies and directed agents for applications of cancer imaging. *Journal of Clinical Oncology*, **24**, 3299.

52. Weissleder, R. (2006). Molecular imaging in cancer. *Science*, **312**, 1168–1171.

53. Wang, X., Yang, L., Chen, Z. G., & Shin, D. M. (2008). Application of nanotechnology in cancer therapy and imaging. *CA: A Cancer Journal for Clinicians*, **58**, 97–110.

54. Cai, W., Shin, D. W., Chen, K., *et al.* (2006). Peptide-labeled near-infrared quantum dots for imaging tumor vasculature in living subjects. *Nano Letters*, **6**, 669–676.

55. Chan, W., & Nie, S. (1998). Quantum dot bioconjugates for ultrasensitive nonisotopic detection. *Science*, **281**, 2016.

56. Gao, X., Cui, Y., Levenson, R. M., Chung, L., & Nie, S. (2004). *In vivo* cancer targeting and imaging with semiconductor quantum dots. *Nature Biotechnology*, **22**, 969–976.

57. Kim, S., Lim, Y. T., Soltesz, E. G., *et al.* (2004). Near-infrared fluorescent type II quantum dots for sentinel lymph node mapping. *Nature*, **22**, 93–97.

58. Sjögren, C. E., Johansson, C., Naevestad, A., *et al.* (1997). Crystal size and properties of superparamagnetic iron oxide (SPIO) particles. *Magnetic Resonance Imaging*, **15**, 55–67.

59. Lee, J. H., Huh, Y. M., Jun, Y., *et al.* (2006). Artificially engineered magnetic nanoparticles for ultra-sensitive molecular imaging. *Nature Medicine*, **13**, 95–99.

60. Wei, W., Ma, G., Hu, G., *et al.* (2008). Preparation of hierarchical hollow $CaCO_3$ particles and the application as anticancer drug carrier. *Journal of the American Chemical Society*, **130**, 15808–15810.

61. Wei, W., Yue, Z. G., Qu, J. B., *et al.* (2010). Galactosylated nanocrystallites of insoluble anticancer drug for liver-targeting therapy: An *in vitro* evaluation. *Nanomedicine*, **5**, 589–596.

62. Bianco, A., Sainz, R., Li, S., *et al.* (2008). Biomedical applications of functionalised carbon nanotubes. *Medicinal Chemistry and Pharmacological Potential of Fullerenes and Carbon Nanotubes*, 23–50.

63. Feazell, R. P., Nakayama-Ratchford, N., Dai, H., & Lippard, S. J. (2007). Soluble single-walled carbon nanotubes as longboat delivery systems for platinum (IV) anticancer drug design. *Journal of the American Chemical Society*, **129**, 8438–8439.

64. Liu, Z., Sun, X., Nakayama-Ratchford, N., & Dai, H. (2007). Supramolecular chemistry on water-soluble carbon nanotubes for drug loading and delivery. *ACS Nano*, **1**, 50–56.

65. Wu, W., Li, R., Bian, X., *et al.* (2009). Covalently combining carbon nanotubes with anticancer agent: Preparation and antitumor activity. *ACS Nano*, **3**, 2740–2750.

66. Welsher, K., Liu, Z., Sherlock, S. P., *et al.* (2009). A route to brightly fluorescent carbon nanotubes for near-infrared imaging in mice. *Nature Nanotechnology*, **4**, 773–780.

67. Brannon-Peppas, L., & Blanchette, J. O. (2004). Nanoparticle and targeted systems for cancer therapy. *Advanced Drug Delivery Reviews*, **56**, 1649–1659.

68. Chen, H., Gao, J., Lu, Y., *et al.* (2008). Preparation and characterization of PE38KDEL-loaded anti-HER2 nanoparticles for targeted cancer therapy. *Journal of Controlled Release*, **128**, 209–216.

69. Li, S. D., & Huang, L. (2006). Targeted delivery of antisense oligodeoxynucleotide and small interference RNA into lung cancer cells. *Molecular Pharmaceutics*, **3**, 579–588.

70. Farokhzad, O. C., Cheng, J., Teply, B. A., *et al.* (2006) Targeted nanoparticle-aptamer bioconjugates for cancer chemotherapy *in vivo*. *Proceedings of the National Academy of Sciences of the United States of America*, **103**, 6315–6320.

71. Li, L., Wartchow, C. A., Danthi, S. N., *et al.* (2004). A novel antiangiogenesis therapy using an integrin antagonist or anti-Flk-1 antibody coated 90Y-labeled nanoparticles. *International Journal of Radiation Oncology, Biology, Physics*, **58**, 1215–1227.

72. Ruoslahti, E. (2002). Specialization of tumour vasculature. *Nature Reviews Cancer*, **2**, 83–90.

73. Laskin, J. J., & Sandler, A. B. (2004). Epidermal growth factor receptor: A promising target in solid tumours. *Cancer Treatment Reviews*, **30**, 1–17.

74. Nobs, L., Buchegger, F., Gurny, R., & Allémann, E. (2006). Biodegradable nanoparticles for direct or two-step tumor immunotargeting. *Bioconjugate Chemistry*, **17**, 139–145.

75. Head, J. F., Wang, F., & Elliott, R. L. (1997). Antineoplastic drugs that interfere with iron metabolism in cancer cells. *Advances in Enzyme Regulation*, **37**, 147–169.

76. Heidel, J. D., Yu, Z., Liu, J., *et al.* (2007). Administration in non-human primates of escalating intravenous doses of targeted nanoparticles containing ribonucleotide reductase subunit M2 siRNA. *Proceedings of the National Academy of Sciences of the United States of America*, **104**, 5715–5721.

77. Rodolfo, M., Melani, C., Zilocchi, C., *et al.* (1998). IgG2a induced by interleukin (IL) 12-producing tumor cell vaccines but not IgG1 induced by IL-4 vaccine is associated with the eradication of experimental metastases. *Cancer Research*, **58**, 5812.

78. Fogueri, L. R., & Singh, S. (2009). Smart polymers for controlled delivery of proteins and peptides: A review of patents. *Recent Patents on Drug Delivery & Formulation*, **3**, 40–48.

79. Van Tomme, S. R., Storm, G., & Hennink, W. E. (2008). *In situ* gelling hydrogels for pharmaceutical and biomedical applications. *International Journal of Pharmaceutics*, **355**, 1–18.

80. Ankareddi, I., & Brazel, C. S. (2007). Synthesis and characterization of grafted thermosensitive hydrogels for heating activated controlled release. *International Journal of Pharmaceutics*, **336**, 241–247.

81. Shen, Z., Terao, K., Maki, Y., *et al.* (2006). Synthesis and phase behavior of aqueous poly (*N*-isopropylacrylamide-co-acrylamide), poly (*N*-isopropylacrylamide-*co*-*N*,*N*-dimethylacrylamide) and poly (*N*-isopropylacrylamide-*co*-2-hydroxyethyl methacrylate). *Colloid & Polymer Science*, **284**, 1001–1007.

82. Li, Y., Pan, S., Zhang, W., & Du, Z. (2009). Novel thermo-sensitive core–shell nanoparticles for targeted paclitaxel delivery. *Nanotechnology*, **20**, 065104.

83. Puri, A., Kramer-Marek, G., Campbell-Massa, R., *et al.* (2008). HER2-specific affibody-conjugated thermosensitive liposomes (Affisomes) for improved delivery of anticancer agents. *Journal of Liposome Research*, **18**, 293–307.

84. Langereis, S., Keupp, J., van Velthoven, J., *et al.* (2009). A temperature-sensitive liposomal 1H CEST and 19F contrast agent for MR image-guided drug delivery. *Journal of the American Chemical Society*, **131**, 1380–1381.

85. Lee, E. S., Gao, Z., & Bae, Y. H. (2008). Recent progress in tumor pH targeting nanotechnology. *Journal of Controlled Release*, **132**, 164–170.

86. Na, K., Seong Lee, E., & Bae, Y. H. (2003). Adriamycin loaded pullulan acetate/sulfonamide conjugate nanoparticles responding to tumor pH: pH-dependent cell interaction, internalization and cytotoxicity *in vitro*. *Journal of Controlled Release*, **87**, 3–13.

87. Na, K., Lee, K. H., & Bae, Y. H. (2004). pH-sensitivity and pH-dependent interior structural change of self-assembled hydrogel nanoparticles of pullulan acetate/oligo-sulfonamide conjugate. *Journal of Controlled Release*, **97**, 513–525.

88. Bae, Y., Jang, W. D., Nishiyama, N., Fukushima, S., & Kataoka, K. (2005). Multifunctional polymeric micelles with folate-mediated cancer cell targeting and pH-triggered drug releasing properties for active intracellular drug delivery. *Molecular BioSystems*, **1**, 242–250.

89. Kang, H. C., & Bae, Y. H. (2007). pH-tunable endosomolytic oligomers for enhanced nucleic acid delivery. *Advanced Functional Materials*, **17**, 1263–1272.

90. Lee, E. S., Na, K., & Bae, Y. H. (2005). Doxorubicin loaded pH-sensitive polymeric micelles for reversal of resistant MCF-7 tumor. *Journal of Controlled Release*, **103**, 405–418.

91. Oh, K. T., Yin, H., Lee, E. S., & Bae, Y. H. (2007). Polymeric nanovehicles for anticancer drugs with triggering release mechanisms. *Journal of Materials Chemistry*, **17**, 3987–4001.

92. Lee, E. S., Na, K., & Bae, Y. H. (2003). Polymeric micelle for tumor pH and folate-mediated targeting. *Journal of Controlled Release*, **91**, 103–113.

93. Chen, H., Gu, Y., Hu, Y., & Qian, Z. (2007). Characterization of pH-and temperature-sensitive hydrogel nanoparticles for controlled drug release. *PDA Journal of Pharmaceutical Science and Technology*, **61**, 303.

94. Alexiou, C., Arnold, W., Klein, R. J., *et al.* (2000). Locoregional cancer treatment with magnetic drug targeting. *Cancer Research*, **60**, 6641.

95. Medarova, Z., Pham, W., Farrar, C., Petkova, V., & Moore, A. (2007). *In vivo* imaging of siRNA delivery and silencing in tumors. *Nature Materials*, **13**, 372–377.

96. Rozhkova, D., Ulasov, I., Kim, D. H., *et al.* (2010). Biofunctionalized magnetic-vortex microdiscs for targeted cancer-cell destruction. *Science Education in the 21st century: Advantages, Pitfalls, and Future Trends*, 38.

97. Gupta, A. K., & Gupta, M. (2005). Synthesis and surface engineering of iron oxide nanoparticles for biomedical applications. *Biomaterials*, **26**, 3995–4021.

98. Komatsu, H., Matsumoto, S., Tamaru, S., *et al.* (2009). Supramolecular hydrogel exhibiting four basic logic gate functions to fine-tune substance release. *Journal of the American Chemical Society*, **131**, 5580–5585.

99. Peng, X. H., Qian, X., Mao, H., & Wang, A. Y. (2008). Targeted magnetic iron oxide nanoparticles for tumor imaging and therapy. *International Journal of Nanomedicine*, **3**, 311.

100. Alexiou, C., Jurgons, R., Schmid, R. J., *et al.* (2003). Magnetic drug targeting-biodistribution of the magnetic carrier and the chemotherapeutic agent mitoxantrone after locoregional cancer treatment. *Journal of Drug Targeting*, **11**, 139–150.

101. Lee, J., Yang, J., Ko, H., *et al.* (2008). Multifunctional magnetic gold nanocomposites: Human epithelial cancer detection via magnetic resonance imaging and localized synchronous therapy. *Advanced Functional Materials*, **18**, 258–264.

Chapter 12

Microspheres for Targeting Delivery to Brain

Chen Jiang, Xinguo Jiang, Yang Liu, Kun Shao, and Rongqin Huang

Key Laboratory of Smart Drug Delivery, Ministry of Education & PLA, Department of Pharmaceutics, School of Pharmacy, Fudan University, Shanghai, P. R. China

jiangchen@shmu.edu.cn

12.1 Introduction

The blood–brain barrier (BBB) protects the central nervous system (CNS) from potentially harmful xenobiotics and endogenous molecules (Mehdipour & Hamidi, 2009). Despite this beneficial role, more than 98% of candidate brain-targeting drugs have been halted mid-development due to the poor permeability of the BBB, presenting a major problem to the pharmaceutical industry (de Boer & Gaillard, 2007). The most important physical structure of BBB is the brain capillary endothelial cells (BCECs) processing unique characteristics such as the presence of tight junctions between the capillary endothelial cells and the lack of paracellular transport (Su & Sinko, 2006). Because of these obstacles, the therapy of brain diseases has been considered as a difficult challenge. A small molecule does not cross the BBB in pharmacologically significant amounts, unless the molecule is both lipid soluble and has a molecular weight (MW) < 400 Da (Pardridge, 2005). Nearly 100% of large-molecule drugs do not cross the BBB (Pardridge, 2003). Gene therapy holds great potential in treating brain diseases such as neurodegenerative diseases and brain tumors (Roy *et al.*, 2008; Brayden, 2003), however limited by the presence of BBB. Therefore, drug delivery systems targeting

Microspheres and Microcapsules in Biotechnology: Design, Preparation, and Applications
Edited by Guanghui Ma and Zhiguo Su

ISBN 978-981-4316-47-7 (Hardcover), 978-981-4364-62-1 (eBook)
www.panstanford.com

brain have been expected to be an effective strategy to deliver drugs across BBB into brain.

Nanotechnology can be applied to develop technologies and approaches for delivering drugs across the BBB (Silva, 2007). Nano-scaled plat can be designed to carry out multiple specific functions potentially for successful drug delivery to the CNS. The use of nanoparticles (NPs) to deliver drugs to the brain across the BBB may provide significant advantages over current strategies (Lockman *et al.*, 2002), and nanoscaled plat can be easily modified to render more functions. The modification provides several approaches for drug-loading NPs crossing the BBB such as adsorptive-mediated endocytosis (AME) (Tamai *et al.*, 1997; Chen *et al.*, 2004), cell-penetrating peptides–mediated transmembrane transport (Foged & Nielsen, 2008; Magzoub & Graslund, 2005), and receptor-mediated endocytosis (Lockman *et al.*, 2002; Friden, 1994) as discussed in this section. Successfully constructed brain-targeting drug delivery systems that can be applied in the therapy for brain diseases such as Parkinson's disease (PD), Alzheimer's disease, and brain tumors will be also be discussed in this section.

12.2 Drug Delivery Systems to Brain

The success of drug delivery to brain depends on the effective carriers, and major drug transfer approaches employ liposomes, polymer, and so on. More suitable approaches to clinical application should be developed. Several kinds of drug carriers applied in brain drug delivery will be discussed in this section briefly.

12.2.1 Liposomes

When liposome was discovered by Bangham in 1960s, it attracted wide attention as a drug carrier. Traditional liposome is made of two layers of phospholipids, whose structure is similar to the cellular membrane, and the particle diameter ranges from 20 nm to 100 μm. Due to qualities such as good biocompatibility and no toxicity and biodegradability, liposomes can be developed as carriers of hydrophobic, hydrophilic, and amphipathic drugs. Because of the high lipophilicity, liposomes can transport their contents into the brain parenchyma in many ways, such as passive diffusion, fusion with the brain capillary endothelial cell, or endocytosis (Tiwari & Amiji, 2006).

Initial research mainly focused on the relatively large size of liposomes (0.2–1.0 μm), which showed that liposomes were not able to deliver drugs across the BBB (Tökés *et al.*, 1980). This may be explained by the relatively large size of the liposomes that could be cleared by the cells in the reticuloendothelial system (RES), particularly liver and spleen, before they reached the target site. Small unilamellar vesicles (SUVs, ranging from 0.025–0.1 μm) were developed subsequently to increase the circulation time *in vivo* (Lasic, 1993).

For example, it was reported that when β-galactosidase-incorporated liposomes were injected into rat tail vein, they penetrated the BBB and reached the lysosomes in the CNS tissue more effectively than the free enzyme itself (Onodera *et al.*, 1983). Also, when liposomes-entrapped γ-amino butyric acid (GABA) was administrated intraperitoneally to Sprague-Dawley rats with penicillin-induced epileptic activity, the formulation decreased or prevented the epileptic activity, whereas no significant changes were seen when free GABA was administered (Loeb *et al.*, 1982). Liposome-mediated BBB transport of thyrotropin-releasing hormone was also reported. In this study, it was shown that following an intravenous (i.v.) injection, compared with free thyrotropin-releasing hormone, the percentage of injected dose of thyrotropin-releasing hormone in the brain increased from 0.12% to 0.28–0.44% following the injection of liposome-entrapped TRH after 3 hours (Postmes *et al.*, 1980).

Systemically injected liposomes, even SUVs, are cleared by RES. Therefore, to reduce the uptake of liposomes by the RES and induce brain target specificity, surface modification of liposomes has been attempted. On one hand, sterically stabilized liposomes, surface modified with polyethylene glycol (PEG), have a distinctly prolonged circulation time in the blood stream compared to conventional liposomes. On the other hand, coupling of vectors or ligands to the surface of liposomes, increases the number of liposomes crossing the BBB through the carrier-mediated or receptor-mediated pathway.

It has been reported that PEG-modified immunoliposomes were employed to target transfect β-galactosidase and luciferase into the brain (Shi *et al.*, 2001). The gene is incorporated into the center of the liposome and the surface of the liposome is then coated with PEG to prolong the circulation time. In addition, a further 2% of the PEG strands have a mAb to the transferrin receptor (TfR) (8D3mAb) attached to them. The mAb interacts with TfRs and effectively targets the immunoliposome to tissues that have a high expression of TfRs such as the liver and the brain. If a brain-specific promoter is encapsulated with the β-galactosidase gene, for example, the promoter for glial fibrillary acidic protein, then the enzyme expression becomes confined to the brain. Similar immunoliposome constructs using OX26 mAb on the surface have been used to delivery digoxin, the substrate of P-gp, to the CNS (Huwyler *et al.*, 2002).

Cereport TM (RMP-7) is an analog of bradykinin which possesses both highly preferential binding to the B bradykinin receptor expressed on the BCECs and agonist properties. When RMP-7 was conjugated to amphotericin-containing liposomes, the concentration of the drug was several times higher than free drug or mixture of drug with RMP-7 (Zhang *et al.*, 2003b).

To achieve target specificity, liposomes can also be combined with virus components (virosomes), which have capability to penetrate BCECs. Jiang *et al.* reported the delivery of glial cell–derived neurotrophic factor (GDNF) across the BBB *in vivo*. The hemagglutination virus of Japan (HVJ)-modified liposomes encapsulating mGDNF gene were administered to rats via the internal carotid artery. ELISA and Western blotting assays suggested that the transfected rats showed a marked increase in the brain level of GDNF on day 3 after the administration, and the level remained significantly elevated for at least 12 days. Immunohistochemical staining revealed an increase in GDNF immunoreactivity throughout the transfected forebrain. These results indicate that the gene was successfully transferred *in vivo* from HVJ liposomes into BCECs, and the gene product was secreted into the brain (Jiang *et al.*, 2000; Jiang *et al.*, 2003). Another study reported the *in vitro* uptake of HVJ liposomes–encapsulated fluorescein isothiocyanate–labeled dextran (FITC-dextran) and FITC-labeled oligodeoxynucleotide (FITC-ODN) by BCECs, indicating the utility of these fusogenic vesicles to transport macromolecules and oligonucleotides to BCECs (Matsuo *et al.*, 2000).

12.2.2 Cationic Lipids

Cationic lipids are one of the most efficient gene transfer vectors. DNA can complex with cationic lipids forming lipoplex and then be transported into cells for gene expression (Yu *et al.*, 2010). A lipoplex-based cationic lipid was used to increase plasmid DNA expression in the brain, and then picomolar amounts of siRNA was directed against gene expression with this lipoplex in CNS (Hassani *et al.*, 2005). The cationic liposomes associated to transferrin lipoplexes (Tf-lipoplexes) carrying siRNA showed efficient knockdown of the transcription factor c-Jun responsible for neuronal cell death (Cardoso *et al.*, 2010). Tf-lipoplex leads to a significant c-Jun knockdown in the mouse hippocampus *in vivo*, resulting in the attenuation of both neuronal death and inflammation following kainic acid–mediated lesion of this region.

12.2.3 Synthetic Cationic Polymers

12.2.3.1 Polyethylenimine

Polyethylenimine (PEI) has been widely studied as gene-delivery carrier both *in vitro* and *in vivo* (Merdan *et al.*, 2003; Jeong *et al.*, 2001; Nimesh *et al.*, 2006). PEGylated technology has been used in PEI/DNA polyplexes for overcoming the rapid blood clearance and accumulation at RES sites upon i.v. administration. Systemic delivery of PEGylated liposomes carrying PEI–oligodeoxynucleotides (ODN) polyplex was also applied. PEI-based nonviral vectors were shown to be well tolerated and effective in transferring genes through a large portion of canine brain. Although the nonviral plasmid vectors were less effective than adenovirus, they still achieved appreciable gene expression levels (Oh *et al.*, 2007). The cytotoxicity of PEI also limits its application *in vivo* and increases as the MW of PEI increases (Guillem & Alino, 2004). Therefore, it should take both efficiency and toxicity aspects into consideration. Abdallah *et al.* showed that the cationic polymer PEI provided unprecedentedly high levels of transgene expression in the mature mouse brain. Three different preparations of PEI (25-, 50-, and 800-kD) were compared for their transfection efficiencies in the brains of adult mice. The highest levels of transfection were obtained with the 25-kD PEI (Abdallah *et al.*, 1996).

PEI, a low-toxicity vector, appears to have potential for fundamental research and genetic therapy of the brain efficiency. PEGylation could improve the solubility of the complexes, minimize their aggregation, and reduce their interaction with proteins in the physiological fluid. *In vivo* application of PEG-modified PEI has been investigated for gene expression in the CNS (Tang *et al.*, 2003).

PEI conjugate with one segment of PEG was able to increase transgene expression in the spinal cord up to 11-fold higher than PEI homopolymer after intrathecal administration of its DNA complexes into the lumbar spinal cord subarachnoid space. Gene expression in the brain was improved after the lumbar injection of this system. The study provides *in vivo* evidence that an appropriate degree of PEG modification is a decisive element in gene transfer by the PEGylated polymers. A novel system, nanogel based on nanoscale network of cross-linked PEG and PEI was proposed for ODN delivery to the brain (Vinogradov *et al.*, 2004). PEG-modified nanogels can bind and encapsulate spontaneously negatively charged ODN, resulting in formation of stable aqueous dispersion of polyelectrolyte complex with particle sizes less than 100 nm. PEG-modified nanogels incorporated ODN could be transported effectively across the bovine brain microvessel endothelial cells monolayers as an *in vitro* model of BBB. On being modified with Tf and insulin, the transport efficacy of nanogels through BBB in both cases was further increased in the transcellular pathway.

The feasibility of achieving gene transfer into CNS neurons has been investigated by peripheral intramuscular injection of plasmid PEI/DNA complexed in the rat hypoglossal system (Wang *et al.*, 2001). The PEI/DNA complexes could migrate through the retrograde axonal transport to neuronal cell bodies in the brain stem after being internalized from nerve terminals in the tongue muscle.

12.2.3.2 Polyamidoamine (PAMAM)

Dendrimers constitute a unique class of hyperbranched macromolecules composed of layers of monomer units radiating from a central core (Tomalia, 2005; Tomalia *et al.*, 2003). They have the potential for use in gene therapy and other therapeutic applications due to their safety and lack of immunogenicity (Dufes *et al.*, 2005). Dendrimer is highly inoculated by polymer molecules, and the diameter of the dendrimer is similar to polymer micelles. For example, PAMAM is a typical kind of dendrimer, and its diameter ranges from 1.5–14.5 nm. A drug can enter the inner cavity of PAMAM through hydrophobic interaction. In physiological conditions, as the dendrimer degrades, the drug is released. The surface of the dendrimer can be conjugated with hydrophilic or hydrophobic polymer block, making a stable system with longer circulation time *in vivo*.

As a brain-targeting delivery system, dendrimers are used to treat brain tumor after combining them with antitumor drugs (Wu *et al.*, 2006). Also, Huang *et al.* (2007b) used PAMAM surface modified with Tf to deliver the gene into the brain. Hildgen's team members prepared an intellectual brain-targeting delivery system with dendrimers (Dhanikula *et al.*, 2008). They synthesized PEPE, whose surface is modified with D-aminoglucose entrapped with antitumor drug methotrexate. When PEPE is modified with D-aminoglucose, it can not only promote the NPs crossing the BBB, but also make them more effective for tumor targeting, which may be related to facilitative glucose transporters expressed on the surface of tumor cells. The antineoplasmic activity of PEPE entrapped with methotrexate was evaluated on neurogliocytoma cell and on the model of human's neurogliocytoma cells, respectively. It is reported that, on the two cell models, the quantity of internalization by glycosylated PEPE is always larger than nonglycosylated PEPE. Incubating bEnd.3 cells and U373 MG cells as BBB model *in vitro*, the quantity of methotrexate entrapped by glycosylated PEPE across the BBB is three to five times compared with nonglycosylated ones. It means that aminoglucose can not only be an effective neurogliocytoma-targeted ligand but also promote the NPs crossing the BBB.

PAMAM dendrimer emerged a new class of nanoscopic, spherical polymer with a high degree of monodispersity, polyvalency, and controlled molecular architecture (Luo *et al.*, 2002; Kitchens *et al.*, 2007). PAMAM owning the above characteristic features have been considered appropriate as delivery vehicles for gene and drug (Kitchens *et al.*, 2007, 2008; Saovapakhiran *et al.*, 2009; Gillies & Frechet, 2005; Asthana *et al.*, 2005). It is believed that the primary amine groups on the surface of PAMAM could bind and compact DNA to form the complex, and then promote DNA cellular uptake. Meanwhile, the buried tertiary amino groups of PAMAM act as a proton sponge in endosomes, thus enhancing the release of DNA into the cytoplasm (Yoon *et al.*, 2000; Santos *et al.*, 2010). In addition, the primary amines located on the surface of PAMAM make it possible for the dendrimer to be well modified with functional molecules. Because of its relatively high transfection efficiency, PAMAM is the one of the most commonly used dendrimers for gene delivery (Bowen *et al.*, 2002). On the other hand, there are conflicting reports regarding their biological safety (Malik *et al.*, 2000; Roberts *et al.*, 1998). PEG has been conjugated on the surface of PAMAM to engineer a nontoxic, highly transfection effective and biocompatible gene carrier (Wang *et al.*, 2009). Brain-targeting ligands–modified PEGylated PAMAM have been designed as nonviral gene vectors, shown in Fig. 12.1 (Huang *et al.*, 2007b; Ke *et al.*, 2009; Liu *et al.*, 2009; Huang *et al.*, 2008). The transfection efficiency in brain of modified PAMAM-PEG/DNA complex was higher than that of PAMAM/DNA complex without modification both *in vitro* and *in vivo*. Meanwhile, the toxicity of modified PAMAM-PEG/DNA complex decreased (Huang *et al.*, 2009b).

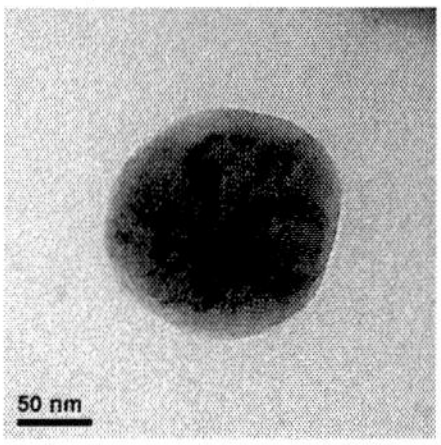

Figure 12.1 Transmission electron microscope result of PAMAM-PEG/DNA NPs.

12.2.3.3 Polyamino acids

Degradable polyamino acids are made of amino acids, most of which are cationic to facilitate DNA binding. The amido linkage between the monomer of amino acids is thought to be hydrolyzed by enzymes (Tsogas *et al.*, 2007). Poly-L-lysine (PLL) was one of the first gene transfer polymers to be used (Wu & Wu, 1987; Pouton *et al.*, 1998). A self-assembled nonviral gene carrier, PLL-modified iron oxide NPs (IONP-PLL), which is formed by modifying PLL to the surface of IONP has been developed (Xiang *et al.*, 2003). After i.v. injection, IONP-PLL, transferred reporter gene EGFP-C2 to lung, brain, spleen, and kidney. Furthermore, IONP-PLL transferred exogenous DNA across the BBB to the glial cells and neuron of brain. Dendrigraft poly-L-lysine (DGL) has emerged as a new kind of synthetic polymer consisting of lysine (Tsogas *et al.*, 2007; Klok *et al.*, 2002; Cottet *et al.*, 2007) and has been employed as gene vector due to its degradability and rich external amino groups. Besides their biodegradability (Tsogas *et al.*, 2007), the external amino groups could encapsulate DNA through electric interactions. Furthermore, they could be modified with PEG and targeting ligand, rendering vectors long circulation and targeting properties.

DGL-PEG-Leptin30 was complexed with plasmid DNA yielding NPs (Liu *et al.*, 2010). It has been proved that targeted NPs can be effectively transported across the BBB in the *in vitro* BBB model, and most of them can be accumulated in brain after an i.v. injection, resulting in relatively high gene transfection efficiency both *in vitro* and *in vivo*.

12.2.4 Polymeric Micelles

A stable core–shell structure of polymeric micelles is formed when amphiphilic copolymers associate spontaneously in aqueous phase. Self-assembly starts when the copolymer concentration reaches a threshold value known as the critical micelle concentration (CMC). The core of polymeric micelles is usually composed of hydrophobic polymer [such as poly–DL lactic acid (PLA), poly(caprolactone), and hydrophobic saturated aliphatic chain], and the shell is composed of hydrophilic blocks (commonly PEG). Most of them are biodegradable and biocompatible (Denora *et al.*, 2009). Polymeric micelles are characterized by a smaller size, usually between 10 and 100 nm. Small size is considered as one of the major requirements for effective delivery of NPs to brain (Koziara *et al.*, 2003). This is possibly because smaller particles are not taken up by RES *in vivo*, prolonging their circulation time in the bloodstream which provides the necessary condition for effective drug delivery to the target site.

Pluronic® block copolymers are commonly used materials for preparing polymeric micelles. These block copolymers consist of hydrophilic ethylene oxide and hydrophobic propylene oxide blocks. Pluronic® block copolymer micelles can penetrate the membrane of BCECs and inhibit loaded drug efflux by P-gp (Batrakova *et al.*, 2005). Therefore using Pluronic® block copolymers as materials for incorporating a variety of hydrophobic P-glycoprotein substrate drugs can increase the accumulation of drugs in the brain. A recent study has shown that polymeric micelles self-assembled from the transcriptional activator peptide, transcriptional activator protein–PEG-block-cholesterol (TAT–PEG–b–Chol) can be used as carrier for drug delivery across BBB with ciprofloxacin as a model drug. The uptake of micelles with TAT by human brain endothelial cells was greater than those without TAT. Micelles with TAT can be used as promising carriers for delivering antibiotics across the BBB to treat brain infections (Liu *et al.*, 2008a,b). PE-PEG-based micelles as nanoscaled drug carriers (Fig. 12.2) were investigated by incorporating a poorly water-soluble antibiotic amphotericin B with 80% drug-entrapping efficiency and 12% drug-loading efficacy. Angiopep-2 was conjugated to the polymeric micelles surface and the penetration efficiency of the brain-targeting delivery was evaluated *in vitro* and *in vivo*. Angiopep-2-modified PE-PEG-based polymeric micelles have the capacity to increase the internalization of AmB by BCECs and to reduce AmB cytotoxicity to mammalian cells. Angiopep-2 can improve the penetration of micelles into the brain. The mechanism of Angiopep-2-modified micelles penetrating into the brain has been investigated. Rhodamine 123, a substrate of P–gp, was incorporated into Angiopep-2-modified micelles and was administered to mice through tail vein. The brain distribution of Rhodamine 123 was observed by fluorescence microscope. Significantly high accumulation of Rhodamine 123–incorporated Angiopep-2-modified micelles was observed in the mouse brain, whereas unmodified micelles were observed to be accumulated only in some regions of the mouse brain. It probably indicated that incorporation of drugs into the micelles decreases the possibility of drugs recognized by P-gp, and the transcellular traffic of angiopep-2 modified micelles could probably bypass the P-gp systems (Shao *et al.*, 2010).

Figure 12.2 AFM image of AmB-incorporated angiopep-PEG-PE micelles.

12.2.5 Polymeric Nanoparticles

Polymeric NPs are solid colloid particles made of polymer materials and with diameters ranging from 1 nm to 1000 nm. These polymer materials include poly-methylacrylic acid, poly-alkylcyanoacrylate, poly-D,L-lactide-glycolide, and polylactic acid. We can also use natural proteins (such as albumin and gelatin) and polyose (polyglucosan, starch, and chitosan) as polymer materials.

Poly(D,L-lactide-co-glycolide) (PLGA) and polylactide (PLA) are two of the few polymers officially approved for clinical use and have several qualities that make them appealing for brain delivery. PLGA and PLA also easily form hydrolysable bonds with drugs (Dechy-Cabaret *et al.*, 2004; Leonard *et al.*, 2009; Lu *et al.*, 2009) and targeting ligands such as lectin (Lu *et al.*, 2009). Drug loading into PLGA/PLA nanocarriers modified with brain-targeting ligand could achieve drug delivery into brain. Multiple-fold increases in the concentration of the brain drug, which was administered via intranasal route, were indeed observed in PLGA/PLA systems (Gao *et al.*, 2009; Kim & Martin, 2006). Even large molecules such as peptides have been shown to be capable of crossing the BBB in animal models (Gao *et al.*, 2009).

Surfactant-coated poly(butyl)cyanoacrylate (PBCA) NPs have been widely used to deliver drugs to the CNS. The notable example is the delivery of hexapeptide dalargin. Dalargin (Tyr-D-Ala-Gly-Phe-Leu-Arg), a leuenkephalin analog, is an artificial endorphin possessing opioid activity. The drug is not effective when administered intravenously, but exhibits its analgesic effect only after direct injection into the brain. Coumarin-6- and docetaxel-encapsulated PLGA NPs were prepared, and the NPs' surfaces were further modified with polysorbate-80 (Tween® 80), poloxamer 188 (F68), or poloxamer 407 (F127). As a result, the F68-coated PLGA NPs demonstrated the greater cellular uptake and achieved higher fluorescence concentration in the brain tissues over those with T80 and F127 surface modification. Therefore, surface modification is a feasible and efficient strategy for NPs made of biodegradable polymers to deliver diagnostic and therapeutic agents across the BBB (Kulkarni & Feng, 2011). When the drug was bound to PBCA NPs and coated with polysorbate 80, a strong analgesic effect was observed both during hot plate test after i.v. injection and tail-flick test (Ramge *et al.*, 1999). The transport of tubocurarine-loaded NPs across the BBB was shown by rat perfusion experiments (Alyautdin *et al.*, 1998). The results suggested that polysorbate 80-coated PBCA NPs can transport tubocurarine across the BBB.

The NPs can get through the BBB not only by coating surfactant on the surface but also by conjugating PEG. Calvo *et al.* prepared poly(methoxy-PEG cyanoacrylate-*co*-*n*-hexadecyl cyanoacrylate) (PEG-PHDCA) NPs, and results demonstrated that NPs can penetrate into brain. Results showed that the ability of PEG-PHDCA NPs penetrating into brain is stronger than that of Tween 80-PHDCA, Poloxamer 98-PHDCA, and naked PHDCA. It was observed through the fluorescence microscope that these NPs distributed in the ependymal cells of choroid plexus and endothelial cells of Pia Mater and cerebral ventricle, while there were only a little of them in the vascular endothelial cells of the BBB (Calvo *et al.*, 2001).

12.2.6 Solid Lipid Nanoparticles

Solid lipid nanoparticles (SLNs) are dispersions of solid lipids stabilized with emulsifier or emulsifier/co-emulsifier complex in water. Solid lipids used to prepare SLN include widely used food lipids and waxes (e.g., cetyl palmitate) and commonly used emulsifiers include different kinds of poloxamers, polysorbates, lecithins, and bile salts.

SLNs can incorporate an antitumor drug, camptothecin with poloxamer 188. The biological distribution of camptothecin was studied in mice after i.v. administration (Yang *et al.*, 1999). The results showed that the area under the curve (AUC) of concentration versus time profiles normalized to the dose administered, while the mean residence times of the SLN formulation were much higher than those of free drug solution especially in brain. The mean residence time of camptothecin increased about 18 times in plasma compared with the same dose that was camptothecin free. It is suggested that the surface of the SLN coating of Poloxamer-188 caused the sustained release of the drug from SLN. Drug-incorporated SLN concentrated in brain and showed the highest AUC among the tested organs (10.4 folds). Similar results have been shown with oral administration of camptothecin in SLN formulation (Yang *et al.*, 1999). Thus, SLN may be an advantageous carrier system for targeting lipophilic anticancer drugs to the CNS when effective drug concentrations in tumor mass are desired for a prolonged period of time.

Emulsifying wax NPs were coated with thiamine for efficient cell binding to achieve specific targeting to brain. The results showed that surface modification with thiamine enhanced the interaction of the NPs with the cells due to specific association with the BBB thiamine transporter and increased the cell binding (Lockman *et al.*, 2003).

12.3 Brain-Targeting Strategies for Drug Delivery

Several strategies have been developed for nonviral gene delivery systems to overcome the BBB. First, as the membrane of BCECs is negatively charged, the most direct way is to deliver the positively charged gene loading NPs to brain through AME. Second, cell-penetrating peptides–modified NPs have the potent ability to translocate across the BBB (Lo *et al.*, 2008) and have been studied intensively during the last decades as materials well suitable for the development of delivery vehicles (Mathupala *et al.*, 2009; Belting & Wittrup, 2009). However, because of lack of selectivity, the above discussed drug delivery systems might cause unnecessary uptake by disease-unrelated cells or tissues. Third, receptor-mediated endocytosis is one of the major mechanisms by which various agents can cross the BBB. The BCECs can uptake the drug delivery systems modified with ligands or receptor monoclonal antibodies and can cross BBB via endocytosis pathway mediated by the specific ligand-receptor binding. In addition, the blood-to-brain influx transporters could be exploited as a new brain-targeting delivery approach for nonviral gene delivery systems.

12.3.1 Adsorptive-Mediated Endocytosis

AME is triggered by an electrostatic interaction between a positively charged moiety of the substances and a negatively charged plasma membrane surface region. Adsorptive-mediated transcytosis provides a means for brain delivery of medicines across the BBB. The concept of drug delivery utilizing AME through the BBB was originally proposed for cationized albumin. Nevertheless, specific targeting by AME to the CNS cannot be expected because of nonspecific operation in other tissues, and the lower affinity and higher capacity of AME compared with those of receptor-mediated endocytosis should be favorable for the delivery of drug to the brain (Pardridge, 1995). Lu *et al.* developed and evaluated a novel drug carrier for brain delivery, cationic bovine serum NP (CBSA–NP) (Lu *et al.*, 2005). Native bovine serum albumin (BSA) was cationized and thiolated, followed by conjugation through the maleimide function located at the distal end of PEG surrounding the NP's surface. The qualitative and quantitative results of CBSA–NP uptake experiment compared with those of BSA–NP showed that rat BCECs took in much more CBSA–NP than BSA–NP at 37°C and at different concentrations and time incubations. After injections of CBSA-NP and BSA-NP in mice caudal veins, fluorescent microscopy of brain coronal sections showed a higher accumulation of CBSA–NP in the lateral ventricle, third ventricle, and periventricular region than that of BSA–NP. In the same group, plasmid pORF-hTRAIL (pDNA) was incorporated into CBSA-NP and the resulting CBSA-NP-hTRAIL was evaluated as a nonviral vector for gene therapy of gliomas (Lu *et al.*, 2006).

12.3.2 Cell-Penetrating Peptides–Mediated Transmembrane Transport

Cell-penetrating peptides are short peptides that facilitate cellular uptake of various molecular cargos. Since the first cell-penetrating peptides TAT were discovered by Frankel *et al.* (1988), a lot of cell-penetrating peptides have been developed and applied in targeting drug delivery systems, especially for macromolecular drugs which cannot cross the membrane barrier easily. Recently, researches on the cell-penetrating peptides–modified brain-targeting drug delivery strategy have increased. Till now, several cell-penetrating peptides including TAT, Syn-B, and Penetratin (Schwarze *et al.*, 1999; Rousselle *et al.*, 2001) have been proven to mediate the drugs or drug delivery systems into the brain (Table 12.1).

Table 12.1 Several brain-targeting cell-penetrating peptides

Peptide	Origin	Length (a.a.)	Sequence	Reference
TAT	HIV-1	11	YGRKKRRQRRR	Torchilin, 2008 Liu *et al.*, 2008b
Antp	Antennapedia	16	RQIKIWFQNRRMKWKK	Huang *et al.*, 2006a Rousselle *et al.*, 2000
LMWP	Protamine	14	VSRRRRRRGGRRRR	Mathupala *et al.*, 2009

LMWP: Low molecular weight protamine.

TAT peptide (TAT sequence: YGRKKRRQRRR, residues 47–57) was the first cell-penetrating peptide that could be efficiently uptaken by cells (Torchilin, 2008). The TAT peptide is effective in binding to negatively charged gene medicine through charge interaction and condensing with six arginines and two lysines in its eleven amino acid residues (Yoon *et al.*, 2004).

Polymeric micelles self-assembled from cholesterol-conjugated PEG and anchored with transcriptional activator TAT peptide (TAT-PEG-b-Col) were fabricated for delivery of antibiotics across the BBB (Liu *et al.*, 2008b). The results showed that the presence of TAT promoted an enhanced uptake of micelles by human astrocytes (ACs). Their distribution in hippocampus brain sections of rats was proved at 2 h after i.v. injection of micelles.

Syn-B peptide family is derived from the natural antimicrobial peptide protegrin 1, originally isolated from porcine leukocytes. They are able to interact with the cell surface and efficiently cross the plasma membrane. Syn-B1 (RGGRLSYSRRRFSTSTGR) and Syn-B3 (RRLSYSRRRF) are usually applied in brain-targeting drug delivery. Doxorubicin was coupled covalently to small peptide vectors: L-SynB1 (18 amino acids), L-SynB3 (10 amino acids), and its enantio form D-SynB3. *In situ* mouse brain perfusion method was used to evaluate the brain uptake vectorization with either L-SynB1, L-SynB3, or D-SynB3, and the result demonstrated that all the vector peptides could significantly increase the brain uptake of doxorubicin (about 30 folds). Besides, the inhibition effect of poly (L-lysine) and protamine, the endocytosis inhibitors, on the transport across the brain was also investigated. Both inhibitors reduced the brain uptake of vectorized doxorubicin in a dose-dependent manner (Rousselle *et al.*, 2001).

The exact mechanism of cell-penetrating peptides translocation is still not clear. Early research intended to believe that its entry to cells is energy-independent and has nothing to do with endocytosis. However, recently more and more researches have indicated that most of the cell-penetrating peptides entered the cells in an energy- and temperature-dependent manner. Their translocation may relate to the negative components in a membrane and be through the adsorptive-mediated endocytosis. The TAT–nucleic acid conjugate was shown to be time-, temperature- and concentration-dependent. Besides, caveolae involved in the process of cellular uptake. The cellular uptake of Syn-B and (Arg)9 also related to the adoptive-mediated endocytosis. The characteristics of Syn-B-mediated cellular uptake are that it is energy-dependent and saturable (Rousselle *et al.*, 2001, Drin *et al.*, 2002).

12.3.2.1 Antennapedia peptide (Antp)

Antennapedia, a Drosophila transcription factor and membrane-penetrating peptide, has been reported to bypass the endosome–

lysosome pathway and directly enter the cells via a fusion mechanism. Antp, a 16-amino acid–long homeodomain corresponding to the third helix of Antennapedia, possesses translocating properties comparable to that of the entire Antennapedia (Derossi *et al.*, 1996). Antp has been successfully used to directly modify drugs and increase their amount in the brain, across BBB (Rousselle *et al.*, 2000).

Huang *et al.* (2006b) compared the Antp-modified chitosan, PEI, and PAMAM DNA loading NPs for brain-targeting delivery (Fig. 12.3). Lipofectamine2000 served as control. The incorporation efficiency was determined by agarose gel electrophoresis and PicoGreen assay. The expression efficiency was qualified with fluorescence microscopy and quantified with the luciferase assay system. Heparin and DNase I were employed to investigate the stability. The results indicated that the gene products of chitosan/DNA NPs and DNA/Lipofectamine 2000 complexes whether modified or unmodified with Antp could not be observed within BCECs. The PEI/DNA NPs and PAMAM/DNA NPs could be internalized into BCECs to a certain extent, and the expression efficiency was enhanced significantly with the modification of Antp. The NPs could protect themselves from the displacement of anionic substances and the digestion of DNase I. These observations demonstrated that polymers and DNA can form stable NPs, and the modification of Antp can enhance the expression efficiency of DNA-loaded NPs in BCECs.

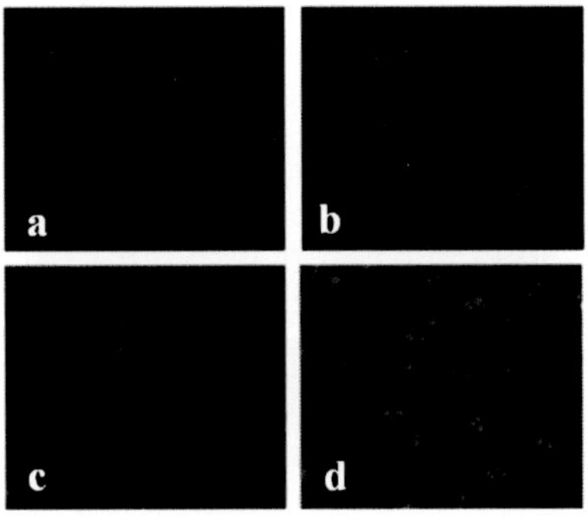

Figure 12.3 Confocal images of BCECs incubated with DNA/PEI NPs containing EMA-DNA at 4°C (a, b) and 37°C (c without Antp, d with Antp). Red: EMA-DNA. Original magnification: 200×.

In addition, an initial *in vivo* experiment in Drosophila showed that abdominal injection of biotin-tagged penetratin permeated the

BBB. The same effect was observed for biotin-tagged penetratin fused with apoE mimetic peptide which demonstrated anti-inflammatory and neuroprotective activities (Eguchi *et al.*, 2001).

12.3.2.2 Low molecular weight protamine

Several nontoxic, arginine-rich peptides termed LMWP are formed by enzymatic digestion of natural protamine (Byun *et al.*, 1999; Chang *et al.*, 2001a). These LMWPs were derived directly from native protamine by enzymatic digestion with thermolysin. It is most important to note that unlike other cationic proteins/peptides, these LMWPs elicited only minimal complement activation and no detectable hypotensive or toxic responses in dogs (Mathupala *et al.*, 2009). Furthermore, *in vitro* and animal studies also demonstrated that LMWPs were neither antigenic (Tsui *et al.*, 2001) nor mutagenic (Liang *et al.*, 2003). As these LMWP peptides possess high arginine content and carry a significant sequence similarity (Chang *et al.*, 2001a; Park *et al.*, 2003) to that of TAT47–57 (Schwarze *et al.*, 1999), the most extensively studied PTD to date, it is hypothesized that they possess a similar cell-translocating, potency-like TAT. In addition, the toxicity profiles of LMWP have already been thoroughly established (Chang *et al.*, 2001b; Tsui *et al.*, 2001; Liang *et al.*, 2003). Also, unlike the other PTDs, LMWP could be produced from native protamine in mass quantities with relative ease and low costs. LMWP have been studied intensively to improve the uptake and transfection efficiency of nonviral gene delivery systems. LMWP peptides demonstrated excellent DNA condensation ability and could form very compact DNA condensates with particles of small size (about 100 nm) (Kharidia *et al.*, 2008). More interestingly, LMWP peptide-mediated *in vitro* gene delivery showed prolonged (up to 12 days) gene expression. Results from the study suggest that designing DNA condensers with appropriate and tunable DNA binding strengths and condensation abilities would be an effective means to improve gene expression and thus gene therapy efficiency. Park *et al.* (2003) reported that the LMWP prepared by their group showed similar transcellular localization behavior and kinetics to those of TAT and efficiently transferred the pDNA into nucleus and cytoplasm in a short time period. The size and zeta potential of the pDNA/LMWP complex were adequately suitable for cellular uptake. After forming the complex, LMWP appeared to effectively protect pDNA against DNase I attack. The pDNA/LMWP complex showed significantly enhanced gene transfer than both naked pDNA and the pDNA/PEI complex and exhibited a markedly reduced cytotoxicity than that of the pDNA/PEI complex. The application of LMWP in brain-targeting gene delivery systems has not been reported yet.

12.3.3 Receptor-Mediated Endocytosis

Certain large molecule peptides or plasma proteins are selectively transported across the BBB via receptor-mediated transport systems, including the insulin receptor, TfR, leptin receptor, LDL-related protein receptor (LPR), and so on. This indicates that the BBB does possess specific receptor-mediated transport mechanisms that can potentially be exploited as a means to target drugs to brain. This part summarizes main researches on a variety of receptors which have now been exploited as targets in the field of brain drug delivery (Table 12.2).

Table 12.2 Receptor-mediated brain-targeting drug delivery system

Targeting site	Ligand	Drug delivery system	Reference
TfR	Tf	Liposome	Mishra *et al.*, 2006
		NPs	Huang *et al.*, 2007b
		Linkage-drug	Zhang *et al.*, 2001
	OX26 MAb	albumin NPs	Zhang *et al.*, 2003a
		Dendrimer NPs	Ulbrich *et al.*, 2009
		Liposome	Zhang *et al.*, 2006
	8D3MAb	Liposome	Zhang *et al.*, 2002a Shi *et al.*, 2001
LRP	ApoE	NPs	Michaelis *et al.*, 2006
	Angiopep-2	NPs	Ke *et al.*, 2009
	Polysobate80	NPs	Kurakhmaeva *et al.*,2009
Insulin receptors	HIRMab	Fusion protein	Boado *et al.*, 2008b
Leptin receptors	leptin30	NPs	Liu *et al.*, 2010
	Lep70-89	Liposomes	Tamaru *et al.*, 2010
Nicotinic acetylcholine receptor (nAchR)	RVG29	Complex	Mathupala *et al.*, 2009
		NPs	Liu *et al.*, 2009

12.3.3.1 Transferrin receptor

The TfR is present at a relatively high concentration on the vascular endothelium of the brain capillaries. The natural ligand to TfR is

Tf, which has been considered as a classic brain-targeting ligand. A large number of researches were carried out using Tf for modified brain drug delivery systems.

Huang *et al.* (2007b) synthesized a Tf-modified PEGylated PAMAM dendrimer-based vector, PAMAM-PEG-Tf, for improving *in vitro* and *in vivo* transfection efficiency of exogenous genes. The results showed that the transfection efficiency of PAMAM-PEG-Tf/DNA complex was much higher than that of PAMAM/DNA and PAMAM-PEG/DNA complexes in BCECs. The *in vivo* transfection potency of complexes was evaluated via frozen sections and measurement of tissue luciferase activity in Balb/c mice after i.v. administration. Results suggested that with a PAMAM to DNA weight ratio at 10:1, the brain gene expression of PAMAM-PEG-Tf/DNA complex was about two-fold higher than that of PAMAM/DNA and PAMAM-PEG/DNA complexes (Fig. 12.4). These results indicated that PAMAM-PEG-Tf can be exploited as a potential nonviral gene vector targeting to brain via i.v. administration.

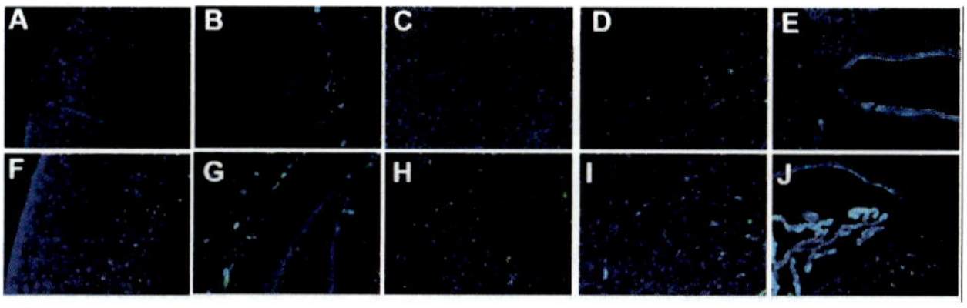

Figure 12.4 Distribution of gene expression in the brains of Balb/c mice treated with PAMAM-PEG/DNA nanoparticles (panel A-E) and PAMAM-PEG-Tf/DNA nanoparticles (panel F-J) two days after i.v. administration. Frozen sections (20 μm thick) of cortical layer (A, F), hippocampus (B, G), caudate putamen (C, H), substantia nigra (D, I) and 4th ventricle (E, J) were examined by fluorescent microscopy. The sections were stained with 300 nM DAPI for 10 min at room temperature. Green: GFP. Blue: cell nuclei. Original magnification: 100×.

Cardoso *et al.* (2007) reported that Tf-modified cationic lipid-based vectors can enhance siRNA delivery to neurons both *in vitro* and *in vivo*. The optimized Tf-lipoplexes were applied to transfer siRNA targeting the firefly luciferase reporter gene into primary neuronal cultures and to evaluate Tf-lipoplex-mediated siRNA delivery as well as luciferase and c-Jun silencing *in vivo*, after stereotactic injection in the striatum of the transgenic luciferase reporter mice. The results demonstrated that Tf-lipoplexes could efficiently promote siRNA delivery to neuronal cells both *in vitro* and *in vivo*, resulting in enhanced siRNA internalization and pronounced gene silencing effects, proposing a great potential for therapeutic applications.

Besides gene medicine, Tf was also used to modify brain delivery systems loading hydrophilic drugs/peptides, which have poor BBB permeability (Soni *et al.*, 2005, 2008; Gupta *et al.*, 2007; Mishra *et al.*, 2006; Gan & Feng *et al.*, 2010; Jain *et al.*, 2011). Soni *et al.* reported a Tf-coupled liposomal system for brain delivery of 5-florouracil (Soni *et al.*, 2005, 2008). 5-florouracil and (99m)Tc-DTPA bearing noncoupled liposomes were prepared by cast film method, which were coupled with the Tf by incubating these liposomes with Tf in the presence of the 1-ethyl-3-(3-dimethylaminopropyl) carbodiimide hydrochloride in saline phosphate buffer (pH 7.4). The size of the liposomes increased after coupling with Tf while the percentage of drug entrapment reduced. The results of the *in vitro* release profile demonstrated that noncoupled liposomal formulation releases a comparatively higher percent (i.e., 74.8+/−3.21%) of drug than coupled liposomes. Results of *in vivo* study suggested a selective uptake of the Tf-coupled liposomes from the BCECs. In case of coupled liposomes, the level of radioactivity was 17-fold more compared with the free radioactive agent and 13 times more with the noncoupled liposomes. Therefore, it could be concluded that using Tf-coupled liposomes the brain uptake of the drug could be enhanced. Furthermore, the potential of engineered albumin NPs was evaluated for brain-specific delivery after i.v. administration by Mishra and coworkers (2006). Long circulatory PEGylated albumin NPs encapsulating water-soluble antiviral drug azidothymidine (AZT) were prepared by ultra-emulsification method using chemical cross-linking by glutaraldehyde. The surface of the PEGylated NPs was modified by anchoring Tf as a ligand for brain targeting. Fluorescence studies revealed the enhanced uptake of Tf-anchored NPs in the brain tissues when compared with unmodified NPs. *In vivo* evaluation was carried out on albino rats to evaluate tissue distribution of engineered NPs after i.v. administration. A significant enhancement of brain localization of AZT was observed for Tf-anchored PEGylated albumin nanoparticles (Tf-PEG-NPs). Hence, the specific role of Tf ligand on NPs for brain targeting was confirmed. Gan and Feng (2010) also developed a NP system for targeted drug delivery across the BBB, which consists of the Tf-conjugated NPs of poly(lactide)-D-alpha-tocopheryl PEG succinate (PLA-TPGS) diblock copolymer. Cellular uptake and cytotoxicity of the Tf-conjugated PLA-TPGS NPs formulation of coumarin 6 as a model imaging agent or Docetaxel as a model drug were investigated in close comparison with those for the PLGA NPs formulation, the bare PLA-TPGS NPs formulation as well as with the clinical Taxotere. The Tf-conjugated PLA-TPGS NPs formulation demonstrated great advantages over the other two NPs formulations and the original imaging/therapeutic agents. IC_{50} data showed that the Tf-conjugated PLA-TPGS NPs formulation of Docetaxel could be 23.4%, 16.9%, and 229% more efficient than the PLGA NPs, the PLA-TPGS NPs formulations and Taxotere after 24 h treatment, respectively. Moreover, the preliminary *ex vivo* biodistribution investigation demonstrated that although not as satisfactory, the Tf-conjugated PLA-TPGS NPs formulation could deliver imaging/therapeutic agents across the BBB.

However, it has been reported that TfR are almost saturated under physiologic conditions because of high endogenous plasma concentration of Tf (Pardridge *et al.*, 1987). Thus, Tf itself is limited as a brain drug transport ligand (Aktas *et al.*, 2005). Another candidate, a mouse monoclonal antibody to the rat Tf receptor (OX26), has been studied and has shown some ability in the delivery of therapeutic agents to the brain (Qian *et al.*, 2002; Gosk *et al.*, 2004; Aktas *et al.*, 2005). Pang *et al.* (2008) prepared PEG-poly(caprolactone) (PEG-PCL) polymersomes conjugated with OX26, OX26-PO, and evaluated its brain-delivery property. The enzyme-linked immunosorbant assay result indicated that the surface OX26 densities ranged from 5 to 92. The optimized OX26 number conjugated per polymersome was 34, which can acquire the greatest BBB permeability surface area product and percentage of injected dose per gram brain (%ID/g brain). Furthermore, NC-1900, as a model peptide, was encapsulated into OX26-PO. Improved scopolamine-induced learning and memory impairments in a water maze task were observed via i.v. administration. These results indicated that OX26-PO is a promising carrier for peptide brain delivery. Olivier *et al.* (2002) synthesized immunonanoparticles by conjugation of OX26 to maleimide-grafted PEGylated NPs based on PLA and bi-functional PEG. The number of OX26 MAb molecules conjugated per individual PEGylated NP was 67 ± 4. The same group described the expression of DNA-loaded PEGylated immunoliposomes conjugated with the OX26 in brain after i.v. administration (Shi & Pardridge, 2000). This enabled targeting of DNA to brain via the endogenous BBB TfR. Unlike cationic liposomes, the neutral liposome formulation is stable in blood and does not result in selective entrapment in the lung. The gene expression in the brain peaks at 48 h after a single i.v. administration of 10 μg of DNA per adult rat, a dose that is 30- to 100-fold lower than that used for gene expression in rodents with cationic liposomes. They concluded that widespread gene expression in the brain can be achieved by using a formulation that does not employ viruses or cationic liposomes, but uses endogenous receptor-mediated transport pathways at the BBB. Furthermore, Ulbrich's group also manufactured human serum albumin (HSA) NPs by desolvation (Ulbrich *et al.*, 2009). Tf or Tf monoclonal antibodies (OX26 or R17217) were covalently coupled to the HSA NPs via bifunctional PEG. Loperamide was used as a model drug since it normally does not cross the BBB. The results showed that loperamide-loaded HSA NPs with covalently bound Tf or the OX26 or R17217 antibodies induced significant anti-nociceptive effects in the tail-flick test in ICR mice after i.v. injection, while control counterpart with IgG2a antibodies yielded only marginal effects. These results demonstrated that Tf or the antibodies covalently coupled to HSA NPs are able to transport loperamide and possibly other drugs across the BBB.

Nevertheless, Ji *et al.* argued that there was only a slight amount of OX26 in the brain after i.v. administration, and the brain uptake of OX26 and Tf was not significantly different (Ji *et al.*, 2006). Furthermore, commercially available OX26 antibodies are less prone to uptake into the brain tissue, unlike the one purified by the researchers themselves (Moos & Morgan, 2001). Thus, other monoclonal antibodies such as rat 8D3 mAb were also developed for brain-targeting drug delivery. Zhang and Pardridge (2005) directly conjugated bacterial beta-galactosidase to the rat 8D3 mAb via a streptavidin-biotin linkage. The results showed that unconjugated beta-galactosidase was rapidly removed from the blood compartment owing to avid uptake by liver and spleen and minimal uptake was observed by brain. Following conjugation of the enzyme to the 8D3 TfRmAb, there was a 10-fold increase in brain uptake of the enzyme. The capillary depletion technique showed that more than 90% of the enzyme-8D3 conjugate that entered into the endothelial compartment of brain passed through the BBB to enter brain parenchyma.

12.3.3.2 Lactoferrin receptors

Lactoferrin (Lf) is an iron-binding protein involved in host defense against infection and severe inflammation; it accumulates in the brain during neurodegenerative disorders (Fillebeen *et al.*, 1999). It is becoming increasingly evident that Lf is a multifunctional protein to which several physiological roles have been attributed (Ward *et al.*, 2002). The multiple biological activities of Lf are mediated by Lf receptors (Suzuki *et al.*, 2005). Several lines of evidence including the results of immunohistochemistry and reverse transcriptase–polymerase chain reaction (RT-PCR) have indicated the presence of specific Lf receptors in the brain (Talukder *et al.*, 2003; Suzuki & Lonnerdal, 2004). Low density lipoprotein receptor–related protein (LRP) might be involved as Lf receptor because its specific antagonist, the receptor-associated protein (RAP), inhibits 70% of Lf transport (Fillebeen *et al.*, 1999). Recently, the Lf receptors were characterized, exhibiting two classes of binding sites with high affinity (K_d = 6.8 nM) and low affinity (K_d = 4815 nM) (Huang *et al.*, 2007a). As known, the plasma concentration of endogenous Lf is approximately 5 nM (Talukder *et al.*, 2003), much lower than K_d of Lf receptors in the BBB. The Lf receptor–mediated transcytosis through the BBB has also been demonstrated (Fillebeen *et al.*, 1999). All these results suggest the promising use of Lf as a ligand for facilitating drug delivery system into the brain as the Lf receptors are less prone to be saturated. Moreover, an increased expression of Lf receptors has been reported in patients with PD (Faucheux *et al.*, 1995; Grau *et al.*, 2001), which showed further potential of Lf for clinical therapy of neurodegenerative diseases.

Hu *et al.* constructed Lf-conjugated PEG-poly(lactide) NP (Lf-NP) as a novel biodegradable brain drug delivery system (Hu *et al.*, 2009). The Lf ELISA results confirmed the biorecognitive activity of Lf after the coupling procedure and suggested that the average number of Lf conjugated on each NP was around 55. To evaluate the brain delivery properties of the Lf-NP, a fluorescent probe, coumarin-6 was incorporated into it. The uptake of Lf-NP by bEnd.3 cells was shown significantly higher than that of unconjugated NP. Following an i.v. administration, nearly three folds of coumarin-6 were found in the mice brain carried by Lf-NP compared to that carried by NP. Cell viability experiment results confirmed good safety of the biodegradable Lf-NP. The significant *in vitro* and *in vivo* results suggest that Lf-NP is a promising brain drug delivery system with low toxicity. In another research, Lf-modified procationic liposome (Lf-PCL) was developed as a new drug carrier for brain delivery (Chen *et al.*, 2010). The procationic liposomes (PCLs) were neutral or negatively charged at physiological pH, and when they touched BCECs with the help of Lf, they were changed into cationic liposomes (CL). The PCLs and Lf-PCLs with different CHETA/Lf ratio were prepared and characterized. An *in vitro* model of BBB, developed by the co-culture of BCECs and ACs, was employed to evaluate the ability and mechanisms of liposomes to cross endothelial cells. The liposome uptake by the mouse brain *in vivo* was detected by HPLC-fluorescence analysis. The results indicated that compared with the conventional liposomes and CLs, PCL and Lf-PCLs showed an improved performance in the uptake efficiency and cytotoxicity. Besides the uptake mediated by clathrin-dependent endocytosis of PCL, Lf-PCL crossed the BCECs through lipid raft/caveloae-mediated endocytosis. The endocytosis involved in the transport of Lf-PCL crossing BBB was mediated by both receptor- and absorption-mediated transcytosis. Compared with the conventional liposomes, PCL and Lf-PCL-8 (CHETA/Lf ratio = 1:8, w/w) were observed to show much improved characteristics of the localization in the brain. This study suggested that Lf-PCL was an available brain drug delivery carrier with potential future application.

Some studies compared the ability of Lf and Tf as brain-targeting ligands. Huang *et al.* (2008) applied Lf as a brain-targeting ligand in the design of PAMAM-based nonviral gene vector to brain, using Tf as a positive control ligand. Using PEG as a spacer, PAMAM-PEG-Lf and PAMAM-PEG-Tf was successfully synthesized. This vector showed a concentration-dependent manner in the uptake in BCECs. The brain uptake of PAMAM-PEG-Lf was 2.2 folds compared to that of PAMAM-PEG-Tf *in vivo*. The transfection efficiency of PAMAM-PEG-Lf/DNA complex was higher than that of PAMAM-PEG-Tf/DNA complex *in vitro* and *in vivo*. The results of frozen sections showed the widespread expression of an exogenous gene in mouse brain via i.v. administration. With a PAMAM/DNA weight ratio of 10:1, the brain gene expression of the PAMAM-PEG-Lf/DNA complex was about 2.3 folds when compared with that of the PAMAM-PEG-Tf/DNA complex. These results indicated that PAMAM-PEG-Lf could be exploited as a potential nonviral gene vector targeting to the brain via noninvasive administration. Furthermore, the mechanisms of Lf-modified NPs to brain were systematically investigated by the same group (Huang *et al.*, 2009a). The uptake of Lf-modified vectors and NPs by BCECs was related to clathrin-dependent endocytosis, caveolae-mediated endocytosis, and macropinocytosis. The intracellular trafficking results showed that Lf-modified NPs could rapidly enter the acidic endolysosomal compartments within 5 min and then partly escape within 30 min. Both Lf-modified vectors and NPs showed higher BBB-crossing efficiency than unmodified counterparts (Fig. 12.5). All the results suggest that both receptor- and adsorptive-mediated mechanisms contribute to the cellular uptake of Lf-modified vectors and NPs. Enhanced brain-targeting delivery could be achieved through the synergistic effect of the macromolecular polymers and the ligand. This effect was also verified *in vivo* (Huang *et al.*, 2010a).

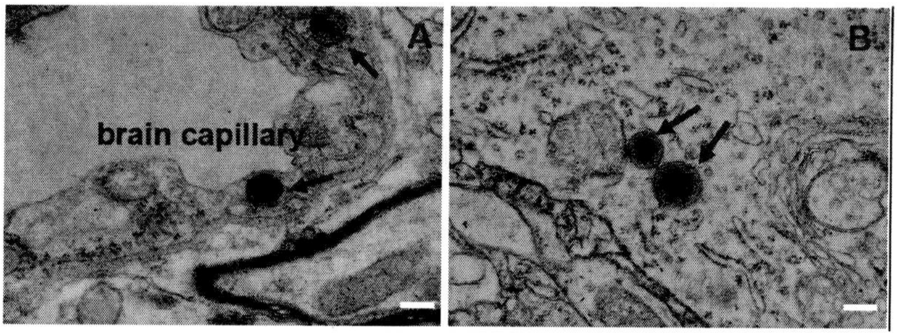

Figure 12.5 Observation of Lf-modified NPs using analytical transmission electron micrographs. The brain capillary (A) and brain tissues (B) of balb/c mice intravenously injected with copper–chlorophyll-labeled Lf-modified NPs are shown. The NPs are indicated by black arrows. Bar = 200 nm.

However, a different conclusion was given by Gao *et al.* (2010). They evaluated the effect of Lf and Tf in brain targeting, using polymersomes as vectors. It was shown that the uptake of Lf-polymersomes and Tf-polymersomes by bEnd.3 cells was time-, temperature-, and concentration-dependent. Both Lf and Tf could increase the cell uptake of polymersomes at 37°C, but the uptake of Tf-polymersomes was significantly greater than that of Lf-polymersomes. *In vivo* tissue distribution and pharmacokinetics in mice revealed higher brain uptake and distribution of Tf-polymersomes than Lf-polymersomes, which was in accordance with *in vitro* uptake results. The drug-targeting index of Tf-polymersomes with regard to Lf-polymersomes was 1.51.

12.3.3.3 Insulin receptors

Circulating insulin crosses the BBB into the CNS. There are many insulin receptors in various areas of the brain; they are expressed by both ACs and neurons (Laron, 2009). Certain brain tumors over-express insulin receptors.

The BBB expresses insulin receptors that mediate the uptake of endogenous insulin (Duffy & Pardridge, 1987). The BBB insulin receptors also transport certain peptidomimetic monoclonal antibodies, which bind an exofacial epitope on the BBB insulin receptors, and this binding enables the mAb to "piggy-back" across the BBB via the endogenous brain capillary endothelial insulin receptor (Pardridge, 2002). Earlier work showed that the murine 83-14 MAb to the human insulin receptor (HIRMab) was taken up by isolated human brain capillaries, was used as an *in vitro* model system of the human BBB, and that this uptake was mediated by the BBB HIR (Prigent *et al.*, 1990). In addition, the murine HIRMAb bound the BBB insulin receptor of Old World primates, such as the rhesus monkey. Approximately 2% of the injected dose of the HIRMAb was rapidly taken up by the rhesus monkey brain *in vivo* following an i.v. administration (Pardridge, 1995). Both chimeric and humanized forms of HIRMAb have been genetically engineered (Paddison *et al.*, 2002; Spagnou *et al.*, 2004).

The HIRMAb could be used to deliver drugs to brain. Hui *et al.* re-engineered human tumor necrosis factor-alpha receptor extracellular domain (TNFR ECD) by fusion of the receptor protein to the carboxyl terminus of the chimeric HIRMAb (Hui *et al.*, 2009). The HIRMAb-TNFR fusion protein is bifunctional and binds both the HIR, to trigger receptor-mediated transport across the BBB, and TNFα, to sequester this cytotoxic cytokine. COS cells were dual transfected with the heavy-chain (HC) and light-chain fusion protein expression plasmids, and the HC of the fusion protein was immunoreactive with antibodies to both human IgG and TNFR. The HIRMAb-TNFR fusion protein bound to the ECD of the HIR with an affinity comparable to the HIRMAb, and bound TNFα with a K_d of 0.34 +/− 0.17 nM. Both the TNFR:Fc fusion protein and the HIRMAb-TNFR fusion protein blocked the cytotoxic actions of TNFα on human cells in a bioassay. In conclusion, these studies describe the re-engineering of the TNFR ECD to make this decoy receptor transportable across the human BBB. Boado *et al.* also described the genetic engineering, expression, and validation of a fusion protein of avidin (AV) and HIRMAb (Boado *et al.*, 2008b). The 15 kDa AV monomer was fused to the carboxyl terminus of the HC of the HIRMAb. The fusion protein HC reacted with antibodies specific for human IgG and AV, and had the same affinity for binding to the HIR ECD as the original chimeric HIRMAb. The fusion protein qualitatively bound biotinylated ligands, but was secreted fully saturated with biotin by COS cells, owing to the high level of biotin in tissue culture medium. Chinese hamster ovary (CHO) cells were permanently transfected with a tandem vector expressing the fusion protein genes, and high expression cell lines were isolated by methotrexate amplification and dilutional cloning. The product expressed by CHO cells had high binding to the HIR and migrated as a homogeneous species in size exclusion HPLC and native polyacrylamide gel electrophoresis. The HIRMAb-AV increased biotin uptake by human cells by >15 folds and mediated the endocytosis of fludrescein-biotin, as demonstrated by confocal microscopy. In summary, the HIRMAb-AV fusion protein is a new drug-targeting system for humans that can be adapted to monobiotinylated drugs or nucleic acids. The same group further reported another research relating to HIRMAb using re-engineering technique (Boado *et al.*, 2008a). Human GDNF was re-engineered by fusing the mature GDNF protein to the carboxyl terminus of HIRMAb. The HIRMAb-GDNF fusion protein is bifunctional, and both bind the HIR, to trigger receptor-mediated transport across the BBB, and bind the GDNF receptor (GFR)-alpha 1, to activate GDNF neuroprotection pathways behind the BBB. COS cells were dual transfected with the HC and light-chain fusion protein expression plasmids, and the HC of the fusion protein was immunoreactive with antibodies to both human IgG and GDNF. The HIRMAb-GDNF fusion protein bound with high affinity to the ECD of both the HIR, ED_{50} = 0.87 +/− 0.13 nM, and the GFR alpha 1, ED_{50} = 1.68 +/− 0.17 nM. The HIRMAb-GDNF fusion protein activated luciferase gene expression in human neural SK-N-MC cells dual transfected with the c-ret kinase and a luciferase reporter gene under the influence of the rat tyrosine hydroxylase promoter, and the ED_{50} 11.68 +/− 0.45 nM was identical to the ED_{50} in the GFR alpha 1 binding assay. The fusion protein was active *in vivo* in a rat middle cerebral artery occlusion model, where the stroke volume was reduced 77% ($P < 0.001$). In conclusion, these studies describe the re-engineering of GDNF, to make this neurotrophin transportable across the human BBB. Similar research was carried out to transfer brain-derived neurotrophic factor (BDNF) across the

BBB (Boado *et al.*, 2007). Other than fusion technique, HIBMAb could be used to modify brain delivery systems. In another study, plasmids were encapsulated in the interior of an "artificial virus" comprised of an 85 nm PEGylated immunoliposome, which was targeted to the rhesus monkey brain *in vivo* with HIRMAb (Zhang *et al.*, 2003a). The HIRMAb enables the liposome carrying the exogenous gene to undergo transcytosis across the BBB and endocytosis across the neuronal plasma membrane following i.v. injection. The level of luciferase gene expression in the brain was 50-fold higher in the rhesus monkey compared with the rat. Widespread neuronal expression of the beta-galactosidase gene in primate brain was demonstrated by both histochemistry and confocal microscopy. This approach makes feasible reversible adult transgenics in 24 hours.

Moreover, Ulbrich *et al.* applied insulin or an anti-insulin receptor monoclonal antibody (29B4) to covalently couple to HSA NPs, using the NHS-PEG-MAL-5000 cross-linker. Loperamide loaded HSA NP with covalently bound insulin or the 29B4 antibodies induced significant antinociceptive effects in the tail-flick test in ICR (CD-1) mice after i.v. injection, demonstrating that insulin or these antibodies, covalently coupled to HSA NP, are able to transport loperamide across the BBB which it normally is unable to cross. Control loperamide-loaded HSA NP with IgG antibodies yielded only marginal effects. The loperamide transport across the BBB using the NP with covalently attached insulin could be totally inhibited by pretreatment with the antibody 29B4 (Ulbrich *et al.*, 2011).

12.3.3.4 Low-density lipoprotein receptor family

The low-density LRP is one of the low-density lipoprotein (LDL) receptor family and is highly expressed on not only the BBB but also glioma cells. LRP could bind numerous ligands, including proteinases, proteinase-inhibitor complexes, RAP, certain apoE- and lipoprotein lipase–enriched lipoproteins. Furthermore, LRP could mediate the cellular internalization of the ligands and their transport across the BBB (Bell *et al.*, 2006; Ito *et al.*, 2006; Shibata *et al.*, 2000). LRP is also reported to play important roles in the transport of some proteins across the BBB, including Lf and melanotransferrin (Fillebeen *et al.*, 1999). The feature that LRP exhibits strong transcytosis capacity and parenchymal accumulation allows people to consider it for receptor-mediated drug targeting to the brain.

Aprotinin, a 6500 Da protease inhibitor is a LRP, is a megalin (LRP2) ligand, and possesses a Kunitz protease inhibitor domain (Demeule *et al.*, 2008; Moestrup *et al.*, 1995; Hussain *et al.*, 1999). Alignment of the amino acid sequence of aprotinin with the Kunitz domains of human proteins allowed the identification and design of a family of peptides, named Angiopeps (Demeule *et al.*, 2008). *In vitro* and *in vivo* results showed that these peptides, and in particular Angiopep-2 (sequence: TFFYGGSRGKRNNFKTEEY), have a higher ability to accumulate in the brain than other proteins such as Tf and RAP, suggesting that the Kunitz-derived peptide could be advantageously used as a new brain-targeting ligand for pharmacological agents that do not readily enter the brain.

In some work, Angiopep-2 was linked to drugs directly for brain delivery. The uptake of angiopep-2 paclitaxel conjugate, ANG1005, into brain and brain metastases of breast cancer in rodents, was evaluated (Thomas *et al.*, 2009). The BBB K(in) for ^{125}I-ANG1005 uptake (7.3 +/− 0.2 × 10(−3) mL/s/g) exceeded that for 3H-paclitaxel (8.5 +/− 0.5 × 10(−5)) by 86 folds. Over 70% of ^{125}I-ANG1005 tracer stayed in brain after capillary depletion or vascular washout. Brain ^{125}I-ANG1005 uptake was reduced by unlabeled angiopep-2 vector and by LRP ligands, consistent with receptor transport. *In vivo* uptake of ^{125}I-ANG1005 into vascularly corrected brain and brain metastases was 4–54 folds more than that of ^{14}C-paclitaxel. The results demonstrate that ANG1005 shows significantly improved delivery to brain and brain metastases of breast cancer compared with free paclitaxel. Ché *et al.* described the synthesis and preliminary biological characterization of ANG1007 and ANG1009, two new chemical entities under development for the treatment of primary and secondary brain cancers (Ché *et al.*, 2010). ANG1007 consists of three doxorubicin molecules conjugated to Angiopep-2 which could cross the BBB by an LRP1 receptor-mediated transcytosis mechanism. ANG1009 has a similar structure, with the exception that three etoposide moieties are conjugated to Angiopep-2. Both agents killed cancer cell lines *in vitro* with similar IC_{50} values and with apparently similar cytotoxic mechanisms as unconjugated doxorubicin and etoposide. ANG1007 and ANG1009 exhibited dramatically higher BBB influx rate constants than unconjugated doxorubicin and etoposide and pooled within brain parenchymal tissue. Passage through the BBB was similar in Mdr1a (−/−) and wild type mice. These results provide further evidence of the potential of this drug development platform in the isolation of novel therapeutics with increased brain penetration.

In some other researches, Angiopep-2 was used to modify drug delivery systems to brain. Ke *et al.* (2009) exploited Angiopep-2 as a ligand for effective brain-targeting gene delivery. PAMAM dendrimers were modified with angiopep-2 through bifunctional PEG, then complexed with DNA, yielding PAMAM-PEG-Angiopep-2/DNA NPs. The angiopep-2-modified NPs showed higher efficiency in crossing BBB than unmodified NPs in an *in vitro* BBB model and accumulated in brain more *in vivo*. The angiopep-2-modified NPs also showed higher efficiency in gene expressing in brain than the unmodified NPs, indicating that PAMAM-PEG-Angiopep-2 held great potential to be applied in designing brain-targeting gene delivery system. The mechanism study showed that the angiopep-2-modified NPs were internalized by BCECs through a clathrin- and caveolae-mediated energy-depending endocytosis, also partly through marcopinocytosis. The cellular uptake of the angiopep-modified NPs was competed by angiopep-2, RAP, and Lf, indicating that LRP1-mediated endocytosis may be the main mechanism of cellular internalization of angiopep-2-modified NPs. A tandem dimer sequence of apoprotein-E (apoE) conjugated to poly-lysine sequence was used as a novel DNA delivery vector for transfection of brain cells either *in vitro* or *in vivo*. The vector helps to carry DNA through the blood stream to reach BBB and transport it to brain cells and eventually help DNA expression in target cells (Salehi, 2007). Shao *et al.* (2010) also exploited Angiopep-2 as brain-targeting ligand. Angiopep-2 modified PE-PEG-based micellar drug-delivery system loaded with the antifungal drug AmB to evaluate the efficiency of AmB accumulating into the brain. PE-PEG-based micelles as nanoscaled drug carriers were investigated by incorporating AmB with high drug entrapping efficiency, improving solubilization of AmB and reducing its toxicity to mammalian cells. The AmB-incorporated angiopep-2-modified micelles showed highest penetrating efficiency across the BBB than unmodified micelles and Fungizone (deoxycholate amphotericin B) *in vitro* and *in vivo*. Meanwhile, contrary to the free Rho 123, the enhancement of Rho 123-incorporated angiopep-2-modified micelles across the BBB can be explained by angiopep-2-modified polymeric micelles that have a potential to overcome the activity of efflux proteins expressed on the BBB such as P-glycoprotein. Therefore, angiopep-2-modified polymeric micelles could be developed as a novel drug delivery system for brain targeting. Due to the overexpression of LRP on both BBB and glioma cells, Angiopep-2 has dual-targeting effects when used for glioma therapy (Xin *et al.*, 2011).

12.3.3.5 Leptin receptors

Leptin is a peptide hormone produced primarily by adipose tissue that acts as a major regulator of food intake and energy homeostasis. Impaired transport of leptin across the BBB contributes to leptin resistance, which is a cause of obesity (Price *et al.*, 2010). Thus leptin itself is a candidate for the treatment of this obesity. In the brain, leptin acts on numerous different cell types via the long-form leptin receptor (LepRb) to elicit its effects. Substantial LepRb mRNA expression in hypothalamic and extrahypothalamic sites has been described before, including in the dorsomedial nucleus of the hypothalamus, ventral premammillary nucleus, ventral tegmental area, parabrachial nucleus, and the dorsal vagal complex. Expression in insular cortex, lateral septal nucleus, medial preoptic area, rostral linear nucleus, and in the Edinger-Westphal nucleus was also observed and has been previously unreported (Scott *et al.*, 2009). Leptin receptors are also highly expressed in some brain tumors including glioblastomas and anaplastic astrocytomas (Riolfi *et al.*, 2010).

Some researches focus on the brain drug delivery based on leptin receptors. Considering the 16 kDa MW of leptin preclude the possibility of passive diffusion across the BBB, a process that provides limited entry for some relatively small molecules, the leptin receptor–mediated transport is believed to be one of the possible mechanisms besides leptin receptor independent pathway. Barrett's group has identified several leptin-derived peptides that are taken up by the brain (Barrett *et al.*, 2009). Among these, peptide corresponding to positions 61 to 90 (leptin30, 30 amino acids) were observed to possess the highest brain:plasma ratio, which was equivalent to that of leptin. For leptin30, the majority of the radioactivity was localized more to parenchyma than capillaries. As the regions conferring brain uptake correspond closely to the sequences that are necessary for receptor binding, the results reinforce the view that the receptor pathway plays a major role in leptin uptake (Barrett *et al.*, 2009). Since leptin is known to be taken up into all regions of the brain (Banks *et al.*, 1996; Tang *et al.*, 2007), leptin30 could be exploited as the brain-targeting ligand modified on the surface of brain delivery systems.

Furthermore, it could be deduced that after transcytosis across the BBB, leptin30 could enhance internalization to the brain parenchyma cells by the specific ligand-receptor-mediated endocytosic pathway. Liu *et al.* (2010) reported a leptin30-modified nonviral gene delivery system. DGL is used as nonviral gene vector due to its rich external amino groups and biodegradability in the study. DGL-PEG-leptin30 was complexed with plasmid DNA yielding NPs. The targeted NPs were proved to be transported across *in vitro* BBB model effectively and accumulate more in brains after i.v. administration, resulting in relatively high gene transfection efficiency both *in vitro* and *in vivo*. Besides, the MTT assay revealed that the cytotoxicity of NPs at low, medium, and high concentration in BCECs was similar and relatively low which proved its safety in cellular level. The above results demonstrate that leptin30 modified nonviral gene vectors provided a safe and noninvasive approach for the delivery of gene across the BBB.

Another group investigated the function of a leptin-derived peptide (Lep70-89) as a ligand for mouse brain–derived endothelial cells (MBEC4) (Tamaru *et al.*, 2010). Lep70-89-modified liposomes, prepared with a PEG spacer (Lep70-89-PEG-LPs), exhibited a significantly higher cellular uptake than unmodified liposomes (PEG-LPs). Furthermore, the cellular uptake was inhibited by amiloride, while no significant inhibitory effect was observed by the presence of chlorpromazine and filipin III, suggesting that macropinocytosis largely contributed to the cellular uptake of Lep70-89-PEG-LPs. Imaging studies revealed that Lep70-89-PEG-LPs were not colocalized with endosome/lysosomes, whereas neutral dextran was predominantly colocalized with these compartments. This indicates that Lep70-89-PEG-LPs are taken up via macropinocytosis and are subject to nonclassical intracellular trafficking, resulting in the circumvention of lysosomal degradation in endothelial cells.

12.3.3.6 Nicotinic acetylcholine receptor

To date, cDNA studies and purification of nAchRs have shown that there are many different subtypes of the receptors in the central and peripheral nervous systems (Nakayama *et al.*, 1994). This could be used as the receptor-mediated basis for brain delivery. Kumar *et al.* (2007) reported that a 29-amino-acid peptide derived from rabies virus glycoprotein (RVG) could bind to the acetylcholine receptor (AchR) and be used for siRNA delivery across the BBB to neuronal cells. SiRNA was complexed with RVG peptide though electrostatic interaction by the nonamer arginine residues at the carboxy terminus of RVG peptide (RVG-9R). As the nAchR is present both on neurons and the vascular endothelium of the BBB, the siRNA/RVG-9R complex was effectively internalized across the BBB by receptor-mediated endocytosis and then targeted to the neurons in the CNS by i.v. administration, resulting in specific gene silencing within the brain. However, the exact mechanism by which RVG penetrates the BBB, and the cell-specific targeting within the CNS, remain to be elucidated (as cells other than neurons also express the AchR) (Mathupala *et al.*, 2009). In another experiment (Liu *et al.*, 2009), RVG peptide (defined as RVG29) was modified on the surface of PAMAM by covalent linkage bond through bifunctional PEG, yielding PAMAM-PEG-RVG29. The PAMAM-PEG-RVG29/plasmid DNA complex was nanoscaled. The NPs could be uptaken by BCECs and accumulate in brain efficiently. The PAMAM-PEG-RVG29/DNA NPs showed higher BBB-crossing efficiency than PAMAM/DNA NPs in an *in vitro* BBB model. Gene expression of the PAMAM-PEG-RVG29/DNA NPs was observed in the brain and was significantly higher than unmodified NPs (Fig. 12.6).

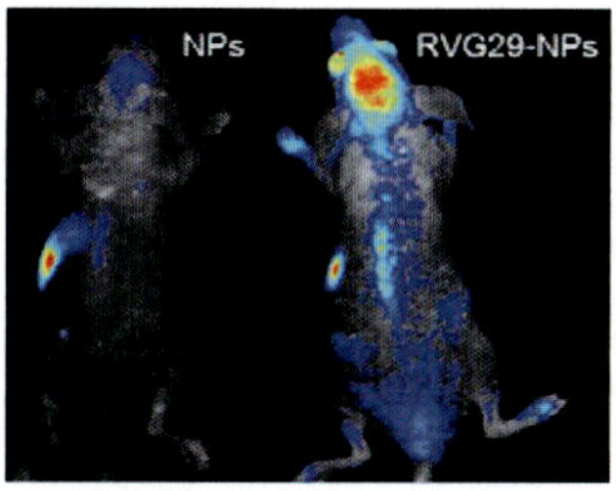

Figure 12.6 *In vivo* distribution of NPs after i.v. administration. Images were taken at 80 min after injection.

The mechanism of RVG29-mediated endocytosis was further explored. The cellular internalization of PAMAM-PEG-RVG29/DNA NPs could not be blocked by the different agonists or antagonists of nAchR in experiment. Though it is possible that the binding sites of the above agonists and antagonists are different from that of RVG29 on nAChR, we hypothesize that the α7 subunit of the nAchR seems not the only way to mediate transcytosis of the PAMAM-PEG-RVG29/DNA NPs across the BBB. As in the physiological condition, the nAchR, an ion channel receptor, can interact with nicotine leading to the ion current, such as Na+/K+, but has rarely been reported to involve in the endocytosis process. However GABA, the endogenic ligand of GABAB, could apparently inhibit the cellular uptake of PAMAM-PEG-RVG29/DNA NPs. It is suggested that the GABAB receptor possibly was involved in the mechanism of RVG29-mediated endocytosis pathway. It was reasonable that the GABAB receptor was demonstrated as a heterodimeric G-protein-coupled receptor (GPCR) involving clathrin-mediated endocytosis (Laffray *et al.*, 2007). The detailed mechanism of RVG peptide and its receptor-mediated endocytosis seems to require further researches. Nevertheless, RVG29 as an efficient brain-targeting ligand was undoubted and could facilitate noninvasive approach for gene delivery across the BBB.

12.4 Drug Delivery Systems Applied in Brain Diseases

While CNS drug therapy for inherited diseases still remains an extremely difficult goal to achieve, far more promise has been shown toward the reversal of course of acquired CNS diseases such as degenerative (Parkinson's and Alzheimer's) diseases and brain cancers. This can be achieved by introduction of transgenes encoding proteins which augment the natural survival and repair systems of the CNS, such as neurotransmitters, neurotrophic factors (NTFs) and their receptors, cytokines, and neuronal-survival (anti-apoptotic) agents by gene delivery systems (Roy *et al.*, 2008).

12.4.1 Parkinson's Disease

PD is a progressive, neurodegenerative disorder for which there is currently no effective neuroprotective therapy. It is typified by the progressive loss of substantia nigra pars compacta dopamine neurons and the consequent decrease in the neurotransmitter dopamine. Patients exhibit a range of clinical symptoms, with the most common affecting motor function and including resting tremor, rigidity, akinesia, bradykinesia, and postural instability (Feng & Maguire-Zeiss, 2010). The gold standard for treatment of both familial and sporadic PD is the peripheral administration of the dopamine precursor, levodopa. However, many patients gradually develop levodopa-induced dyskinesias and motor fluctuations. In addition, dopamine enhancement therapies are most useful when a portion of the nigrostriatal pathway is intact. Consequently, as the number of substantia nigra dopamine neurons and striatal projections decrease, these treatments become less efficacious. Therefore, research interests have been shifted to other therapeutic means including delivering therapeutic proteins or genes to diseased region via drug delivery systems.

12.4.1.1 Therapy with proteins

NTFs are a class of molecules that influence a number of neuronal functions, including cell survival and axonal growth. Experimental studies in animal models suggest that members of neurotrophin family and GDNF family of ligands (GFLs) have the potent ability to protect degenerating dopamine neurons as well as promote regeneration of the nigrostriatal dopamine system (Rangasamy *et al.*, 2010). However, these NTFs must be administered by intracerebral injection, because they could not cross the BBB. In the work done by Fu *et al.*, GDNF was re-engineered to enable receptor-mediated transport across the BBB following fusion of GDNF to the HC of a chimeric monoclonal antibody against the mouse TfR, and this fusion protein is designated as cTfRMAb-GDNF (Fu *et al.*, 2010). This fusion protein has previously shown to retain low nM-binding constants for both the GDNF receptor and the mouse TfR, and to rapidly enter the mouse brain *in vivo* following i.v. administration. Experimental PD in mice was induced by the intrastriatal injection of 6-hydroxydopamine, and mice were treated with either saline or the cTfRMAb-GDNF fusion protein every other day for 3 weeks, starting 1 h after toxin injection. Fusion protein treatment caused a 44% decrease in apomorphine-induced rotation, a 45% reduction in amphetamine-induced rotation, a 121% increase in the vibrissae-elicited forelimb placing test, and a 272% increase in striatal tyrosine hydroxylase enzyme activity at 3 weeks after toxin injection. Fusion protein treatment caused no change in tyrosine hydroxylase enzyme activity in either the contralateral striatum or the frontal cortex. Therefore, following fusion of GDNF to a BBB molecular Trojan horse, GDNF trophic effects in brain in experimental PD are observed following i.v. administration. In another study, Garbayo *et al.* (2009) investigated the neurorestorative effect of controlled GDNF delivery using biodegradable microspheres in an animal model with partial dopaminergic lesion. Microspheres were loaded with *N*-glycosylated recombinant GDNF and were prepared using the Total Recirculation One-Machine System. GDNF-loaded microparticles were unilaterally injected into the rat striatum by stereotaxic surgery two weeks after a unilateral partial 6-OHDA nigrostriatal lesion. Animals were tested for amphetamine-induced rotational asymmetry at different times and were sacrificed two months after microsphere implantation for immunohistochemical analysis. The putative presence of serum IgG antibodies against rat glycosylated GDNF was analyzed for addressing safety issues. The results demonstrated that GDNF-loaded microspheres improved the rotational behavior induced by amphetamine of the GDNF-treated animals together with an increase in the density of tyrosine hydroxylase positive fibers at the striatal level. The developed GDNF-loaded microparticles proved to be suitable to release biologically active GDNF over up to 5 weeks *in vivo*. Furthermore, none of the animals developed antibodies against GDNF demonstrating the safety of glycosylated GDNF use.

12.4.1.2 Gene therapy

Dopamine is synthesized in a two-step reaction from dietary tyrosine via L-dopa by the enzymes tyrosine hydroxylase and aromatic L-amino acid decarboxylase (AADC). Obviously, the most direct approach is to reconstitute the enzymes in the Parkinsonian striatum that are required for dopamine synthesis. It is reported that tyrosine hydroxylase gene delivery alone could increase dopamine synthesis in the partially denervated rodent striatum (Bjorklund & Kordower, 2010). Gene delivery of AADC has been pursued to increase the conversion efficacy of peripheral L-dopa to dopamine in the striatum. The rationale for this approach is that the decreased AADC levels in the parkinsonian striatum would render this enzyme rate-limiting in L-dopa pharmacotherapy. Xia *et al.* (2007a,b) reported that tyrosine hydroxylase expression plasmids were delivered to either cultured cells or to rat brain *in vivo* with Trojan horse liposomes, which target the nonviral plasmid DNA to cells via cell membrane receptors following i.v. administration. The pattern of tyrosine hydroxylase gene expression in cell culture and *in vivo* was similar: the cDNA form of the tyrosine hydroxylase gene was fast-acting with short duration of action, and the genomic form of the tyrosine hydroxylase gene was slow-acting with longer duration of action. The most-sustained replacement of striatal tyrosine hydroxylase enzyme activity in experimental PD was produced by combination gene therapy where both the cDNA and the genomic forms of the tyrosine hydroxylase gene were administered simultaneously.

As mentioned above, GDNF is generally believed to possess most potent trophic effects on dopaminergic neurons (Kordower *et al.*, 2000). Therefore, gene therapy of PD using therapeutic genes encoding GDNF family has been extensively studied in animals, even advanced to clinical testing (Hong *et al.*, 2008; Mochizuki *et al.*, 2008). Zhang and Pardridge (2009) reported that rats with experimental PD are treated with intravenous GDNF plasmid DNA (designated pTHproGDNF) and nonviral gene therapy using Trojan horse liposomes targeted with a MAb to the rat TfR. The GDNF transgene expression is under the influence of the rat tyrosine hydroxylase promoter. Rats were treated with 3 weekly injections of Trojan horse liposomes starting 1 week after the intracerebral injection of 6-hydroxydopamine. The dose of the pTHproGDNF was 10 μg/rat/week injection. Rats were tested with three assays of neurobehavior, and terminal striatal tyrosine hydroxylase enzyme activity was measured at 6 weeks following toxin administration, which is 3 weeks following the last administration of Trojan horse liposomes. Apomorphine-induced contralateral rotation was reduced 87% by Trojan horse liposomes gene therapy; amphetamine-induced ipsilateral rotation was reduced 90% by Trojan horse liposomes gene therapy; and whisker-induced forelimb placement abnormalities were reduced 77% with Trojan horse liposomes gene therapy. The improvement in neurobehavior correlated with a lasting 77% increase in striatal tyrosine hydroxylase enzyme activity, relative to saline-treated rats. The results demonstrated that near-complete abrogation of the neurotoxin effects are achieved with multiple intravenous dosing of GDNF plasmid DNA gene therapy, using receptor-targeted Trojan horse liposomes and a region-specific promoter. Huang *et al.* (2009b) also reported a multiple intravenous dosing administration study choosing human GDNF gene (hGDNF) as the therapeutic gene to be encapsulated in Lf-modified brain-targeting PAMAM-based NPs. The neuroprotective effects were examined in a rat PD model constructed by the unilateral lesion of striatum using 6-hydroxydopamine (6-OHDA), one of the most popular tools for PD studies (Deumens *et al.*, 2002). The results showed that increasing the number of injections of Lf-modified NPs loading hGDNF improved locomotor activity, reduced dopaminergic neuronal loss, and enhanced monoamine neurotransmitter levels in PD rats. Five injections of Lf-modified NPs loading hGDNF exhibited much more powerful neuroprotection than a single injection, indicating the effectiveness and feasibility of multiple dosing administrations. The results of toxicity tests demonstrated that the dosage of NPs used in the present study was safe enough for brain gene delivery. However, the strong unilateral lesion caused by 6-OHDA in rats is not consistent with the real progress of PD and cannot show direct relevance to human PD (Huang *et al.*, 2006a). Many studies showed that the chronic exposure of rotenone could cause highly selective lesions of dopaminergic neurons in the whole nigrostriatal system (Sherer *et al.*, 2003). The rotenone model could produce most of the features of PD including Lewy body formation in the nigral neurons, unlike other existing models (Sindhu *et al.*, 2005). Consequently, the rotenone-induced PD model was considered to more adequately mimic the pathogenesis and progress of PD than previously used models (Hirsch *et al.*, 2003). As a result, the therapeutic effect of hGDNF plasmid loading Lf-modified NPs is also tested in a rotenone-induced chronic rat model of PD after treatment with NPs encapsulating human GDNF gene via a regimen of multiple dosing i.v. administration (Huang *et al.*, 2010b). The results showed that multiple injections of Lf-modified NPs obtained higher GDNF expression and this gene expression was maintained for a longer time than the one with a single injection. Multiple dosing i.v. administrations of Lf-modified NPs could significantly improve locomotor activity, reduce dopaminergic neuronal loss, and enhance monoamine neurotransmitter levels on rotenone-induced PD rats, which indicate its powerful neuro-protective effects.

12.4.2 Alzheimer's Disease

Alzheimer's disease is clinically characterized by progressive cognitive decline, memory loss, and personality change, accompanied with behavioral and psychological symptoms of dementia, such as abnormal behavior, agitation, and mood swings. Till recently, it was believed that accumulation of amyloid beta (Aβ) is the first event in the pathogenesis of Alzheimer's disease (Selkoe, 2001; Small *et al.*, 2001). Pathological characteristics of the Alzheimer's disease brain include the amyloid plaques and neurofibrillary tangles (NFTs), which are formed via aggregation of extracellular amyloid β-peptide and intracellular hyperphosphorylated tau, respectively. Aβ is generated from a type I membrane protein, amyloidprecursor protein, through sequential proteolytic cleavages performed by β-secretase (β-site APP-cleaving enzyme 1) and γ-secretase, which consists of presenilin (PSEN), Aph-1, Pen-2, and nicastrin. β-Secretase-mediated cleavage of amyloid precursor protein gives rise to a soluble *N*-terminal fragment and a membrane-bound C-terminal fragment, which is further cleaved by γ-secretase to release Aβ consisting of 40–43 amino acids (Citron, 2004). Amyloidprecursor protein, which has been well conserved throughout evolution, is present in almost all cells and is abundantly expressed in neural cells. The role of Aβ degradation in the clearance of the Aβ peptide is becoming more broadly understood and appreciated, with the proteases neprilysin, insulin-degrading enzyme, endothelin-converting enzymes 1 and 2, plasmin, and cathepsin B, all capable of regulating Aβ levels *in vivo* (Hemming *et al.*, 2007). The multifactorial causes of Alzheimer's disease offer a variety of possible targets for gene therapy including neurotrophins that sustain growth and synaptic activity of neurons [nerve growth factor (NGF) and BDNF], enzymes directly involved in Aβ degradation (neprilysin, ECE and cathepsin B), Aβ burden-associated factors (APOE), and proteins involved in Aβ generation (β-site APP-cleaving enzyme 1 and amyloidprecursor protein) that are targets for siRNA mediated down-regulation (Nilsson *et al.*, 2010). Most researches concentrated on the viral based gene therapy (Brian *et al.*, 2008; Carty *et al.*, 2008; Perkins, 2007). For example, as ApoE has been shown to influence brain Aβ and amyloid burden, both in humans and in transgenic mice, direct intracerebral administration of lentiviral vectors expressing the three common human apoE isoforms differentially alters hippocampal Abeta and amyloid burden in the PDAPP mouse model of Alzheimer's disease (Dodart *et al.*, 2005). The results demonstrated that gene delivery of apoE2 may prevent or reduce brain Abeta burden and the subsequent development of neuritic plaques. Some viral gene therapy is under clinic trials. Unfortunately, it seems that there are few studies about the nonviral gene therapy for Alzheimer's disease so far. According to de Boer and Gaillard (2007), targeted delivery of the human neprilysin gene after an i.v. injection by novel nonviral drug–targeting technology, 2B-Trans™, might provide a human applicable method to increase Neprilysin activity and to decrease Aβ peptide concentrations in the brain.

In addition, immunotherapy of Alzheimer's disease could also regulate the already deposited or depositing Aβ. Kiyota *et al.* showed expression of the mouse interleukin (IL)-4 gene in beta-amyloid precursor protein + presenilin-1 bigenic mice attenuating Alzheimer's disease pathogenesis (Kiyota *et al.*, 2010). Recently, Aβ DNA vaccines were developed by using virus vectors (Haraa *et al.*, 2004; Zhang *et al.*, 2003c). Although these vaccines effectively decreased Aβ depositions in the brains of model mice, the possibility of viral replication could not be completely excluded. The plasmid vector is safe and has no possibility of viral infection and transformation because it exists as an episome without being built into the chromosome in eukaryotic cells (Nishikawa & Huang, 2001). Another important factor is related to technology. When DNA vaccines are in clinical use, large amounts of vaccines are necessary for treatment of a large number of patients who would be treated for a long period. Nonviral DNA vaccines have an advantage because they can be mass-produced with a high purity at a low price. It was reported that naked plasmid DNAs encoding proteins are taken into cells where they produce proteins in small amounts for a relatively long period when injected into the muscle or skin (Wolff *et al.*, 1990). Okura *et al.* (2006) developed nonviral Aβ DNA vaccines and were able to reduce the amyloid burden in the cerebral cortex and hippocampus of Alzheimer's disease model mice by vaccination. Therapeutic treatment started after Aβ deposition reduced Aβ burden to 50% at the age of 18 months. Importantly, this therapy induced neither neuroinflammation nor T-cell responses to Aβ peptide in both APP23 and wild-type B6 mice, even after long-term vaccination.

The NGF is essential for the survival of both peripheral ganglion cells and central cholinergic neurons in the basal forebrain. The model of acute scopolamine-induced amnesia in rats as well as in the model of 1-methyl-4-phenyl-1,2,3,6-tetrahydropyridine (MPTP)-induced Parkinsonian syndrome were employed to evaluate brain delivery of NGF adsorbed on poly(butyl cyanoacrylate) NPs coated with polysorbate 80 and the pharmacological efficacy of this delivery system (Kurakhmaeva *et al.*, 2009). The i.v. administration of the NP-bound NGF successfully reversed scopolamine-induced amnesia, improved recognition and memory, and reduced the basic symptoms of Parkinsonism (oligokinesia, rigidity, tremor). In addition, the efficient transport of NGF across the BBB was confirmed by direct measurement of NGF concentrations in the murine brain. Wu *et al.* (1997) reported that ^{125}I-Abeta1-40, a peptide radiopharmaceutical which is a potential imaging agent for brain disorders, was monobiotinylated (bio) and conjugated to a BBB drug delivery and brain-targeting system comprising a complex of the 83-14 monoclonal antibody to the human insulin receptor, which is tagged with streptavidin. A marked increase in rhesus monkey brain uptake of the ^{125}I-bio-Abeta1-40 was observed after conjugation to brain-targeting delivery system at 3 h after i.v. injection and was higher than that of unmodified one.

12.4.3 Brain Tumors

Brain tumors include any intracranial tumor created by abnormal and uncontrolled cell division, normally either in the brain itself (neurons, glial cells, lymphatic tissue, blood vessels), in the cranial nerves, in the brain envelopes, skull, pituitary and pineal gland, or spread from cancers primarily located in other organs (metastatic tumors). Glioblastoma multiforme is the deadliest and most common form of malignant brain tumor (Clarke *et al.*, 2010). Even when aggressive multimodality therapy consisting of radiotherapy, chemotherapy, and surgical excision is used, median survival is only 12–17 months. The current standard of care includes maximal safe surgical resection, followed by a combination of radiation and chemotherapy with temozolomide. Despite that, recurrence is quite common and more effective treatments both for initial therapy and at the time of recurrence still need to be worked upon. Gene therapy for brain tumors is one of the promising ways that may halt the progress of the disease or even cure the tumor. However, its application is also limited by the presence of BBB. Therefore, gene delivery systems are needed for gene therapy.

To solve the problem of drug delivery across BBB to brain tumor, quite a few CNS delivery strategies have been developed, among which the most promising approach is the receptor-mediated transport. By coupling drug-loaded vehicles with ligands which specifically recognize receptors on the BBB, the RMT strategy combines the advantages of brain targeting, high incorporation capacity, reduction of side effects, and circumvention of the multidrug efflux system. Some receptors are highly expressed on the endothelial cells forming the BBB, such as the insulin receptor, TfR, LDL receptor and its related protein, and others. There have been many reports about RMT strategy employing Tf, Lf, mouse anti-rat monoclonal antibody OX26 or angiopeps as the specific ligands to enable nanocarriers crossing the BBB (Huwyler *et al.*, 1996; Zhang & Pardridge, 2005; Hu *et al.*, 2009; Demeule *et al.*, 2008). TfR is highly expressed on the brain tumor cells, and mediated uptake of drugs and macromolecules has been studied extensively. Pang *et al.* (2008) synthesized the diblock copolymers of methoxy-PEG-PCL and Maleimide-PEG-PCL and applied them to prepare polymersomes conjugated with OX26. It was shown that OX26-polymersomes significantly enhanced theoretical gene delivery into the brain tumor. Angiopep-2 is a ligand of LDL receptor–related protein expressing on the BBB and glioma cells widely. Angiopep-2 has been used as a potential targeting moiety for enhancing drug delivery system to delivery drug into the brain tumor. Recently, it was proved that Angiopep-2-conjugated PEG-PCL NPs concentrated drug into glioma across the BBB. Chlorotoxin (CTX), which has been demonstrated to bind specifically to receptor expressed in glioma, was exploited as the targeting ligand to conjugate PAMAM via bifunctional PEG and constructing CTX modified NPs. The *in vivo* distribution of CTX-modified NPs in the brain glioma was higher than that of PEG NPs (Fig. 12.7) (Huang *et al.*, 2011).

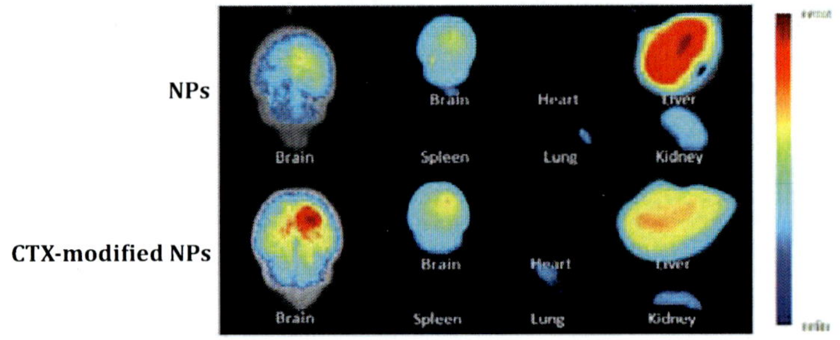

Figure 12.7 The fluorescent distribution of PEG NPs and chlorotoxin-modified NPs (CTX-modified NPs) in the main organs in glioma-bearing mice. NPs were intravenously administered. Intensity of the signal: dark red is the strongest while dark blue is the weakest, as shown by the bar.

12.4.3.1 Antisense/RNAi gene therapy

The therapeutic targets can be similar to that of other periphery tumors. For example, high-grade brain gliomas overexpress the epidermal growth factor receptor (EGFR) and EGFR antisense gene therapy can reduce the growth of EGFR-dependent gliomas. However, the difference between brain tumors and periphery tumors should be considered when designing gene therapy systems. It is still in controversy that the integrity of BBB is interfered by the progress of tumor growth. Some researchers hold the opinion that the integrity of BBB is destroyed at the initial stage of brain tumors (Lee *et al.*, 2006; Chertok *et al.*, 2008). Others believe that the integrity of BBB persists at the early stage of tumor and is destroyed gradually with the tumor processes (Huynh *et al.*, 2006). RNAi has the potential to knock down oncogenes in cancer, including brain cancer (Pardridge, 2004). However, the therapeutic potential of RNAi will not be realized until the rate-limiting step of delivery is solved. The development of RNA-based therapeutics is not practical due to the instability of RNA *in vivo*. Plasmid DNA can be engineered to express short hairpin RNA (shRNA), similar to endogenous microRNAs. RNAi-based gene therapy can be coupled with gene therapy that replaces mutated tumor suppressor genes to build a polygenic approach to the gene therapy of cancer. Viral vectors are considered the most effective and have been used in a number of animal models and the majority of brain tumor gene therapy clinical trials (Lawler *et al.*, 2005). Despite encouraging preclinical results, clinical studies have not shown similar success; few trials have shown substantial extended patient survival. This is thought to be caused by the following reasons: (i) lack of transport of the virus across the BBB; (ii) inflammation in brain following intracerebral administration of a single dose of either adenovirus or herpes simplex virus; (iii) a failure to distribute the viral vector throughout the tumor mass; (iv) a wide variability in the infectivity of brain tumor cells, resulting in an insufficient number of transfected tumor cells (Lawler *et al.*, 2005). As a result, nonviral gene vectors such as lipid- and polymer-based gene therapy strategies have been more extensively studied and have been tested in both preclinical and small clinical trials. Gene-targeting technology should enable the delivery to brain cancer of a nonviral formulation of the therapeutic gene via a simple i.v. injection. These delivery systems are comparatively simple, easy to formulate, and are generally less immunogenic and less neurotoxic compared to viral systems. Well-designed synthetic lipid vectors can be nearly as efficacious as viral vectors (Von Eckardstein *et al.*, 2001). For example, the efficiency of DNA-liposome complexes was similar to that of retrovirus-producing cells for HSV-tk suicide gene therapy (Princen *et al.*, 2000). Though there are quite a lot of researches on the liposome/DNA polyplex (Huynh *et al.*, 2006), there is rapid sequestration of the complex by lung, with no uptake by brain following i.v. injection of cationic lipid/DNA complexes. If antisense gene therapy is to be used for brain cancer, then alternative forms of gene-targeting technology need to be developed such as employing strategies discussed in the third section of the review.

The human EGFR antisense gene encoding plasmid with the SV40 promoter and elements that facilitate extra-chromosomal replication is packaged in the interior of 85 nm PEGylated immunoliposomes that is targeted to human glioma cells with the 83-14 murine MAb to the HIR, designated HIR MAb (Zhang *et al.*, 2002a). The HIR is expressed at both the BBB perfusing human brain gliomas and at the plasma membrane of human glioma cells. Endocytosis followed by entry into the nucleus was demonstrated for the HIR MAb conjugated PEGylated immunoliposomes carrying fluorescein-labeled plasmid DNA. The PEGylated immunoliposomes delivered exogenous genes to virtually all cells in culture, based on β-galactosidase histochemistry. Targeting the EGFR antisense gene to U87 glioma cells with the PILs resulted in more than 70% reduction in (3H) thymidine incorporation into the cells; this was paralleled by a 79% reduction in the level of immunoreactive human EGFR. This study provides a new approach to gene targeting that is effective *in vivo*, following i.v. administration without viral vectors. In another research, the plasmid DNA encoding antisense mRNA against the human EGFR gene is packaged within the interior of PEGylated immunoliposomes and delivered to the brain tumor with MAbs that target the mouse TfR and the insulin receptor (Zhang *et al.*, 2002b). The mouse TfR MAb enables transport across the tumor vasculature, which is of mouse brain origin, and the insulin receptor MAb causes transport across the plasma membrane and the nuclear membrane of the human brain cancer cell. Mice implanted with intracranial U87 human glial brain tumors are treated with plasmid-encapsulating Trojan horse liposomes. The lifespan of the mice treated weekly with an i.v. administration of the EGFR antisense gene therapy is increased 100% relative to mice treated either with a luciferase gene or with saline. Similarly, an expression plasmid encoding a shRNA directed at nucleotides 2529–2557 within the human EGFR mRNA was encapsulated in PEGylated immunoliposomes (Zhang *et al.*, 2004). The PEGylated immunoliposome was targeted to brain cancer with two receptor-specific MAb, the murine HIRMAb and the rat 8D3 MAb to the mouse TfR. The delivery of the RNAi expression plasmid resulted in a 95% suppression of EGFR function, based on measurement of thymidine incorporation or intracellular calcium signaling. Weekly i.v. RNAi gene therapy caused reduced tumor expression of immunoreactive EGFR and an 88% increase in survival time of mice with advanced intracranial brain cancer. In another study, PEI complexation of siRNAs was used as an applicable platform for RNAi *in vitro* and *in vivo*, indicating the potential of PEI/siRNA-mediated secreted growth factor pleiotrophin gene targeting as a novel therapeutic option in glioblastoma multiforme (Grzelinski *et al.*, 2006).

12.4.3.2 Gene-based immunotherapy

Immunotherapy could also be introduced with gene therapy for brain tumors (Dow *et al.*, 1999). The IFN-β gene was transferred by intratumoral injection with cationic liposomes or cationic liposomes associated with anti-glioma monoclonal antibody (immunoliposomes) (Mizuno & Yoshida, 1998). When intratumoral injections (a total of six injections) of liposomes or immunoliposomes containing the human IFN-β gene were given every second day starting 7 days after tumor transplantation (a suspension of human glioma U251-SP cells was injected into the brains of nude mice), complete disappearance of the tumor was observed in six of the seven mice that had received liposomes and in all seven mice receiving immunoliposomes. In addition, experimental gliomas injected with immunoliposomes were much smaller than those injected with ordinary liposomes following delayed injections beginning 14 days after transplantation. An immunohistochemical study of the treated nude mouse brains revealed a remarkable induction of natural killer cells expressing asialoGM1 antigen. Intratumoral administration of cationic liposomes containing pSV2muIFN-β [lip(pSV2muIFN-β)] resulted in prolonged survival time and a 50% tumor-free incidence in the treated mice (Natsume *et al.*, 2000). Besides direct growth-inhibitory effects by the IFN-β gene on the tumor cells, the study also revealed that activation of systemic cellular immunity may participate in antitumor effects *in vivo*, despite the fact that CNS is generally regarded as an immunologically privileged site.

12.4.3.3 Other routes

Besides the above-mentioned intravenous and intratumoral injections, there are some other routes through which nonviral gene therapy could be carried out *in vivo*. For example, electroporation, a standard laboratory method of introducing exogenous molecules into cells, has been gaining importance as a very effective nonviral physical technique of gene delivery. Yoshizato *et al.* injected plasmid DNA encoding luciferase, green fluorescent protein, and diphtheria toxin fragment DT-A into C6 rat glioma tumors and subjected the tumors to square wave pulses from an electro-orator (Yoshizato *et al.*, 2000). Gene expression is found to be several orders of magnitude higher when the tumors are pulsed with the optimized electrical parameters compared with the controls. DT-A shows a massive death in tumor tissue. A special circular array of six needles through which pulses are delivered with rotating electric field is found to be highly efficient in transferring genes inside the tumor. Direct injection of plasmid DNA followed by electroporation allows very high *in vivo* gene transfer and its subsequent expression into tumor tissues.

12.5 Summary

This chapter focused on the development and application of brain-targeting NP-based platform that carries therapeutic agents. Some targeting ligands and polymers have been investigated as brain-targeting delivery systems. These NPs are one of the most recent and attractive tools developed to combat neurodegenerative diseases and brain cancer. Although remarkable advances have been made during the past two decades, extensive developments available in polymer chemistry and targeting mechanisms will likely lead to development of lower-toxicity NPs for clinical use.

References

Abdallah, B., Hassan, A., Banjoist, C., Gouda, D., Behr J. P., & Desenex, B. A. (1996). A powerful nonviral vector for *in vivo* gene transfer into the adult mammalian brain: Polyethylenimine. *Human Gene Therapy*, **7**(16), 1947–1954.

Aktas, Y., Yenisei, M., Adieux, K., Gorsy, R. N., Alonso, M. J., Fernandez-Mega, E., *et al.* (2005). Development and brain delivery of chitosan-PEG nanoparticles functionalized with the monoclonal antibody OX26. *Bioconjugate Chemistry*, **16**(6), 1503–1511.

Alyautdin, R. N., Tezikov, E. B., Ramge, P., Kharkevich, D. A., Begley, D. J., & Kreuter J. (1998) Significant entry of tubocurarine into the brain of rats by absorption to polysorbate 80-coated polybutyl-cyanoacrylate nanoparticles: An *in situ* brain perfusion study. *Journal of Microencapsulation*, **15**, 67–74.

Asthana, A., Chauhan, A. S., Diwan, P. V., & Jain, N. K. (2005). Poly (amidoamine)(PAMAM) dendritic nanostructures for controlled site-specific delivery of acidic anti-inflammatory active ingredient. *AAPS PharmSciTech*, **6**(3), 536–542.

Banks, W. A., Kastin, A. J., Huang, W., Jaspan, J. B., & Maness, L. M. (1996). Leptin enters the brain by a saturable system independent of insulin. *Peptides*, **17**(2), 305–311.

Barrett, G. L., Trieu, J., & Naim, T. (2009). The identification of leptin-derived peptides that are taken up by the brain. *Regulatory Peptides*, **155**(1–3), 55–61.

Batrakova, E. V., Vinogradov, S. V., Robinson, S. M., Niehoff, M. L., Banks, W. A., & Kabanov A. V. (2005). Polypeptide point modifications with fatty acid and amphiphilic block copolymers for enhanced brain delivery. *Bioconjugate Chemistry*, **16**(4), 793–802.

Bell, R. D., Sagare, A. P., Friedman, A. E., Bedi, G. S., Holtzman, D. M., Deane, R., *et al.* (2006). Transport pathways for clearance of human Alzheimer's amyloid β-peptide and apolipoproteins E and J in the mouse central nervous system. *Journal of Cerebral Blood Flow & Metabolism*, **27**(5), 909–918.

Belting, M., & Wittrup, A. (2009). Developments in macromolecular drug delivery. *Methods in Molecular Biology*, **480**, 1–10.

Bjorklund, T., & Kordower, J. H. (2010). Gene therapy for Parkinson's disease. *Movement Disorders*, **25**(Suppl 1), S161–S173.

Boado, R. J., Zhang, Y. F., Zhang, Y., & Pardridge, W. M. (2007). Genetic engineering, expression, and activity of a fusion protein of a human neurotrophin and a molecular Trojan horse for delivery across the human blood–brain barrier. *Biotechnology and Bioengineering*, **97**(6), 1376–1386.

Boado, R. J., Zhang, Y., Zhang, Y. F., Wang, Y. T., & Pardridge, W. M. (2008a). GDNF fusion protein for targeted-drug delivery across the human blood–brain barrier. *Biotechnology and Bioengineering*, **100**(2), 387–396.

Boado, R. J., Zhang, Y. F., Zhang, Y., Xia, C. F., Wang, Y. T., & Pardridge, W. M. (2008b). Genetic engineering, expression, and activity of a chimeric monoclonal antibody-avidin fusion protein for receptor-mediated delivery of biotinylated drugs in humans. *Bioconjugate Chemistry* **19**(3), 731–739.

Bowen, G. P., Borgland, S. L., Lam, M., Libermann, T. A., Wong, N. C. W., & Muruve, D. A. (2002). Adenovirus vector-induced inflammation: Capsid-dependent induction of the C–C chemokine RANTES requires NF-κB. *Human Gene Therapy*, **13**(3), 367–379.

Brayden, D. J. (2003). Controlled release technologies for drug delivery. *Drug Discovery Today*, **8**(21), 976–978.

Brian, S., Robert, M., Edward, R., Leslie, C., Anthony, A., Rewati, P., *et al.* (2008). Long-term neprilysin gene transfer is associated with reduced levels of intracellular Abeta and behavioral improvement in APP transgenic mice. *BMC Neuroscience*, **9**, 109.

Byun, Y., Singh, V. K., & Yang, V. C. (1999). Low molecular weight protamine: A potential nontoxic heparin antagonist. *Thrombosis Research*, 1999, **94**(1), 53–61.

Calvo, P., Gouritin, B., Chacun, H., Desmaële, D., D'Angelo, J., Noel, J. P., *et al.* (2001). Long-circulating PEGylated polycyanoacrylate nanoparticles as new drug carriers for brain delivery. *Pharmaceutical Research*, **18**, 1157–1166.

Cardoso, A. L., Costa, P., de Almeida, L. P., Simoes, S., Plesnila, N., Culmsee, C., *et al.* (2010). Tf-lipoplex-mediated c-Jun silencing improves neuronal survival following excitotoxic damage *in vivo*. *Journal of Controlled Release*, **142**(3), 392–403.

Cardoso, A. L., Simoes, S., Almeida, L. P., Pelisek, J., Culmsee, C., Wagner, E., *et al.* (2007). siRNA delivery by a transferrin-associated lipid-based vector: A non-viral strategy to mediate gene silencing. *Journal of Gene Medicine*, **9**(3), 170–183.

Carty, N. C., Nash, K., Lee, D., Mercer, M., Gottschall, P. E., Meyers, C., *et al.* (2008). Adeno-associated Viral (AAV) Serotype 5 Vector mediated gene delivery of endothelin-converting enzyme reduces Aβ deposits in APP+PS-1 transgenic mice. *Molecular Therapy*, **16**(9), 1580–1586.

Chang, L. C., Lee, H. F., Yang, Z. Q., & Yang, V. C. (2001a). Low molecular weight protamine (LMWP) as nontoxic heparin/low molecular weight heparin antidote (I): Preparation and characterization. *AAPS Journal*, **3**(3), 7–14.

Chang, L. C., Wrobleski, S., Wakefield, T. W., Lee, L. M., & Yang, V. C. (2001b). Low molecular weight protamine as nontoxic heparin/low molecular weight heparin antidote (III): Preliminary *in vivo* evaluation of efficacy and toxicity using a canine model. *AAPS Journal*, **3**(3), 24–31.

Ché, C., Yang, G., Thiot, C., Lacoste, M. C., Currie, J. C., Demeule, M., *et al.* (2010). New Angiopep-modified doxorubicin (ANG1007) and etoposide (ANG1009) chemotherapeutics with increased brain penetration. *European Journal of Medicinal Chemistry*, **53**(7), 2814–2824.

Chen, Y., Dalwadi, G., & Benson, H. A. (2004). Drug delivery across the blood–brain barrier. *Current Drug Delivery*, **1**(4), 361–376.

Chen, H., Tang, L., Qin, Y., Yin, Y., Tang, J., Tang, W., *et al.* (2010). Lactoferrin-modified procationic liposomes as a novel drug carrier for brain delivery. *European Journal of Pharmaceutical Sciences*, **40**(2), 94–102.

Chertok, B., Moffat, B. A., David, A. E., Yu, F., Bergemann, C., Ross, B. D., *et al.* (2008). Iron oxide nanoparticles as a drug delivery vehicle for MRI monitored magnetic targeting of brain tumors. *Biomaterials*, **29**(4), 487–496.

Citron, M. (2004). Strategies for disease modification in Alzheimer's disease. *Nature Reviews Neuroscience*, **5**(9), 677–685.

Clarke, J., Butowski, N., & Chang, S. (2010). Recent advances in therapy for glioblastoma. *Archives of Neurology*, **67**(3), 279–283.

Cottet, H., Martin, M., Papillaud, A., Souaid, E., Collet, H., & Commeyras, A. (2007). Determination of dendrigraft poly-L-lysine diffusion coefficients by Taylor dispersion analysis. *Biomacromolecules*, **8**(10), 3235–3243.

de Boer, A. G., & Gaillard, P. J. (2007). Drug targeting to the brain. *Annual Review of Pharmacology and Toxicology*, **47**, 323–355

Dechy-Cabaret, O., Martin-Vaca, B., & Bourissou, D. (2004). Controlled ring-opening polymerization of lactide and glycolide. *Chemical Reviews* **104**, 6147–6176.

Demeule, M., Régina, A., Ché, C, Poirier, J., Nguyen, T., Gabathuler, R., *et al.* (2008). Identification and design of peptides as a new drug delivery system for the brain. *Journal of Pharmacology and Experimental Therapeutics*, **324**(3), 1064–1072.

Denora, N., Trapani, A., Laquintana, V., Lopedota, A., & Trapani, G. (2009). Recent advances in medicinal chemistry and pharmaceutical technology-strategies for drug delivery to the brain. *Current Topics in Medicinal Chemistry*, **9**(2), 182–196.

Derossi, D., Calvet, S., Trembleau, A., Brunissen, A., Chassaing, G., & Prochiantz, A. (1996). Cell internalization of the third helix of the Antennapedia homeodomain is receptor-independent. *Journal of Biological Chemistry*, **271**(30), 18188–18193.

Deumens, R., Blokland, A., & Prickaerts, J. (2002). Modeling Parkinson's disease in rats: An evaluation of 6-OHDA lesions of the nigrostriatal pathway. *Experimental Neurology*, **2175**(2), 303–317.

Dhanikula, R. S., Argaw, A., Bouchard, J. F., & Hildgen, P. (2008). Methotrexate loaded polyether-copolyester dendrimers for the treatment of gliomas: Enhanced efficacy and intratumoral transport capability. *Molecular Pharmaceutics*, **5**, 105–116.

Dodart, J. C., Marr, R. A., Koistinaho, M., Gregersen, B. M., Malkani, S., Verma, I. M., *et al.* (2005). Gene delivery of human apolipoprotein E alters brain Aβ burden in a mouse model of Alzheimer's disease. *Proceedings of the National Academy of Sciences of the United States of America*, **102**(4), 1211–1216.

Dow, S. W., Fradkin, L. G., Liggitt, D. H., Willson, A. P., Heath, T. D., & Potter T. A. (1999). Lipid-DNA complexes induce potent activation of innate immune responses and antitumor activity when administered intravenously. *Journal of Immunology*, **163**(3), 1552–1561.

Drin, G., Rousselle, C., Scherrmann, J. M., Rees, A. R., & Temsamani, J. (2002). Peptide delivery to the brain via adsorptive-mediated endocytosis: Advances with SynB vectors. *AAPS PharmSci*, **4**(4), E26.

Dufes, C., Uchegbu, I. F., & Schatzlein, A. G. (2005). Dendrimers in gene delivery. *Advanced Drug Delivery Reviews*, **57**(15), 2177–2202.

Duffy, K. R., & Pardridge, W. M. (1987). Blood–brain barrier transcytosis of insulin in developing rabbits. *Brain Research*, **420**(1), 32–38.

Eguchi, A., Akuta, T., Okuyama, H., Senda, T., Yokoi, H., Inokuchi, H., *et al.* (2001). Protein transduction domain of HIV-1 Tat protein promotes efficient delivery of DNA into mammalian cells. *Journal of Biological Chemistry*, **276**(28), 26204–26210.

Faucheux, B. A, Nillesse, N., Damier, P., Spik, G., Mouatt-Prigent, A., Pierce, A., *et al.* (1995). Expression of lactoferrin receptors is increased in the mesencephalon of patients with Parkinson disease. *Proceedings of the National Academy of Sciences of the United States of America*, **92**(21), 9603–9607.

Feng, L. R., & Maguire-Zeiss, K. A. (2010). Gene therapy in Parkinson's disease: Rationale and current status. *CNS Drugs*, **24**(3), 177–192.

Fillebeen, C., Descamps, L., Dehouck, M. P., Fenart, L., Benaissa, M., Spik, G., *et al.* (1999). Receptor-mediated transcytosis of lactoferrin through the blood–brain barrier. *Journal of Biological Chemistry*, **74**(11), 7011–7017.

Foged, C., & Nielsen, H. M. (2008). Cell-penetrating peptides for drug delivery across membrane barriers. *Expert Opinion on Drug Delivery*, **5**(1), 105–117.

Frankel, A. D., Bredt, D. S., & Pabo, C. O. (1988). Tat protein from human immunodeficiency virus forms a metal-linked dimer. *Science*, **240**(4848), 70–73.

Friden, P. M. (1994). Receptor-mediated transport of therapeutics across the blood–brain barrier. *Neurosurgery*, **35**(2), 294–298.

Fu, A., Zhou, Q. H., Hui, E. K., Lu, J. Z., Boado, R. J., & Pardridge, W. M. (2010). Intravenous treatment of experimental Parkinson's disease in the mouse with an IgG-GDNF fusion protein that penetrates the blood–brain barrier. *Brain Research*. **1352**, 208–213.

Gan, C. W., & Feng, S. S. (2010). Transferrin-conjugated nanoparticles of poly(lactide)-D-alpha-tocopheryl polyethylene glycol succinate diblock copolymer for targeted drug delivery across the blood–brain barrier. *Biomaterials*, **31**(30), 7748–7757.

Gao, H. L., Pang, Z. Q., Fan, L., Hu, K. L., Wu, B. X., & Jiang, X. G. (2010). Effect of lactoferrin- and transferrin-conjugated polymersomes in brain targeting: *In vitro* and *in vivo* evaluations. *Acta Pharmacologica Sinica*, **31**(2), 237–243.

Gao, X., Wu, B., Zhang, Q., Chen, J., Zhu, J., Zhang, W., *et al.* (2007). Brain delivery of vasoactive intestinal peptide enhanced with the nanoparticles conjugated with wheat germ agglutinin following intranasal administration. *Journal of Controlled Release*, **121**, 156–167.

Garbayo, E., Montero-Menei, C. N., Ansorena, E., Lanciego, J. L., Aymerich, M. S., & Blanco-Prieto, M. J. (2009). Effective GDNF brain delivery using microspheres–A promising strategy for Parkinson's disease. *Journal of Controlled Release*, **135**(2), 119–126.

Gillies, E. R., & Frechet, J. M. (2005). Dendrimers and dendritic polymers in drug delivery. *Drug Discovery Today*, **10**(1), 35–43.

Gosk, S., Vermehren, C., Storm, G., & Moos, T. (2004). Targeting anti-transferrin receptor antibody (OX26) and OX26-conjugated liposomes to brain capillary endothelial cells using *in situ* perfusion. *Journal of Cerebral Blood Flow & Metabolism*, **24**(11), 1193–1204.

Grau, A. J., Willig, V., Fogel, W., & Werle, E. (2001). Assessment of plasma lactoferrin in Parkinson's disease. *Movement Disorders*, **16**(1), 131–134.

Grzelinski, M., Urban-Klein, B., Martens, T., Lamszus, K., Bakowsky, U., Höbel, S., *et al.* (2006). RNA interference-mediated gene silencing of pleiotrophin through polyethylenimine-complexed small interfering RNAs *in vivo* exerts antitumoral effects in glioblastoma xenografts. *Human Gene Therapy*, **17**(7), 751–766.

Guillem, V. M., & Alino, S. F. (2004). Transfection pathways of nonspecific and targeted PEI-polyplexes. *Gene therapy & Molecular Biology*, **8**, 369–384.

Gupta, Y., Jain, A., & Jain, S. K. (2007). Transferrin-conjugated solid lipid nanoparticles for enhanced delivery of quinine dihydrochloride to the brain. *Journal of Pharmacy and Pharmacology*, **59**(7), 935–940.

Haraa, H., Monsonegob, A., Yuasac, K., Adachia, K., Xiaod, X., Takedac, S., *et al.* (2004). Development of a safe oral A vaccine using recombinant adeno-associated virus vector for Alzheimer's disease. *Journal of Alzheimer's Disease*, **6**(5), 483–488.

Hassani, Z., Lemkine, G. F., Erbacher, P., Palmier, K., Alfama, G., Giovannangeli, C., *et al.* (2005). Lipid-mediated siRNA delivery down-regulates exogenous gene expression in the mouse brain at picomolar levels. *Journal of Gene Medicine*, **7**(2), 198–207.

Hemming, M. L., Patterson, M., Reske-Nielsen, C., Lin, L., Isacson, O., & Selkoe, D. J. (2007). Reducing amyloid plaque burden via *ex vivo* gene delivery of an Abeta-degrading protease: A novel therapeutic approach to Alzheimer disease. *PLOS Medicine*, **4**(8), e262.

Hirsch, E. C., Hoglinger, G., Rousselet, E., Breidert, T., Parain, K., Feger, J., *et al.* (2003). Animal models of Parkinson's disease in rodents induced by toxins: An update. *Journal of Neural Transmission Supplementa*, **65**, 89–100.

Hong, M., Mukhida, K., & Mendez, I. (2008). GDNF therapy for Parkinsons disease. *Expert Review of Neurotherapeutics*, **8**(7), 1125–1139.

Hu, K., Li, J., Shen, Y., Lu, W., Gao, X., Zhang, Q., *et al.* (2009). Lactoferrin-conjugated PEG-PLA nanoparticles with improved brain delivery: *In vitro* and *in vivo* evaluations. *Journal of Controlled Release*, **134**(1), 55–61.

Huang, R. Q., Ke, W. L., Liu, Y., Jiang, C., & Pei, Y. (2008). The use of lactoferrin as a ligand for targeting the polyamidoamine-based gene delivery system to the brain. *Biomaterials*, **29**(2), 238–246.

Huang, R. Q., Ke, W. L., Qu, Y. H., Zhu, J. H., Pei, Y. Y., & Jiang, C. (2007a). Characterization of lactoferrin receptor in brain endothelial capillary cells and mouse brain. *Journal of Biomedical Science*, **14**(1), 121–128.

Huang, J., Liu, H., Gu, W., Yan, Z., Xu, Z., Yang, Y., *et al.* (2006a). A delivery strategy for rotenone microspheres in an animal model of Parkinson's disease. *Biomaterials*, **27**(6), 937–946.

Huang, R. Q., Qu, Y. H., Ke, W. L., Zhu, J. H., Pei, Y. Y., & Jiang, C. (2007b). Efficient gene delivery targeted to the brain using a transferrin-conjugated polyethyleneglycol-modified polyamidoamine dendrimers. *FASEB Journal.*, **21**, 1117–1125.

Huang, R. Q., Yang, W., Jiang, C., & Pei, Y. Y. (2006b). Gene delivery into brain capillary endothelial cells using Antp-Modified DNA-loaded nanoparticles. *Chemical & Pharmaceutical Bulletin (Tokyo)*, 2006, **54**(9), 1254–1258.

Huang, R., Han, L., Li, J., Ren, F., Ke, W., Jiang, C., *et al.* (2009a). Neuroprotection in a 6-hydroxydopamine-lesioned Parkinson model using lactoferrin-modified nanoparticles. *Journal of Gene Medicine*, **11**(9), 754–763.

Huang, R., Ke, W., Han, L., Liu, Y., Shao, K., Ye, L., *et al.* (2009b). Brain-targeting mechanisms of lactoferrin-modified DNA-loaded nanoparticles. *Journal of Cerebral Blood Flow & Metabolism*, **29**(12), 1914–1923.

Huang, R., Ke, W., Han, L., Liu, Y., Shao, K., Jiang, C., *et al.* (2010a). Lactoferrin-modified nanoparticles could mediate efficient gene delivery to the brain *in vivo*. *Brain Research Bulletin*, **81**(6), 600–604.

Huang, R., Ke, W., Liu, Y., Wu, D., Feng, L., Jiang, C., *et al.* (2010b). Gene therapy using lactoferrin-modified nanoparticles in a rotenone-induced chronic Parkinson model. *Journal of the Neurological Sciences*, **290** (1–2), 123–130.

Huang, R. Q., Ke, W. L., Han, L., Li, J. F., Liu, S. H., & Jiang, C., (2011). Targeted delivery of chlorotoxin-modified DNA-loaded nanoparticles to glioma via intravenous administration. *Biomaterials*, **32**, 2399–2406.

Hui, E. K., Boado, R. J., & Pardridge, W. M. (2009). Tumor necrosis factor receptor-IgG fusion protein for targeted drug delivery across the human blood–brain barrier. *Molecular Pharmaceutics*, **6**(5), 1536–1543.

Hussain, M. M., Strickland, D. K., & Bakillah, A. (1999). The mammalian low-density lipoprotein receptor family. *Annual Review of Nutrition*, **19**(1), 141–172.

Huwyler, J., Cerletti, A., Fricker, G., Eberle, A. N., & Drewe, J. (2002). By-passing of P-glycoprotein using immunoliposomes. *Journal of Drug Targeting*, **10**, 73–79.

Huwyler, J., Wu, D., & Pardridge, W. M. (1996). Brain drug delivery of small molecules using immunoliposomes. *Proceedings of the National Academy of Sciences of the United States of America*, **93**, 14164–14169.

Huynh G. H., Deen D. F., & Szoka F. C. (2006). Barriers to carrier mediated drug and gene delivery to brain tumors. *Journal of Controlled Release*, **110**(2), 236–259.

Ito, S., Ohtsuki, S., & Terasaki, T. (2006). Functional characterization of the brain-to-blood efflux clearance of human amyloid-[beta] peptide (1-40) across the rat blood–brain barrier. *Neuroscience Research*, **56**(3), 246–252.

Jain, A., Chasoo, G., Singh, S. K., Saxena, A. K., & Jain, S. K. (2011). Transferrin-appended PEGylated nanoparticles for temozolomide delivery to brain: *In vitro* characterization. *Journal of Microencapsulation*, **28**(1), 21–28.

Jeong, J. H., Song, S. H., Lim, D. W., Lee, H., & Park, T. G. (2001). DNA transfection using linear poly (ethylenimine) prepared by controlled acid hydrolysis of poly (2-ethyl-2-oxazoline). *Journal of Controlled Release*, **73**(2–3), 391–399.

Ji, B., Maeda, J., Higuchi, M., Inoue, K., Akita, H., & Harashima, H., (2006). Pharmacokinetics and brain uptake of lactoferrin in rats. *Life Science*, **78**(8), 851–855.

Jiang, C., Koyabu, N., Yonemitsu, Y., Shimazoe, T., Watanabe, S., Naito, M., *et al.* (2003). *In vivo* delivery of glial cell-derived neurotrophic factor across the blood–brain barrier by gene transfer into brain capillary endothelial cells. *Human Gene Therapy*, **14**(12), 1181–1191.

Jiang, C., Matsuo, H., Koyabu, N., Ohtani, H., Fujimoto, H., Yonemitsu, Y., *et al.* (2000). Efficient introduction of macromolecules and oligonucleotides into brain capillary endothelial cells using HVJ-liposomes. *Journal of Drug Targeting*, **8**(4), 207–216.

Ke, W. L., Shao, K., Huang, R. Q., Han, L., Liu, Y., Li, J. F., *et al.* (2009). Gene delivery targeted to the brain using an Angiopep-conjugated polyethyleneglycol-modified polyamidoamine dendrimer. *Biomaterials*, **30**(36), 6976–6985.

Kharidia, R., Friedman, K. A., & Liang, J. F. (2008). Improved gene expression using low molecular weight peptides produced from protamine sulfate. *Biochemistry* (*Mosc*), **73**(10), 1162–1168.

Kim, D. H., & Martin, D. C. (2006). Sustained release of dexamethasone from hydrophilic matrices using PLGA nanoparticles for neural drug delivery. *Biomaterials*, **27**, 3031–3037.

Kitchens, K. M., Foraker, A. B., Kolhatkar, R. B., Swaan, P. W., & Ghandehari, H. (2007). Endocytosis and interaction of poly (amidoamine) dendrimers with Caco-2 cells. *Pharmaceutical Research*, **24**(11), 2138–2145.

Kitchens, K. M., Kolhatkar, R. B., Swaan, P. W., & Ghandehari, H. (2008). Endocytosis inhibitors prevent poly (amidoamine) dendrimer internalization and permeability across Caco-2 cells. *Molecular Pharmaceutics*, **5**(2), 364–369.

Kiyota, T., Okuyama, S., Swan, R. J., Jacobsen, M. T., Gendelman H. E, & Ikezu, T. (2010). CNS expression of anti-inflammatory cytokine interleukin-4 attenuates Alzheimer's disease-like pathogenesis in APP+PS1 bigenic mice. *FASEB Journal*, **24**, 1–10.

Klok, H. A., & Rodriguez-Hernandez, J. (2002). Dendritic-graft polypeptides. *Macromolecules*, **35**(23), 8718–8723.

Kordower, J. H., Emborg, M. E., Bloch, J., Ma, S. Y., Chu, Y., Leventhal, L., *et al.* (2000). Neurodegeneration prevented by lentiviral vector delivery of GDNF in primate models of Parkinson's disease. *Science*, **290**(5492), 767–773.

Koziara, J. M., Lockman, P. R., Allen, D. D., & Mumper, R. J. (2003). *In situ* blood–brain barrier transport of nanoparticles. *Pharmaceutical Research*, **20**(11), 1772–1778.

Kulkarni, S. A., & Feng, S. S. (2011). Effects of surface modification on delivery efficiency of biodegradable nanoparticles across the blood–brain barrier. *Nanomedicine (Lond)*, **6**(2), 377–394.

Kumar, P., Wu, H., McBride, J. L., Jung, K. E., Kim, M. H., Davidson, B. L., Lee, S. K., *et al.* (2007). Transvascular delivery of small interfering RNA to the central nervous system. *Nature*, **448**(7149), 39–43.

Kurakhmaeva, K. B., Djindjikhashvili, I. A., Petrov, V. E., Balabanyan, V. U., Voronina, T. A., Trofimov, S. S., *et al.* (2009). Brain targeting of nerve growth factor using poly(butyl cyanoacrylate) nanoparticles. *Journal of Drug Targeting*, **17**(8), 564–574.

Laffray, S., Tan, K., Dulluc, J., Bouali-Benazzouz, R., Calver, A. R., Nagy, F., *et al.* (2007). Dissociation and trafficking of rat GABAB receptor heterodimer upon chronic capsaicin stimulation. *European Journal of Neuroscience*, **25**(5), 1402–1416.

Laron, Z. (2009). Insulin and the brain. *Archives of Physiology and Biochemistry*. **115**(2), 112–116.

Lasic, D. D. (1993). *Liposomes: From Physics to Applications[M]*. Amsterdam, the Netherlands: Elsevier Science Publishers Raven Press.

Lawler, S. E., Peruzzi, P. P., & Chiocca, E. A. (2005). Genetic strategies for brain tumor therapy. *Cancer Gene Therapy*, **13**(3), 225–233.

Lee, S. W., Kim, W. J., Park, J. A., Choi, Y. K., Kwon, Y. W., & Kim, K. W. (2006). Blood–brain barrier interfaces and brain tumors. *Archives of Pharmacal Research*, **29**(4), 265–275.

Leonard, D. J., Pick, L. T., Farrar, D. F., Dickson, G. R., Orr, J. F., & Buchanan, F. J., (2009). The modification of PLA and PLGA using electron-beam radiation. *Journal of Biomedical Materials Research, Part A* **89**, 567–574.

Liang, J. F., Zhen, L., Chang, L. C., & Yang, V. C. (2003). A less toxic heparin antagonist—low molecular weight protamine. *Biochemistry* (*Mosc*), **68**(1), 116–120.

Liu, L., Guo, K., Lu, J., Venkatraman, S. S., Luo, D., Ng, K. C., *et al.* (2008a). Biologically active core/shell nanoparticles self-assembled from cholesterol-terminated PEG-TAT for drug delivery across the blood-brain barrier. *Biomaterials*, **29**, 1509–1517.

Liu, Y., Huang, R. Q., Han, L., Ke, W. L., Shao, K., Ye, L., *et al.* (2009). Brain-targeting gene delivery and cellular internalization mechanisms for modified rabies virus glycoprotein RVG29 nanoparticles. *Biomaterials*, **30**(25), 4195–4202.

Liu, Y., Li, J. F., Shao, K., Huang, R. Q., Ye, L., Lou, J., & Jiang, C. (2010). A leptin derived 30-amino-acid peptide modified PEGylated poly-L-lysine dendrigraft for brain targeted gene delivery. *Biomaterials*, **31**(19), 5246–5257.

Liu, L., Venkatraman, S. S., Yang, Y. Y., Guo, K., Lu, J., He, B., *et al.* (2008b). Polymeric micelles anchored with TAT for delivery of antibiotics across the blood–brain barrier. *Biopolymers*, **90**(5), 617–623.

Lo, S. L., & Wang, S. (2008). An endosomolytic Tat peptide produced by incorporation of histidine and cysteine residues as a nonviral vector for DNA transfection. *Biomaterials*, **29**(15), 2408–2414.

Lockman, P. R., Mumper, R. J., Khan, M. A., & Allen, D. D. (2002). Nanoparticle technology for drug delivery across the blood–brain barrier. *Drug Development and Industrial Pharmacy*, **28**(1), 1–13.

Lockman, P. R., Oyewumi, M. O., Koziara, J. M., Roder, K. E., Mumper, R. J., & Allen, D. D. (2003). Brain uptake of thiamine-coated nanoparticles. *Journal of Controlled Release*, **93**(3), 271–282.

Loeb, C., Benassi, E., Besio, G., Maffini, M., & Tanganelli, P. (1982). Liposme entrapped GABA modifies behavioral and electrographic changes of penicillin-induced epileptic activity. *Neurology*, **32**, 1234–1238.

Lu, W., Sun, Q., Wan, J., She, Z., & Jiang, X. G. (2006). Cationic albumin-conjugated PEGylated nanoparticles allow gene delivery into brain tumors via intravenous administration. *Cancer Research*, **66**(24), 11878.

Lu, J. M., Wang, X., Marin-Muller, C., Wang, H., Lin, P. H., Yao, Q., *et al.* (2009). Current advances in research and clinical applications of PLGA-based nanotechnology. *Expert Review of Molecular Diagnostics*, **9**, 325–341.

Lu, W., Zhang, Y., Tan, Y. Z., Hu, K. L., Jiang, X. G., & Fu, S. K. (2005). Cationic albumin-conjugated PEGylated nanoparticles as novel drug carrier for brain delivery. *Journal of Controlled Release*, **107**(3), 428–448.

Luo, D., Haverstick, K., Belcheva, N., Han, E., & Saltzman, W. M. (2002). Poly (ethylene glycol)-conjugated PAMAM dendrimer for biocompatible, high-efficiency DNA delivery. *Macromolecules*, **35**(9), 3456–3462.

Magzoub, M., & Graslund, A. (2005). Cell-penetrating peptides: Small from inception to application. *Quarterly Reviews of Biophysics*, **37**(02), 147–195.

Malik, N., Wiwattanapatapee, R., Klopsch, R., Lorenz, K., Frey, H., Weener, J. W., *et al.* (2000). Dendrimers relationship between structure and biocompatibility *in vitro*, and preliminary studies on the biodistribution of 125I-labelled polyamidoamine dendrimers *in vivo*. *Journal of Controlled Release*, **65**(1–2), 133–148.

Mathupala, S. P. (2009). Delivery of small-interfering RNA (siRNA) to the brain. *Expert Opinion on Therapeutic Patents*, **19**(2), 137–140.

Matsuo, H., Okamura, T., Chen. J., Takanaga, H., Ohtani, H., Kaneda, Y., *et al.* (2000). Efficient introduction of macromolecules and oligonucleotides into brain capillary cells using HVJ-liposomes. *Journal of Drug Targeting*, **8**(4), 207–16.

Mehdipour, A. R., & Hamidi, M. (2009). Brain drug targeting: A computational approach for overcoming blood–brain barrier. *Drug Discovery Today*, **14**(21–22), 1030–1036

Merdan, T., Callahan, J., Petersen, H., Kunath, K., Bakowsky, U., Kopec kova, P., *et al.* (2003). PEGylated Polyethylenimine-'Fab antibody fragment conjugates for targeted Gene delivery to human ovarian carcinoma cells. *Bioconjugate Chemistry*, **14**(5), 989–996.

Michaelis, M. L., Chen, Y. X., Hill, S., Reiff, E., Georg, G., Rice, A., *et al.* (2006). Amyloid peptide toxicity and microtubule-stabilizing drugs. *Journal of Molecular Neuroscience*, **19**(1–2), 101–105.

Mizuno, M., & Yoshida, J. (1998). Effect of human interferon beta gene transfer upon human glioma, transplanted into nude mouse brain, involves induced natural killer cells. *Cancer Immunology, Immunotherapy*, **47**(4), 227–232.

Mishra, V., Mahor, S., Rawat, A., Gupta, P. N., Dubey, P., Khatri, K., *et al.* (2006). Targeted brain delivery of AZT via transferrin anchored PEGylated albumin nanoparticles. *Journal of Drug Targeting*, **14**(1), 45–53.

Mochizuki, H., Yasuda, T., & Mouradian, M. M. (2008). Advances in gene therapy for movement disorders. *Neurotherapeutics*, **5**(2), 260–269.

Moestrup, S. K., Cui, S., Vorum, H., Bregengard, C., Bjorn, S. E., Norris, K., *et al.* (1995). Evidence that epithelial glycoprotein 330/megalin mediates uptake of polybasic drugs. *Journal of Clinical Investigation*, **96**(3), 1404–1413.

Moos, T., & Morgan, E. H. (2001). Restricted transport of anti-transferrin receptor antibody (OX26) through the blood–brain barrier in the rat. *Journal of Neurochemistry*, **79**(1), 119–129.

Nakayama, H., Okuda, H., & Nakashima, T. (1994). Molecular diversity and properties of brain nicotinic acetylcholine receptor. *Nippon Yakurigaku Zasshi*, **104**(3), 241–249.

Natsume, A., Tsujimura, K., Mizuno, M., Takahashi, T., & Yoshida, J. (2000). IFN-β gene therapy induces systemic antitumor immunity against malignant glioma. *Journal of Neuro-Oncology*, **47**(2), 117–124.

Nilsson, P., Iwata, N., Muramatsu, S. I., Tjernberg, L. O., Winblad, B., & Saido, T. C. (2010). Gene therapy in Alzheimer's disease -potential for disease modification. *Journal of Cellular and Molecular Medicine*, **14**(4), 741–757.

Nimesh, S., Goyal, A., Pawar, V., Jayaraman, S., Kumar, P., Chandra, R., *et al.* (2006). Polyethylenimine nanoparticles as efficient transfecting agents for mammalian cells. *Journal of Controlled Release*, **110**(2), 457–468.

Nishikawa, M., & Huang, L. (2001). Nonviral vectors in the new millennium: Delivery barriers in gene transfer. *Human Gene Therapy*, **12**(8), 861–870.

Oh, S., Pluhar, G. E., McNeil, E. A., Kroeger, K. M., Liu, C., Castro, M. G., *et al.* (2007). Efficacy of nonviral gene transfer in the canine brain. *Journal of Neurosurgery*, **107**(1), 136–144.

Okura, Y., Miyakoshi, A., Kohyama, K., Park, I. K., Staufenbiel, M., & Matsumoto, Y. (2006). Nonviral Aβ DNA vaccine therapy against Alzheimer's disease: Long-term effects and safety. *Proceedings of the National Academy of Sciences of the United States of America*, **103**(25), 9619–9624.

Olivier, J. C., Huertas, R., Lee, H. J., Calon, F., & Pardridge, W. M. (2002). Synthesis of PEGylated immunonanoparticles. *Pharmaceutical Research*, **19**(8), 1137–1143.

Onodera, H., Takada, G., Tada, K., Desnick, R. J. (1983). Microautoradiographic study on the tissue localization of liposome-entrapped or unentrapped 3H-labeled beta-galactosidase injected into rats. *Tohoku Journal of Experimental Medicine*, **140**(1), 1–13.

Paddison, P. J., Caudy, A. A., Bernstein, E., Hannon, G. J., & Conklin, D. S. (2002). Short hairpin RNAs (shRNAs) induce sequence-specific silencing in mammalian cells. *Genes & Development*, **16**(8), 948–958.

Pang, Z., Lu, W., Gao, H., Hu, K., Chen, J., Zhang, C., *et al.* (2008). Preparation and brain delivery property of biodegradable polymersomes conjugated with OX26. *Journal of Controlled Release*, **128**(2), 120–127.

Pardridge, W. M. (1995). Transport of small molecules through the blood-brain barrier: Biology and methodology. *Advanced Drug Delivery Reviews*, **15**(1–3), 5–36.

Pardridge, W. M. (2002). Targeting neurotherapeutic agents through the blood–brain barrier. *Archives of Neurology*, **59**(1), 35–40.

Pardridge, W. M. (2003). Blood–brain barrier drug targeting: The future of brain drug development. *Molecular Interventions*, **3**(2), 90–105.

Pardridge, W. M. (2004). Intravenous, non-viral RNAi gene therapy of brain cancer. *Expert Opinion on Biological Therapy*, **4**(7), 1103–1113.

Pardridge, W. M. (2005). The blood–brain barrier: Bottleneck in brain drug development. *NeuroRx*, **2**(1), 3–14.

Pardridge, W. M., Eisenberg, J., & Yang, J. (1987). Human blood–brain barrier transferrin receptor. *Metabolism*, **36**(9), 892–895.

Park, Y. J., Liang, J. F., Ko, K. S., Kim, S. W., & Yang, V. C. (2003). Low molecular weight protamine as an efficient and nontoxic gene carrier: *In vitro* study. *Journal of Gene Medicine*, **5**(8), 700–711.

Perkins, D. (2007). Targeting apoptosis in neurological disease using the herpes simplex virus. *Journal of Cellular and Molecular Medicine*, **6**(3), 341–356.

Prigent, S. A., Stanley, K. K., & Siddle, K. (1990). Identification of epitopes on the human insulin receptor reacting with rabbit polyclonal antisera and mouse monoclonal antibodies. *Journal of Biological Chemistry*, **265**(17), 9970–9977.

Princen, F., Lechanteur, C., Lopez, M., Gielen, J., Bours, V., & Merville, M. P. (2000). Similar efficiency of DNA-liposome complexes and retrovirus-producing cells for HSV-tk suicide gene therapy of peritoneal carcinomatosis. *Journal of Drug Targeting*, **8**(2), 79–89.

Postmes, T. J., Hukkelhoven, M., van den Bogaard, A. E., Halders, S. G., & Coenegracht, J. (1980). Passage through the blood–brain barrier of thyrotropin-releasing hormone encapsulated in liposomes. *Journal of Pharmacy and Pharmacology*, **32**, 722–724.

Pouton, C. W., Lucas, P., Thomas, B. J., Uduehi, A. N., Milroy, D. A., & Moss, S. H. (1989). Polycation-DNA complexes for gene delivery: A comparison of the biopharmaceutical properties of cationic polypeptides and cationic lipids. *Journal of Controlled Release*, **53**(1–3), 289–299.

Price, T. O., Farr, S. A., Yi, X., Vinogradov, S., Batrakova, E., Banks, W. A., *et al.* (2010). Transport across the blood–brain barrier of pluronic leptin. *Journal of Pharmacology and Experimental Therapeutics*, **333**(1), 253–263.

Qian, Z. M., Li, H., Sun, H., & Ho, K. (2002). Targeted drug delivery via the transferrin receptor-mediated endocytosis pathway. *Pharmacol Rev*, **54**(4), 561–587.

Rangasamy, S. B., Soderstrom, K., Bakay, R. A., & Kordower, J. H. (2010). Neurotrophic factor therapy for Parkinson's disease. *Progress in Brain Research*, **184**, 237–264.

Ramge, P., Kreuter, J., & Lemmer, B. (1999). Circadian phase-dependent antinociceptive reaction in mice after i.v. injection of dalargin-loaded nanoparticles determined by the hot-plate test and the tail flick test. *Chronobiology International*, **16**, 767–777.

Riolfi, M., Ferla, R., Del Valle, L., Piña-Oviedo, S., Scolaro, L., Micciolo, R., *et al.* (2010). Leptin and its receptor are overexpressed in brain tumors and correlate with the degree of malignancy. *Brain Pathology*, **20**(2), 481–489.

Roberts, J. C., Bhalgat, M. K., & Zera, R. T. (1998). Preliminary biological evaluation of polyamidoamine (PAMAM) StarburstTM dendrimers. *Journal of Biomedical Materials Research, Part A*, **30**(1), 53–65.

Rousselle, C., Clair, P., Lefauconnier, J. M., Kaczorek, M., Scherrmann, J. M., & Temsamani, J. (2000). New advances in the transport of doxorubicin through the blood–brain barrier by a peptide vector-mediated strategy. *Molecular Pharmacology*, **57**(4), 679.

Rousselle, C., Smirnova, M., Clair, P., Lefauconnier, J. M., Chavanieu, A., Calas, B., *et al.* (2001). Enhanced delivery of doxorubicin into the brain via a peptide-vector-mediated strategy: Saturation kinetics and specificity. *Journal of Pharmacology and Experimental Therapeutics*, **296**(1), 124–131.

Roy, I., Stachowiak, M. K., & Bergey, E. J. (2008). Nonviral gene transfection nanoparticles: Function and applications in the brain. *Nanomedicine*, **4**(2), 89–97.

Salehi, M. (2007). Gene delivery to brain cells with apoprotein E derived peptide conjugated to polylysine (apoEdp-PLL). *Journal of Drug Targeting*, **15**(3), 226–230.

Santos, J. L., Pandita, D., Rodrigues, J. C., Pego, A. P., Granja, P., Balian, G., *et al.* (2010). Receptor-mediated gene delivery using PAMAM dendrimers conjugated with peptides recognized by mesenchymal stem cells. *Molecular Pharmaceutics*, **7**(3), 763–774.

Saovapakhiran, A., D'Emanuele, A., Attwood, D., & Penny, J. (2009). Surface modification of PAMAM dendrimers modulates the mechanism of cellular internalization. *Bioconjugate Chemistry*, **20**(4), 693–701.

Scott, M. M., Lachey, J. L., Sternson, S. M., Lee, C. E., Elias, C. F., Friedman, J. M., *et al.* (2009). Leptin targets in the mouse brain. *Journal of Comparative Neurology*, **514**(5), 518–532.

Schwarze, S. R., Ho, A., Vocero-Akbani, A., & Dowdy, S. F. (1999). *In vivo* protein transduction: Delivery of a biologically active protein into the mouse. *Science*, **285**(5433), 1569–1572.

Selkoe, D. J. (2001). Alzheimer's disease results from the cerebral accumulation and cytotoxicity of amyloid beta-protein. *Journal of Alzheimer's Disease*, **3**(1), 75–80.

Shao, K., Huang, R. Q., Li, J. F., Han, L., Ye, L. Y., Lou, J. N., *et al.* (2010). Angiopep-2 modified PE-PEG based polymeric micelles for amphotericin B delivery targeted to the brain. *Journal of Controlled Release*, **147**(1), 118–126.

Sherer, T. B., Betarbet, R., Kim, J. H., & Greenamyre, J. T. (2003). Selective microglial activation in the rat rotenone model of Parkinson's disease. *Neuroscience Letters*, **341**(2), 87–90.

Shi, N., & Pardridge, W. M. (2000). Noninvasive gene targeting to the brain. *Proceedings of the National Academy of Sciences of the United States of America*, **97**(13), 7567–7572.

Shi, N., Zhang, Y., Zhu, C., Boado, R. J., & Pardridge, W. M. (2001). Brain-specific expression of an exogenous gene after i.v. administration. *Proceedings of the National Academy of Sciences of the United States of America*, **98**, 12754–12759.

Shibata, M., Yamada, S., Kumar, S. R., Calero, M., Bading, J., Frangione, B., *et al.* (2000). Clearance of Alzheimer's amyloid-β1-40 peptide from brain by LDL receptor–related protein-1 at the blood–brain barrier. *Journal of Clinical Investigation*, **106**(12), 1489–1499.

Sindhu, K. M., Saravanan, K. S., & Mohanakumar, K. P. (2005). Behavioral differences in a rotenone-induced hemiparkinsonian rat model developed following intranigral or median forebrain bundle infusion. *Brain Research*, **1051**(1–2), 25–34.

Silva, G. A. (2007). Nanotechnology approaches for drug and small molecule delivery across the blood brain barrier. *Surg Neurol*, **67**(2), 113–116.

Small, D. H., San Mok, S., & Bornstein, J. C. A. (2001). Alzheimer's disease and Aβ toxicity: From top to bottom. *Nature Reviews Neuroscience*, **2**(8), 595–598.

Spagnou, S., Miller, A. D., & Keller, M. (2004). Lipidic carriers of siRNA: Differences in the formulation, cellular uptake, and delivery with plasmid DNA. *Biochemistry (Mosc)*, **43**(42), 13348–13356.

Soni, V., Kohli, D. V., & Jain, S. K. (2005). Transferrin coupled liposomes as drug delivery carriers for brain targeting of 5-florouracil. *Journal of Drug Targeting*, **13**(4), 245–250.

Soni, V., Kohli, D. V., & Jain, S. K. (2008). Transferrin-conjugated liposomal system for improved delivery of 5-fluorouracil to brain. *Journal of Drug Targeting*, **16**(1), 73–78.

Su, Y., & Sinko, P. J. (2006). Drug delivery across the blood–brain barrier: Why is it difficult? How to measure and improve it? *Expert Opinion on Drug Delivery*, **3**(3), 419–435.

Suzuki, Y. A., & Lonnerdal, B. (2004). Baculovirus expression of mouse lactoferrin receptor and tissue distribution in the mouse. *Biometals*, **17**(3), 301–309.

Suzuki, Y. A., Lopez, V., & Lonnerdal, B. (2005). Mammalian lactoferrin receptors: Structure and function. *Cellular and Molecular Life Sciences*, **62**(22), 2560–2575.

Talukder, M. J. R, Takeuchi, T., & Harada, E. (2003). Receptor-mediated transport of lactoferrin into the cerebrospinal fluid via plasma in young calves. *Journal of Veterinary Medical Science*, **65**(9), 957–964.

Tamai, I., Sai, Y., Kobayashi, H., Kamata, M., Wakamiya, T., & Tsuji, A. (1997). Structure-internalization relationship for adsorptive-mediated endocytosis of basic peptides at the blood–brain barrier. *Journal of Pharmacology and Experimental Therapeutics*, **280**(1), 410–415.

Tamaru, M., Akita, H., Fujiwara, T., Kajimoto, K., & Harashima, H. (2010). Leptin-derived peptide, a targeting ligand for mouse brain-derived endothelial cells via macropinocytosis. *Biochemical and Biophysical Research Communications*, **394**(3), 587–592.

Tang, C. H., Lu, D. Y., Yang, R. S., Tsai, H. Y., Kao, M. C., Fu, W. M., *et al.* (2007). Leptin-induced IL-6 production is mediated by leptin receptor, insulin receptor substrate-1, phosphatidylinositol 3-kinase, Akt, NF-{kappa} B, and p300 pathway in microglia. *Journal of Immunology*, **179**(2), 1292–1302.

Tang, G. P., Zeng, J. M., Gao, S. J., Ma, Y. X., Shi, L., Li, Y., *et al.* (2003). Polyethylene glycol modified polyethylenimine for improved CNS gene transfer: Effects of PEGylation extent. *Biomaterials*, **24**(13), 2351–2362.

Thomas, F. C., Taskar, K., Rudraraju, V., Goda, S., Thorsheim, H. R., Gaasch, J. A., *et al.* (2009). Uptake of ANG1005, a novel paclitaxel derivative, through the blood–brain barrier into brain and experimental brain metastases of breast cancer. *Pharmaceutical Research*, **26**(11), 2486–2494.

Tiwari, S. B., & Amiji, M. M. (2006). A review of nanocarrier-based CNS delivery systems. *Current Drug Delivery*, **3**(2), 219–232.

Tökés, Z. A., St Péteri, A. K., & Todd, J. A. (1980). Availability of liposome content to the nervous system. Liposomes and the blood–brain barrier. *Brain Research*, **188**(1), 282–286.

Torchilin, V. P. (2008). Tat peptide-mediated intracellular delivery of pharmaceutical nanocarriers. *Advanced Drug Delivery Reviews*, **60**(4–5), 548–558.

Tsogas, I., Theodossiou, T., Sideratou, Z., Paleos, C. M., Collet, H., Rossi, J. C., *et al.* (2007). Interaction and transport of poly (L-lysine) dendrigrafts through liposomal and cellular membranes: The role of generation and surface functionalization. *Biomacromolecules*, **8**(10), 3263–3270.

Tsui, B., Singh, V. K., Liang, J. F., & Yang, V. C. (2001). Reduced reactivity towards anti-protamine antibodies of a low molecular weight protamine analogue. *Thrombosis Research*, **101**(5), 417–420.

Tomalia, D. A., Naylor, A. M., & Goddard, W. A. (2003). Starburst dendrimers: Molecular-level control of size, shape, surface chemistry, topology, and flexibility from atoms to macroscopic matter. *Angewandte Chemie International Edition England*, **29**(2), 138–175.

Tomalia, D. A. (2005). Birth of a new macromolecular architecture: Dendrimers as quantized building blocks for nanoscale synthetic polymer chemistry. *Progress in Polymer Science*, **30**(3–4), 294–324.

Ulbrich, K., Hekmatara, T., Herbert, E., & Kreuter, J. (2009). Transferrin- and transferrin-receptor-antibody-modified nanoparticles enable drug delivery across the blood–brain barrier (BBB). *European Journal of Pharmaceutics and Biopharmaceutics*, **71**(2), 251–256.

Ulbrich, K., Knobloch, T., & Kreuter, J. (2011). Targeting the insulin receptor: Nanoparticles for drug delivery across the blood–brain barrier (BBB). *Journal of Drug Targeting*, **19**(2), 125–132.

Vinogradov, S. V., Batrakova, E. V., & Kabanov, A. V. (2004). Nanogels for oligonucleotide delivery to the brain. *Bioconjugate Chemistry*, **15**(1), 50–60.

Von Eckardstein, K. L., Patt, S., Zhu, J., Zhang, L., Cervos-Navarro, J., & Reszka, R. (2001). Short-term neuropathological aspects of *in vivo* suicide gene transfer to the F98 rat glioblastoma using liposomal and viral vectors. *Histology and Histopathology*, **16**(3), 735–745.

Wang, S., Ma, N., Gao, S. J., Yu, H., & Leong, K. W. (2001). Transgene expression in the brain stem effected by intramuscular injection of polyethylenimine/DNA complexes. *Molecular Therapy*, **3**(5), 658–664.

Wang, W., Xiong, W., Wan, J., Sun, X., Xu, H., & Yang, X. (2009). The decrease of PAMAM dendrimer-induced cytotoxicity by PEGylation via attenuation of oxidative stress. *Nanotechnology*, **20**(10), 105103.

Ward, P. P., Uribe-Luna, S., & Conneely, O. M. (2002). Lactoferrin and host defense. *Biochemistry & Cell Biology*, **80**(1), 95–102.

Wolff, J. A., Malone, R. W., Williams, P., Chong, W., Acsadi, G., Jani, A., *et al.* (1990). Direct gene transfer into mouse muscle *in vivo*. *Science*, **247**(4949), 1465–1468.

Wu, D., Yang, J., & Pardridge, W. M. (1997). Drug targeting of a peptide radiopharmaceutical through the primate blood–brain barrier *in vivo* with a monoclonal antibody to the human insulin receptor. *Journal of Clinical Investigation*, **100**(7), 1804–1812.

Wu, G., Barth, R. F., Yang, W., Kawabata, S., Zhang, L., & Green-Church, K. (2006). Targeted delivery of methotrexate to epidermal growth factor receptor-positive brain tumors by means of cetuximab (IMC-C225) dendrimer bioconjugates. *Molecular Cancer Therapeutics*, **5**, 52–59.

Wu, G. Y., & Wu, C. H. (1987). Receptor-mediated *in vitro* gene transformation by a soluble DNA carrier system. *Journal of Biological Chemistry*, **262**(10), 4429.

Xia, C. F., Boado, R. J., Zhang, Y., Chu, C., & Pardridge, W. M. (2007a). Intravenous glial-derived neurotrophic factor gene therapy of experimental Parkinson's disease with Trojan horse liposomes and a tyrosine hydroxylase promoter. *Journal of Gene Medicine*, **10**(3), 306–315.

Xia, C. F., Zhang, Y., Zhang, Y., Boado, R. J., & Pardridge, W. M. (2007b). Intravenous siRNA of brain cancer with receptor targeting and avidin-biotin technology. *Pharmaceutical Research*, **24**(12), 2309–2316.

Xiang, J. J., Tang, J. Q., Zhu, S. G., Nie, X. M., Lu, H. B., Shen, S. R., *et al.* (2003). IONP-PLL: A novel non-viral vector for efficient gene delivery. *Journal of Gene Medicine*, **5**(9), 803–817.

Xin, H., Jiang, X., Gu, J., Sha, X., Chen, L., Law, K., *et al.* (2011). Angiopep-conjugated poly(ethylene glycol)-co-poly(ε-caprolactone) nanoparticles as dual-targeting drug delivery system for brain glioma. *Biomaterials*, **32**(18), 4293–4305.

Yang, S. C., Lu, L. F., Cai, Y., Zhu, J. B., Liang, B. W., & Yang, C. Z. (1999). Body distribution in mice of intravenously injected camptothecin solid lipid nanoparticles and targeting effect on brain. *Journal of Controlled Release*, **59**(3), 299–307.

Yang, S., Zhu, J., Lu, Y., Liang, B., & Yang, C. (1999). Body distribution of camptothecin solid lipid nanoparticles after oral administration. *Pharmaceutical Research*, **16**, 751–757.

Yoon, H. C., Hong, M. Y., & Kim, H. S. (2000). Functionalization of a poly (amidoamine) dendrimer with ferrocenyls and its application to the construction of a reagentless enzyme electrode. *Analytical Chemistry*, **72**(18), 4420–4427.

Yoon, J. S., Jung, Y. T., Hong, S. K., Kim, S. H., Shin, M. C., Lee, D. G., *et al.* (2004). Characteristics of HIV-Tat protein transduction domain. *Journal of Microbiology*, **42**, 328–335.

Yoshizato, K., Nishi, T., Goto, T., Dev, S. B., Takeshima, H., Kino, T., *et al.* (2000). Gene delivery with optimized electroporation parameters shows potential for treatment of gliomas. *International Journal of Oncology*, **16**(5), 899–904.

Yu, X., Zhang, Y., Chen, C., Yao, Q., & Li, M. (2010). Targeted drug delivery in pancreatic cancer. *Biochimica et Biophysica Acta*, **1805**(1), 97–104.

Zhang, X. P., Jiang, F., Kalkanis, S. N., Yang, H. Y., Zhang, Z. G., Katakowski, M., *et al.* (2006). Combination of surgical resection and photodynamic therapy of 9L gliosarcoma in the nude rat. *Photochemistry and Photobiology*, **82**(6), 1704–1711.

Zhang, Y., Lee, H. J., Boado, R. J., & Pardridge, W. M. (2002a). Receptor-mediated delivery of an antisense gene to human brain cancer cells. *Journal of Gene Medicine*, **4**(2), 183–194.

Zhang, Y., & Pardridge, W. M. (2001). Conjugation of brain-derived neurotrophic factor to a blood–brain barrier drug targeting system enables neuroprotection in regional brain ischemia following intravenous injection of the neurotrophin. *Brain Research*, **889**(1–2), 49–56.

Zhang, Y., & Pardridge, W. M. (2005). Delivery of b-galactosidase to mouse brain via the blood–brain barrier transferrin receptor. *Journal of Pharmacology and Experimental Therapeutics*, **313**, 1075–1081.

Zhang, Y., & Pardridge, W. M. (2009). Near complete rescue of experimental Parkinson's disease with intravenous, non-viral GDNF gene therapy. *Pharmaceutical Research*, **26**(5), 1059–1063.

Zhang, Y., Schlachetzki, F., & Pardridge, W. M. (2003a). Global non-viral gene transfer to the primate brain following intravenous administration. *Molecular Therapy*, **7**(1), 11–18.

Zhang, X., Xie, J., Li, S., Wang, X., & Hou, X. (2003b). The study on brain targeting of the amphotericin B liposomes. *Journal of Drug Targeting*, **11**(2), 117–122

Zhang, J., Wu, X., Qin, C., Qi, J., Ma, S., Zhang, H., *et al.* (2003c). A novel recombinant adeno-associated virus vaccine reduces behavioral impairment and [beta]-amyloid plaques in a mouse model of Alzheimer's disease. *Neurobiology of Disease*, **14**(3), 365–379.

Zhang, Y., Zhang, Y., Bryant, J., Charles, A., Boado R. J., & Pardridge, W. M. (2004). Intravenous RNA interference gene therapy targeting the human epidermal growth factor receptor prolongs survival in intracranial brain cancer. *Clinical Cancer Research*, **10**(11), 3667–3677.

Zhang, Y., Zhu, C., & Pardridge, W. M. (2002b). Antisense gene therapy of brain cancer with an artificial virus gene delivery system. *Molecular Therapy*, **6**(1), 67–7

Chapter 13

Nano/Microspheres in Bioimaging and Medical Diagnosis

Xiaohui Li and Chunying Chen

Key Lab for Biomedical Effects of Nanomaterials and Nanosafety of Chinese Academy of Sciences, National Center for Nanoscience and Technology, Beijing, P. R. China

chenchy@nanoctr.cn

13.1 Introduction

The rapid development of nanoscience and the application of nanotechnology are changing the foundations of bioimaging and medical diagnosis. With the process of nanotechnology, different nano- and microparticles function as imaging agents, target probes, and therapeutic carriers and have an exciting potential for imaging and diagnosis. The application of nanotechnology to molecular imaging and diagnosis has lots of advantages, such as (a) monitoring the primary intervention procedure to assess desirable distribution and localization of delivered therapeutics at the targets, (b) enhancing the effectiveness of the delivered therapeutics to achieve a sufficient level of therapeutic effect at the targets, and (c) monitoring the function and effect period of the delivered therapeutics at the targets (1–3).

Among different types of nano- and microparticles, polymeric nano/microsphere is one of the most important numbers. Compared to conventional materials, polymeric nano/microsphere and microcapsule have several advantages such as biocompatibility, diversity of chemical structure, large specific surface area, stable dispersion and uniform size, surface chemistry, and morphology.

Microspheres and Microcapsules in Biotechnology: Design, Preparation, and Applications
Edited by Guanghui Ma and Zhiguo Su

ISBN 978-981-4316-47-7 (Hardcover), 978-981-4364-62-1 (eBook)
www.panstanford.com

(4). These unique chemical and physical properties are useful for developing multifunctional nanomaterials. Modification of the these imaging particles, such as conjugation of ligands to the particle surface, enables generation of site-specific ligand molecule interaction, which therefore permits visualization of the targets and assessment of the functionality of target sites at the molecular level, termed molecular imaging (5). Furthermore, loading of therapeutics, such as genes and drugs into the target nano/microsphere leads to the generation of new approaches for site-specific delivery of therapeutics, and thus, target-specific therapies for diseases (6–8).

This chapter provides an overview of the current status of the application of nano/microsphere in medical diagnosis and bioimaging. First, we address how nanotechnology is used in the field medical diagnosis and bioimaging then, how nano/microsphere can be used for bioimaging; and finally, how nanoparticles are being developed as medical diagnosis agents.

13.2 Nano/Microspheres in Bioimaging

The goal of bioimaging is to understand the components, processes, and dynamics of a disease from a molecular perspective. The research fields of bioimaging include the synthesis of bioimaging probes, live cell imaging, early diagnosis, high throughput drug screening, and the development of various imaging methods. Appropriate bioimaging methods and corresponding bioimaging probes are the main tools for bioimaging (9,10).

Presently, the most common bioimaging methods based on noninvasive technology used in current clinical practice are optical imaging, magnetic resonance imaging (MRI), positron emission tomography (PET), and ultrasound imaging (11–14). Compared with biopsy, surgery, or other invasive techniques, these methods have enormous potential for the early detection and treatment of diseases (9–12). The imaging capability of nano/microsphere involves the construction of particles using imaging contrast agents, such as optically active compounds for optical imaging, paramagnetic or superparamagnetic metals for MRI, or gases for ultrasonographic imaging. For MRI, the most common bioimaging probes are metals such as gadolinium (Gd^{3+}) or iron. For PET, the most common bioimaging probes are isotopes such as ^{18}F, ^{11}C, ^{15}O, ^{99m}Tc, and ^{111}In, which can be very expensive for clinical use. Beside these methods, the advent of optical strategies holds great promise for imaging applications. For example, studies by the laboratories of Weissleder (15,16) have shown that it is possible to image fluorescent dye–labeled nanoparticles *in vivo* with high sensitivity and specificity by using near-infrared fluorescent (NIRF) probes. The regular bioimaging systems are reviewed in Table 13.1 (2).

Table 13.1 Overview of bioimaging systems

System	Penetration depth	Resolution	Imaging probes	Clinical application
Optical imaging	<10 cm	1–3 mm	Fluorochromes, photoproteins	Under development
MRI	No limit	10–100 μm	Gadolinium, iron oxide particles	Yes
PET/SPECT	No limit	1–2 mm	^{18}F, ^{11}C, ^{15}O, ^{99m}Tc, ^{111}In	Yes
Ultrasound	<20 cm	50 μm	Microbubbles	Yes

SPECT: Single-photon emission computed tomography.
Source: Reprinted from Ref. 2, with permission from Elsevier.

These imaging methods are fairly generalized, using nonspecific dyes or image contrast agents, while this lack of specificity and ease of excretion have made it difficult to achieve enough sensitivity necessary for high-quality imaging, particularly in the early stages of a disease (17). Recent advances in polymer-based nano/microsphere have increased the bioimaging specificity and therapeutic potential (18). First, the polymeric materials often exhibit a prolonged circulation time in the bloodstream by avoiding the reticuloendothelial system (RES) and increasing cellular-uptake properties (2,19). Second, nano/microsphere can combine multivalent imaging sensors. By connecting specific ligands on the microsphere, which have an affinity with target tissues of interest, specifically localized molecular events can be visualized in body in real time. These new imaging sensors open the door for the development of bioimaging technology.

13.2.1 Nano/Microspheres for Optical Imaging

13.2.1.1 Optical imaging

Optical imaging involves the use of light interactions, one of the most versatile and noninvasive imaging radiations, which create a contrast due to changes in their intensity, wavelength, polarization, coherence, interference (20–24), lifetime, and nonlinear effects. Among the various optical imaging methods, fluorescence imaging has been believed to be the most important one. Fluorescence technology is particularly suited for bioimaging, as fluorescent probes are safe and sensitive and can be specifically labeled to biomacromolecules and biomaterials. The fundamental barriers to fluorescence imaging of tissue are light scattering, autofluorescence, and absorption by tissues in the visible range (25). In contrast to excitation light in the visible region, which cannot reach the imaging target in tissue, near-infrared (NIR) light of 650–900 nm, which has a higher tissue penetration due to minimal absorbance by the surface tissue, avoids some of these limitations.

13.2.1.2 Materials used for preparing optical imaging nano/microspheres

To achieve NIRF imaging, a large number of NIRF agents have been developed to examine fundamental processes at the organ, tissue, cellular, and molecular levels. These fluorescent agents should be physico-chemically stable, nontoxic, highly biocompatible, biodegradable or excretable (26), and they should fulfill as many of the following requirements as possible (27,28):

1. *Be excitable in visible or NIR.* NIRF agents are particularly promising for bioimaging since there is less interference due to autofluorescence and scattering and their two-photon excitation characteristic.
2. *Have high brightness.* Brightness is commonly defined as a product of molar absorption coefficient ε and luminescence quantum yield QY. ε and QY should both be strong to be suitable for any practical use.
3. *Respond to the analyte of interest and preferably be inert to other species.* By appropriate chemical modification of an indicator or by making use of permeation-selective properties of a polymer, NIRF agents should aim at the analyte of interest in order to minimize some of the undesirable effects.

4. *Have good photostability.* Photostability is particularly critical if high light intensities are used for interrogation (such as in microscopy or fiber-optic micro-sensors) or if the measurements are performed over a long time.
5. *Possess good solubility in the polymer or have functional groups for immobilization.* Generally, suitable NIRF agents are either lipophilic enough or are rendered lipophilic. It should be ensured that NIRF agents remain in polymeric nano/microspheres in monomeric form and neither aggregate nor leach out into the surrounding medium.
6. *Be nontoxic. Low cytotoxicity is desirable but not essential for the choice of NIRF agents.* Evidently, cytotoxicity is greatly reduced upon immobilization of agents into polymeric nano/microspheres. But, if even one agent molecule leaches out of the nano/microspheres, it may undesirably affect a biological system.

Virtually, a lot of luminescent dyes can be potentially used for staining polymeric nano/microspheres. Several common dyes are described in Table 13.2, which contain some fluorescent organic dyes and phosphorescent metal-ligand complexes. Most of these dyes possess good brightness exceeding 40,000. Modifications are done to increase photostability, to render the dyes lipophilic, or to make them available for covalent coupling. For example, fluoresceins, seminaphthorhodafluors, and 1-hydroxy pyrene-3,6,8-trisulfonate (or its alkylsulfonamides) are well-known pH indicators and are used for the preparation of pH-sensitive nano/microspheres. Other dye classes can be used for the preparation of fluorescent labels and tracers.

Currently, the widely used fluorescent agents such as QDs have several limitations for bioimaging, such as lack of specificity for target cells, tissues, or organs; rapid aggregation; photobleaching; and a short half-life in blood (29–32). Nano/microsphere bioimaging probes, which consist of polymer-conjugated nano/microsphere and fluorescent agents, have been developed to avoid these problems. They take advantage of polymer therapeutics, which have prolonged plasma half-lives, enhanced stability, reduced toxicity, improved targeting, and reduced background. Over the past several years, a number of polymer-based nano/microsphere optical bioimaging probes have been successfully developed (30,33–35).

Table 13.2 Chemical structures and typical spectral properties of the common fluorophores that can be used for staining polymeric microspheres

Dyes	λ_{max} (abs) nm	ε ($M^{-1}cm^{-1}$)
Coumarins	350–580	40,000–50,000
Fluoresceins	490–520	80,000
Rhodamines	510–600	80,000–140,000
Seminaphthorhodafluors	560	50,000
Phenoxazines	500–630	40,000–80,000
Naphthalimides	420	15,000
Perylene	450–580	50,000
Boradipyrroles	500–670	80,000
Squaraines	630–670	300,000
Cyanine	550–750	200,000
Ruthenium(II) polypyridyl complexes	450–470	20,000–30,000
Palladium(II) and platinum(II) Polypyridyl complexes	380–430 (Soret) 500–550 (Q-bands)	150,000–300,000 (Soret) 15,000–25,000 (Q-bands)
Cyclometallated iridium(III) complexes	400–500	5,000
Europium(III) β-diketonates	300–380 400–430 (with antenna)	70,000
Palladium(II) and platinum(II) ketoporphyrin complexes	390–410 (Soret) 590–605 (Q-bands)	90,000 (Soret) 55,000 (Q-bands)
Iridium(III) coumarin complexes	450–490	90,000
Palladium(II) and platinum(II) tetrobenzoporphyrin complexes	420–450 (Soret) 600–635 (Q-bands)	200,000–400,000 (Soret) 130,000–180,000 (Q-bands)

Source: Adapted from Ref. 27.

A variety of polymers have been used for the preparation of nano/microsphere imaging probes. Some widely used polymers are described in Table 13.3. Among different kinds of polymers, nonpolar polymers such as polystyrene (PS) are excellent matrices for immobilization of lipophilic indicators (36). Such nano/microspheres are often provided with charged groups which dramatically increase their stability in aqueous media. Notably, the surface of such nano/microspheres is also available for further functionalization.

Table 13.3 Chemical structures of the polymers commonly used for the preparation of optical imaging microspheres

Polymer	Type
PS	Neutral
Poly(methyl methacrylate acid) (PMMA)	Neutral
PVC	Neutral
Polyacrylonitrile (PAN)	Neutral
Poly(styrene-co-maleic acid) (PS-MA)	Charged
Poly(methyl methacrylate-co-methacrylic acid) (PMMA-MA)	Charged
RL100 and RS100	Charged
Poly(acrylonitrile-co-acrylic acid) (PAN-AA)	Charged
Poly(styrene-block-vinylpyrrolidone) (PS-PVP)	Core–shell

Source: Adapted from Ref. 27.

Polar polymers can be employed to design particles for sensing pH, ions, and basic and acidic gases. For example, polar polyacrylamide, polyurethane hydrogels are polymers of choice for designing the pH-sensing nano/microspheres. Plasticized polyvinylchloride (PVC) was found to be very suitable for preparing ion-sensitive nano/microspheres. Block polymers are able to form core–shell nano/microspheres in aqueous media, where the neutral lipophilic shell is stable in water and does not aggregate even at very high ionic strength.

13.2.1.3 Polymer-based nano/microspheres used for imaging probes

One of the unique capabilities of NIR global imaging is the ability to perform high-throughput applications, using the parallel capabilities of the technique to evaluate multiple samples simultaneously. To realize the clinical potential of nano/microsphere bioimaging probes, effective *in vivo* delivery technologies are obviously required. One way to achieve the aim is to apply the advantages of nanomedicine platform to bioimaging systems. To implement this function, the physical and chemical properties of nano/microsphere, usually their surfaces, need to be modified with target ligands. These ligands can be peptides, monoclonal antibodies or their fragments, small molecular peptidomimetics (small protein-like chains that contain both natural and non-natural amino acids), vitamins, or aptamers (small molecules that can bind to another molecule) (25).

Fluorescent microspheres with carboxyl, amine/hydrazide, and maleimide groups for covalent coupling of proteins, peptides, oligonucleotides, and polysaccharides can be purchased. Non-covalent linkages are useful for some applications. Species-specific anti-IgGs coupled to microspheres facilitate direct binding of immunoglobulins from ascites fluids or concentrated hybridoma supernatants when further purification is not practical. High- and low-density streptavidin-coated fluorospheres with the capacity to bind biotinylated analytes are available for adjusting the dynamic range of quantitative assays and signal-to-noise ratios. Qiang Ma (37) and coworkers have prepared multicolor quantum QD-encoded microspheres via the layer-by-layer (LbL) assembly approach. PS microspheres of 3 μm diameter were used as templates for the deposition of different-sized CdTe QDs/polyelectrolyte multilayers via electrostatic interactions. Two kinds of biofunctional multicolor microspheres with two different antibodies, anti-human IgG, and anti-rabbit IgG were prepared. Human IgG and rabbit IgG can be detected as target antigens in the multiplexed fluoroimmunoassays. Furthermore, a novel microfluidic on-chip device was developed to detect two kinds of antigen-conjugated multicolor QD-encoded microspheres; the microspheres can be distinguished from each other based on their fluorescence signals.

Qian *et al.* (38) synthesized successfully a kind of monodisperse luminescent PS-based microspheres with diameters of 200–400 nm encapsulated with rare earth ion–doped $NaYF_4$ upconversion nanoparticles by a microemulsion polymerization process. The microspheres emit strong NIR-to-visible upconversion fluorescence upon excitation by a 980 nm laser (Fig. 13.1).

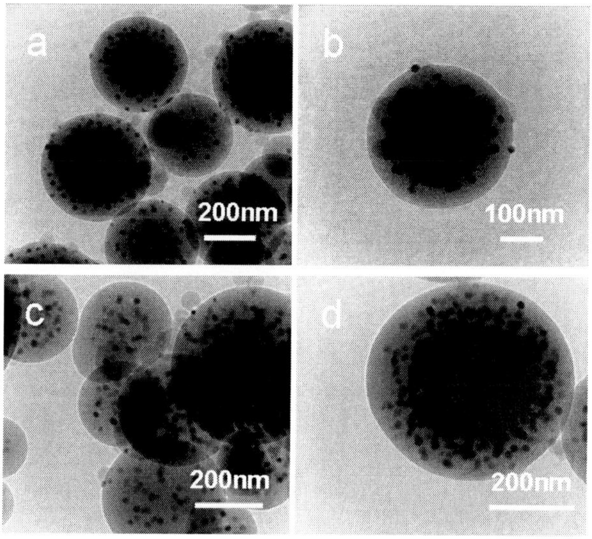

Figure 13.1 TEM images of monodispersed $NaYF_4$:Yb/Er nanoparticles (a,b) and $NaYF_4$:Yb/Tm nanoparticles (c,d). Reproduced from Ref. 38, with permission from Springer Science+Business Media.

Karthik Vishwanath and coworkers (39) described how PS microspheres can reduce fluorophore quantum yield via collision quenching. They reported studies on UV-visible (fluorescein-based) and NIR (IR125-based) phantoms with different fluorophore and scatterer concentrations, as well as different microsphere sizes. Results suggest that the fluorophore's intrinsic excited-state lifetime decreased due to collisional quenching from PS microspheres and that the quenching efficiency was dependent on the concentration ratio of fluorophores to microspheres. Lifetime decreases ranging from 10–35% (average in 20%) were measured for fluorescein-based (IR125-based) phantoms. Since PS microspheres are commonly used in tissue-simulating phantoms for quantitative studies of fluorescence light propagation, their quenching effects on fluorescence intensities may be difficult to separate from intensity loss attributed to optical absorption and scattering in the phantom unless fluorescence lifetime measurements are performed simultaneously.

Optical imaging probes mainly include various organic fluorescent molecules, quantum dots (QDs), transition/lanthanide metal ions–doped luminescent compounds. The poor photostability of the fluorescent dyes and the extreme toxicity of the QDs always make them suffer from some limitations in biomedical applications. For many samples and applications, nano/microsphere bioimaging probes will provide suitable method to avoid these problems.

The combination of luminescent properties into a single micro- or nanocomposite system through polymer matrix will allow the development of a novel multifunctional biomedical platform for multimodal imaging and simultaneous diagnosis. Till now, the varieties of polymer matrix are still limited. It is an important job to fabricate different fluorescence microspheres with stable fluoresence and higher quantum efficiency. At the same time, there still are the following challenges in the fabrication. How to increase the photostability of the fluorescent probe complexes and to improve biocompatibility of the entire particles? How to avoid the fluorescent quenching of the fluorophore on the surface of the particle caused by the interaction of a number of organic dye molecules and by solvent molecules?

While it is still in its developing phase, nano- and microparticle-based optical imaging has already presented the exciting potential to imaging of different biology samples. With the three-in-one functions of imaging, targeting, and transporting, plays a central role in optical imaging.

13.2.2 Nano/Microspheres for Magnetic Resonance Imaging

MRI is used primarily in medical settings to produce high quality images of the inside of the body. The technique measures the change in the characteristics of hydrogen nuclei in water and other nuclei with similar chemical shifts across the image slice. The characteristics of hydrogen nuclei depend on their position in the molecule. Rather than providing information about chemical shifts and coupling constants, MRI gives a spatial distribution of the density of water protons (40).

The advantages of MRI are that it is noninvasive, non-ionizing radiation, and high soft-tissue resolution and discrimination in any imaging plane. MRI is designed to allow noninvasive bioimaging in the human body and therefore has been widely employed by clinicians. In 1988, the advent of MRI bioimaging probes (contrast agents) made it possible to use MRI for angiography and perfusion imaging. The main difficulty in this regard is to accumulate enough bioimaging probes in the area of interest while maintaining the agent at a minimum level in other organs and tissues. The market for MRI bioimaging probe is approximately a quarter of that of anti-cancer agents and is expected to grow continuously (41). The most commonly used clinical MRI bioimaging probes are low–molecular weight–based chelates of metal ions such as Mn^{2+} and Gd^{3+} and iron oxide particles.

However, the transient tissue retention time, toxicities, and unfavorable pharmacokinetic profiles of the above-mentioned metal ions result in short imaging-time window, low signal-to-noise ratio, and limited clinical applications. Various polymer -based approaches have been proposed to overcome this limitation and thus improve its therapeutic value. Polymer-based nano/microsphere MRI bioimaging probes have some advantages, such as the ability to reduce the toxicity of the metals, persistence and long circulation in the blood, significantly increased contrast, and the possibility to attach various functionalities to the metal particles. Below, several of the polymer-based nano/microspheres that have shown promise for use in MRI bioimaging probes are discussed in detail.

13.2.2.1 Polymer-based nano/microspheres for MRI bioimaging probes

MRI contrast agents play a significant role in meeting the healthcare needs. The spinel metal ferrites and metal-doped iron oxides, which are important MRI contrast agents due to their enhanced magnetic properties, have been developed greatly. The spinel metal ferrites with a composition of MFe_2O_4, where M is +2 cation of Mn, Fe, Co, or Ni, have been fabricated by various methods to tune specific magnetic properties.

To improve magnetic properties and decrease toxic side effect for applications, Lee *et al.* (42) reported the synthesis and characterization of $MnFe_2O_4$, $FeFe_2O_4$, $CoFe_2O_4$, and $NiFe_2O_4$ by high-temperature reaction between divalent metal chloride and iron tris-2,4-pentadioate. Through comparison of various metal-doped ferrite nanoparticles, this group has demonstrated that $MnFe_2O_4$ nanoparticles are nontoxic *in vitro* and possess higher magnetic susceptibility than magnetite nanoparticles, suggesting that they may be used as an ultra-sensitive MRI probe. In addition to this, encapsulation of paramagnetic gadolinium into polymeric microspheres enables detection of the microspheres at MRI (Fig. 13.2). Table 13.4 reviewed polymer-conjugated Gd chelate MRI probes.

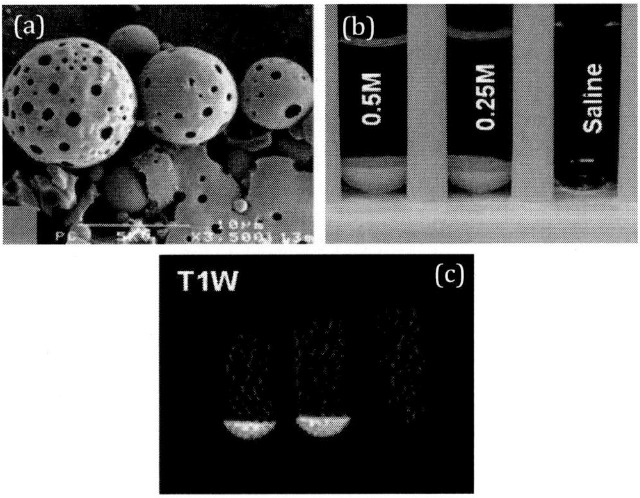

Figure 13.2 (a) Electron microscopic image of polymeric microspheres. Multiple micropores on the wall of polymeric microspheres allow exchange between interior (encapsulated) MR contrast agents and exterior free water, generating MR signals. (b) Polymeric microspheres encapsulated with gadolinium (gadopentetate dimeglumine) at 0.5 and 0.25 mol/L solution and saline. The gadolinium microspheres are aggregated in a pellet at the tube bottom after centrifugation. (c) Subsequent *in vitro* T1-weighted MR image shows hyperintense signals produced by gadolinium microspheres; no such signal is detected from saline. Reprinted with permission from Ref. 43, © John Wiley & Sons, Inc.

Table 13.4 Polymer-conjugated Gd chelate MRI probes

Structure or name	Body distribution	Application
Liposome, Paramagnetic	RES-directed	Hepatectomy
Polylysine-(Gd-DTPA)*x*-dextran Lymphangiography	Lymph nodes	

Structure or name	Body distribution	Application
Diphenylcyclohexyl phophodiester-Gd-DTPA, MS325 EPIX	Blood pool	MR angiograpy (MRA)
4-Pentyl-bicyclo[2.2.2] octan-1-carboxyl-di-L-aspartyllysine-DTPA, MP 2269	Blood pool	MRA
(Gd-DTPA)-17, 24 cascade polymer	Blood pool	MRA
Gd-DTPA-PEG-polymers	Blood pool	MRA
(Gd-DTPA)*n*-albumin	Blood pool	MRA
(Gd-DOTA)*n*-albumin		Breast imaging
(Gd-DTPA)*n*-polylysine	Blood pool	MRA
(Gd-DTPA)*n*-dextran	Blood pool	MRA
(Gd-DTPA)*n* l-cystine bisamide copolymers, PGA-cystamine-Gd-DO3A	Blood pool	MRA

Source: Reprinted from Ref. 2, with permission from Elsevier.

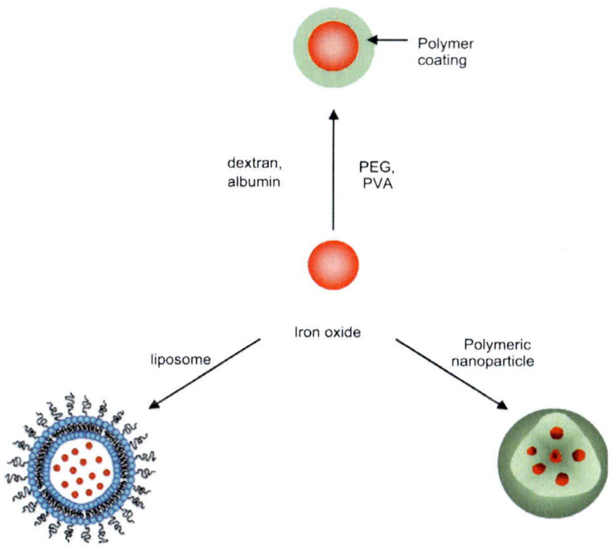

Figure 13.3 Polymer-based iron oxide contrast agents. Reprinted from Ref. 2, with permission from Elsevier.

Colloidal iron oxide particles, such as superparamagnetic iron oxide (SPIO) and ultrasmall superparamagnetic iron oxide (USPIO), have been the most extensively investigated MRI materials for biomedical applications due to their excellent biocompatibility and ease of synthesis. They are typically composed of nanocrystalline magnetite (Fe_3O_4) or maghemite (γFe_2O_3) protected with a polymeric coating. In the case of Fe_3O_4, magnetization arises from electron hopping between the Fe^{2+} and Fe^{3+} ions that coexist at the octahedral sites. In addition to magnetic properties, favorable biocompatibility and biodegrade ability of these MNPs have contributed greatly to their widespread use in biomedical applications (Fig. 13.3). Currently, the SPIO and USPIO being utilized or under investigation for clinical application as MRI contrast agents are predominately synthesized by an aqueous co-precipitation process in the presence of the coating material (43,44).

13.2.2.2 Magnetic dye-doped polymeric nano/microspheres

Dye-doped polymeric nano/microspheres with magnetic properties (either para- or ferromagnetic) became increasingly popular in the past decades. Compared to conventional nano/microspheres, these particles can be magnetically guided to a place of interest by an external magnetic field or even rotated by an alternating field. On one hand, the magnetic separation from the medium can increase the signal strength, speed up separation in sedimentation arrays, or localize an effect of the particle to a specific point of a biological material (e.g., targeted drug delivery). On the other hand, the modulated rotation of such particles can greatly extend the signal-to-noise ratio. The magnetic nano/microspheres usually consist of inorganic or metallic nanoparticles which are either homogeneously or inhomogeneously (e.g., core–shell) distributed in the nano/microspheres.

Magnetic fluorescent particles have some main advantages over their nonmagnetic counterparts. First of all, their separation is usually faster than plain sedimentation and does not require filtration or centrifugation steps. The magnetic properties of the microspheres can be utilized as contrast agent for MRI, and the incorporated dyes are used as labels for affordable real-time tracking of particles *in vivo* (16). A careful selection of both the surface groups and the dye is particularly important for designing multimodal probes. The surface property affects the cycle lifetime *in vivo*, the potential cellular-uptake rate, and also the contrasting efficiency in MRI (45). NIR dyes are usually selected in order to ensure maximal penetration depth in human tissue. Besides the dual mode imaging of tumor tissue, it is also possible to magnetically guide and deliver a drug there. For example, Deng *et al.* (46) incorporated doxorubicin, a common anticancer drug, in a stimuli-responsive poly(*N*-isopropylacrylamide) shell, which was polymerized around a core of magnetic fluorescent nano/microspheres. It should be mentioned here that magnetic fluorescent silica microspheres (47) are used far more often for *in vivo* applications (such as multimodal imaging, magnetically guided drug delivery, and stimuli-mediated drug release) than the polymeric microspheres, probably due to their excellent stability, simple modification chemistry, and biocompatibility.

One particularly interesting report on polymeric microspheres with magnetic, luminescent, and cell-targeting properties was published by Kopelman (48). They constructed polyacrylamide microspheres with magnetic nanoparticles and a photosensitizer in the core. The surface was modified with an RGD peptide for specific cell targeting and polyethylene glycol for increasing the plasma residence time. These particles attached to the cell surface of tumor cells, enhancing the contrast in an MRI image and finally producing reactive oxygen species (ROS) in the close vicinity of the tumor cells.

13.2.3 Nano/Microspheres for Radionuclide Imaging

13.2.3.1 Radionuclide imaging

Because of high sensitivity and little quantity of tracer molecules required, radionuclide imaging methods are widely used for clinical applications. Radionuclide imaging commonly contains three general imaging techniques: (i) planar gamma scintigraphy, (ii) three-dimensional imaging by SPECT, and (iii) PET. Planar gamma scintigraphy uses different kinds of single-photon radionuclides which emit gamma (γ) rays in the energy range of approximately 75 to 360 keV (Table 13.5) to achieve quantification of tissue distribution as a percentage of the injected dose and compresses the complex anatomical structure of the organs into only two dimensions. SPECT uses the same radionuclides as planar gamma scintigraphy, but it can be used for acquisition of three-dimensional data. PET offers the most accurate images, however, the short half-lives of PET radionuclides (e.g., ^{11}C, ^{18}F, and ^{64}Cu) sometimes limit their applications (49).

Table 13.5 Radionuclides for tumor imaging

Radionuclide	Emission type	Half-life	E_{max} (kev)	Production
^{131}I	γ (81.2%), β	8.0 days	(γ): 284, 364, 637	$^{131}Te(\beta)$ ^{131}I
^{67}Ga	γ	78.3 h	93, 134, 300, 393	$^{68}Zn(n,p)$ ^{67}Ga
^{111}In	γ	67.2 h	171, 245	$^{111}Cd(p,n)$ ^{111}In
^{123}I	γ	13.2 h	159	$^{121}Sn(\alpha,2n)$ ^{123}I
^{99m}Tc	γ	6.0 h	140	$^{99}Mo/^{99m}Tc$-generator
^{18}F	Positron	1.83 h	640	$^{18}O(p,n)$ ^{18}F
^{11}C	Positron	20.4 min	960	$^{14}N(p,\alpha)$ ^{11}C
^{13}N	Positron	9.96 min	1190	$^{16}O(p,\alpha)$ ^{13}N
^{15}O	Positron	2.07 min	1720	$^{14}N(d,n)$ ^{15}O

Source: Reprinted from Ref. 2, with permission from Elsevier.

In most cases, pure β-emitting radionuclides such as ^{32}P and $^{90}\gamma$ are generally preferable, as other β-emitting radionuclides produce relatively small fraction of their energy γ-rays, imposing the problem of exposing normal tissues to undesirable irradiation due to the long range of γ-rays. However, in many cases, the latter radionuclides might be advantageous, as they allow for a scintigraphic follow-up of their biodistribution via gamma cameras.

Nano/microsphere plays an important role in nuclear imaging because of their improved specificity, prolonged circulation time, amplified signals and ease of modification. In this chapter, we focus microsphere encapsulating radionuclide imaging probes mainly on planar gamma scintigraphy and briefly on SPECT and PET.

13.2.3.2 Nano/microspheres for SPECT and PET

In the last few decades, the use of radiolabeled nano/microspheres in imaging studies has been widely reported. They have played a key role in the emergence of SPECT and PET techniques (50). In local radiotherapy in hepatic tumors, these microspheres provide the extra benefit of avoiding gastrointestinal toxicity resulting from extrahepatic radioactivity diffusion due to arteriovenous shunting or intrahepatic vascular alterations (51,52).

SPECT imaging is performed by using a gamma camera to acquire two-dimensional images from multiple angles. The images are then digitally processed using a tomographic reconstruction algorithm. The SPECT technique has been clinically used for the imaging of myocardial perfusion, tumors, and brain. Zielhuis and coworkers demonstrated that the incorporation of holmium into alginate microspheres allows their visualization with a gamma camera. For the purpose of imaging-guided radioembolization of tumors, these microspheres are labeled with radioactive ^{166}Ho (53). Intra-arterially-injected microspheres of ^{99m}Tc macroaggregated albumin (MAA) microspheres have been widely clinically used in the detection of arteriovenous shunting in various cancer cases like hepatic tumors and malignant gliomas (54,55). These MAA microspheres with diameters of 25 to 50 μm are trapped in the capillary bed of the organs under physiological conditions (56,57). The utilization of these radioisotopes-loaded microspheres appears to be very advantageous as a diagnostic tool for local tumors therapy.

PET is a nuclear imaging technique that produces a three-dimensional image or map of functional processes in the body. The method utilizes a metabolically active compound labeled with an isotope with a short half-life that decays by emitting a positron. There is a waiting period while the metabolically active molecule, most commonly fluorodeoxyglucose (FDG), becomes concentrated in the tissues of interest. The PET technique has been clinically used for oncology, neurology, cardiology, and pharmacology. One important indication of microspheres includes the elaboration of Gallium-68-labeled MAA (58) and human serum albumin (HSA) for PET imaging. Okada *et al.* developed a kit for labeling HSA microspheres with ^{68}Ga for liver anomalies and tumor scanning (59). The kit was clinically tested by intravenously injecting a dose of 5 μg/kg of these labeled microspheres to 5 normal volunteers and 25 patients, 11 of them had chronic hepatitis and 14 had liver cirrhosis, for PET imaging of both liver and spleen. The microsphere uptake in these organs was evaluated for clinical usefulness as a function test of the RES. Through PET imaging, it was found that the liver volume and dimensionality decreased while the spleen volume increased, which was in accordance with the degree of the chronic liver disease.

13.2.4 Nano/Microspheres for Ultrasound Imaging

13.2.4.1 Ultrasound imaging

Ultrasound is the most widely used imaging technique. Ultrasound imaging is a versatile, noninvasive, low-risk, cost-effective, and portable real-time diagnostic modality, but its use has been limited by the lack of effective bioimaging probes (60,61). Ultrasound bioimaging probes can improve imaging by introducing a material with different acoustic properties from that of tissue. The most common approach is the intravenous injection of small air or gas bubbles (microbubbles), which is based on the principle of using sound waves to detect a difference in density between the probe (gas) and the surrounding medium (blood or soft tissue) at different time points during the examination.

Microbubbles are a kind of hollow microspheres which are typically used as additives to lower the density of a material (62). These microbubbles have diameters of 1–10 mm and can pass through the lung capillaries; thus, they act as true red blood cell tracers. Two key aspects of microbubbles are their filling with a gas of low diffusivity and blood solubility and the inclusion of a coating material that can stabilize the microbubble. Gases with high molecular weight and density, such as perfluorochemicals and sulfur hexafluoride, are widely used as microbubble-based contrast agents. In particular, perfluorocarbon tends to remain biologically inert in the bloodstream after destruction of the microbubble, and it is gradually eliminated after multiple passes through the lungs. A shell encapsulating the gas bubble is essential for the longevity of the agent (61).

The shell has a role in countering the surface tension at the gas–liquid interface, and thus stabilizing the bubble against dissolution during its transit in the bloodstream. Also, the material properties of the shell influence the nonlinear dynamic behavior of the microbubbles. Current shell materials include proteins (Albunex and Optisons), lipids (SonoVues, Definity^{TM}, and Imagents), or a surfactant (ST68) (63,64); however, most of these imaging probes lack stability and are often destroyed under routine conditions such as during continuous imaging mode or at higher power outputs.

Polymer shell–based microbubbles have many advantages. They are more stable in plasma under ultrasonic waves than monomolecular layers of proteins, lipids, or surfactants because they have better mechanical strength (65). The elasticity of the shells can be controlled by adjusting the chemical composition and the molecular weight of the polymer. Also, the shell provides a good surface for conjugating with target-specific ligands. When combined with specific targeted imaging, polymer shell-based agents can allow the detection of a single microbubble in targeted organs or cells (66). Their controlled release characteristics and biocompatibility also allows them to be used as carriers for targeted drug delivery or gene therapy (67).

13.2.4.2 Microbubbles for ultrasound imaging

Microbubble imaging agents enhance visualization of blood flow and thus improve our ability to see changes in blood flow associated with therapy. A number of ablation techniques are currently in use to treat regional cancer disease, including alcohol ablation, cryosurgery, radiofrequency (RF) ablation and high-intensity focused ultrasound (HIFU) (68–70). All of these techniques have the primary goal of ablating a well-defined volume of tissue comprising a tumor and a surrounding margin of tumor-free tissue. Concomitant with ablation, there is devascularization of the coagulated tissue.

Real-time ultrasound can show these changes in blood flow associated with coagulation. Figure 13.4 shows a liver tumor pre- and post-ablation with RF energy. The vascular region is clearly defined post-contrast. HIFU is another ablation technique, which unlike the others may not require that a probe or needle be inserted into the tissue. HIFU relies upon focused high-energy ultrasound to rapidly heat a region of tissue to produce coagulative necrosis. Since the ultrasound can be applied across the intact skin and focused to a given depth within the tissues, HIFU may afford advantages of being less invasive than the techniques such as RF ablation, which requires that one or more probes be inserted into the tissue. We hypothesized that since microbubbles increased the attenuation in a medium, they would increase the rate of heating with therapeutic ultrasound.

Recent experimental and modeling experiments indicate this to be true (71). Microbubbles increase the rate of heating from HIFU and may shorten the procedure for HIFU by up to a factor of 4. This is potentially important clinically. HIFU is approved by the regulatory authorities in Europe for treating prostate cancer. The procedure takes several hours to complete. Use of microbubbles may shorten the procedure significantly. Additionally, therapeutic imaging with microbubbles may improve visualization of blood flow and monitoring of the procedures so that margins may be made even more accurate. Improved definition of margins may be important in decreasing complications. For prostate ablation, this translates to better tumor control with improved maintenance of potency and decreased incidence of incontinence. Lindner (72) has developed a number of different targeted microbubbles to targets such as pSelectin and $\alpha_V\beta_3$. Figure 13.4 shows an image of tumor angiogenesis pre- and post-enhancement with a microbubble targeted to integrin. Such a microbubble might be used to detect tumors or monitor response to therapy with agents such as angiogenesis inhibitors. Also such an agent might be used as a targeted enhancer in HIFU. HIFU is used to ablate tumors via coagulative necrosis. A targeted tumor imaging agent might be used to visualize tumor margins to more accurately define margins for ablation and also to enhance uptake of ultrasound energy to improve coagulation and shorten the procedure.

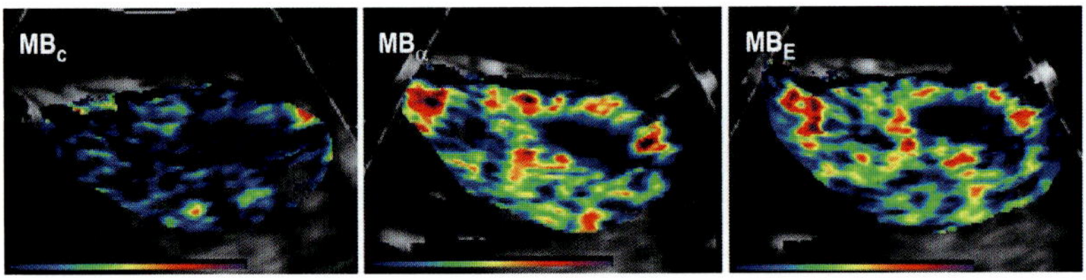

Figure 13.4 Contrast-enhanced ultrasound (CEU) assessment of microbubble retention in matrigel. Examples of background-subtracted, color-coded CEU images reflecting signal from microbubbles retained in matrigel plugs 15 min after intravenous injection of microbubbles. The color-coding scale for background-subtracted acoustic intensity is shown at the bottom of each image. MB$_c$, control microbubble; MB$_\alpha$, microbubbles with monoclonal antibody against murine αv conjugated on surface; MB$_E$, microbubbles with echistain conjugated to surface. Reproduced with permission from Ref. 73. Copyright © 2012 by American Heart Association, Inc.

Microbubbles represent a new class of drug delivery/therapeutic products. The microbubbles are activated locally by ultrasound energy. The occurrence of cavitation creates a local shock wave that can be exploited for *in vivo* therapeutic effects. Potential applications of this technology include treatment of thrombosis and drug delivery. Drug carrying bubbles such as AALs might be used for local delivery of system agents. In the brain, systemic delivery of microbubbles and application of ultrasound across the intact skull can potentially be used to open the blood–brain barrier (BBB) for local delivery of therapeutic agents. For delivery of drugs to the brain, bubbles similar to the current FDA-approved lipid-coated microbubble product might be used to open the BBB. In this case, a drug might be co-administered with the bubbles rather than entrapped in the bubbles itself. Similarly, microbubbles not carrying drugs, but acting as a drug themselves in concert with ultrasound may be used in SonoLysisk therapy. Drug-carrying microbubbles can be developed as "magic bullets" to deliver therapeutic agents to the target region using ultrasound to activate release. This has the potential to open a new field of targeted gene therapy.

13.3 Nano/Microspheres in Medical Diagnosis

In medical diagnostics, polymer microspheres have found various applications particularly in the aggregation tests, detection of antigens/antibodies, bacteria, viruses, and tissue receptors. The nano/microspheres are used in diagnostics as "reagents" and as elements of diagnostic devices. Microspheres with attached antigens (or antibodies) could be used for the detection of complementary antibodies (or antigens) in tests based on principles other than simple aggregation. In the book, we described several examples of applications of nano/microspheres in medical diagnostic tests including microsphere-based aggregation tests, nucleic acid analysis, flow cytometric analysis, and microsphere-based immunoassay (74).

13.3.1 Nano/Microspheres-Based Aggregation Tests

The simplest is the agglutination test, which consists of mixing a drop of a suspension of nano/microspheres containing immobilized antibodies with a drop of an analyzed liquid. In the presence of antigens specific to the attached antibodies, the nano/microspheres aggregate. The aggregation of nano/microspheres is a result of the binding of antigens to antibodies immobilized onto different particles. Antigens in this process act as cross linking agents. The presence of aggregates can be observed even with the naked eye. Thus, nano/microspheres in agglutination tests are used as "magnifying agents." Visual observation of aggregation, especially at low concentrations of antigens, is quite subjective.

However, the agglutination tests can indicate only whether an antigen is present (or absent) in an analyzed liquid. Quantitative determination of a "target" compound in a liquid is possible by monitoring the kinetics of aggregation of primary particles or the size of the aggregates formed. In these tests, a variety of analytical instruments based on light scattering, ultrasound measurements, and measurements of rates of sedimentation are used in order to improve sensitivity or to extend the concentration range over which a given compound can be detected.

13.3.2 Nano/Microspheres-Based Nucleic Acid Analysis

An important area where nano/microspheres are making an impact is in the area of nucleic acid analysis. For genetic and genomic analysis, the number of molecular features targeted for analysis can be quite large, tens of thousands in the case of gene expression analysis and hundreds of thousands in the case of single nucleotide polymorphisms. The flat DNA microarray, or DNA chip, developed from the need to make highly parallel molecular measurements from a single sample has emerged as a workhorse analysis platform (73).

However, for studies involving large numbers of samples, the flat chip format is not especially well-suited to high throughput analysis. Optically encoded arrays of nano/microspheres offer numerous advantages in terms of preparation and use the conventional flat microarray. Multiplexed nucleic acid applications can be grouped into two main categories: sequence detection and sequence analysis. Sequence detection applications include gene expression analysis and the detection of PCR products. Sequence analysis is generally concerned with variations in nucleic acid sequence. The analysis of single nucleotide polymorphisms (or

SNPs) has important implications for areas ranging from disease diagnostics and drug discovery to bacterial identification and forensic analysis, and the combination of a high level of multiplexing with the high serial throughput of flow cytometry make nano/microsphere arrays an attractive format for these applications.

The primary step in the preparation of nano/microsphere arrays for nucleic acid analysis is the functionalization of the nano/microsphere surface with a nucleic acid. The nucleic acid is most often a synthetic oligonucleotide, but could also be an oligonucleotide analogue or natural DNA or RNA, depending on the application. As for immobilization of antibodies for immunoassays, nucleic acids may be immobilized on the nano/microsphere surface using non-covalent physical adsorption, affinity binding methods, or covalent conjugation. Common methods are the use of affinity tags to tightly, but non-covalently, attach nucleic acid to a surface, especially biotin-modified nucleic acids to avidin- or streptavidin-functionalized surfaces. Such affinity interactions are generally tight, specific, and fairly stable, and have been used in a number of multiplexed applications.

However, in terms of stability and specificity, the most widely used nucleic acid immobilization methods are involved in covalent attachment. The most popular approach is to couple amino-modified nucleic acids to carboxylated surfaces using carbodiimide chemistry. Synthetic oligonucleotides are readily obtained with an amino-modification at one or both ends or at an internal position. Longer nucleic acids can be produced by enzymatic methods (i.e., PCR) by using amino-modified oligonucleotide primers or amino-modified nucleotide analogues. Sulthydral-modified nucleic acids can be prepared in similar manners for attachment via maleimide chemistry or for binding to gold surfaces. For both covalent and affinity-based non-covalent attachment, the inclusion of a multi-carbon linker to serve as a spacer seems to improve the efficiency of hybridization to a surface immobilized nucleic acid, though the degree of improvement can vary.

13.3.3 Nano/Microspheres-Based Flow Cytometric Analysis

The unique properties of nano/microspheres, which include their range of sizes, stability, uniformity, and capacity to absorb and retain fluorescent dye, were readily adapted to flow cytometric analysis of multiplexed reactions. Some of the immunoassays detected antibodies to cytomegalovirus and herpes simplex virus, different *Candida albicans* preparations, hepatitis C core and nonstructural proteins, and four recombinant DNA-produced HIV proteins (75–78) after capture on antigen-coated microspheres of different sizes. Proteins, oligonucleotides, polysaccharides, lipids, or small peptides have been previously adsorbed or chemically coupled to the surface of microspheres to capture analytes that are subsequently measured by a fluorochrome-conjugated detection molecule. More recently, assays for similar analytes have been multiplexed, or analyzed in the same assay volume, by performing each reaction on a set of microspheres that are dyed to different fluorescent intensities and, therefore, are spectrally distinct. Some recent applications with fluorescent microspheres have expanded to include cytokine quantification, single nucleotide polymorphism genotyping, phosphorylated protein detection, and characterization of the molecular interactions of nuclear receptors.

13.3.4 Other Nano/Microspheres-Based Analysis

The test based on measurements of electrophoretic mobility was developed for the detection and determination of antibodies against *Helicobacter pylori* (anti-*H. pylori*) in blood serum, a bacterium that causes stomach ulcers and cancers. Poly(styrene/polyglycidol) microspheres with immobilized *H. pylori* antigens were used in the test. In contact with blood sera, the particles were slightly swollen but did not aggregate. However, their electrophoretic mobility did change significantly after exposure to diluted sera from patients infected with *H. pylori*. Independent measurements based on ELISA assays (enzyme linked immunosorbent assay) were used for calibration of the test. The test allowed detection of antibodies in the sera of patients with titer (maximum degree of dilution at which ELISA gives a positive result) from 1:500 to 1:32,000, that is, in a similar range to ELISA, but in contrast to ELISA, its accuracy was the largest over the low range of concentrations of anti-*H. pylori* antibodies. New types of ELISA tests were developed using microspheres with magnetite nanoparticles in the cores and hydrophilic shells allowing for covalent immobilization of antibodies. Furthermore, ^{111}In-oxine, ^{99m}Tc-hexamethylpropyleneamine oxime, and ^{99m}Tc-stannous fluoride colloid–labeled white blood cells are good tools for the detection of inflammatory diseases and infections. In addition, one further application is labeling platelets with ^{111}In for the diagnosis of vascular thrombosis and the detection of infection regions (79).

Theragnostic multifunctional nanoparticles hold great promise in simultaneous diagnosis of disease, targeted drug delivery with minimal toxicity, and monitoring of treatment. One of the current challenges in cancer treatment is enhancing the tumor-specific targeting of both imaging probes and anticancer agents. Herein, we report tumor-homing chitosan-based nanoparticles (CNPs) that simultaneously execute cancer diagnosis and therapy (cancer theragnosis). These CNPs are unique for their three distinctive characteristics, such as stability in serum, deformability, and rapid uptake by tumor cells. These properties are critical in increasing their tumor targeting specificity and reducing their nonspecific uptake by normal tissues. To develop these CNPs into novel theragnostic nanoparticles, we labeled them with Cy5.5, a NIRF dye, for imaging and also loaded them with paclitaxel (PTX-CNPs), an anticancer drug for cancer treatment. Cy5.5 labeled PTX-CNPs exhibited significantly increased tumor-homing ability with low nonspecific uptake by other tissues in SCC7 tumor-bearing mice. Theragnostic nanoparticles, Cy5.5 labeled PTX-CNPs, are highly useful for simultaneous diagnosis of early-stage cancer and drug delivery (80,81).

Furthermore, hybrid structures that combine polymers with imaging moieties will begin to make polymer-based imaging probes a credible option for theragnosis. Theragnosis means molecular therapeutic and diagnostic approaches to treat diseases with the help of biotechnology, nanotechnology, and information technology, especially in the fields of bioimaging and nanomedicine. Theragnosis will offer the unprecedented and paradigm changing opportunity to study and diagnose various diseases at the molecular and cellular level. In the near future, "theragnosis" will begin to replace conventional *in vivo* diagnostics and drug delivery systems, a change that promises to revolutionize the treatment of diseases and disabilities.

13.4 Conclusion and Prospects

Nano/microspheres have shown promising potential in different fields of life sciences, such as biotechnology, drug carriers, biodetection, bioimaging and diagnosis and so on. This chapter has summarized the current status of the application of nano/microsphere in the medical diagnosis and bioimaging. The nano- and microparticles can be constructed as a multifunctional cargo to carry imaging agents, such as paramagnetic or superparamagnetic metals for MRI, optically active compounds for optical imaging, or microbubble imaging agents and radiolabeled agents used in SPECT and PET techniques. Even the capability to have both imaging and therapeutic agents incorporated in the same particle has been established. However, application of polymer-based nano/microsphere for bioimaging and diagnosis is still in its developing phase and quite a lot of issues need to be resolved, including application of new types of imaging probes, preparation of new nano/microspheres, improvements in image and diagnosis analysis and computational simulation, and evaluation of the accuracy or reproducibility of imaging and diagnostic methods and techniques. However, one thing for sure is that polymer-based nano/microsphere for bioimaging and diagnosis have already presented the exciting potential to monitor primary interventional procedures for precise therapeutic delivery, enhance the effectiveness of delivered therapeutics, and monitor therapeutic efficiency. In future, it is believed that rapid progresses in nano/microsphere will provide new opportunities for bioimaging and medical diagnosis.

References

1. Yang, X. (2007). Nano- and microparticle-based imaging of cardiovascular interventions: Overview. *Radiology*, **243**, 340–347.
2. Kim, J. H., Park K., Nam, H. Y., Lee, S., Kim, K., & Kwon, I. C. (2007). Polymers for bioimaging. *Progress in Polymer Science*, **32**, 1031–1053.
3. Guccione, S., Li, K. C., & Bednarski, M. D. (2004). Vascular-targeted nanoparticles for molecular imaging and therapy. *Methods in Enzymology*, **386**, 219–236.
4. Kawaguchi, H. (2000). Functional polymer microspheres. *Progress in Polymer Science*, **25**, 1171–1210.

5. Faraasen, S., Vörös, J., Csúcs, G., Textor, M., Merkle, H. P., & Walter, E. (2003). Ligand-specific targeting of microspheres to phagocytes by surface modification with poly (L-lysine)-grafted poly(ethylene glycol) conjugate. *Pharmaceutical Research*, **20**, 237–246.
6. Keegan, M. E., Royce, S. M., Fahmy, T., & Saltzman, W. M. (2006). *In vitro* evaluation of biodegradable microspheres with surface-bound ligands. *Journal of Controlled Release*, **110**, 574–580.
7. Pappo, J., Ermak, T. H., & Steger, H. J. (1991). Monoclonal antibody-directed targeting of fluorescent polystyrene microspheres to Peyer's patch M cells. *Immunology*, **73**, 277–280.
8. Gref, R., Couvreur, P., Barratt, G., & Mysiakine, E. (2003). Surface-engineered nanoparticles for multiple ligand coupling. *Biomaterials*, **24**, 4529–4537.
9. Zhao, M., Beauregard, D. A., Loizou, L., Davletov, B., & Brindle, K. M. (2001). Non-invasive detection of apoptosis using magnetic resonance imaging and a targeted contrast agent. *Nature Medicine*, **7**, 1241–1244.
10. Behrens, T. E. J., Johansen-Berg, H., Woolrich, M. W., Wheeler-Kingshoot, C. A. M., Boulby, P. A., Barker, G. J., *et al.* (2003). Non-invasive mapping of connections between human thalamus and cortex using diffusion imaging. *Nature Neuroscience*, **6**, 750–757.
11. Andrews, J., Al-Nahhas, A., Pennell, D. J., Hossain, M. S., Davies, K. A., Haskard, D. O., *et al.* (2004). Non-invasive imaging in the diagnosis and management of Takayasu's arteritis. *Annals of the Rheumatic Diseases*, **63**, 995–1000.
12. Prati, F., Mallus, M. T., Broglia, L., & Albertucci, M. (2010). Integrated non-invasive imaging techniques. *EuroIntervention*, 6 Suppl G, 161–168.
13. Boppart, S. A., Oldenburg, A. L., Xu, C. Y., & Marks, D. L. (2005). Optical probes and techniques for molecular contrast enhancement in coherence imaging. *Journal of Biomedical Optics*, **10**, 41208.
14. Morawski, A. M., Lanza, G. A., & Wickline, S. A. (2005). Targeted contrast agents for magnetic resonance imaging and ultrasound. *Current Opinion in Biotechnology*, **16**, 89–92.
15. Weissleder, R., & Imhof, H. (2007). Molecular imaging — A new emphasis in radiology. *Der Radiologe*, **47**, 6–7.
16. Weissleder, R., & Pittet, M. J. (2008). Imaging in the era of molecular oncology. *Nature*, **452**, 580–589.

17. Wu, K. C., Kim, R. J., Bluemke, D. A., Rochitte, C. E., Zerhouni, E. A., Becker, L. C., *et al.* (1998). Quantification and time course of microvascular obstruction by contrast-enhanced echocardiography and magnetic resonance imaging following acute myocardial infarction and reperfusion. *Journal of the American College of Cardiology*, **32**, 1756–1764.
18. Ghoroghchian, P. P., Therien, M. J., & Hammer, D. A. (2009). *In vivo* fluorescence imaging: A personal perspective. *Wiley Interdisciplinary Reviews. Nanomedicine and Nanobiotechnology*, **1**, 156–167.
19. Yockman, J. W., Kastenmeier, A., Erickson, H. M., Brumbach, J. G., Whitten, M. G., Albanil, A., *et al.* (2008). Novel polymer carriers and gene constructs for treatment of myocardial ischemia and infarction. *Journal of Controlled Release*, **132**, 260–266.
20. Ghoroghchian, P. P., Frail, P. R., Susumu, K., Blessington, D., Brannan, A. K., Bates, F. S., *et al.* (2005). Near-infrared-emissive polymersomes: Self-assembled soft matter for *in vivo* optical imaging. *Proceedings of the National Academy of Sciences of the United States of America*, **102**, 2922–2927.
21. Kim, Y. E., Chen, J., Chan, J. R., & Langen, R. (2010). Engineering a polarity-sensitive biosensor for time-lapse imaging of apoptotic processes and degeneration. *Nature Methods*, **7**, 67–73.
22. Langlois, A. J., Holder Jr., W. D., Iglehart, J. D., Nelson-Rees, W. A., Wells Jr., S. A., & Bolognesi, D. P. (1979). Morphological and biochemical properties of a new human breast cancer cell line. *Cancer Research*, **39**, 2604–2613.
23. Ji, H., Ohmura, K., Mahmood, U., Lee, D. M., Hofhuis, F. M. A., Boackle, S. A., *et al.* (2002). Arthritis critically dependent on innate immune system players. *Immunity*, **16**, 157–168.
24. Chen, J., Tung, C. H., Mahmood, U., Ntziachristos, V., Gyurko, R., Fishman, M. C., *et al.* (2002). *In vivo* imaging of proteolytic activity in atherosclerosis. *Circulation*, **105**, 2766–2771.
25. Licha, K., Hessenius, C., Becker, A., Henklein, P., Bauer, M., Wisniewski, S., *et al.* (2001). Synthesis, characterization, and biological properties of cyanine-labeled somatostatin analogues as receptor-targeted fluorescent probes. *Bioconjugate Chemistry*, **12**, 44–50.
26. Funovics, M., Weissleder, R., & Tung, C. H. (2003). Protease sensors for bioimaging. *Analytical and Bioanalytical Chemistry*, **377**, 956–963.
27. Borisov, S. M., Mayr, T., & Klimant I. (2010). Dye-doped polymeric particles for sensing and imaging. *Advanced Fluorescence Reporters*

in Chemistry and Biology II, Springer Series on Fluorescence, **9**, 193–228.

28. Gouanve, F., Schuster, T., Allard, E., Meallet-Renault, R., & Larpent, C. (2007). Fluorescence quenching upon binding of copper ions in dye-doped and ligand-capped polymer nanoparticles: A simple way to probe the dye accessibility in nano-sized templates. *Advanced Functional Materials*, **17**, 2746–2756.
29. Jamieson, T., Bakhshi, R., Petrova, D., Pocock, R., Imani, M., & Seifalian, A. M. (2007). Biological applications of quantum dots. *Biomaterials*, **28**, 4717–4732.
30. Jin, T., Tiwari, D. K., Tanaka, S., Inouye, Y., Yoshizawa, K., & Watanabe, T. M. (2010). Antibody-protein A conjugated quantum dots for multiplexed imaging of surface receptors in living cells. *Molecular BioSystems*, **6**, 2325–2331.
31. Bentolila, L. A., Ebenstein, Y., & Weiss, S. (2009). Quantum dots for *in vivo* small-animal imaging. *J Nucl Med*, **50**, 493–496.
32. Hardman, R. (2006). A toxicologic review of quantum dots: Toxicity depends on physicochemical and environmental factors. *Environmental Health Perspectives*, **114**, 165–172.
33. Du, J., Yu, C., Pan, D., Li, J., Chen, W., Yan, M., *et al.* (2010). Quantum-dot-decorated robust transductable bioluminescent nanocapsules. *Journal of the American Chemical Society*, **132**, 12780–12781.
34. Sigot, V., Arndt-Jovin, D. J., & Jovin, T. M. (2010). Targeted cellular delivery of quantum dots loaded on and in biotinylated liposomes. *Bioconjugate Chemistry*, **21**, 1465–1472.
35. Zrazhevskiy, P., Sena, M., & Gao, X. (2010). Designing multifunctional quantum dots for bioimaging, detection, and drug delivery. *Chemical Society Reviews*, **39**, 4326–4354
36. Sanchez-Martin, R. M., Alexander, L., & Bradley, M. (2008). Multifunctionalized biocompatible microspheres for sensing. *Annals of the New York Academy of Sciences*, **1130**, 207–217.
37. Ma, Q., Wang, X. Y., Li, Y. B., Su, X. G., & Jin, Q. H. (2007). The use of COP quantum dot fluorescent microspheres in fluoro-immunoassays and a microfluidic chip system. *Luminescence*, **22**, 438–445.
38. Qian, H. S., Li, Z. Q., & Zhang, Y. (2008). A facile synthesis of multicolor polystyrene microspheres encapsulating upconversion fluorescent nanoparticles. *IFMBE Proceedings*, **21**, 73–76.
39. Vishwanath, K., Zhong, W., Close, M., & Mycek, M. A. (2006). Fluorescence quenching by polystyrene microspheres in UV-visible and NIR tissue-simulating phantoms. *Optics Express*, **14**, 7776–7788.

40. Kobayashi, H., & Brechbiel, M. W. (2005). Nano-sized MRI contrast agents with dendrimer cores. *Advanced Drug Delivery Reviews*, **57**, 2271–2286.

41. Calabi, L., Alfieri, G., Biondi, L., Miranda, M. D., Paleari, L., & Ghelli, S. (2002). Application of high-resolution magic-angle spinning NMR spectroscopy to define the cell uptake of MRI contrast agents. *Journal of Magnetic Resonance*, **156**, 222–229.

42. Lee, J. H., Huh, Y. M., Jun, Y. W., Seo, J. W., Jang, J. T., Song, H. T., *et al.* (2007). Artificially engineered magnetic nanoparticles for ultra-sensitive molecular imaging. *Nature Medicine*, **13**, 95–99

43. Chen, H. H., Visage, L. C., Qiu, B. S., Du, X. Y., Ouwerkerk, R., Leong, K. W., *et al.* (2005). MR imaging of biodegradable polymeric microparticles: A potential method of monitoring local drug delivery. *Magnetic Resonance in Medicine*, **53**, 614–620.

44. Shen, T., Weissleder, R., Papisov, M., Bogdanov, A., & Brady, T. J. (1993). Monocrystalline iron-oxide nanocompounds (Mion) - Physicochemical Properties. *Magnetic Resonance in Medicine*, **29**, 599–604.

45. Fang, C., & Zhang, M. (2009). Multifunctional magnetic nanoparticles for medical imaging applications. *Journal of Materials Chemistry*, **19**, 6258–6266.

46. Deng, Y. H., Wang, C. C., Shen, X. Z., Yang, W. L., Jin, L., Gao, H., *et al.* (2005). Preparation, characterization, and application of multistimuli-responsive microspheres with fluorescence-labeled magnetic cores and thermoresponsive shells. *Chemistry*, **11**, 6006–6013.

47. Lee, J. H., Jun, Y. W., Yeon, S. I., Shin, J. S., & Cheon, J. (2006). Dual-mode nanoparticle probes for high-performance magnetic resonance and fluorescence imaging of neuroblastoma. *Angewandte Chemie*, **45**, 8160–8162.

48. Kopelman, R., Lee Koo, Y. E., Philbert, M., Moffat, B. A., Reddy, G. R., McConville, P., *et al.* (2005). Multifunctional nanoparticle platforms for *in vivo* MRI enhancement and photodynamic therapy of a rat brain cancer. *Journal of Magnetism and Magnetic Materials*, **293**, 404–410.

49. Hamoudeh, M., Kamleh, M. A., Diab, R., & Fessi, H. (2008). Radionuclides delivery systems for nuclear imaging and radiotherapy of cancer. *Advanced Drug Delivery Reviews*, **60**, 1329–1346.

50. Nijsen, J. F., Krijger, G. C., & van Het Schip, A. D. (2007). The bright future of radionuclides for cancer therapy. *Anti-Cancer Agents in Medicinal Chemistry*, **7**, 271–290.

51. Kaplan, W. D., Ensminger, W. D., Come, S. E., Smith, E. H., D'Orsi, C. J., Levin, D. C., *et al.* (1980). Radionuclide angiography to predict patient response to hepatic artery chemotherapy. *Cancer Treatment Reports*, **64**, 1217–1222.

52. Ensminger, W., Niederhuber, J., Dakhil, S., Thrall, J., & Wheeler, R. (1981). Totally implanted drug delivery system for hepatic arterial chemotherapy. *Cancer Treatment Reports*, **65**, 393–400.

53. Zielhuis, S. W., Seppenwoolde, J. H., Bakker, C. J. G., Jahnz, U., Zonnenberg, B. A., van het Schip, A. D., *et al.* (2007). Characterization of holmium loaded alginate microspheres for multimodality imaging and therapeutic applications. *Journal of Biomedical Materials Research, Part A*, **82**, 892–898.

54. Mariani, L., Schroth, G., Wielepp, J. P., Haldemann, A., & Seiler, R. W. (2001). Intratumoral arteriovenous shunting in malignant gliomas. *Neurosurgery*, **48**, 353–358.

55. Walser, R. H., Haldemann, A. R., Rösler, H., Koranda, P., Kinser, J. A., & Triller, J. (1996). Diagnostic angioscintigraphic evaluation of malignant hepatic tumors before catheter embolization: Determination of shunt, flow distribution, and reflux. *Cardiovascular and Interventional Radiology*, **19**, 77–81.

56. Moreno, A. J., Rodriguez, A. A., Spicer, M. J., Jackson, R. E., & Turnbull, G. L. (1986). The value of obtaining renal images with brain scintigraphy when using Tc-99m glucoheptonate. *Journal of Nuclear Medicine*, **27**, 315–315.

57. Briz Kanafani, S., & Lazos Constantino, G. (1969). Perfusion I-131 macroaggregate brain scanning; a clinical evaluation of its diagnostic efficiency. *American Journal of Roentgenology, Radium Therapy, and Nuclear Medicine*, **106**, 333–339.

58. Even, G. A., & Green, M. A. (1989). Gallium-68-labeled macroaggregated human serum albumin, 68Ga-MAA. *International Journal of Radiation Applications and Instrumentation — Part B*, **16**, 319–321.

59. Okada, S., Ohto, M., Kuniyasu, Y., Higashi, S., Arimizu, N., & Uematsu, S. (1990). [Estimation of the reticuloendothelial function by positron emission computed tomography (PET) study in chronic liver disease]. *Nihon Shokakibyo Gakkai Zasshi*, **87**, 90–99.

60. Unger, E. C., Porter, T., Culp, W., Labell, R., Matsunaga, T., & Zutshi, R. (2004). Therapeutic applications of lipid-coated microbubbles. *Advanced Drug Delivery Reviews*, **56**, 1291–1314.

61. Schutt, E. G., Klein, D. H., Mattrey, R. M., & Riess, J. G. (2003). Injectable microbubbles as contrast agents for diagnostic ultrasound imaging: The key role of perfluorochemicals. *Angewandte Chemie, International Edition*, **42**, 3218–3235.

62. Raisinghani, A., & DeMaria, A. N. (2002). Physical principles of microbubble ultrasound contrast agents. *American Journal of Cardiology*, **90**, 3–7.

63. Nanda, N. C., Kitzman, D. W., Dittrich, H. C., & Hall, G. (2003). Imagent improves endocardial border delineation, inter-reader agreement, and the accuracy of segmental wall motion assessment. *Echocardiography*, **20**, 151–161.

64. Forsberg, F., Wu, Y., Makin, I. R., Wang, W., & Wheatley, M. A. (1997). Quantitative acoustic characterization of a new surfactant-based ultrasound contrast agent. *Ultrasound in Medicine & Biology*, **23**, 1201–1208.

65. Pisani, E., Tsapis, N., Paris, J., Nicolas, V., Nicolas, L., & Fattal, E. (2006). Polymeric nano/microcapsules of liquid perfluorocarbons for ultrasonic imaging: Physical characterization. *Langmuir*, **22**, 4397–4402.

66. Cavalieri F., Hamassi, A. E., Chiessi, E., & Paradossi, G. (2006). Tethering functional ligands onto shell of ultrasound active polymeric microbubbles. *Biomacromolecules*, **7**, 604–611.

67. Uhrich, K. E., Cannizzaro, S.M., Langer, R. S., & Shakesheff, K. M. (1999). Polymeric systems for controlled drug release. *Chemical Reviews*, **99**, 3181–3198.

68. Miller, D. L., & Song, J. M. (2003). Tumor growth reduction and DNA transfer by cavitation-enhanced high-intensity focused ultrasound *in vivo*. *Ultrasound in Medicine & Biology*, **29**, 887–893.

69. Kohrmann, K. U., Michel, M. S., Steidler, A., Marlinghaus, E., Kraut, O., & Alken, P. (2002). Technical characterization of an ultrasound source for noninvasive thermoablation by high-intensity focused ultrasound. *BJU International*, **90**, 248–252.

70. Wu, F., Chen, W. Z., Bai, J., Zou, J. Z., Wang, Z. L., Zhu, H., *et al.* (2002). Tumor vessel destruction resulting from high-intensity focused ultrasound in patients with solid malignancies. *Ultrasound in Medicine & Biology*, **28**, 535–542.

71. Hilgenfeldt, S., Lohse, D., & Zomack, M. (2000). Sound scattering and localized heat deposition of pulse-driven microbubbles. *The Journal of the Acoustical Society of America*, **107**, 3530–3539.

72. Lindner, J. R., Song, J., Xu, F., Klibanov, A. L., Singbartl, K., Ley, K., *et al.* (2000). Noninvasive ultrasound imaging of inflammation using microbubbles targeted to activated leukocytes. *Circulation*, **102**, 2745–2750.

73. Leong-Poi, H., Christiansen, J., Klibanov, A. L., Kaul, S., & Linsner, J. R. (2003). Noninvasive assessment of angiogenesis by ultrasound and microbubbles targeted to av-Integrins. *Circulation*, **107**, 455– 460.

74. Jain, R., Dandekar, P., & Patravale, V. (2009). Diagnostic nanocarriers for sentinel lymph node imaging. *Journal of Controlled Release*, **138**, 90–102.

75. McHugh, T. M., Miner, R. C., Logan, L. H., & Stites, D. P. (1988). Simultaneous detection of antibodies to cytomegalovirus and herpes simplex virus by using flow cytometry and a microsphere-based fluorescence immunoassay. *Journal of Clinical Microbiology*, **26**, 1957–1961.

76. McHugh, T. M., Wang, Y. J., Chong, H. O., Blackwood, L. L., & Stites, D. P. (1989). Development of a microsphere-based fluorescent immunoassay and its comparison to an enzyme immunoassay for the detection of antibodies to three antigen preparations from Candida albicans. *Journal of Immunological Methods*, **116**, 213–219.

77. McHugh, T. M., Viele, M. K., Chase, E. S., & Recktenwald, D. J. (1997). The sensitive detection and quantitation of antibody to HCV by using a microsphere-based immunoassay and flow cytometry. *Cytometry*, **29**, 106–112.

78. Scillian, J. J., Mchugh, T. M., Busch, M. P., Tam, M., Fulwyler, M. J., Chien, D. Y., *et al.* (1989). Early detection of antibodies against rDNA-produced HIV proteins with a flow cytometric assay. *Blood*, **73**, 2041–2048.

79. Basinska, T. (2005). Hydrophilic core–shell microspheres: A suitable support for controlled attachment of proteins and biomedical diagnostics. *Macromolecular Bioscience*, **5**, 1145–1168.

80. Weissleder, R., Tung, C. H., Mahmood, U., & Bogdanov, A. Jr. (1999). *In vivo* imaging of tumors with protease-activated near-infrared fluorescent probes. *Nature Biotechnology*, **17**, 375–378.

81. Morimoto, S. (2007). *In vivo* imaging of tumors with protease-activated near-infrared fluorescent probes. *Nature Biotechnology*, **17**, 375–378.

Chapter 14

Affinity Nanoparticles for Detection

Haruma Kawaguchi[a] **and Hiroshi Handa**[b]

[a]*Graduate School of Engineering, Kanagawa University, Yokohama, Japan*
[b]*Graduate School of Bioscience and Biotechnology, Tokyo Institute of Technology, Yokohama, Japan*
haruma@kanagawa-u.ac.jp

14.1 Introduction

14.1.1 Definition and Feature of Affinity Latex

There are a number of biospecific couples which form hybrids or supramolecules via their bioaffinity. The couples include antibody/antigen, DNA double strand, sequence-specific DNA/DNA-binding protein, sequence-specific peptide/peptide-binding protein, etc. When either component of the couple is immobilized on latex particles, the hybrid particles thus obtained are referred to as "affinity latex." Affinity latex can catch only its complementary "target" molecule onto the particles from a mixture of biomolecules such as miscellaneous proteins and DNAs. Therefore, it contributes to bioassay and diagnosis if the target molecule is concerned with any disease. This chapter deals with what the key points for designing affinity latex are, how to prepare affinity latex, and what the affinity latex can do.

14.1.2 Affinity Chromatography

Affinity latex system is a kind of affinity chromatography. Affinity chromatography is an established bioseparation technique based

Microspheres and Microcapsules in Biotechnology: Design, Preparation, and Applications
Edited by Guanghui Ma and Zhiguo Su

ISBN 978-981-4316-47-7 (Hardcover), 978-981-4364-62-1 (eBook)
www.panstanford.com

on biorecognition (1). Affinity column chromatography has come into wide use (2). It is demonstrated using a column packed with macro- or micro-porous gel beads, e.g., agarose beads, of about 10 μm, which carry biospecific molecules (ligands or probe) on and in them. A sample to be examined is run through the affinity column. The efficiency for the ligands to catch the target molecules is not so high. It is because of two reasons. One is the effect of complicated network structure which is shown in the TEM view of Fig. 14.1. The ligands are immobilized in different microenvironment in the network structure. They are surrounded by various atmospheres that cause different affinity-binding force between the ligand and the target. This prevents sharp affinity-based separation. The gel's complicated network structure is another obstacle for sharp performance of beads; the network structure disturbs smooth flow of eluent in the column and, therefore, clear-cut separation becomes difficult. The second disadvantage of affinity column chromatography is the fact that ligands are on stationary phase. In affinity column chromatography, the ligands are at a loss as they have no other option than to wait for coming targets. In contrast, affinity latex overcomes these drawbacks of affinity column. The active reaction site of affinity latex is only the surface of the latex particle. This means the reaction

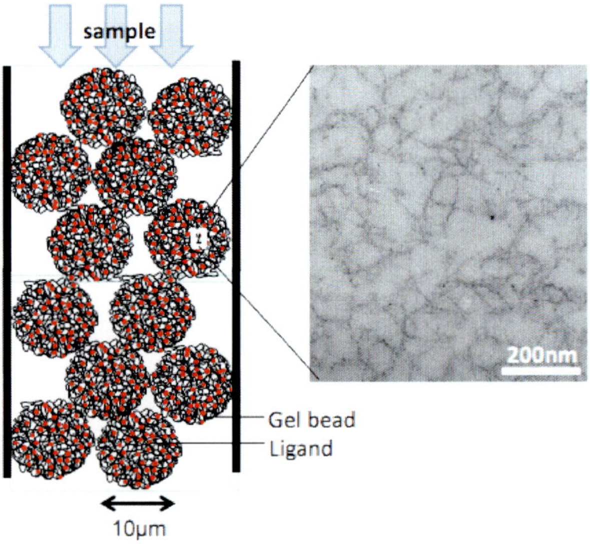

Figure 14.1 Affinity column chromatograph and a close-up of gel structure.

site is uniform and homogeneous. Submicron-sized latex particles have such surface in a very huge scale. In addition, the reaction site of affinity latex is mobile. This means the ligands on the latex particle have frequent chances to meet target molecules in the medium. While these are advantages of affinity latex, its disadvantage is that it needs a particle recovery process after the operation. Recovery can be done by centrifugation but, more conveniently by application of magnetic force for magnetic nanoparticle-containing affinity latex.

14.1.3 History of Affinity Latex

The term "affinity latex" has been coined by us, but affinity latex has a fairly long history. The first affinity latex appeared as a latex diagnosis in a literature of 1956 (3). It was a dispersion of microparticles composed of poly(styrene derivative) carrying anti-rheumatoid antibody. The affinity latex particles aggregate if the blood sample of a patient has rheumatoid factor, which can be detected by antigen–antibody binding as shown in Fig. 14.2. Thus, the extent of aggregate formation is regarded as the measure of rheumatism.

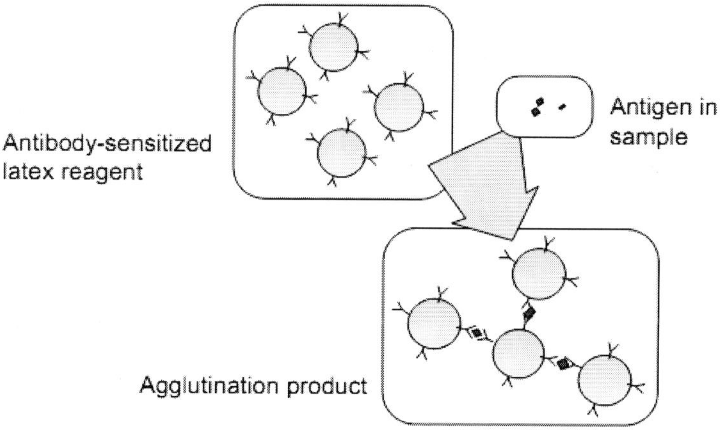

Figure 14.2 Typical latex diagnosis based on latex agglutination caused by antigen–antibody reaction.

The second innovation in affinity latex was brought by Rembaum, who used latex particles as a cell marker (4). The amount and distribution of specific receptors on the cell surface were clearly detected by the attached affinity latex particles that carry a specific molecule complementary to the receptor. Ugelstad was another key person to render distinguished service for the development of affinity particles (5). His particles were micron-sized and monodispersed, and they carried anticancer antibodies. They were employed for cancer cell separation (6). The aggregates of cancer cells and polymer beads were removed from the body fluid with the aid of magnetic force. Rembaum and Ugelstad were special invited speakers for the 2nd Microsphere Symposium at Fukui in 1982. Seven years later, Uhlen *et al.* proposed new methodology for specific DNA separation using DNA-carrying particles (7) and, in the same year, Kawaguchi and Handa *et al.* published their first paper on affinity latex (8).

14.2 Design of Affinity Latex Particle

A simplified figure of affinity latex particle is presented in Fig. 14.3. In this section, important points for the preparation of affinity latex system are presented.

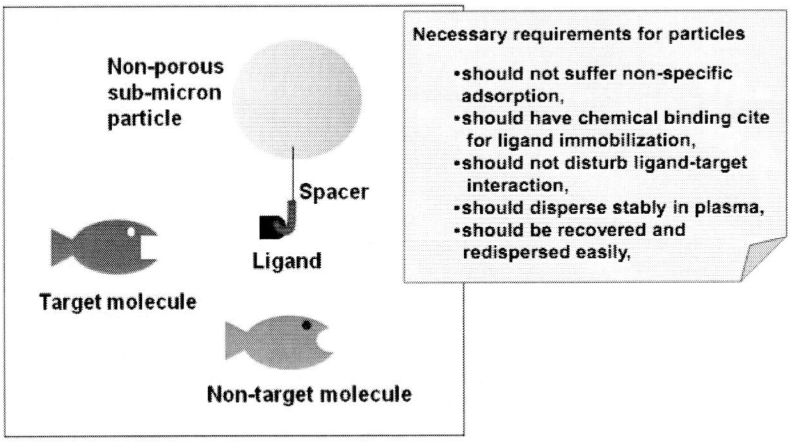

Figure 14.3 Construction of affinity latex particle.

14.2.1 Design of Particle

The particle should selectively catch the target molecule (fish) with immobilized ligand (bait). Therefore, the particle itself should prevent nonspecific adsorption because nonspecific adsorption, corresponding to randomly occurring adsorption, spoils the affinity-based selection. In order to realize the satisfying condition, the surface of latex particle must be moderately hydrophilic. After careful experiment, poly(styrene-co-glycidyl methacrylate) particle was chosen as the best one for our purpose (9). The particles were prepared by soap-free emulsion polymerization of styrene and glycidyl methacrylate using potassium persulfate as an initiator. The resulting particles were monodispersed and had a diameter of 200–300 nm. The surface of the particle was covered with glycidyl and glycelol groups as well as persulfate groups. Glycelol groups were prepared *in situ* from glycidyl groups by hydrolysis. The ratio of glycidyl to glycol groups was dependent on the reaction conditions, such as pH and temperature. Glycidyl and glycelol groups are modelately hydrophilic and reject nonspecific adsorption of proteins and other biomolecules. The amount of protein adsorption onto the poly(styrene-co-glycidyl methacrylate) particle was one fiftieth of that of polystyrene particle. Glycidyl group contributes not only to reject nonspecific adsorption but also to serve as a binding site of ligands. Ligands having amino, hydroxyl, thiol groups are capable to attach poly(styrene-co-glycidyl methacrylate) particle via chemical reaction between those functional groups and glycidyl group on the surface.

Another important characteristic of the particle is that it neither has micropores nor has macropores inside. Due to the nonporous structure of latex particles, ligands are immobilized only on the surface of latex particles. All such ligands have equal or similar ability to form supramolecules with target molecules. This is the advantage of nonporous particles compared to macro- and microporous particles that carry ligands with different activity to catch the targets and, therefore, suppress clear-cut bioseparation.

14.2.2 Ligand

Ligands to be immobilized onto affinity latex particle are proteins, DNA, drug, etc. Proteins include enzyme, antibody, and other functional proteins. In addition to these molecules, a part of such molecules can be immobilized. For example, an antibody is replaced with an antigen-binding fragment of antibody (Fab) as a ligand for latex diagnosis (10). Fab was separated from Fc fragment and used solely as a ligand to catch antigen. This method gives a merit that Fc fragment that is useless for affinity interaction is eliminated.

Another example is to use only an essential part as ligand instead of the whole molecule in cell adhesive peptide, arginine–glycine–aspartic acid (RGD). Cell adhesive molecules have RGD peptide, which contributes to cell adhesion via interaction between RGD in the molecule and integlin (GP IIb/IIIa) on cell membrane. This means affinity latex can demonstrate affinity binding with integlin if the latex carries RGD as the ligand. This was confirmed using RGDS-carrying particles in cell extract (Fig. 14.4a) (11) and in cell membrane (Fig. 14.4b) (12) for integlin separation and cell activation. It is worth mentioning that [Mg^{2+}] (magnesium ion concentration) must be carefully checked in this system because the special structure of RGD-binding site of GPIIb is controlled by [Mg^{2+}].

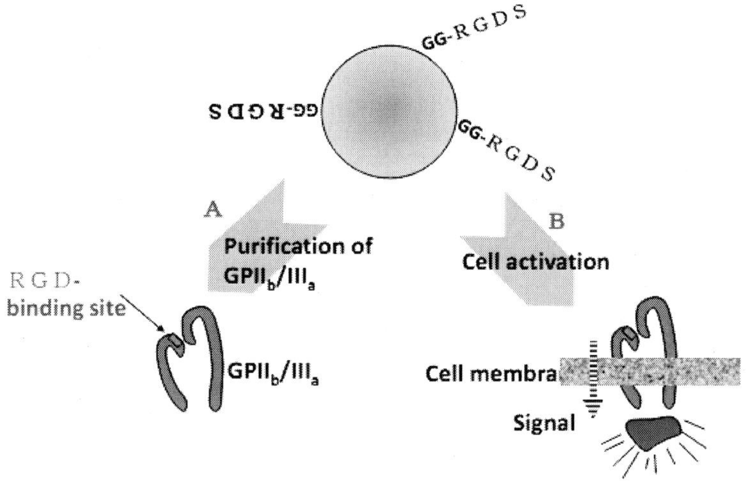

Figure 14.4 RGDS-immobilized affinity latex and two applications.

14.2.3 Immobilization of Ligand

14.2.3.1 Chemical bonding or physical adsorption

It was described in the previous section that affinity latex particles should not be susceptible to nonspecific adsorption of biocomponents. This does not necessarily mean that the particles themselves never be susceptible to nonspecific adsorption. Highly hydrophobic polymer particles adsorb proteins fast and irreversibly. Thus, adsorption is one of the easiest methods to immobilize a ligand protein and, in fact, polystyrene latex particles have been used as an affinity carrier of early latex diagnosis due to easy immobilization of ligand. In this case, the remaining exposed hydrophobic surface of the ligand-carrying particle must be masked with hydrophilic polymers such as polyethyleneglycol and albumin in order to avoid further nonspecific adsorption of proteins (13). Immobilization of ligand by physical adsorption on hydrophobic particle has sometimes severe disadvantages. When strand-type ligands were immobilized onto hydrophobic particles, it was quite difficult to bind them at the end of the strand. They rather stick to the surface in train or loop modes. These conformations do not result in efficient affinity performance. Therefore, latex particles having hydrophobic surface cannot be used for binding DNA molecules.

14.2.3.2 DNA immobilization via chemical bonding using protruding chain end

DNA are immobilized on particles having hydrophilic surface. An efficient method for immobilizing long double-strand DNA onto hydrophilic surface is presented below.

A protruding chain end plays an important role in the immobilization of DNA at the chain end, because the protruding DNA has an exposed base which can combine with glycidyl group on particle surface. The length of protruding chain end decided the amount of immobilization as shown in Fig. 14.5 (14). These results support the above-mentioned reaction scheme. After immobilization, the remaining glycidyl groups were masked with tris(hydroxymethyl) amminomethane or mercaptoethanol. Thus, further binding of DNA chains was suppressed and the DNA was immobilized only at the chain end without train or loop conformations. The same methodology can be applied when single-strand DNA is bound at the tail to the particle (15). In this case, double-strand DNA is immobilized using a protruding end and melted to two single-strand DNAs, one of which is bound to the surface and the other dissolves into water.

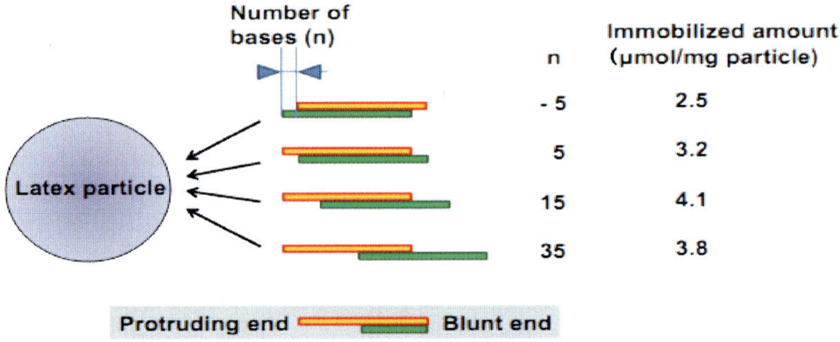

Figure 14.5 Effect of protruding chain length on the efficiency of DNA immobilization.

14.2.3.3 Position of active site of on-surface ligand

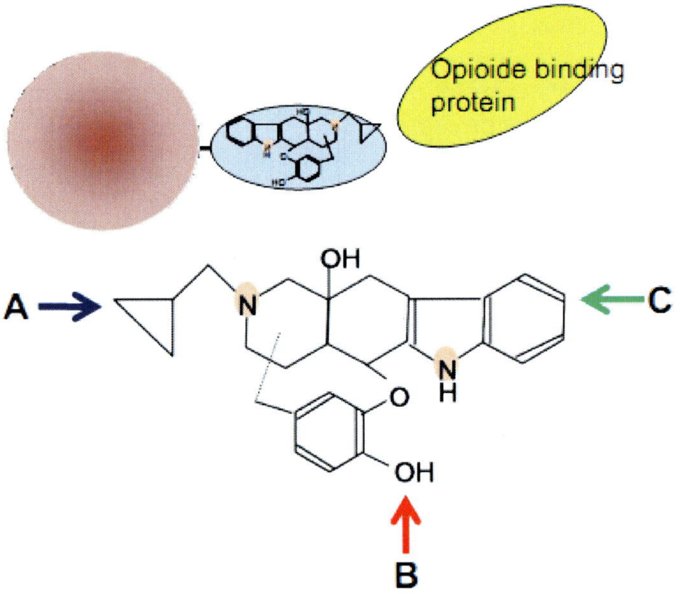

Figure 14.6 Affinity particle to catch antagonist of opioid receptor, which is a cell membrane protein concerning pain relief.

The direction of the active site of ligand on a particle is another important factor deciding the efficiency of affinity binding. If whole molecule of the antibody is immbilized onto the particle, it must be bound to the particle at Fc fragment. It is because the affinity recognition is done by Fab fragment, which must face the aqueous phase. Not only high molecular weight protein but also low molecular weight drug must be immobilized properly in order to exhibit efficient affinity with target protein. If the target protein is known, the ligand drug is bound at a position that does not affect hybridization. But, if the target protein is unknown, trial-and-error examination is necessary to decide the binding site of the drug. Such examination was done to immobilize an antagonist of the opioid receptor whose molecular structure is shown in Fig. 14.6 (16). The molecule was bound to the latex particle at A, B, or C and incubated with a mixture of proteins. As a result, only the particle immobilizing the antagonist at A showed affinity binding with an aimed protein. Such kind of information suggests that a small part of the drug molecule is not the critical binding site with receptor protein and can be modified for improving the drug function.

14.2.3.4 Spacer

The efficiency of on-surface ligands to recognize and catch the targets significantly depends on the free space around the ligand. In order to maintain such condition, a certain distance between the ligand molecule and the particle surface is kept by a tie molecule, called "spacer." Ethyleneglycol diglycidyl ether and glycyl-glycine are representative spacers (17). They are moderately hydrophilic and have a suitable chain length. In a study on sequence-specific DNA carrying affinity latex for transcription factor collection, it was found that a part of the DNA serves as the spacer.

Transcription factor is an important protein to regulate transcription according to the process shown in Fig. 14.7. First, transcription factor attaches to a specific base sequence on a double-strand DNA (7,14). The attachment brings about binding of polymerase at the downstream of DNA. Then transcription, that is, mRNA formation starts, followed by translation and protein formation. Therefore, transcription factor is a key molecule to regulate RNA formation and protein formation as well. An affinity latex to collect transcription factor from the mixture was created. Fourteen-base, sequence-specific, double-strand DNA was immobilized onto poly(styrene-glycidyl methacrylate) particles using protruding end–binding method. But this trial was in vain. It was attributed to steric hindrance near the immobilized DNA. To resolve this problem, the DNA was polymerized in tandem and then attached to the particle. This particle successfully gathered the transcription factor. These experimental results showed that the DNA near the surface of particle is regarded as a kind of spacer.

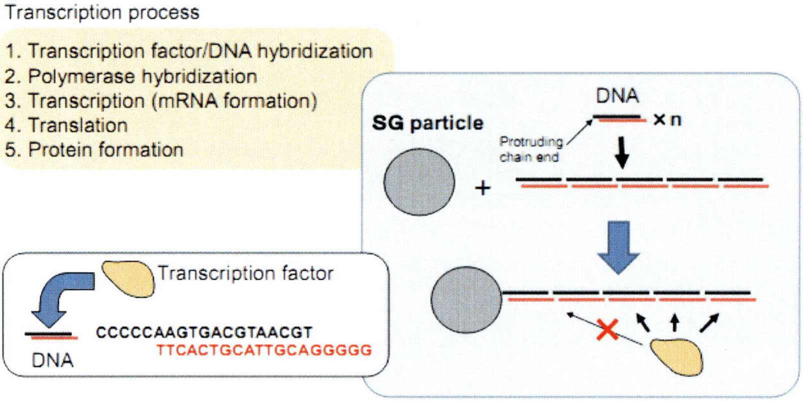

Figure 14.7 Preparation of sequence-specific DNA-carrying affinity particle for separation of transcription factor.

14.3 Applications

14.3.1 DNA Diagnosis

A large number of human genetic diseases and single nucleotide polymorphisms (SNPs) are caused by gene mutations including single-base substitutions and small additions or deletions (Fig. 14.8). Therefore, it is very meaningful to detect such gene mutations as early as possible, and rapid, easy to handle, and high-throughput mismatch detection methods are urgently demanded. Here, two types of affinity latexes for DNA diagnosis are introduced. One is single-strand DNA–carrying latex and the other is enzyme MutS–carrying latex.

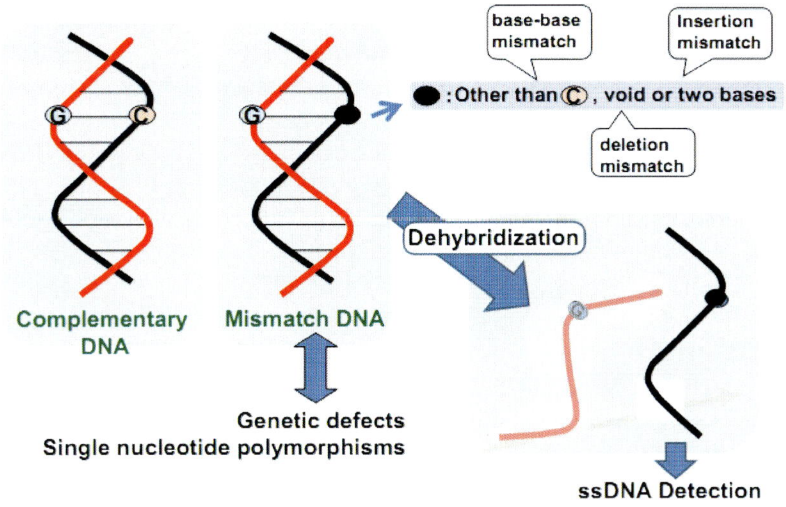

Figure 14.8 Types of SNPs and their detection.

14.3.1.1 Separation of normal and point mutant DNAs using DNA-carrying particles

This method is applied to the de-hybridized DNA solution (right bottom of Fig. 14.8) (18,19). The affinity particle carries single-strand DNA complementary to point mutant DNA (right bottom of Fig. 14.8). The ability of the affinity latex to distinguish complementary DNA from noncomplementary one was examined using the latex particles that carry 5′-CCTACGCCACCAGCTCCAAC (pDNA) via 10G spacer and two 20-base DNA solutions: GGATGCGGTGGTCGAGGTTG (complementary to the DNA attached to particle, cDNA) and GGATGCGGTCGTCGAGGTTG (one-point mismatch to the DNA attached to particle, mDNA). The two DNAs are different from each other by one base only. Two methods were employed; one is individual hybridization in which affinity latex was mixed separately with cDNA solution or mDNA solution. The progress of hybridization was measured as a function of time and the result is shown in Fig. 14.9a. As can be seen in the figure, cDNA hybridized quickly with pDNA on affinity latex whereas hybridization is very slow for mDNA. In the second method, cDNA and mDNA solutions with different molar ratio (1:1 ~ 1:100) are simultaneously mixed with affinity latex. Note that a fluorescent probe was combined to cDNA to identify which DNA hybridized with pDNA. Competitive hybridization was expected in this system. But this is not the case. The result shown in Fig. 14.9b indicated that hybridization of cDNA was absolutely not hindered by mDNA even if mDNA exists 100 times more than cDNA. Such phenomena were not observed in solution. The extraordinary phenomena in latex system is attributed to the restricted mobility of pDNA immobilized on the particle. That is, conformation of pDNA hardly adapts to that of mDNA due to the anchored chain end. Therefore, the results in Fig. 14.9 strongly suggest that affinity latex is a promising tool for discrimination of one-point mismatch DNA.

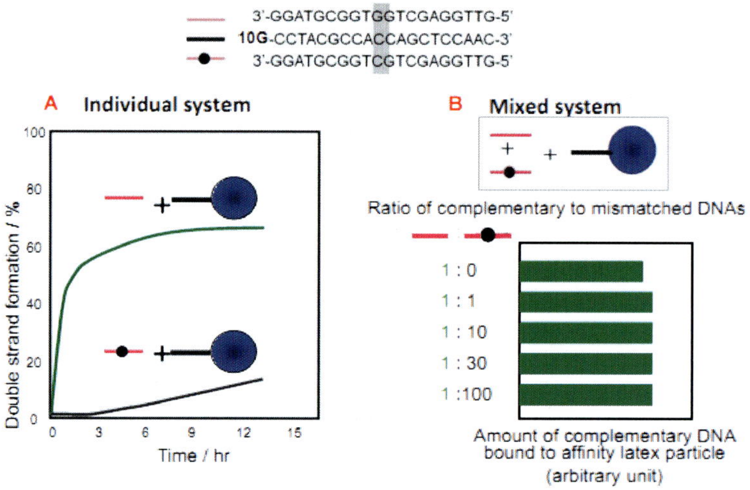

Figure 14.9 Detection of point mismatch DNA using ssDNA-immobilized affinity particle in individual and mixed systems.

14.3.1.2 Enzyme-carrying affinity latex as diagnostic reagent

This affinity latex enabled direct determination of mismatch DNA (20). The latex particle carries two key materials. One is MutS and the other is anthraquinone (AQ). MutS protein is a crucial element of DNA mismatch repair (MMR) system in many organisms such as *Escherichia coli* (21). MutS is a 97 kDa protein that can specifically recognize and bind all possible single-base mismatches as well as

1–4 base insertion or deletion loops (22) but it releases complementary double-strand DNA. AQ derivative) is a light-activated agent that cleaves double-stranded DNA (23) with the aid of UV irradiation. Use of affinity latex made it possible to discriminate complementary and mismatched DNAs and photocleaved mismatched DNA selectively.

The experimental result is summarized in Fig. 14.10. The DM value indicates the degree of mismatched DNA discrimination. The high value is an evidence of the high ability of MutS to selectively bind mismatched DNA. This resulted from the difference in dissociation constants (K_d) of G/C homoduplex (K_d = 1000 nM) and G/T heteroduplex (K_d = 68.5 nM). The affinity latex carrying MutS but no AQ showed no change in the amount of bound DNA even after UV irradiation. In contrast, affinity latex having both MutS and AQ showed significant change in the amount of bound DNA with UV irradiation. The efficiency of photocleavage of mismatched DNA bound by the affinity particles increased with UV irradiation time. It was also found that the functional particles were reusable. Using the functional particles and a dsDNA-binding fluorescent dye, SYBR-Gold, a system was constructed for detection of mismatched DNA that was 40 base pairs. The system proved useful and, therefore, this MutS- and AQ-carrying affinity latex system is a promising tool for mismatch detection in DNA diagnosis.

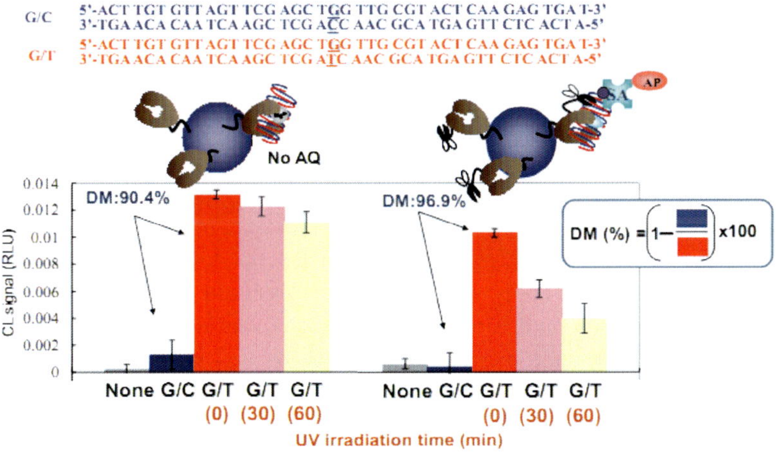

Figure 14.10 Activity of AQ- and MutS-carrying latex in discrimination of one-point mismatch DNA.

14.3.2 Drug-Carrying Affinity Latex

One of the valuable applications of affinity latex is examination of drug activity and identification of drug receptor. Drug-carrying affinity latex can select and catch the receptor protein among various proteins in serum or on cell membrane. Cisplatin-binding proteins were identified using cisplatin-immobilized particles (24). The structure of drug-protein couple formed on latex particle was analyzed in relation to the medical effect. If it is found through this examination that the complex formed between the drug and the protein causes side effect, then the structure of drug is modified.

14.3.3 Affinity Latex to Determine Telomerase Activity

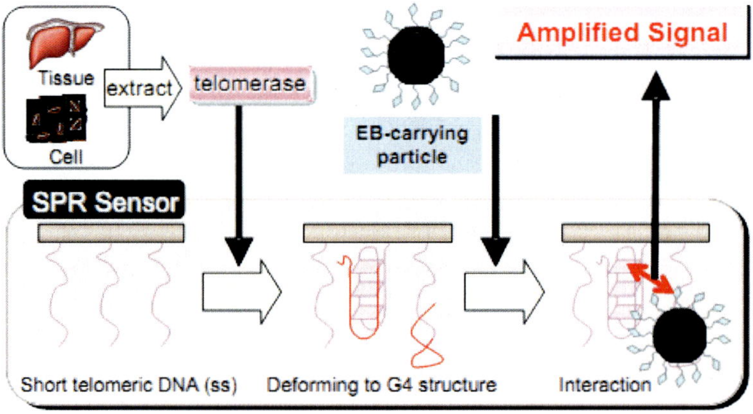

Figure 14.11 Schematic representation of telomerase detection by using G4-DNA binder conjugated polymeric microsphere.

Interests are focused on detection of telomerase, because highly sensitive detection of telomerase enables to detect many kinds of cancers in their early stage and also gives us much important information about cell biology (25). A new method of telomerase detection using polymeric microspheres on surface plasmon resonance (SPR)-based sensor was proposed (26), referring to Maezawa's work (27). The mechanism is presented in Fig. 14.11. A short telomere was immobilized on an SPR sensor chip. Then the sample was added. If the sample included telomerase, it elongated short telomere to G4-DNA (28). The particles which carried ethidium bromide (EB), a G4-DNA recognizable probe, were passed through the SPR sensor and attached to G4-DNA in proportional to the amount of G4-DNA. Therefore, SPR signal should be proportional to the telomerase activity. The EB-carrying particles could contribute to not only DNA structure recognition but also SPR signal amplification. The particles could amplify the signal through the interaction with G4-DNA by about 760 times compared with free EB system. This method enabled the detection of telomerase that was extracted from less than 1 cell. The results suggest that this assay of EB-carrying latex in SPR system is suitable for detection of telomerase activity with ultrahigh sensitivity.

14.3.4 Multiplex Assay Using Affinity Latex

14.3.4.1 Two multiplex assay systems

Multiplex assay is defined as the assay for simultaneous detection of multiple targets by using multiple ligands. This assay needs only very small amount of samples and, therefore, patients suffer little anguish. Speed is another merit of multiplex assay. There are two types of muassay systems. One is composed of chip and the other is composed of dispersing particles. In the former, each ligand is identified by the position on a chip. In contrast, in multiplex assay using affinity particles, called as suspension array or beads array, each particle in the dispersion has to carry its own sign, or bar code, to indicate its identity as shown in Fig. 14.12.

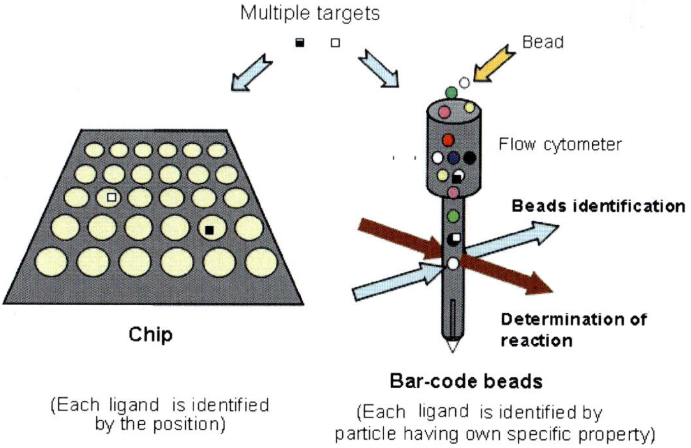

Figure 14.12 Two types of multiplex assay systems.

14.3.4.2 Coding of particles

Color, size, density, electric charge, etc. have been used as the bar code. A series of multiple-colored particles were obtained by mixing multiple dyes or quantum dots (QD) with different mixing ratio (29). For example, a hundred solutions of two fluorescent dyes were prepared by using ten different concentrations of each dye in different combinations, and microbeads of 5 μm diameter were coded with these solutions. Such beads were identified in a flow cell by the specific fluorescence emitted from each dye embedded in the beads. In some QD-assisted multiplex assay systems, leakage of QD made trouble. To prevent the leakage of QDs, silica nanoparticles were deposited on the polymer particles to form silica shell (30).

14.3.4.3 Practice of multiplex beads assay

Multiplex beads assay is carried out as follows: when the particles having sequence-specific DNAs are mixed with a labeled sample of DNA, only the particle having complementary DNA hybridizes with the labeled sample and the identification of the hybridized particle informs us about the DNA sequence of the sample. Then identification is done by flow cytometer as shown in the right of Fig. 14.11. The cytometer is interfaced with a personal computer carrying data processor and each DNA is quantified. This assay system prefers to chip system in terms of speed and easy processing.

Improvement of sensitivity is a keen requirement not only in single-plex assays but also in multiplex assays. If the sensitivity is high enough, no PCR amplification followed by troublesome processes is required. Several trials are going on in this subject. Zheng *et al.* developed a new method to enable accurate and simple quantification of mRNA in whole blood (31).

14.4 Postface

Design, preparation, and application of affinity latex were described in this chapter, focusing especially on bioseparation and diagnosis. Affinity latex has some advantages compared with affinity column chromatography regarding speed, purity, and concentration of recovered target molecules. The drawback of hybrid particle is difficult recovery after operation. One of the solutions to this problem is the use of magnetic nanoparticle-containing particles (32). A semiautomatic affinity latex bioseparator using magnetic system was put on the market.

References

1. Cuatrecasas, P., Wilchek, M., & Anfinsen, C. B. (1968). Selective enzyme purification by affinity chromatography. *Proceedings of the National Academy of Sciences of the United States of America*, **61**, 636–643.
2. Kadonaga, J. T., & Tjian, R. (1986). Affinity purification of sequence-specific DNA binding proteins. *Proceedings of the National Academy of Sciences of the United States of America*, **83**, 5889–5893.
3. Singer, J. M., & Plotz, C. M. (1956). The latex fixation test: I. Application to the serologic diagnosis of rheumatoid arthritis. *American Journal of Medicine*, **21**, 888–892.
4. Rembaum, A., Molday, R. S., & Yen, S. P. S. (1977). Application of magnetic microspheres in label and separation of cells. *Nature*, **268**, 437–438.
5. Ugelstad, J., Berge, A., Ellingsen, T., Shmid, R., Nilsen, T. N., Mork, P. C., *et al.* (1992). Preparation and application of new monosized polymer particles. *Progress in Polymer Science*, **17**, 87–161.
6. Treleaven, J. G., Ugelstad, J., Rembaum, A., Caine, G. D., & Kemshead, J. T. (1984). Removal of neuroblastoma cells from bone marrow with monoclonal antibodies conjugated to magnetic microspheres. *Lancet*, **323**, 70–73.
7. Uhlen, M. (1989). Magnetic separation of DNA. *Nature*, **340**, 733–734.
8. Kawaguchi, H., Asai, A., Ohtsuka, Y., Watanabe, H., Wada, T., & Handa, H. (1989). Purification of DNA-binding transcription factors by their selective adsorption on the affinity latex particles. *Nucleic Acids Research*, **17**, 6229–6240
9. Inomata, Y., Kawaguchi, H., Wada, T., & Handa, H. (1991). Improvement of DNA immobilized latex for the purification of transcription factors. *Kobunshi Ronbunshu*, **48**, 335–340.
10. Kawaguchi, H., Sakamoto, K., Ohtsuka, Y., Ohtake, T., Sekiguchi, H., & Iri, H. (1989). Fundamental-study on latex reagents for agglutination tests. *Biomaterials*, **10**, 225–229.
11. Inomata, Y., Kasuya, Y., Fujimoto, K., Handa, H., & Kawaguchi, H. (1995). Purification of membrane-receptors with peptide-carrying affinity latex-particles. *Colloids and Surfaces, B: Biointerfaces*, **4**, 231–241.

12. Kasuya, Y., Fujimoto, K., Kawaguchi, H., & Miyamoto, M. (1994). Influence of the structure of support microspheres on leucocyte activation by RGDS-carrying microspheres. *Biomaterials*, **15**, 570–576.

13. Akazawa, T., Fujimoto, K., & Kawaguchi, H. (1993). Application of hydrogel microspheres to latex diagnosis. *Kobunshi Ronbunshu*, **50**, 431–435.

14. Inomata, Y., Kawaguchi, H., Hiramoto, M., Wada, T., & Handa, H. (1992). Direct purification of multiple Atf/E4tf3 polypeptides from hela-cell crude nuclear extracts using dna affinity latex-particles. *Analytical Biochemistry*, **206**, 109–114.

15. Imai, T., Sumi, Y., Hatakeyama, M., Fujimoto, K., Kawaguchi, H., Hayashida, *et al.* (1996). Selective isolation of DNA or RNA using single-stranded DNA affinity latex particles. *Journal of Colloid and Interface Science*, **177**, 245–249.

16. Hasegawa, M., Ohno, H., Tanaka, H., Hatakeyama, M., Kawaguchi, H., Takahashi, T., *et al.* (2006). Affinity identification of delta-opioid receptors using latex nanoparticles. *Bioorganic & Medicinal Chemistry Letters*, **16**, 158–161.

17. Shimizu, N., Sugimoto, K., Tang, J., Nishi, T., Sato, I., Hiramoto, M., *et al.* (2000). High-performance affinity beads for identifying drug receptors. *Nature Biotechnology*, **18**, 877–881.

18. Hatakeyama, M., Iwato, S., Fujimoto, K., Handa, H., & Kawaguchi, H. (1998). DNA-carrying latex particles for DNA diagnosis. 1. Separation of normal and point mutant DNAs according to the difference in hybridization efficiency. *Colloids and Surfaces, B: Biointerfaces*, **10**, 161–169.

19. Hatakeyama, M., Nakamura, K., Iwato, S., Fujimoto, K., Handa, H., & Kawaguchi, H. (1998). DNA-carrying latex particles for DNA diagnosis 2. Distinction of normal and point mutant DNA using S1 nuclease. *Colloids and Surfaces, B: Biointerfaces*, **10**, 171–178.

20. Oba, S., Hatakeyama, M., Handa, H., & Kawaguchi, H. (2005). Development of polymer latex particles for selective cleavage of mismatched DNA and their application for DNA diagnosis. *Bioconjugate Chemistry*, **16**, 551–558.

21. Marti, T. M., Kunz, C., & Fleck, O. (2002). DNA mismatch repair and mutation avoidance pathways. *Journal of Cellular Physiology*, **191**, 28–41.

22. Obmolova, G., Ban, C., Hsieh, P., & Yang, W. (2000). Crystal structures of mismatch repair protein MutS and its complex with a substrate DNA. *Nature*, **407**, 703–710.

23. Armitage, B., Yu, C., Devadoss, C., & Schuster, G. B. (1994). Cationic anthraquinone derivatives as catalytic DNA photonucleases: Mechanisms for DNA damage and quinone recycling. *Journal of the American Chemical Society*, **116**, 9847–9859.
24. Tomohiro, T., Sawada, J., Sawa, C., Nakura, H., Yoshida, S., Kodaka, M., *et al.* (2002). Total analysis and purification of cellular proteins binding to cisplatin-damaged DNA using submicron beads. *Bioconjugate Chemistry*, **13**, 163–166.
25. Shay, J. W., & Wright, W. E. (2002). Telomerase: A target for cancer therapeutics. *Cancer Cell*, **2**, 257–265.
26. Hara, R., & Kawaguchi, H. (2012) Highly sensitive detection of telomerase by using G-Quartet DNA binder conjugated polymeric microspheres. *Kobunshi Ronbunshu*, **69**, 122–128.
27. Maesawa, C., Inaba, T., Sato, H., Iijima, S., Ishida, & K., Terashima, M., *et al.* (2003). A rapid biosensor chip assay for measuring of telomerase activity using surface plasmon resonance. *Nucleic Acids Research*, **31**, e4.
28. Parkinson, G. N., Lee, M. P. H., & Neidle, S. (2002). Crystal structure of parallel quadruplexes from human telomeric DNA. *Nature*, **417**, 876–880.
29. Fulton, R. J., McDade, R. L., Smith, P. L., Kienker, L. J., & Kettman Jr, J. R. (1997). Advanced multiplexed analysis with the FlowMetrixTM system. *Clinical Chemistry*, **43**, 1749–1756.
30. Cao, Y. C., Huang, Z. L., Liu, T. C., Wang, H. Q., Zhu, X. X., Wang, Z., *et al.* (2006). Preparation of silica encapsulated quantum dot encoded beads for multiplex assay and its properties. *Analytical Biochemistry*, **351**, 193–200.
31. Zheng, Z., Luo, Y., & McMaster, G. K. (2006). Sensitive and quantitative measurement of gene expression directly from a small amount of whole blood. *Clinical Chemistry*, **52**, 1294–1302.
32. Nishio, K., Masaike,Y., Ikeda, M., Narimatsu, H., Gokon, N., Tsubouchi, S., *et al.* (2008). Development of novel magnetic nano-carriers for high-performance affinity purification. *Colloids and Surfaces, B: Biointerfaces*, **64**, 162–169.

Index